5/22/97

D0139654

This is the third edition of the principal text on the finite element method, which has become a major solution technique for engineering electromagnetics. It presents the method in a mathematically undemanding style, accessible to undergraduates who may be encountering it for the first time. Like the earlier editions, it begins by deriving finite elements for the simplest familiar potential fields, then builds on these to formulate finite elements for a wide range of applied electromagnetics problems. These include wave propagation, diffusion, and static fields; open-boundary problems and nonlinear materials; axisymmetric, planar and fully three-dimensional geometries; scalar and vector fields.

This new edition is more than half as long again as its predecessor and includes much new material. Exterior problems of wave propagation, edge elements, and many other recent developments are now included. The original material has been extensively revised and in many cases entirely new treatments have been substituted for the old. A wide selection of fully documented demonstration programs is included with the text, allowing the reader to see how the methods are put to work in practice.

This graduate-level textbook is easily accessible to the senior undergraduate for it requires neither previous experience with finite elements, nor any advanced mathematical background. It will also be a valuable reference text for professional engineers and research students to make full use of finite elements in electrical engineering.

From reviews of the second edition:

Highly recommended for senior electrical engineering students and professionals in the field of modern electromagnetics.

Engineering

This book provides a clear, detailed introduction to finite element analysis ... it is suitable for use as a text in a graduate electrical engineering course. It is also an excellent reference for all electrical engineers who want to understand finite-element analysis well enough to write or modify their own finite-element codes.

IEEE Antennas and Propagation Magazine

Highly recommended for engineering libraries serving senior and graduate electrical engineering students, as well as for professionals in the field of electromagnetics.

American Library Association

Finite elements for electrical engineers

Finite elements for
electrical engineers
Third edition

PETER P. SILVESTER
University of British Columbia and McGill University

RONALD L. FERRARI
University of Cambridge

CAMBRIDGE
UNIVERSITY PRESS

Published by the Press Syndicate of the University of Cambridge
The Pitt Building, Trumpington Street, Cambridge CB2 1RP
40 West 20th Street, New York, NY 10011–4211, USA
10 Stamford Road, Oakleigh, Melbourne 3166, Australia

© Cambridge University Press 1983, 1990, 1996

First published 1983
Reprinted 1986
Second edition 1990
Third edition 1996

Printed in Great Britain at the University Press, Cambridge

A catalogue record for this book is available from the British Library

Library of Congress cataloguing in publication data
Silvester, P.P. (Peter Peet)
    Finite elements for electrical engineers/Peter P. Silvester,
Ronald L. Ferrari.–3rd ed.
        p.   cm.
    Includes bibliographical references.
    ISBN 0 521 44505 1 (hc) – ISBN 0 521 44953 7 (pb)
    1. Electric engineering–Mathematics. 2. Finite element method.
I. Ferrari, R.L. (Ronald L.) II. Title.
TK153.S53   1996
621.3'01'515353–dc20   95-32756 CIP

ISBN 0 521 44505 1 hardback
ISBN 0 521 44953 7 paperback

KW

Contents

Machine-readable programs

All programs in this book are available in machine-readable form. Users of the World-Wide-Web will find them available from the server maintained by Cambridge University Press:

http://www.cup.cam.ac.uk/onlinepubs/FinElements/FinElementstop.html

Users of the Compuserve network can download them directly from:
Go Science; Library 5, Mathematics; file FE4EE3

The material is also available by anonymous FTP as follows:
Address: club.eng.cam.ac.uk
Path: /pub/fe4ee3/*

The FTP directory fe4ee3 contains an information file README.1ST and a number of subdirectories. The information file gives details of the subdirectory structure of fe4ee3/* and notes on how the files contained within the structure may be downloaded by means of FTP operations. The program files contain characters only, no pictorial material.

The files at the WWW, Compuserve and anonymous FTP locations described above may be updated occasionally. Similar files may be found, from time to time, at other Internet FTP sites, but these are not under the control of Cambridge University Press, of Compuserve, or of the authors.

Some readers may prefer to purchase the programs in this book on diskette. For current prices, formats and shipping times, write to Dr R.L. Ferrari, University Engineering Department, University of Cambridge, Trumpington Street, Cambridge CB2 1PZ, UK.

Note

The programs and related data in this book are copyright material, like all other parts of the book. The files provided at the anonymous FTP site and on the MS-DOS diskette obtainable from the authors, are provided there solely as a convenience for the reader; they remain copyright materials and must be treated on the same basis as this book, except that a security back-up copy may be made. In other words, they may be

Preface to the third edition

Six years have passed since the second edition of this book was published. In the meantime, it has been joined by other books that deal with electrical applications of finite elements, and many more specialities within electrical engineering have espoused the finite element method in that time. In the well-established application areas, topics at the advanced research forefront then have now become routine industrial practice. A new edition must therefore cover a larger span of material and cover it more extensively. This book is therefore about half again as large as the second edition. Readers acquainted with the earlier edition will find the text enhanced by new applications, new mathematical methods, and new illustrations of programming techniques.

Although a wide selection of commercial finite element programs has become available in the past decade, the range of finite element applications in electrical engineering has grown at least equally fast. Interest in the creation of sophisticated special-purpose programs has thus not slackened, on the contrary it has grown rapidly as commercial programs have taken over the more routine tasks. To answer the needs of this more demanding audience, the authors have not only broadened the range of material included in this volume, but widened its scope to include topics a bit closer to current or recent research than in past editions. However, introductory material has not been reduced; on the contrary, there is a wider range of simple computer programs suited to modification and experimentation, programs mainly intended to be easy to read, only secondarily (if at all) to be computationally efficient.

Both authors wish to acknowledge the many contributions that their students, colleagues, and collaborators have made to this volume. These have ranged from broad suggestions about content to contributed illustrations and specific mathematical proofs. To all, the authors express their gratitude and appreciation.

PPS and RLF
Victoria and Cambridge
1996

Preface to the first edition

Although there are now many research papers in the literature that describe the application of finite element methods to problems of electro-magnetics, no textbook has appeared to date in this area. This is surprising, since the first papers on finite element solution of electrical engineering problems appeared in 1968, about the same time as the first textbook on finite element applications in civil engineering.

The authors have both taught courses in finite elements to a variety of electrical engineering audiences, and have sorely felt the lack of a suitable background book. The present work is intended to be suitable for advanced undergraduate students, as well as for engineers in professional practice. It is undemanding mathematically, and stresses applications in established areas, rather than attempting to delineate the research forefront.

Both authors wish to thank the many people who have helped shape this book – especially their students.

PPS and RLF
Montreal and Cambridge
June 1983

1

Finite elements in one dimension

1. Introduction

This chapter introduces the finite element method by showing how a simple one-dimensional problem can be solved. Subsequent chapters of this book then generalize that elementary technique to cover more complicated situations.

The illustrative problem considered in this beginning chapter is that of a lossy direct-current transmission line, a problem not only physically simple but analytically solvable that permits direct comparison of the finite element approximation with known exact solutions. Two- and three-dimensional cases are dealt with in later chapters, where both the mathematical formulation of finite elements and the relevant numerical solution techniques are dealt with.

2. A direct-current transmission line

Suppose a pair of buried pipes is used for transmission of electrical signals, as in Fig. 1.1(a). Such a situation may arise, for example, if gas pipelines are used to carry signals for flow measurement, flaw detection, or other communications. It will be assumed that only very low frequencies are of interest, but that the pipes are resistive so there will be some longitudinal voltage drop as well as current leakage through the earth between pipes. The transmission line formed by the pair of pipes is assumed to be driven by a voltage source (e.g., an amplifier connected to thermocouples); the receiving end is taken to be very lightly loaded, practically an open circuit. The problem is to determine the distribution of signal voltage along the line, and the voltage at the receiving end.

To solve the pipeline problem analytically, consider a short length $\mathrm{d}x$ of the line, as in Fig. 1.1(b). Its longitudinal resistance $r\,\mathrm{d}x$ causes a longitudinal voltage drop $\mathrm{d}v$, while a current $\mathrm{d}i$ flows through the earth

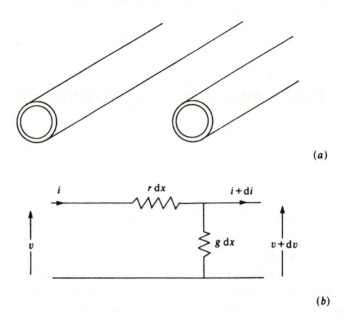

$$(a)$$

$$(b)$$

Fig 1.1   (*a*) Buried pipes used to transmit instrumentation signals as well as fluids. (*b*) Equivalent circuit of a short length dx of the pipeline.

between pipes, represented by the shunt conductance $g\,$dx. The differences dv and di in voltages and currents at the two ends x and $x + dx$ are

$$\mathrm{d}v = ir\,\mathrm{d}x \tag{2.1}$$

$$\mathrm{d}i = (v + \mathrm{d}v)g\,\mathrm{d}x. \tag{2.2}$$

Rewriting, then discarding second-order terms, these equations become

$$\frac{\mathrm{d}v}{\mathrm{d}x} = ri \tag{2.3}$$

$$\frac{\mathrm{d}i}{\mathrm{d}x} = gv. \tag{2.4}$$

Differentiating with respect to x,

$$\frac{\mathrm{d}^2 v}{\mathrm{d}x^2} = r\frac{\mathrm{d}i}{\mathrm{d}x}, \tag{2.5}$$

$$\frac{\mathrm{d}^2 i}{\mathrm{d}x^2} = g\frac{\mathrm{d}v}{\mathrm{d}x}, \tag{2.6}$$

cross-substitution then produces the pair of differential equations

$$\frac{d^2v}{dx^2} = rgv \tag{2.7}$$

$$\frac{d^2i}{dx^2} = rgi. \tag{2.8}$$

Clearly, both voltage and current are governed by the same second-order differential equation. It should be physically evident that two boundary conditions apply here: first, that the voltage has a prescribed value, say v_0, at the sending end $x = L$,

$$v|_{x=L} = V_0, \tag{2.9}$$

and second, that the current must vanish at the other, $i = 0$. But Eq. (2.3) clearly implies that vanishing current is equivalent to a vanishing derivative of voltage, so

$$\frac{dv}{dx}\bigg|_{x=0} = 0. \tag{2.10}$$

The latter two equations provide all the boundary conditions necessary for solving the second-order equation in voltage v. To do so, substitute the trial solution

$$v = V_1 \exp(+px) + V_2 \exp(-px) \tag{2.11}$$

into (2.7), to yield

$$p^2\{V_1 \exp(+px) + V_2 \exp(-px)\}$$
$$= rg\{V_1 \exp(+px) + V_2 \exp(-px)\} \tag{2.12}$$

This equation can hold only if p has the fixed value $p = \sqrt{rg}$. Therefore the trial solution must be amended to read

$$v = V_1 \exp(+\sqrt{rg}x) + V_2 \exp(-\sqrt{rg}x). \tag{2.13}$$

The values of V_1 and V_2 still remain to be determined from boundary conditions. At the open (receiving) end of the line, $x = 0$, these require the first derivative of voltage to vanish. Differentiating Eq. (2.13),

$$\frac{dv}{dx}\bigg|_{x=0} = \sqrt{rg}(V_1 - V_2), \tag{2.14}$$

which implies that there is actually only one unknown constant V, $V = V_1 = V_2$, not two distinct constants. Hence the desired solution is

$$v = V\{\exp(+\sqrt{rg}x) + \exp(-\sqrt{rg}x)\}. \tag{2.15}$$

The value of v is easily determined from the requirement that at $x = L$, the voltage is fixed at V_0,

$$V\{\exp(+\sqrt{rg}L) + \exp(-\sqrt{rg}L)\} = V_0. \tag{2.16}$$

Solving for V and substituting into Eq. (2.15),

$$v = V_0 \frac{\exp(+\sqrt{rg}x) + \exp(-\sqrt{rg}x)}{\exp(+\sqrt{rg}L) + \exp(-\sqrt{rg}L)} \tag{2.17}$$

finally emerges as the complete analytic solution of the lossy transmission line problem. This solution will be studied in further detail later, in conjunction with its approximate counterpart obtained by a finite element method.

3.   The finite element approach

In the finite element method, the differential equations that describe the transmission line are not tackled directly. Instead, this method takes advantage of the equivalent physical principle that the voltage distribution along the line will always so adjust itself as to minimize power loss. When coupled with the familiar idea of piecewise approximation, the minimum power principle directly leads to finite element methods, as will be shown in this section.

3.1   *Power redistribution and minimization*

While it is sometimes very difficult to express the distribution of power in closed mathematical form, it is not hard to formulate this minimum principle in terms of functional approximations. The steps in this process are the following.

(1) Express the power W lost in the line in terms of the voltage distribution $v(x)$:

$$W = W\{v(x)\}. \tag{3.1}$$

   The precise form of this expression is unimportant for the moment; it will be developed further below.

(2) Subdivide the domain of interest (the entire transmission line) into K finite sections or *elements*.

(3) Approximate the voltage $v(x)$ along the line, using a separate approximating expression in each element, of the form

$$v(x) = \sum_{i=1}^{M} v_i f_i(x) \tag{3.2}$$

where the $f_i(x)$ are some convenient set of known functions. These expressions will necessarily include M constant but as yet unknown coefficients v_i on each element, for the voltage along the line is not yet known.

(4) Express power in each element in terms of the approximating functions $f_i(x)$ and their M undetermined coefficients v_i. Because the functions $f_i(x)$ are chosen in advance and therefore known, the power becomes a function of the coefficients only:

$$W = W(v_1, v_2, \ldots v_M). \tag{3.3}$$

(5) Introduce constraints on the MK coefficients so as to ensure the voltage is continuous from element to element. Thus constrained, the ensemble of all elements will possess some N degrees of freedom, $N \le MK$.

(6) Minimize the power by varying each coefficient v_i in turn, subject to the constraint that voltage along the line must vary in a continuous fashion:

$$\frac{\partial W}{\partial v_i} = 0, \quad i = 1, \ldots, N. \tag{3.4}$$

This minimization determines the coefficients and thereby produces an approximate expression for the voltage along the line.

This relatively abstract prescription opens the door to a wide variety of methods. How many elements should be used and should they be of the same size? What sort of approximating functions are best? How should constraints be introduced? What type of minimization technique should be used? These and other questions all admit various answers and thereby define whole families of finite element methods. Only one combination of choices will be sketched out in this chapter; some others follow in later chapters of this book.

There is one key in the finite element process which does not brook much ambiguity: the quantity to be minimized is power, so an expression for the power dissipated by the transmission line must be the starting point. It is therefore appropriate to begin work by expressing the power in a form suited to the task.

3.2 *Explicit expressions for power*

To find an expression for total power in terms of the line voltage, consider a short section dx of the line, as in Fig. 1.1(b). The power entering this section at its left end is given by

$$W_{\text{in}} = vi \tag{3.5}$$

while the power leaving on the right is

$$W_{\text{out}} = (v + dv)(i + di).$$ (3.6)

The difference between the incoming and outgoing amounts must represent the power lost in the section dx. Neglecting the second-order term $dv\,di$, this difference is given by

$$dW = v\,di + i\,dv.$$ (3.7)

The first term on the right may be thought to represent the power dissipated in the earth between the two conductors, the second term to represent the power loss in the conductors themselves. To write the power entirely in terms of voltage, substitute the two expressions

$$di = -g\,dx\,v$$ (3.8)

$$i = -\frac{1}{r}\frac{dv}{dx}$$ (3.9)

into Eq. (3.7). The power loss per unit length then reads

$$\frac{dW}{dx} = -gv^2 - \frac{1}{r}\left(\frac{dv}{dx}\right)^2$$ (3.10)

and the total loss for the whole line is

$$W = -\int_0^L \left\{ gv^2 + \frac{1}{r}\left(\frac{dv}{dx}\right)^2 \right\} dx.$$ (3.11)

This result represents a law of physics, as it were, and is not open to choice. The techniques for approximation and minimization, on the other hand, are various. They will be developed next.

3.3   *Piecewise approximation*

To describe the voltage $v(x)$ along the transmission line, the simple but very common artifice will now be employed of constructing a *piecewise-straight* approximation. The entire line, spanning $0 \le x \le L$, is subdivided into K segments or *finite elements* and the voltage is assumed to vary linearly between endpoint values within each element. Where the segments join, voltage must be continuous; but because its variation is assumed to be piecewise straight, it clearly cannot have continuous slope as well. Continuity of v is necessary to ensure the differentiation in Eq. (3.11) can be carried out; continuity of its slope is not.

To organize the work in a manner easy to implement by computer, let the K elements be numbered in order, beginning at the left (receiving) end. The left and right ends of each element will be denoted by the suffixes l and r, as illustrated in Fig. 1.2(*a*) for five elements. The voltages and currents at the two ends of element k will be identified in two ways, the choice depending on context. One useful form of identification will be by simply assigning each node a number, as in Fig. 1.2(a), and referring to voltages and currents as v_3 or i_9. The other approach, preferable when examining a single element, is to label nodes by element number, with suffixes 1 and r to denote the left and right ends respectively. In this notation, voltages and currents might be labelled as $v_{(k)\mathrm{l}}$ or $i_{(k)\mathrm{r}}$, and the corresponding positions of the two element ends as $x_{(k)\mathrm{l}}$ and $x_{(k)\mathrm{r}}$. The assumption that voltage varies linearly with distance x within any one element may be described in this way by the equation

$$v = \frac{x_{(k)\mathrm{r}} - x}{x_{(k)\mathrm{r}} - x_{(k)\mathrm{l}}} v_{(k)\mathrm{l}} + \frac{x - x_{(k)\mathrm{l}}}{x_{(k)\mathrm{r}} - x_{(k)\mathrm{l}}} v_{(k)\mathrm{r}}. \tag{3.12}$$

For simplicity, the parenthesized subscript will be omitted in the following; no ambiguity can arise because only a single element is under discussion anyway. It is convenient to rewrite (3.12) as

$$v = \alpha_\mathrm{l}(x)v_\mathrm{l} + \alpha_\mathrm{r}(x)v_\mathrm{r} \tag{3.13}$$

where the position functions

$$\alpha_\mathrm{l}(x) = \frac{x_\mathrm{r} - x}{x_\mathrm{r} - x_\mathrm{l}} \tag{3.14}$$

and

$$\alpha_\mathrm{r}(x) = \frac{x - x_\mathrm{l}}{x_\mathrm{r} - x_\mathrm{l}} \tag{3.15}$$

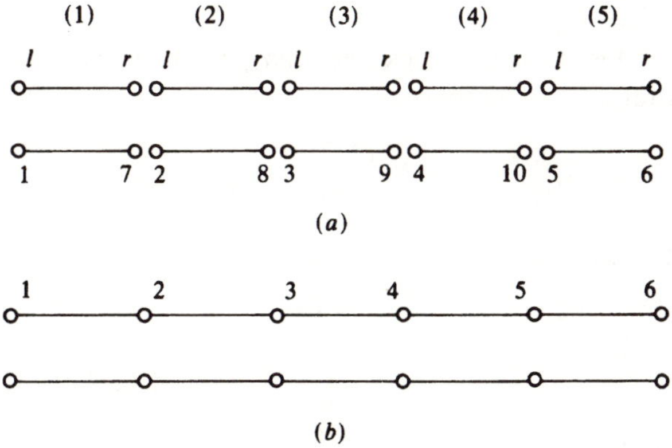

Fig 1.2   The transmission line is regarded as a cascade of five short line elements. (a) Element numbering and node identification, with elements considered separately. (b) Numbering of nodes after connection.

have been introduced in the interests of brevity. A similar expression applies in each finite element.

The total power lost in the entire line is easily calculated as the sum of powers in the individual elements,

$$W = \sum_{k=1}^{K} W_k, \tag{3.16}$$

where the power in element k is given by

$$W_k = -\int_{x_{(k)\mathrm{l}}}^{x_{(k)\mathrm{r}}} \left\{ gv^2 + \frac{1}{r} \left(\frac{\mathrm{d}v}{\mathrm{d}x} \right)^2 \right\} \mathrm{d}x. \tag{3.17}$$

Substituting the approximate expression (3.13) for voltage in each element, the power in the kth element is found to be

$$W_k = -\frac{1}{r_k} \int_{x_{(k)\mathrm{l}}}^{x_{(k)\mathrm{r}}} \left(v_\mathrm{l} \frac{\mathrm{d}\alpha_\mathrm{l}}{\mathrm{d}x} + v_\mathrm{r} \frac{\mathrm{d}\alpha_\mathrm{r}}{\mathrm{d}x} \right)^2 \mathrm{d}x - g_k \int_{x_{(k)\mathrm{l}}}^{x_{(k)\mathrm{r}}} (v_\mathrm{l}\alpha_\mathrm{l} + v_\mathrm{r}\alpha_\mathrm{r})^2 \mathrm{d}x. \tag{3.18}$$

Here and in the following it will be assumed that the resistance and conductance per unit length, r and g, are constant within any one element, though they do not necessarily have to be equal in all elements. In other words, r and g are piecewise constant and the element subdivision must be made in such a way that the line properties only change at element endpoints. This restriction considerably simplifies work organization and programming, though it is not in principle obligatory. In Eq. (3.18), for example, it permits moving the line parameters r and g outside the integral signs.

The power lost in a single element may be stated very neatly as a matrix quadratic form, as follows:

$$W_k = -[v_\mathrm{l} v_\mathrm{r}] \left[\frac{1}{r_k} \mathbb{S} + g_k \mathbb{T} \right] \begin{bmatrix} v_\mathrm{l} \\ v_\mathrm{r} \end{bmatrix}. \tag{3.19}$$

Here $\mathbb{S}$ and $\mathbb{T}$ are 2×2 matrices with entries

$$\mathbb{S}_{ij} = \int_{x_\mathrm{l}}^{x_\mathrm{r}} \frac{\mathrm{d}\alpha_i}{\mathrm{d}x} \frac{\mathrm{d}\alpha_j}{\mathrm{d}x} \, \mathrm{d}x, \tag{3.20}$$

$$\mathbb{T}_{ij} = \int_{x_\mathrm{l}}^{x_\mathrm{r}} \alpha_i \alpha_j \, \mathrm{d}x, \tag{3.21}$$

with the indices i and j assuming both values l and r. More compactly, define a combined matrix $\mathbb{M}$ by

$$\mathsf{M} = \frac{1}{r_k}\mathsf{S} + g_k\mathsf{T}. \tag{3.22}$$

Then the power associated with a typical finite element is simply a quadratic form in the voltages associated with that element,

$$W_k = \mathbf{V}_{(k)}^{\mathrm{T}}\mathsf{M}\mathbf{V}_{(k)}, \tag{3.23}$$

where $\mathbf{V}_{(k)}$ is the vector of left and right end voltages of element k as, in Eq. (3.19) above. This form is not only elegant but also suggests a computational implementation of the method, by emphasizing that the required calculations are similar for each element. It is therefore only necessary to define and program them in detail once, for a typical element.

3.4   *Finite element matrices*

Finite element calculations are usually best carried out by developing the matrices of a single element in detail, then repeating the calculation, without modification, for each element. This approach suggests that the work should be carried out, so far as feasible, in terms of normalized variables independent of the particular element. To this end, let a local coordinate ξ be introduced on the typical element. If

$$L_k = x_{(k)\mathrm{r}} - x_{(k)\mathrm{l}}, \tag{3.24}$$

denotes the length of element k, the normalized coordinate ξ may be defined within the kth element by

$$\xi = \frac{x - x_1}{L_k}. \tag{3.25}$$

The two approximating functions α_{l} and α_{r} applicable to each element can be written in terms of its local coordinate ξ, as

$$\alpha_{\mathrm{l}}(\xi) = 1 - \xi \tag{3.26}$$

$$\alpha_{\mathrm{r}}(\xi) = \xi \tag{3.27}$$

so that the approximate expression for voltage, Eq. (3.13), now reads

$$v = v_{\mathrm{l}}\alpha_{\mathrm{l}}(\xi) + v_{\mathrm{r}}\alpha_{\mathrm{r}}(\xi). \tag{3.28}$$

All the integrations and differentiations necessary for finding the power in an element, as well as other calculations required in due course, are easy to carry out in the local coordinate. For example, the first derivative of the voltage is easily found to be

$$\frac{\mathrm{d}v}{\mathrm{d}x} = \frac{\mathrm{d}v}{\mathrm{d}\xi}\frac{\mathrm{d}\xi}{\mathrm{d}x} = \frac{1}{L_k}(v_{\mathrm{r}} - v_{\mathrm{l}}). \tag{3.29}$$

This value is constant within the element, a direct consequence of the piecewise-straight approximation assumed for the voltage itself.

The derivative of voltage, (3.29) is now expressed in a form applicable to any element whatever its length. The integrations required for evaluating the power in each element, as given by the quadratic form (3.19),

$$W_k = -[v_l\, v_r]\left[\frac{1}{r_k}\mathbb{S} + g_k\mathbb{T}\right]\begin{bmatrix} v_l \\ v_r \end{bmatrix} \tag{3.19}$$

can be put into a length-independent form also. In (3.19), the matrices $\mathbb{S}$ and $\mathbb{T}$ refer to an element of length L_k. Let the corresponding (normalized) matrices be denoted by **S** and **T**, so that

$$\mathbb{S} = \frac{1}{L_k}\mathbf{S} \tag{3.30}$$

$$\mathbb{T} = L_k\mathbf{T}. \tag{3.31}$$

Because the normalized coordinate ξ ranges over $0 \le \xi \le 1$ in any finite element, length normalization makes calculation of the matrices **S** and **T** particularly easy. They are given by expressions that actually involve no lengths at all, for the normalized coordinate ξ enters only as a dummy variable of integration:

$$\mathbf{S} = \begin{bmatrix} \displaystyle\int_0^1 \frac{d\alpha_l}{d\xi}\frac{d\alpha_l}{d\xi}\,d\xi & \displaystyle\int_0^1 \frac{d\alpha_l}{d\xi}\frac{d\alpha_r}{d\xi}\,d\xi \\ \displaystyle\int_0^1 \frac{d\alpha_r}{d\xi}\frac{d\alpha_l}{d\xi}\,d\xi & \displaystyle\int_0^1 \frac{d\alpha_r}{d\xi}\frac{d\alpha_r}{d\xi}\,d\xi \end{bmatrix} \tag{3.32}$$

$$\mathbf{T} = \begin{bmatrix} \displaystyle\int_0^1 \alpha_l\alpha_l\,d\xi & \displaystyle\int_0^1 \alpha_l\alpha_r\,d\xi \\ \displaystyle\int_0^1 \alpha_r\alpha_l\,d\xi & \displaystyle\int_0^1 \alpha_r\alpha_r\,d\xi \end{bmatrix}. \tag{3.33}$$

The matrix entries may look formidable, but their values are quickly worked out because the integrals only involve simple polynomials:

$$\int_0^1 \alpha_l\alpha_r\,d\xi = \int_0^1 (1-\xi)\xi\,d\xi = +\frac{1}{6} \tag{3.34}$$

$$\int_0^1 \alpha_l^2\,d\xi = \int_0^1 \xi^2\,d\xi = +\frac{1}{3} \tag{3.35}$$

$$\int_0^1 \frac{d\alpha_l}{d\xi}\frac{d\alpha_r}{d\xi}\,d\xi = \int_0^1 (-1)(1)\,d\xi = -1 \tag{3.36}$$

$$\int_0^1 \left(\frac{d\alpha_1}{d\xi}\right)^2 d\xi = \int_0^1 (1)^2 d\xi = +1. \tag{3.37}$$

All the matrix entries are constant pure numbers! In fact, **S** and **T** simplify to

$$\mathbf{S} = \begin{bmatrix} +1 & -1 \\ -1 & +1 \end{bmatrix}, \tag{3.38}$$

$$\mathbf{T} = \frac{1}{6}\begin{bmatrix} 2 & 1 \\ 1 & 2 \end{bmatrix}. \tag{3.39}$$

Hence the expression for power loss in a single element is easily stated as

$$W_k = -[v_1 \; v_r]\left(\frac{1}{r_k L_k}\begin{bmatrix} +1 & -1 \\ -1 & +1 \end{bmatrix} + \frac{g_k L_k}{6}\begin{bmatrix} 2 & 1 \\ 1 & 2 \end{bmatrix}\right)\begin{bmatrix} v_1 \\ v_r \end{bmatrix}. \tag{3.40}$$

Having found the approximate power loss in a single element, it remains to find the total power by putting together the individual power values for all the elements, as in Eq. (3.16), and then to minimize power with respect to the nodal voltages, v_i. This task will be addressed in the following section.

4.   Global minimization

The matrix equations that describe individual finite elements are essential, but not in themselves sufficient to solve the problem. It is also necessary to know how the elements are interconnected, and these interconnections must be described next. In this respect finite element analysis somewhat resembles circuit theory: the description of a circuit consists of a set of descriptions of the individual circuit elements (resistors, diodes, etc.), and a set of statements of how their voltages and currents depend on each other, usually obtained by writing Kirchhoff's laws.

4.1   *Joining the elements*

The set of all finite elements that make up the lossy transmission line may be thought of as a cascade of two-port circuits, circuit k being subjected to port voltages $v_{(k)l}$ and $v_{(k)r}$. If the elements are regarded as separate two-port circuits, they must be thought of as electrically interconnected. In other words, it is permissible to assign distinct labels to the right end of one element, and the left end of the next, but the voltage can only have one value because the element ends are connected. To state how they are connected in a manner suited to computation, the finite element nodes (the element endpoints) will be given a globally valid single numbering, as in Fig. 1.2(*b*). This scheme will be referred to as the *connected* numbering, to distinguish it from the *disconnected* numbering of Fig. 1.2(*a*); when it is necessary to distinguish which numbering is meant, a suffix con or dis is appended. For example, the idea that the voltage $v_{(3)r}$ at the right-hand end of element is number 9 in the disconnected numbering, or 4 in the connected numbering, is stated as

$$v_{(3)r} = v_9^{\text{dis}} = v_4^{\text{con}}. \tag{4.1}$$

Two viewpoints are implicitly expressed here. One, illustrated by Fig. 1.2(*a*), regards the elements as fundamental entities. A five-element model is then characterized by ten voltages, two per element, which happen to be constrained by interconnections between element nodes. The other viewpoint, graphically represented in Fig. 1.2(*b*), associates the voltages with nodes, so that a five-element model of the transmission line (which possesses six nodes) forms an indivisible connected whole characterized by the six nodal voltages v_1^{con} through v_6^{con}. In the set of five finite elements of Fig. 1.2, the ten *disconnected* element voltages are related to the six *connected* nodal voltages by the matrix rule

$$
\begin{bmatrix} v_1 \\ v_2 \\ v_3 \\ v_4 \\ v_5 \\ v_6 \\ v_7 \\ v_8 \\ v_9 \\ v_{10} \end{bmatrix}_{\text{dis}} =
\begin{bmatrix}
1 & & & & & \\
& 1 & & & & \\
& & 1 & & & \\
& & & 1 & & \\
& & & & 1 & \\
& & & & & 1 \\
& 1 & & & & \\
& & 1 & & & \\
& & & 1 & & \\
& & & & 1 &
\end{bmatrix}
\begin{bmatrix} v_1 \\ v_2 \\ v_3 \\ v_4 \\ v_5 \\ v_6 \end{bmatrix}_{\text{con}} . \tag{4.2}
$$

This relationship says, in essence, that the six nodal voltages associated with five interconnected finite elements may also be regarded as the ten terminal voltages of the elements, with one numbering mapped onto the other by a certain coefficient matrix. This relationship between connected and disjoint voltage numberings, which may be restated in symbolic matrix form as

$$\mathbf{V}_{\text{dis}} = \mathbf{C}\mathbf{V}_{\text{con}}, \tag{4.3}$$

is usually called the *connection transformation* or *constraint transformation*. It states how the disconnected element voltages $\mathbf{V}_{\text{dis}}$ would have to be constrained in order to make their electrical behaviour identical to that of connected elements subjected to voltages $\mathbf{V}_{\text{con}}$. $\mathbf{C}$ is usually called the *connection matrix*.

A relationship can next be derived between the total power loss expressed in terms of element voltages on the one hand, or in terms of nodal voltages on the other. The total power may always be written as the sum of element powers, as in Eq. (3.16), or equivalently as the quadratic form

$$
W = -\mathbf{V}_{\text{dis}}^{\text{T}}
\begin{bmatrix}
\dfrac{\mathbf{S}}{r_1 L_1} + g_1 L_1 \mathbf{T} & & & \\
 & \dfrac{\mathbf{S}}{r_2 L_2} + g_2 L_2 \mathbf{T} & & \\
 & & \ddots & \\
 & & & \dfrac{\mathbf{S}}{r_5 L_5} + g_5 L_5 \mathbf{T}
\end{bmatrix}
\mathbf{V}_{\text{dis}}.
$$

$$(4.4)$$

The coefficient matrix of this quadratic form is block diagonal. It consists of zeros everywhere, except for 2×2 submatrices placed along its principal diagonal, one such submatrix corresponding to each finite element. If $\mathbf{M}_{\text{dis}}$ represents this coefficient matrix (a 10×10 matrix in the five-element example), then the power W may be written more briefly as

$$
W = -\mathbf{V}_{\text{dis}}^{\text{T}} \mathbf{M}_{\text{dis}} \mathbf{V}_{\text{dis}}.
\tag{4.5}
$$

The disconnected voltages, however, may be replaced by the connected set, in accordance with Eq. (4.3),

$$
W = -\mathbf{V}_{\text{con}}^{\text{T}} \mathbf{C}^{\text{T}} \mathbf{M}_{\text{dis}} \mathbf{C} \mathbf{V}_{\text{con}},
\tag{4.6}
$$

suggesting that the matrix $\mathbf{M}$,

$$
\mathbf{M} = \mathbf{C}^{\text{T}} \mathbf{M}_{\text{dis}} \mathbf{C},
\tag{4.7}
$$

may be viewed as the coefficient matrix associated with the connected set of elements.

If the elements are chosen to be all of equal length L_e, and the transmission line is uniform so that the resistance r and conductance g per unit length are equal in all elements, then the final (connected) matrices of the five-element model are

$$
\frac{1}{r} \mathbf{C}^{\text{T}} \mathbf{S} \mathbf{C} = \frac{1}{L_e r}
\begin{bmatrix}
1 & -1 & & & & \\
-1 & 2 & -1 & & & \\
 & -1 & 2 & -1 & & \\
 & & -1 & 2 & -1 & \\
 & & & -1 & 2 & -1 \\
 & & & & -1 & 1
\end{bmatrix},
\tag{4.8}
$$

$$gC^{T}\mathbb{T}C = \frac{L_{e}g}{6} \begin{bmatrix} 2 & 1 & & & & \\ 1 & 4 & 1 & & & \\ & 1 & 4 & 1 & & \\ & & 1 & 4 & 1 & \\ & & & 1 & 4 & 1 \\ & & & & 1 & 2 \end{bmatrix}. \tag{4.9}$$

The associated total power loss is

$$W = -V_{con}^{T} \left(\frac{1}{r} C^{T} \mathbb{S}C + gC^{T}\mathbb{T}C \right) V_{con} \tag{4.10}$$

or

$$W = -V_{con}^{T} \mathbb{M} V_{con}. \tag{4.11}$$

It now remains to minimize this loss, so as to determine the actual values of nodal voltages.

4.2   *Total power minimization*

The final step in obtaining a finite element solution of the lossy transmission line problem is the minimization of power with respect to the nodal voltage values. In effect, all steps up to this point have produced nothing more than an expression for the power loss in terms of a finite number of voltage values, making it possible to apply the rule that *the voltage must be adjusted so as to minimize power*. The minimization itself is mathematically uncomplicated.

Every node voltage along the line is free to vary, except for the voltage at the sending end of the line; that one is determined independently by the source (generator). Minimization of power can therefore be performed with respect to all nodal voltages except this one. Thus, with N nodes differentiation can only take place with respect to those $N - 1$ variables which are free to vary. Hence power is minimized by setting

$$\frac{\partial W}{\partial v_{k}} = 0, = 1, 2, \ldots, N - 1. \tag{4.12}$$

On substituting Eq. (4.11), power minimization thus produces a matrix equation of N columns, one for each node, but only $N - 1$ rows corresponding to the $N - 1$ nodes whose voltages are free to vary. For the five-element example $N = 6$, so the differentiation results in

$$\begin{bmatrix} M_{11} & M_{12} & \cdots & & \\ M_{21} & M_{22} & & & \\ \vdots & & \ddots & & \vdots \\ & & \cdots & M_{45} & M_{46} \\ & & & M_{55} & M_{56} \end{bmatrix} \begin{bmatrix} v_1 \\ v_2 \\ \vdots \\ v_5 \\ v_6 \end{bmatrix} = 0. \tag{4.13}$$

This equation is in a rather unconventional form. As written, it suggests that all six voltages are unknown. That, however, is not true; v_6 is the sending-end voltage and is therefore known. Transposing this known voltage value to the right-hand side results in

$$\begin{bmatrix} M_{11} & M_{12} & \cdots & \\ M_{21} & M_{22} & & \\ \vdots & & \ddots & \vdots \\ & & \cdots & M_{55} \end{bmatrix} \begin{bmatrix} v_1 \\ v_2 \\ \vdots \\ v_5 \end{bmatrix} = \begin{bmatrix} -M_{16}v_6 \\ -M_{26}v_6 \\ \vdots \\ -M_{56}v_6 \end{bmatrix}. \tag{4.14}$$

This matrix equation represents exactly as many simultaneous algebraic equations as there are free nodes. The unknown nodal voltages are obtained as the solution of this equation, a task easily programmed for a digital computer.

5.   A Fortran program

Coding the procedure discussed above is a straightforward matter in any appropriate programming language. Fortran, Pascal, or C are all suitable for the task, permitting a complete program to be constructed in a few pages of code. An illustrative program LOSSYLIN is listed at the end of this chapter, complete except for a few service routines which appear in Appendix 4: simultaneous equation solving, error handling, and similar general services.

5.1   *Program structure*

Like most finite element programs, LOSSYLIN is organized as a loosely coupled set of subprograms, united by a common pool of data. In Fortran, shared variables are made accessible to several subprograms by placing them in storage areas declared common. In Pascal or C programs, the same result is obtained by making the variables global to the procedures that actually do the work. In LOSSYLIN most variables are regarded as global, so every part of the program (save for a few general-purpose service routines) begins with the same common declarations.

Program LOSSYLIN begins by requesting subroutines matini and vecini to initialize matrices **S**, **T**, and the right-hand side vector to zeros. Next, subroutine `prblem` reads the input data file. The main work of the actual finite element solution is carried out by subroutines `makec`, which makes up the connection matrix **C**; `disjnt`, which creates the disjoint coefficient matrix $\mathbf{M}_{dis}$ (it is called `stdis` in the program); and `connct`, which constraints the disjoint matrix so as to make up the connected coefficient matrix **M**. After transposing the sending-end voltage to the right-hand side with subroutine `rtsd`, the general-purpose equation solver subroutine `eqsolv` is called to determine the nodal voltages. Finally, subroutine `output` calculates both the approximate (finite element) and exact (analytic) solution values at a set of equispaced points along the line, as well as the percentage error in the approximate solution; this result is either displayed on the screen or written to a file. The output file is arranged to permit the results to be imported easily into any popular spreadsheet program (such as Lotus 1-2-3), so many further calculations can be done and graphs or plots can be obtained without any additional programming.

The various subroutines attached to LOSSYLIN perform the expected actions. They do so in the most straightforward way possible, even where much more efficient ways might be found; simplicity rather than computational efficiency has been stressed in program design. For example, subroutine `disjnt` explicitly creates the disjoint coefficient matrix $\mathbf{M}_{dis}$ (called `stdis` in the program), by placing the individual element matrices into an initially null matrix. This procedure creates a large, very sparsely populated matrix which is only required temporarily; a finite element program designed for large-scale application would use much more memory-efficient techniques, but these are usually a good deal less easy to read. As a second example, subroutine `connct` contains nothing but a Fortran version of the matrix multiplication of Eq. (4.7) in explicit form. It creates each entry of the connected matrix **M** by

$$M_{ij} = \sum_{k=1}^{2N-2} \sum_{l=1}^{N-2} C_{ij} M_{(dis)kl} C_{jl}, \qquad (5.1)$$

even though this process is grossly wasteful of time — it involves a number of multiplications proportional to the *fourth* power of the number of nodes! In practical programs, much more efficient procedures are used. Again, **C** appears as a full floating-point matrix, even though it never contains any entries other than 1 or 0 and therefore could be stored in much more compact ways. LOSSYLIN, and indeed most of the other programs in this book, are intended mainly to be read by people, only secondly to be effective in computation; wherever a program design deci-

sion involved a compromise between legibility and efficiency, legibility
was declared the winner.

Subroutine `prblem` reads input data from the standard input
stream, which may be a file or (rarely) the keyboard. It is set up to permit
experimentation with the program, to explore error behaviour and the
effect of different subdivision sizes. `prblem` begins by reading one input
line containing the sending-end voltage and the line parameters
(resistivity and conductivity). Any number of lines then follow, each
describing one section of line. Each section is specified by giving the
number of elements it is to be represented by and its total length. For
example, a data file `SAMPL.DAT` may contain

```
2.0          1.0   0.5
2      2.0
1      0.5
```

This file specifies that a sending-end voltage of 2.0 is applied to a transmission line with a relative resistivity value of 1.0 and a relative conductivity per unit length of 0.5. The line has a total length of 2.5 units; the
portion of length 2.0 nearest the sending end is to be represented by two
finite elements, the remaining length of 0.5 (near the receiving end) is to
be modelled by one. The spacing of data in each line is not critical; list-
directed input is used so formatting is unimportant.

Input and output are handled by `LOSSYLIN` through the standard
input and standard output streams, so in the absence of any other information the program sends all output to the screen and expects input from
the keyboard. Data movement to and from files is handled by redirecting
the standard data streams. How this is done, depends on the operating
system used. Under the Unix and MS-DOS systems, input from file
`SAMPL.DAT` and output to file `SAMPL.OUT` would be requested by

```
LOSSYLIN < SAMPL.DAT > SAMPL.OUT
```

where the $<$ and $>$ signs denote redirection to and from files. Omitting
any redirection instruction causes the default device to be used, so that

```
LOSSYLIN < SAMPL.DAT
```

corresponds to the extremely common combination of taking input from
the file but sending the output to the terminal screen.

5.2   *Solutions and errors*

Solutions obtained using program `LOSSYLIN` have comparatively high accuracy. Subroutine `output` interpolates between element
endpoints and evaluates the analytic solution of Eq. (2.17) as well. A

direct comparison of approximate and exact solutions is thus produced, along with a percentage error. For a transmission line with $\sqrt{rg} = 1$, two length units long and modelled by five finite elements, partial results are shown in Table 1.1. The errors are clearly of the order of 1–2% locally. At the sending end, the two solutions coincide exactly because the sending-end voltage is imposed on both solutions. A graph of another but coarser finite element solution (and for comparison, the analytic solution) is given in Fig. 1.3; here the problem is the same, but only three finite elements have been used. Evidently the approximate solution lies above the exact one in mid-element but dips below it at the nodes. A similar plot for the five-element solution has the same general character but the error

Table 1.1 Error in finite element solution

x	approximate	exact	% error
0.00	0.262344	0.265802	−1.301066
0.02	0.263422	0.265855	−0.915272
0.04	0.264500	0.266015	−0.569405
0.06	0.265578	0.266281	−0.263808
0.08	0.266656	0.266653	0.001207
0.10	0.267735	0.267132	0.225448
0.12	0.268813	0.267718	0.408799
0.14	0.269891	0.268411	0.551209
0.16	0.270969	0.269212	0.652722
0.18	0.272047	0.270120	0.713471
0.20	0.273125	0.271136	0.733660
0.22	0.274203	0.272261	0.713553
⋮	⋮	⋮	⋮
1.82	0.857346	0.841780	1.849189
1.84	0.873197	0.857924	1.780218
1.86	0.889047	0.874411	1.673851
1.88	0.904898	0.891247	1.531575
1.90	0.920748	0.908441	1.354777
1.92	0.936598	0.925997	1.144844
1.94	0.952449	0.943924	0.903108
1.96	0.968299	0.962229	0.630872
1.98	0.984150	0.980918	0.329439
2.00	1.000000	1.000000	0.000000

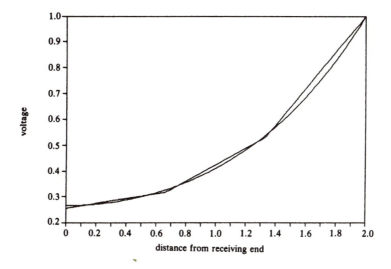

Fig 1.3 Approximate and exact solutions for a three-element model of the transmission line. Values are made to coincide at the sending end.

is much smaller and the difference between solutions is quite hard to see on a graph.

To investigate the error behaviour of finite element solutions, the same problem may be solved using several different finite element models. For the transmission line described above ($l = 2.0$, $\sqrt{rg} = 1$), Fig. 1.4 shows error behaviour for $N = 3, 4, 5, 6, 8, 10, 12, 16, 20, 24$ elements. What is plotted here is the *logarithm* of the root-mean-square error E, as a function of the logarithm of N. Clearly, the result is a straight line whose slope is -2. (A straight-line fit to the data points actually yields -2.018.) Thus

$$\log E = -2\log\frac{N}{a},\qquad(5.2)$$

where a is some constant. Hence

$$E = \exp\left(-2\log\frac{N}{a}\right) = \left(\frac{a}{N}\right)^2.\qquad(5.3)$$

The experimental curve of Fig. 1.4 shows that the root-mean-square error E falls as the *square* of the number of elements used. There are good theoretical reasons for this square-law dependence; these will be discussed later in the wider connection of multidimensional finite elements. The number of elements used in the problem more or less covers the entire

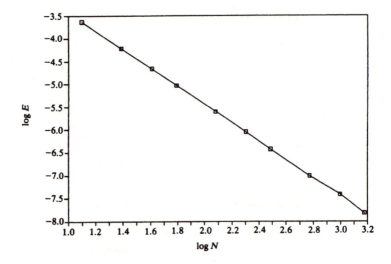

Fig 1.4   Error in the finite element solution falls as the square of the number of elements, as shown by experimental evidence: the slope of the log *E* curve is −2.

useful range. In fact, with single-precision computation (which carries about 7 significant decimal figures) a precision increase much beyond 24 elements is not warranted. With a larger number of elements, Fig. 1.4 implies that solutions will quickly become stable to 5 or 6 significant figures, so that numerical roundoff error arising from finite word-length in computation, rather than finite element discretization error, becomes the limiting factor.

6.   Further reading

Of the many textbooks now available on the finite elements, several introduce the method by treating one-dimensional examples in considerable detail. No doubt for good historical reasons, the major portion of finite element literature deals with applications to structural engineering and related branches of applied elasticity. Hence the main difficulty for the electrical engineer starting out in this field lies less in the availability of books — of which there are many — but in choosing books that allow finite elements to be studied without first undertaking a course in applied elasticity! The text by Bickford (1990) is a model of its genre. Its second chapter, which extends to about 150 pages, introduces finite elements, then explores its physical, mathematical, and computational aspects in great detail, using only one-dimensional examples; the treatment is complete enough to form almost a small book on one-dimensional elements. Two and three dimensions are treated later in this book, which requires little knowledge of applied mechanics and is therefore eminently accessible to electrical engineers. Rao (1989) takes a full chapter to give a similar, if somewhat shorter, treatment of one-dimensional problems and provides the added benefit of a good bibliography; unfortunately, the remaining chapters of Rao's book require a good understanding of structural engineering. Pepper and Heinrich (1992) deal with one-dimensional problems in a fashion similar to this book, but apply the technique to heat conduction, an area probably easier for electrical engineers to grasp than elasticity. Jin (1993) also introduces finite elements through one-dimensional examples; his treatment is less complete and presented in a more abbreviated form than Bickford's. However, it has the advantage of being written for an electrical engineering audience. Mori (1986) emphasizes the mathematical physics of finite elements a little more strongly than the books already mentioned, thus avoiding any need for knowledge about structures; the book is very readable and can be recommended. The volume by Beltzer (1990) devotes a considerable amount of space to one-dimensional problems, but approaches the matter a little differently from the others: all computations are carried out using a symbolic algebra system (MACSYMA). As a result, the book will probably appeal to readers acquainted with MACSYMA.

7.   Bibliography

Beltzer, A. (1990), *Variational and Finite Element Methods: a Symbolic Computation Approach*. Berlin: Springer-Verlag. xi + 254 pp.

Bickford, W.B. (1990), *A First Course in the Finite Element Method*. Homewood, Illinois: Irwin. xix + 649 pp.

Jin, J.-M. (1993), *The Finite Element Method in Electromagnetics*. New York: John Wiley. xix + 442 pp.

Mori, M. (1986), *The Finite Element Method and its Applications*. New York: Macmillan. xi + 188 pp.

Pepper, D.W. and Heinrich, J.C. (1992), *The Finite Element Method: Basic Concepts and Applications*. Washington: Hemisphere. xi + 240 pp. + disk.

Rao, S.S. (1989), *The Finite Element Method in Engineering*, 2nd edn. Oxford: Pergamon. xxvi + 643 pp.

8.   Programs

The LOSSYLIN program which follows is accompanied by the two-line data set used to produce Fig. 1.3.

```
C*************************************************************
C**********                              **********
C********** One-dimensional demonstration program **********
C**********                              **********
C*************************************************************
C       Copyright (c) 1995  P.P. Silvester and R.L. Ferrari
C*************************************************************
C
C       The subroutines  that make up this program  communicate
C       via named common blocks.  The variables in commons are:
C
C       Problem definition and solution
C       common /problm/
C          nodes   =  number of nodes in problem
C          resist  =  resistance per unit length
C          conduc  =  conductance per unit length
C          x       =  nodal coordinates
C          voltag  =  nodal voltage values
C
C       Global matrices and right-hand side
C       common /matrix/
C          stcon   =  global coefficient matrix (connected)
C          rthdsd  =  right-hand side of system of equations
C          c       =  connection matrix for whole problem
C          stdis   =  global coefficient matrix (disconnected)
C
C       Predefined problem size parameter (array dimensions):
C          MAXNOD  =  maximum number of nodes in problem
C
C=============================================================
C     Global declarations -- same in all program segments
C=============================================================
        parameter (MAXNOD = 30)
        common /problm/ nodes, resist, conduc,
     *                  x(MAXNOD), voltag(MAXNOD)
        common /matrix/ stcon(MAXNOD,MAXNOD), rthdsd(MAXNOD),
     *      c(MAXNOD,2*MAXNOD-2), stdis(2*MAXNOD-2,2*MAXNOD-2)
C=============================================================
C
C       Initialize matrices to zeros.
        call matini(stcon, MAXNOD, MAXNOD)
        call matini(stdis, 2*MAXNOD-2, 2*MAXNOD-2)
        call vecini(rthdsd, MAXNOD)
C
C       Problem definition:
C            Sending-end voltage, resistivity, conductivity
C            Node number, x-coordinate (1 or more times)
        call prblem
C
C       Create connection matrix
        call makec
C
C       Construct the disjoint coefficient matrix
        call disjnt
```

```
C
C       Construct connected coefficient matrix
        call connct
C
C       Construct right-hand side vector
        call rtsd
C
C       Solve the assembled finite element equations
        call eqsolv(stcon, voltag, rthdsd, nodes-1, MAXNOD)
C
C       Write the resulting voltage values to output file
        call output
C
        stop
        end
C
C***************************************************************
C
        Subroutine prblem
C
C***************************************************************
C
C       Reads in problem details from input file.
C
C===============================================================
C       Global declarations -- same in all program segments
C===============================================================
        parameter (MAXNOD = 30)
        common /problm/ nodes, resist, conduc,
     *                  x(MAXNOD), voltag(MAXNOD)
        common /matrix/ stcon(MAXNOD,MAXNOD), rthdsd(MAXNOD),
     *        c(MAXNOD,2*MAXNOD-2), stdis(2*MAXNOD-2,2*MAXNOD-2)
C===============================================================
C
C       Set up initial values
        nodes = 1
        do 10 i = 1,MAXNOD
   10   x(i) = 0.
C
C       Read problem data
C          (1) sending voltage, resistivity, conductivity
        read (*, *) sendgv, resist, conduc
C          (2) one or more nodes:  node number, x-coordinate
   30 read (*, *, end=60) node, xnode
        if (node .le. 0 .or. nodes+node .gt. MAXNOD) then
          call errexc('PRBLEM', 1)
        else
          xleft = x(nodes)
          nleft = nodes
          xnode = xleft + xnode
          nodes = nodes + node
          segms = nodes - nleft
          do 40 node=nleft,nodes
            x(node) = (nodes-node)*xleft - (nleft-node)*xnode
```

```
   40       x(node) = x(node) / segms
          endif
          go to 30
C
C     Data successfully read.  Set up problem values.
   60 if (nodes .le. 1) then
          call errexc('PRBLEM', 2)
          endif
          if (sendgv .eq. 0.) sendgv = 1.
          if (resist .eq. 0.) resist = 1.
          if (conduc .eq. 0.) conduc = 1.
          do 70 i = 1,nodes
   70     voltag(i) = 0.
          voltag(nodes) = sendgv
C
          return
          end
C
C***************************************************************
C
          Subroutine makec
C
C***************************************************************
C
C     Establishes connection matrix c.
C
C==============================================================
C     Global declarations -- same in all program segments
C==============================================================
          parameter (MAXNOD = 30)
          common /problm/ nodes, resist, conduc,
      *                   x(MAXNOD), voltag(MAXNOD)
          common /matrix/ stcon(MAXNOD,MAXNOD), rthdsd(MAXNOD),
      *        c(MAXNOD,2*MAXNOD-2), stdis(2*MAXNOD-2,2*MAXNOD-2)
C==============================================================
C
C     Create connection matrix:
          do 60 i = 1,nodes
            do 50 j = 1,2*nodes-2
              if (j/2+1 .eq. i) then
                c(i,j) = 1.
              else
                c(i,j) = 0.
              endif
   50       continue
   60     continue
C
          return
          end
C
C***************************************************************
C
          Subroutine disjnt
C
```

```
C*************************************************************
C
C      Constructs the disjoint coefficient matrix
C
C================================================================
C      Global declarations -- same in all program segments
C================================================================
       parameter (MAXNOD = 30)
       common /problm/ nodes, resist, conduc,
      *                x(MAXNOD), voltag(MAXNOD)
       common /matrix/ stcon(MAXNOD,MAXNOD), rthdsd(MAXNOD),
      *      c(MAXNOD,2*MAXNOD-2), stdis(2*MAXNOD-2,2*MAXNOD-2)
C================================================================
C
       do 70 i = 1,2*nodes-3,2
         j = i + 1
C
C        Determine element length, quit if nonpositive
         ellng = x(j/2+1) - x(j/2)
         if (ellng .le. 0.) then
           call errexc('DISJNT', i)
         else
C          Fit element s and t into disjoint global matrix
           stdis(i,i) = + 1./(resist*ellng) + conduc*ellng/3.
           stdis(i,j) = - 1./(resist*ellng) + conduc*ellng/6.
           stdis(j,i) = - 1./(resist*ellng) + conduc*ellng/6.
           stdis(j,j) = + 1./(resist*ellng) + conduc*ellng/3.
         endif
   70  continue
C
       return
       end
C
C*************************************************************
C
       Subroutine connct
C
C*************************************************************
C
C      Connection transformation:    stcon = c' stdis c
C
C================================================================
C      Global declarations -- same in all program segments
C================================================================
       parameter (MAXNOD = 30)
       common /problm/ nodes, resist, conduc,
      *                x(MAXNOD), voltag(MAXNOD)
       common /matrix/ stcon(MAXNOD,MAXNOD), rthdsd(MAXNOD),
      *      c(MAXNOD,2*MAXNOD-2), stdis(2*MAXNOD-2,2*MAXNOD-2)
C================================================================
C
C      Set connected coefficient matrix to zero:
       do 50 k = 1,nodes
         do 40 l = 1,nodes
```

```
            sum = 0.
            do 30 j = 1,2*nodes-2
              do 20 i = 1,2*nodes-2
                sum = sum + c(k,i) * stdis(i,j) * c(l,j)
   20         continue
   30       continue
            stcon(k,l) = sum
   40     continue
   50   continue
C
      return
      end
C
C*************************************************************
C
      Subroutine rtsd
C
C*************************************************************
C
C     Transposes sending-end voltage to right-hand side
C
C=============================================================
C     Global declarations -- same in all program segments
C=============================================================
      parameter (MAXNOD = 30)
      common /problm/ nodes, resist, conduc,
     *                x(MAXNOD), voltag(MAXNOD)
      common /matrix/ stcon(MAXNOD,MAXNOD), rthdsd(MAXNOD),
     *     c(MAXNOD,2*MAXNOD-2), stdis(2*MAXNOD-2,2*MAXNOD-2)
C=============================================================
C
      do 10 i = 1,nodes-1
        rthdsd(i) = -stcon(i,nodes) * voltag(nodes)
   10   continue
      return
      end
C*************************************************************
C
      Subroutine output
C
C*************************************************************
C
C     Outputs problem and results to standard output stream,
C     accompanied by exact analytic solution.
C
C=============================================================
C     Global declarations -- same in all program segments
C=============================================================
      parameter (MAXNOD = 30)
      common /problm/ nodes, resist, conduc,
     *                x(MAXNOD), voltag(MAXNOD)
      common /matrix/ stcon(MAXNOD,MAXNOD), rthdsd(MAXNOD),
     *     c(MAXNOD,2*MAXNOD-2), stdis(2*MAXNOD-2,2*MAXNOD-2)
C=============================================================
```

```
C
      p = sqrt(resist * conduc)
      dn = exp(p*x(nodes)) + exp(-p*x(nodes))
C
      nbits = 100
      do 10 i = 1,nbits+1
        z = (i-1) * x(nodes)/nbits
        vap = vapx(z)
        vex = voltag(nodes) * (exp(p*z) + exp(-p*z))/dn
        err = (vap - vex)/vex * 100.
        write (*, 1000, err=890) i, z, vap, vex, err
   10   continue
 1000 format (1x, i3, 4f12.6)
      return
  890 call errexc('OUTPUT', 1)
      end
C
C*************************************************************
C
      Function vapx(z)
C
C*************************************************************
C
C     Returns interpolated approximate solution vapx at x = z
C
C=============================================================
C     Global declarations -- same in all program segments
C=============================================================
      parameter (MAXNOD = 30)
      common /problm/ nodes, resist, conduc,
     *                x(MAXNOD), voltag(MAXNOD)
      common /matrix/ stcon(MAXNOD,MAXNOD), rthdsd(MAXNOD),
     *     c(MAXNOD,2*MAXNOD-2), stdis(2*MAXNOD-2,2*MAXNOD-2)
C=============================================================
C
C     Determine in which interval z lies
      int = 0
      do 10 i = 1,nodes-1
        if (x(i) .le. z .and. x(i+1) .ge. z) int = i
   10   continue
      if (int .eq. 0) then
        call errexc('VAPX', 1)
      else
C        Interpolate within interval to find value
        vapx = (voltag(int+1) * (z - x(int)) +
     *     voltag(int) * (x(int+1) - z)) / (x(int+1) - x(int))
      endif
C
      return
      end


    2.        1.      0.5
  4 2.0
```

2

First-order triangular elements for potential problems

1. Introduction

First-order triangular finite elements made their initial appearance in electrical engineering applications in 1968. They were then used for the solution of comparatively simple waveguide problems, but have since been employed in many areas where two-dimensional scalar potentials or wave functions need to be determined. Because of their relatively low accuracy, first-order elements have been supplanted in many applications by elements of higher orders. However, they continue to find wide use in problems where material nonlinearities or complicated geometric shapes are encountered; for example, in analysing the magnetic fields of electric machines, or the charge and current distributions in semiconductor devices.

The first-order methods using triangular elements may be regarded as two-dimensional generalizations of piecewise-linear approximation, a tool widely used in virtually all areas of electrical engineering. The mathematics required in defining such elements is easily mastered, and computer programming at a very simple level can produce many useful results. There are few methods in electromagnetic field theory for which such sweeping claims can be made, and indeed it is surprising that finite elements have not penetrated into electrical engineering applications even more deeply.

In this chapter, simple triangular finite element methods will be developed for solving two-dimensional scalar potential problems. The construction of these simple elements is useful in its own right; but perhaps more importantly, it will also illustrate by way of example many of the principles involved in all finite element methods.

2.   Laplace's equation

Numerous problems in electrical engineering require a solution of Laplace's equation in two dimensions. This section outlines a classic problem that leads to Laplace's equation, then develops a finite element method for its solution. Subsequent sections then examine how the method can be implemented in practice and how it can be generalized to a wider universe of problems.

2.1   *Energy minimization*

To determine the TEM wave properties of a coaxial transmission line composed of rectangular conductors, as in Fig. 2.1(*a*), requires finding the electric potential distribution in the interconductor space. This potential is governed by Laplace's equation,

$$\nabla^2 u = 0. \tag{2.1}$$

Because the transmission line has two planes of symmetry, only one-quarter of the actual problem region needs to be analysed, as suggested in Fig. 2.1(*a*). Two kinds of boundary conditions therefore arise. The inner and outer conductors are metallic so they must be equipotential surfaces, with known potential values. Potential values can only be symmetric about the symmetry planes if the normal derivative of potential vanishes there. Thus the boundary-value problem to be solved consists of (2.1), subject to the boundary conditions

$$\left.\begin{array}{ll} u = u_0 & \text{on conductor surfaces,} \\[2mm] \dfrac{\partial u}{\partial n} = 0 & \text{on symmetry planes.} \end{array}\right\} \tag{2.2}$$

Fixed-value boundary conditions are often referred to as *Dirichlet* conditions, because Dirichlet first showed how to solve Laplace's equation in region with such conditions on all boundaries. For analogous reasons, boundary conditions that prescribe derivative values are usually called *Neumann* conditions.

The well-known principle of minimum potential energy requires the potential u to distribute itself in the transmission line in such a way as to minimize the stored field energy per unit length. To within a constant multiplier this energy is given by

$$W(u) = \frac{1}{2}\int \nabla u \cdot \nabla u \, dS, \tag{2.3}$$

the integration being carried out over the whole two-dimensional problem region. This minimum energy principle is mathematically equivalent to Laplace's equation, in the sense that a potential distribution

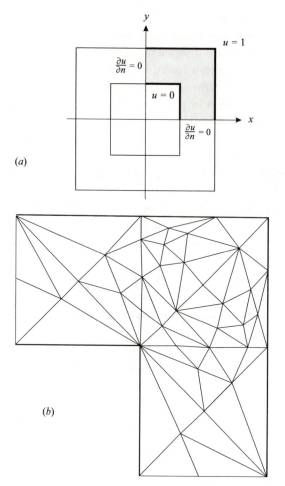

Fig 2.1 (*a*) One-quarter model of a rectangular coaxial line, showing boundary conditions of the problem. (*b*) Finite element mesh for analysis of the line.

which satisfies the latter equation will also minimize the energy, and vice versa. Hence two alternative practical approaches exist for solving the field problem. On the one hand, an approximate solution to Laplace's equation may be sought directly — as it is, for example, in the method of separation of variables, or in finite difference methods. Alternatively, the stored energy may be approximated and minimized. To do so, the potential $u(x, y)$ is assumed to be given by a combination of suitably chosen simple functions with as yet undetermined coefficients. Minimization of the stored energy $W(u)$ associated with the approximate potential $u(x, y)$ then determines the coefficients, and thereby finds an approximation to the potential distribution. Virtually all finite element methods follow this second route, or adaptations of it.

Suppose $u(x,y)$ is the true solution of the problem, while $h(x,y)$ is some sufficiently differentiable function with exactly zero value at every boundary point where the value of u is prescribed by the boundary conditions. The linear combination $u + \theta h$, where θ is a scalar parameter, then has the same prescribed boundary values as u. The energy $W(u + \theta h)$ associated with this incorrect potential distribution is

$$W(u + \theta h) = W(u) + \theta \int \nabla u \cdot \nabla h \, dS + \frac{1}{2} \theta^2 \int \nabla h \cdot \nabla h \, dS. \qquad (2.4)$$

The middle term on the right may be rewritten, using Green's theorem (see Appendix 2), so that

$$W(u + \theta h) = W(u) + \theta^2 W(h) - \theta \int h \nabla^2 u \, dS + \theta \oint h \frac{\partial u}{\partial n} \, ds. \qquad (2.5)$$

Clearly, the third term on the right vanishes because the exact solution u satisfies Laplace's equation. The contour integral term must also vanish because its integrand vanishes everywhere: at each and every Dirichlet boundary point in Fig. 2.1(*a*), h vanishes, while at every Neumann boundary point the normal derivative of u vanishes. Hence,

$$W(u + \theta h) = W(u) + \theta^2 W(h). \qquad (2.6)$$

The rightmost term in this equation is clearly always positive. Consequently, $W(u)$ is indeed the minimum value of energy, reached when $\theta = 0$ for any admissible function h. Admissibility here implies two requirements; h must vanish at Dirichlet boundary points (where u is prescribed), and h must be at least once differentiable throughout the problem region.

It is also evident from Eq. (2.6) that the incorrect energy estimate $W(u + \theta h)$ differs from the correct energy $W(u)$ by an error which depends on the square of θ. If the incorrect potential distribution does not differ very greatly from the correct one — that is to say, if θ is small — the error in energy is much smaller than the error in potential. This point is of very considerable practical importance, for many of the quantities actually required by the engineering analyst are closely related to energy. Impedances, power losses, or the stored energies themselves are often very accurately approximated even if the potential solution contains substantial errors.

2.2   *First-order elements*

To construct an approximate solution by a simple finite element method, the problem region is subdivided into triangular elements, as indicated in Fig. 2.1(*b*). The essence of the method lies in first approximating the potential *u* within each element in a standardized fashion, and thereafter interrelating the potential distributions in the various elements so as to constrain the potential to be continuous across interelement boundaries.

To begin, the potential approximation and the energy associated with it will be developed. Within a typical triangular element, illustrated in Fig. 2.2, it will be assumed that the potential *u* is adequately represented by the expression

$$U = a + bx + cy. \tag{2.7}$$

The true solution *u* is thus replaced by a piecewise-planar function *U*; the smoothly curved actual potential distribution over the *x*–*y* plane is modelled by a jewel-faceted approximation. It should be noted, however, that the potential distribution is continuous throughout the problem region, even across interelement edges. Along any triangle edge, the potential as given by (2.7) is the linear interpolate between its two vertex values, so that if two triangles share the same vertices, the potential will always be continuous across the interelement boundary. In other words, there are no 'holes in the roof', no gaps in the surface *u*(*x*, *y*) which approximates the true solution over the *x*–*y* plane. The approximate solution is piecewise planar, but it is continuous everywhere and therefore differentiable everywhere.

The coefficients *a*, *b*, *c* in Eq. (2.7) may be found from the three independent simultaneous equations which are obtained by requiring

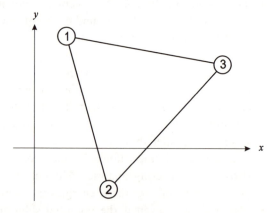

Fig 2.2   Typical triangular finite element in the *x*–*y* plane.

the potential to assume the values U_1, U_2, U_3 at the three vertices. Substituting each of these three potentials and its corresponding vertex location into Eq. (2.7), three equations are obtained. These may be collected to make up the matrix equation

$$\begin{bmatrix} U_1 \\ U_2 \\ U_3 \end{bmatrix} = \begin{bmatrix} 1 & x_1 & y_1 \\ 1 & x_2 & y_2 \\ 1 & x_3 & y_3 \end{bmatrix} \begin{bmatrix} a \\ b \\ c \end{bmatrix}. \tag{2.8}$$

The determinant of the coefficient matrix in Eq. (2.8) may be recognized on expansion as equal to twice the triangle area. Except in the degenerate case of zero area, which is anyway of no interest, the coefficients a, b, c are therefore readily determined by solving this matrix equation. Substitution of the result into Eq. (2.7) then yields

$$U = \begin{bmatrix} 1 & x & y \end{bmatrix} \begin{bmatrix} 1 & x_1 & y_1 \\ 1 & x_2 & y_2 \\ 1 & x_3 & y_3 \end{bmatrix}^{-1} \begin{bmatrix} U_1 \\ U_2 \\ U_3 \end{bmatrix}. \tag{2.9}$$

Inversion of the coefficient matrix in Eq. (2.9) is a little lengthy but quite straightforward. When the inversion is performed and the multiplications are carried out, Eq. (2.9) can be recast in the form

$$U = \sum_{i=1}^{3} U_i \alpha_i(x, y), \tag{2.10}$$

where the position function $\alpha_1(x, y)$ is given by

$$\alpha_1 = \frac{1}{2A} \{(x_2 y_3 - x_3 y_2) + (y_2 - y_3)x + (x_3 - x_2)y\}, \tag{2.11}$$

α_2 and α_3 being obtained by a cyclic interchange of subscripts; A is the area of the triangle. It is easy to verify from (2.11) that the newly defined functions are *interpolatory* on the three vertices of the triangle, i.e., that each function vanishes at all vertices but one, and that it has unity value at that one:

$$\left. \begin{aligned} \alpha_i(x_j, y_j) &= 0 & i \neq j \\ &= 1 & i = j. \end{aligned} \right\} \tag{2.12}$$

The energy associated with a single triangular element may now be determined using Eq. (2.3), the region of integration being the element itself. The potential gradient within the element may be found from Eq. (2.10) as

$$\nabla U = \sum_{i=1}^{3} U_i \nabla \alpha_i, \tag{2.13}$$

so that the element energy becomes

$$W^{(e)} = \frac{1}{2} \int \nabla U \cdot \nabla U \, dS, \tag{2.14}$$

or, from Eq. (2.13),

$$W^{(e)} = \frac{1}{2} \sum_{i=1}^{3} \sum_{j=1}^{3} U_i \int \nabla \alpha_i \cdot \nabla \alpha_j \, dS \, U_j, \tag{2.15}$$

where the superscript (e) identifies the element, and the region of integration is understood to be of that element. For brevity, define matrix elements

$$S_{ij}^{(e)} = \int \nabla \alpha_i \cdot \nabla \alpha_j \, dS, \tag{2.16}$$

where the superscript identifies the element. Equation (2.15) may be written as the matrix quadratic form

$$W^{(e)} = \frac{1}{2} \mathbf{U}^T \mathbf{S}^{(e)} \mathbf{U}. \tag{2.17}$$

Here $\mathbf{U}$ is the column vector of vertex values of potential; the superscript T denotes transposition.

For any given triangle, the matrix $\mathbf{S}^{(e)}$ is readily evaluated. On substitution of the general expression (2.11) into Eq. (2.16), a little algebra yields

$$S_{12}^{(e)} = \frac{1}{4A} \{ (y_2 - y_3)(y_3 - y_1) + (x_3 - x_2)(x_1 - x_3) \}, \tag{2.18}$$

and similarly for the other entries of the matrix $\mathbf{S}$.

2.3   Continuity between elements

For any one triangular element, the approximate element energy may be computed as shown above. The total energy associated with an assemblage of many elements is the sum of all the individual element energies,

$$W = \sum_{e} W^{(e)}. \tag{2.19}$$

Any composite model made up of triangular patches may be built up one triangle at a time. This process is best appreciated by initially considering two triangles only, then generalizing to larger assemblies of elements. To begin, suppose the two elements of Fig. 2.3(*a*) are to be joined to make the composite structure of Fig. 2.3(*b*). Since three potential values are associated with each triangle, all possible states of the pair of uncon-

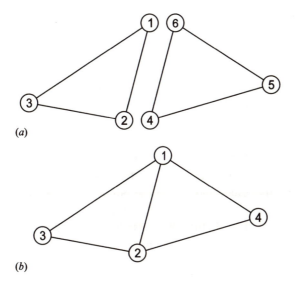

Fig 2.3 Joining of two elements: (*a*) Pair of elements considered electrically disjoint. (*b*) The same elements, with potentials required to be continuous and nodes renumbered.

nected elements can be described by a column vector containing all six vertex potentials.

$$\mathbf{U}_{\text{dis}}^{\text{T}} = \begin{bmatrix} U_1 & U_2 & U_3 & U_4 & U_5 & U_6 \end{bmatrix}_{\text{dis}}, \qquad (2.20)$$

where the subscript dis indicates that disjoint elements (elements not as yet connected electrically) are being considered. The total energy of the pair of elements is then

$$W = \frac{1}{2} \mathbf{U}_{\text{dis}}^{\text{T}} \mathbf{S}_{\text{dis}} \mathbf{U}_{\text{dis}}, \qquad (2.21)$$

where

$$\mathbf{S}_{\text{dis}} = \begin{bmatrix} S_{11}^{(1)} & S_{12}^{(1)} & S_{13}^{(1)} & & & \\ S_{21}^{(1)} & S_{22}^{(1)} & S_{23}^{(1)} & & & \\ S_{31}^{(1)} & S_{32}^{(1)} & S_{33}^{(1)} & & & \\ & & & S_{44}^{(2)} & S_{45}^{(2)} & S_{46}^{(2)} \\ & & & S_{54}^{(2)} & S_{55}^{(2)} & S_{56}^{(2)} \\ & & & S_{64}^{(2)} & S_{65}^{(2)} & S_{66}^{(2)} \end{bmatrix} \qquad (2.22)$$

is the matrix $\mathbf{S}$ (the *Dirichlet matrix*) of the disjoint pair of elements. More briefly, in partitioned matrix form,

$$\mathbf{S}_{\text{dis}} = \begin{bmatrix} \mathbf{S}^{(1)} & \\ & \mathbf{S}^{(2)} \end{bmatrix}. \tag{2.23}$$

In the physical problem, potentials must be continuous across inter-element boundaries. Because the potential in each triangle is approximated by a linear function of x and y, its value varies linearly with distance along any one triangle side. Hence the continuity requirement on potentials is satisfied, provided the potentials at corresponding vertices are identical. That is to say, the potential in Fig. 2.3(*a*) will be continuous everywhere, so long as the potentials at vertices 1 and 6 are forces to be exactly equal, and so are the potentials at vertices 2 and 4. Such equality of potentials is implicit in the node numbering for the quadrilateral region, Fig. 2.3(*b*). Of course there need not be any particular relationship between the node numbers for the disjoint triangles on the one hand, and the quadrilateral on the other; both the disjoint and connected numberings in Fig. 2.3 are quite arbitrary. The equality constraint at vertices may be expressed in matrix form, as a rectangular matrix $\mathbf{C}$ relating potentials of the disjoint elements to the potentials of the conjoint set of elements (also termed the connected system):

$$\mathbf{U}_{\text{dis}} = \mathbf{C}\mathbf{U}_{\text{con}}, \tag{2.24}$$

where the subscripts denote disjoint and conjoint sets of elements, respectively. With the point numberings shown in Fig. 2.3, this equation reads in full

$$\begin{bmatrix} U_1 \\ U_2 \\ U_3 \\ U_4 \\ U_5 \\ U_6 \end{bmatrix}_{\text{dis}} = \begin{bmatrix} 1 & & & \\ & 1 & & \\ & & 1 & \\ & 1 & & \\ & & & 1 \\ 1 & & & \end{bmatrix} \begin{bmatrix} U_1 \\ U_2 \\ U_3 \\ U_4 \end{bmatrix}_{\text{con}}. \tag{2.25}$$

To enhance legibility, all zero matrix elements have been omitted in Eq. (2.25). Substituting Eq. (2.24) into (2.21), the energy for the connected problem becomes

$$W = \tfrac{1}{2}\mathbf{U}_{\text{con}}^{\text{T}}\mathbf{S}\mathbf{U}_{\text{con}}, \tag{2.26}$$

where

$$\mathbf{S} = \mathbf{C}^{\text{T}}\mathbf{S}_{\text{dis}}\mathbf{C} \tag{2.27}$$

represents the assembled coefficient matrix of the connected problem. For the element assembly in Fig. 2.3,

$$\mathbf{S} = \begin{bmatrix} S_{11}^{(1)} + S_{66}^{(2)} & S_{12}^{(1)} + S_{64}^{(2)} & S_{13}^{(1)} & S_{65}^{(2)} \\ S_{21}^{(1)} + S_{46}^{(2)} & S_{22}^{(1)} + S_{44}^{(2)} & S_{23}^{(1)} & S_{45}^{(2)} \\ S_{31}^{(1)} & S_{32}^{(1)} & S_{33}^{(1)} & 0 \\ S_{56}^{(2)} & S_{54}^{(2)} & 0 & S_{55}^{(2)} \end{bmatrix}. \tag{2.28}$$

The disjoint and conjoint numberings are frequently also termed *local* and *global* numberings, respectively. For the assembly of two triangles, all that is required is evidently to add corresponding matrix elements; there is no need to write down the matrix **C** explicitly.

2.4   *Solution of the connected problem*

The energy of a continuous approximate potential distribution was formulated above as a quadratic form involving the column vector of node potentials. To obtain an approximate solution of Laplace's equation, it remains to minimize the stored energy in the connected finite element model. Since the energy expression of Eq. (2.27) is quadratic in the nodal potentials, it must have a unique minimum with respect to each component of the potential vector **U**. Hence, to minimize it is sufficient to set

$$\frac{\partial W}{\partial U_k} = 0. \tag{2.29}$$

Here the index k refers to entries in the connected potential vector $\mathbf{U}_{\text{con}}$, or what is equivalent, to node numbers in the connected model. Differentiation with respect to each and every k corresponds to an unconstrained minimization with the potential allowed to vary at every node. This, however, does not correspond to the boundary-value problem as originally stated in Fig. 2.1, where the potentials at all Dirichlet boundary points were fixed at some prescribed values. Indeed, the unconstrained minimum energy is trivially zero, with exactly zero potential everywhere. Clearly, that subset of the potentials contained in **U** which corresponds to the fixed potential values must not be allowed to vary in the minimization. Suppose the node numbering in the connected model is such that all nodes whose potentials are free to vary are numbered first, all nodes with prescribed potential values are numbered last. In Fig. 2.1(*a*), for example, the nodes in the interconductor space (where the potential is to be determined) would be listed first, and all nodes lying on conductor surfaces (where the potential is prescribed) would be numbered last. Equation (2.29) may then be written with the matrices in partitioned form,

$$\frac{\partial W}{\partial U_k} = \frac{\partial}{\partial [\mathbf{U}_{\text{f}}]_k} [\mathbf{U}_{\text{f}}^{\text{T}} \ \mathbf{U}_{\text{p}}^{\text{T}}] \begin{bmatrix} \mathbf{S}_{\text{ff}} & \mathbf{S}_{\text{fp}} \\ \mathbf{S}_{\text{pf}} & \mathbf{S}_{\text{pp}} \end{bmatrix} \begin{bmatrix} \mathbf{U}_{\text{f}} \\ \mathbf{U}_{\text{p}} \end{bmatrix} = 0, \tag{2.30}$$

where the subscripts f and p refer to nodes with free and prescribed potentials, respectively. Note that the differentiation is carried out only with respect to the kth component of $\mathbf{U}_f$, never with respect to any component of $\mathbf{U}_p$, for the subvector $\mathbf{U}_p$ contains only prescribed potentials which cannot vary. Differentiating with respect to the free potentials only, there results the matrix equation

$$[\mathbf{S}_{ff} \quad \mathbf{S}_{fp}] \begin{bmatrix} \mathbf{U}_f \\ \mathbf{U}_p \end{bmatrix} = 0. \tag{2.31}$$

The coefficient matrix in this equation is rectangular. It contains as many rows as there are unconstrained (free) variables, and as many columns as there are nodel potentials — the total number of free as well as prescribed potential values. Rewriting so as to put all prescribed (and therefore known) quantities on the right, Eq. (2.31) becomes

$$\mathbf{S}_{ff}\mathbf{U}_f = -\mathbf{S}_{fp}\mathbf{U}_p. \tag{2.32}$$

Here the left-hand coefficient matrix is square and nonsingular; a formal solution to the whole problem is therefore given by

$$\mathbf{U} = \begin{bmatrix} -\mathbf{S}_{ff}^{-1}\mathbf{U}_f\mathbf{S}_{fp} \\ \mathbf{U}_p \end{bmatrix}. \tag{2.33}$$

The approximate solution as calculated takes the form of a set of nodal potential values. However, it is important to note that the finite element solution is uniquely and precisely defined everywhere, not only at the triangle vertices, because the energy minimization assumed the solution surface to have a particular shape. The set of nodal potential values is merely a compact representation for the piecewise-planar solution surface which yields minimum energy.

Within each triangle the local potential values are prescribed by Eq. (2.7). Thus, no further approximation is necessary to obtain contour plots of equipotential values, to calculate the total stored energy, or to perform any other desired further manipulations. Since in this method the potential in each element is taken to be the linear interpolate of its vertex values, as in Eq. (2.10), an equipotential plot will necessarily consist of piecewise-straight contour segments. For example, Fig. 2.4 shows equipotential contours for the problem of Fig. 2.1.

It is worth noting in Fig. 2.4 that the Dirichlet boundary conditions (prescribed-value boundary conditions) are exactly satisfied, because the potential values at the boundary nodes are explicitly specified when Eqs. (2.31) and (2.32) are set up. On the other hand, the homogeneous Neumann boundary condition (prescribed zero normal derivative) is not satisfied exactly, but only in a certain mean-value sense which causes

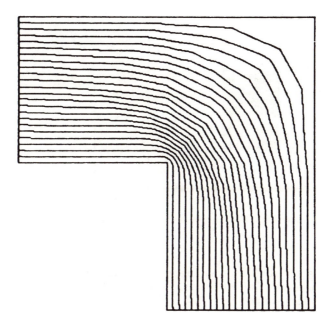

Fig 2.4 Equipotential contours for the coaxial-line problem of Fig. 2.1. Note the sharp bends at interelement boundaries.

the contour integral term in Eq. (2.5) to vanish. This boundary condition could of course be rigidly imposed; but the stored field energy would be raised thereby, so that the overall solution accuracy would in fact be worse. This point will be considered in some detail later. Roughly speaking, the procedure followed here trades error incurred along the Neumann boundary for a global accuracy increase.

3.   Inhomogeneous sources and materials

Many practical field problems involve distributed sources or are otherwise more general than Laplace's equation allows. They can be dealt with using broadly similar techniques. This section widens the scope of solvable problems by including distributed sources and material inhomogeneities.

3.1   *Poisson's equation*

Where distributed sources occur within the field region, an approach similar to the above may be used, but with the difference that the source distributions must be explicitly included. As a simple example, Fig. 2.5 shows an electric machine conductor lying in a rotor slot. The machine rotor and stator are separated by an air gap; the rotor

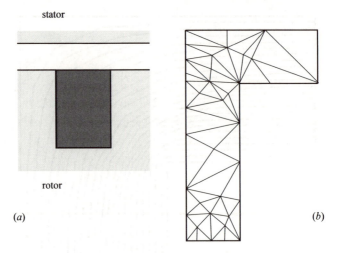

stator

rotor

(*a*)   (*b*)

Fig 2.5 (*a*) Electric machine rotor slot, filled by current-carrying
conductor. (*b*) Finite element mesh for analysis of the fields in the slot
and air gap near the slot.

surface is slotted and each slot contains a current-carrying conductor.
The conductor, which substantially fills the slot, may be assumed to carry
uniformly distributed current. The magnetic field in the air gap and slot
can be described by the magnetic vector potential **A**, which satisfies a
vector form of the Poisson equation. If the slot and conductor are
assumed infinitely long, both the current density **J** and the magnetic
vector potential **A** possess only longitudinally directed components, so
that the vector Poisson equation degenerates to its scalar counterpart:

$$\nabla^2 A = -\mu_o J. \tag{3.1}$$

The machine iron will be assumed infinitely permeable, an assumption
commonly made because its relative permeability is often in the range
10^3–10^4. Under this assumption, the normal derivative of A at all iron
surfaces must be zero. Further, its normal derivative must vanish at the
slot centreline for reasons of symmetry. A short distance from the slot
itself, flux lines must cross the air gap orthogonally; thus the right edge of
the L-shaped region of Fig. 2.5(*b*) represents a flux line, a constant value
of A. Boundary conditions for the problem are thereby clearly defined.

The variational problem equivalent to solving Poisson's equation is
that of minimizing the energy-related functional

$$F(u) = \frac{1}{2}\int \nabla u \cdot \nabla u \, dS - \mu_0 \int u J \, dS. \tag{3.2}$$

To show that this functional does reach a minimum at the true solution of Eq. (3.1), suppose A is the correct solution and h is some differentiable function which vanishes at all boundary points where A is prescribed. Let $F(A + \theta h)$ be evaluated, where θ is a numerical parameter:

$$F(A + \theta h) = F(A) + \theta \int \nabla A \cdot \nabla h \, \mathrm{d}S$$

$$- \theta \mu_0 \int hJ \, \mathrm{d}S + \frac{1}{2} \theta^2 \int \nabla h \cdot \nabla h \, \mathrm{d}S \qquad (3.3)$$

Using Green's theorem once again (Appendix 2), the second term on the right may be changed to read

$$\int \nabla A \cdot \nabla h \, \mathrm{d}S = \oint h \frac{\partial A}{\partial n} \, \mathrm{d}s - \int h \nabla^2 A \, \mathrm{d}S. \qquad (3.4)$$

The contour integral on the right vanishes, since h vanishes at all Dirichlet boundary segments and the normal derivative of A is zero at every remaining boundary point. Since A is the correct solution of Eq. (3.1), the rightmost term of Eq. (3.4) may be written

$$- \int h \nabla^2 A \, \mathrm{d}S = \mu_0 \int hJ \, \mathrm{d}S. \qquad (3.5)$$

This term immediately cancels, so the functional of Eq. (3.3) simplifies to

$$F(A + \theta h) = F(A) + \frac{1}{2} \theta^2 \int \nabla h \cdot \nabla h \, \mathrm{d}S. \qquad (3.6)$$

Since the integral on the right is always positive, it is evident that a minimum will be reached when θ has zero value; and conversely, that $F(u)$ reaches its minimum value for $u = A$, the solution of Eq. (3.1).

It will be noted that the field energy is still calculable by the general expression

$$W = \frac{1}{2} \int \nabla A \cdot \nabla A \, \mathrm{d}S \qquad (3.7)$$

or by evaluating the equivalent expression

$$W = \frac{\mu_0}{2} \int AJ \, \mathrm{d}S. \qquad (3.8)$$

At its minimum value $F(A)$, the functional F evidently has a negative value equal in magnitude to the total stored energy. Once again, the error term in Eq. (3.6) depends on the square of the parameter θ. Near the correct solution, θ is small. The accuracy with which the stored energy can be found is therefore very high, even if the potential values are locally not very accurate.

3.2   *Modelling the source term*

To construct a finite element model of the Poisson-equation problem, a procedure will be employed similar to that used for Laplace's equation. The problem region will again be triangulated, as shown in Fig. 2.5, and initially a typical triangular element will be examined in isolation from the rest. Since the first term in the functional of Eq. (3.2) is identical to the right-hand side of Eq. (2.3), the discretization process follows exactly the same steps, and leads to a matrix representation of the connected finite element indistinguishable from the corresponding result for Laplace's equation, Eq. (2.26). The second term in Eq. (3.2) is new; it requires treatment which is similar in principle, but slightly different in details.

Over any one triangle, the prescribed current density $J(x, y)$ will be approximated in a manner similar to the potential,

$$J = \sum_{i=1}^{3} J_i \alpha_i(x, y), \tag{3.9}$$

where the right-hand coefficients J_i are vertex values of current density within the triangle. These values are of course known, since the current density itself is a prescribed function. The source-term integral in Eq. (3.2) may therefore be written

$$\int AJ\,\mathrm{d}S = \sum_{i=1}^{3}\sum_{j=1}^{3} A_i \int \alpha_i \alpha_i\,\mathrm{d}S\, J_j, \tag{3.10}$$

with the approximated vertex potential values A_i the only unknowns. For each element, let another square matrix of order 3 be defined by

$$T_{ij}^{(e)} = \int \alpha_i \alpha_j\,\mathrm{d}S, \tag{3.11}$$

so that

$$\int AJ\,\mathrm{d}S = \mathbf{A}^{\mathrm{T}}\mathbf{T}^{(e)}\mathbf{J}. \tag{3.12}$$

For the disjoint set of triangular elements, the functional of Eq. (3.2) now becomes

$$F(A) = \frac{1}{2}\mathbf{A}_{\mathrm{dis}}^{\mathrm{T}}\mathbf{S}_{\mathrm{dis}}\mathbf{A}_{\mathrm{dis}} - \mu_0 \mathbf{A}_{\mathrm{dis}}^{\mathrm{T}}\mathbf{T}_{\mathrm{dis}}\mathbf{J}_{\mathrm{dis}}. \tag{3.13}$$

The element interconnection once again expresses itself in the requirement of potential continuity and hence in a constraint transformation like Eq. (2.24). Thus

$$F(A) = \frac{1}{2}\mathbf{A}^{T}\mathbf{S}\mathbf{A} - \mu_0\mathbf{A}^{T}\mathbf{C}^{T}\mathbf{T}_{\text{dis}}\mathbf{J}_{\text{dis}}. \tag{3.14}$$

Minimization of $F(A)$ with respect to each and every unconstrained vertex potential, putting

$$\frac{\partial F}{\partial A_k} = 0, \tag{3.15}$$

then leads to the rectangular matrix equation

$$\mathbf{S}'\mathbf{A} = \mu_0\mathbf{C}^{T}\mathbf{T}_{\text{dis}}\mathbf{J}_{\text{dis}} \tag{3.16}$$

as the finite element model of the boundary-value problem. It should be noted that the current density coefficient vector on the right has not been constrained in the same way as the potentials, but remains discontinuous. This reflects the physics of the problem: there is no need for source densities to be continuous across interelement boundaries. Therefore no further transformations need apply to the right-hand side.

Since the differentiation of Eq. (3.15) cannot be carried out with respect to fixed potentials, the matrix $\mathbf{S}'$ of Eq. (3.16) is rectangular. It possesses as many rows as there are potentials free to vary, while its number of rows equals the total number of nodal potentials (with free as well as prescribed values). Just as in Eq. (2.31), the vector of node potentials will now be partitioned so as to include all unconstrained potentials in the upper, all prescribed potentials in its lower part. The matrix $\mathbf{S}'$ is partitioned comfortably, leading to

$$\begin{bmatrix} \mathbf{S}_{\text{ff}} & \mathbf{S}_{\text{fp}} \end{bmatrix}\begin{bmatrix} \mathbf{A}_{\text{f}} \\ \mathbf{A}_{\text{p}} \end{bmatrix} = \mu_0\mathbf{C}^{T}\mathbf{T}_{\text{dis}}\mathbf{J}_{\text{dis}}. \tag{3.17}$$

As before, the subscripts f and p refer to free and prescribed potential values respectively. Since the latter are known, they will be moved to the right-hand side:

$$\mathbf{S}_{\text{ff}}\mathbf{A}_{\text{f}} = \mu_0\mathbf{C}^{T}\mathbf{T}_{\text{dis}}\mathbf{J}_{\text{dis}} - \mathbf{S}_{\text{fp}}\mathbf{A}_{\text{p}}. \tag{3.18}$$

Solution of this equation determines the unknown nodal potentials, thereby solving the problem.

It is interesting to note that the right-hand side of Eq. (3.18) combines the source term of the differential equation (the inhomogeneous part of the equation) with the effect of prescribed boundary values (the inhomogeneous part of the boundary conditions). Thus there is no fundamental distinction between the representation of a homogeneous differential equation with inhomogeneous boundary conditions on the one hand, and an inhomogeneous differential equation with homogeneous boundary conditions on the other.

A solution of the slot-conductor problem, using the rather simple triangulation of Fig. 2.5, appears in Fig. 2.6. Just as in the case of Laplace's equation, the Dirichlet boundary conditions (flux-line boundary conditions) are rigidly enforced, while the Neumann boundary conditions are not. As can be seen in Fig. 2.6, the latter conditions are therefore locally violated, but satisfied in the mean.

3.3   *Material inhomogeneities*

Many, possibly most, engineering field problems involve multiple physical media. An example is furnished by the shielded microstrip line of Fig. 2.7. It is similar to the coaxial line problem of Fig. 2.1 but contains two dielectric media, typically a ceramic or thermoplastic substrate to support the conductor, and air.

There is really not much difference in principle between the method used for the two-medium problem and the method developed for the single-dielectric coaxial line of Fig. 2.1. In both cases, the governing principle is that of minimum stored energy. Regardless of the dielectric media involved, the stored energy is given by

$$W = \frac{1}{2}\int \mathbf{E} \cdot \mathbf{D} \, dS = \frac{1}{2}\int \epsilon \mathbf{E} \cdot \mathbf{E} \, dS \tag{3.19}$$

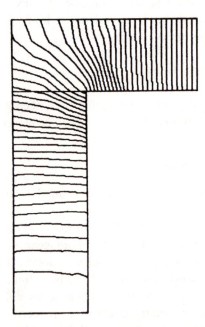

Fig 2.6 Solution of the electric machine slot problem of Fig. 2.5. The equipotentials of A correspond to magnetic flux lines.

Fig 2.7   Shielded microstrip line. The outer shield is
at the same potential as the ground plane.

which may be written in terms of the electric potential u as

$$W = \frac{1}{2} \int \epsilon \nabla u \cdot \nabla u \, \mathrm{d}S. \tag{3.20}$$

Provided that the element subdivision is such that element edges always
follow the medium boundaries, the Dirichlet matrix $\mathbf{S}^{(e)}$ of every element
is formed in the same manner as before, but then multiplied by the locally
applicable permittivity value ϵ before the element is joined to others.

3.4   *Natural interface conditions*

The energy minimization procedure for heterogeneous media is
simple, but a basic question remains unanswered: is it not necessary to
impose the well-known electrostatic boundary conditions that must
always prevail at an interface between media? These require that (1)
the tangential electric field must be continuous across any material interface, (2) the normal electric flux density may be discontinuous if there are
any surface charges, but must otherwise be continuous. The first condition is equivalent to requiring the potential to be continuous, and is
therefore met by any finite element solution. The second is not so
obviously satisfied. To see whether it is, the energy minimization will
now be followed through in detail.

As in the homogeneous-dielectric case, let $U = u + \theta h$ be an approximate solution, u the correct potential value, and h an error function that
vanishes at all Dirichlet boundary points where the correct value of
potential is known from the start. Then the stored energy $W(U)$ may
be written

$$W(u + \theta h) = W(u) + \theta^2 W(h) + \theta \int \epsilon \nabla u \cdot \nabla h \, \mathrm{d}S. \tag{3.21}$$

Suppose for the sake of simplicity that a mesh contains just two elements
L and R, as in Fig. 2.8. The integral in (3.21) is then best evaluated by
integrating over the two elements separately,

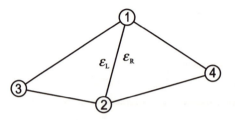

Fig 2.8   Two adjacent elements with different permittivities.

$$W(u + \theta h) = W(u) + \theta^2 W(h)$$
$$+ \theta \left(\int_L \epsilon_L \nabla u \cdot \nabla h \, dS + \int_R \epsilon_R \nabla u \cdot \nabla h \, dS \right). \quad (3.22)$$

Green's theorem once again comes to the rescue, for it allows the integrals to be rewritten,

$$\int \nabla u \cdot \nabla h \, dS = \oint h \frac{\partial u}{\partial n} \, ds - \int h \nabla^2 u \, dS. \quad (3.23)$$

The second term on the right of (3.23) vanishes identically, because u is the correct solution of Laplace's equation, $\nabla^2 u = 0$. Thus (3.22) may be restated as

$$W(u + \theta h) = W(u) + \theta^2 W(h)$$
$$+ \theta \left(\oint_L \epsilon_L h \frac{\partial u}{\partial n} \, ds + \oint_R \epsilon_R h \frac{\partial u}{\partial n} \, ds \right). \quad (3.24)$$

On all external boundaries, the contour integrals in (3.24) have vanishing integrands, for $h = 0$ on all Dirichlet boundaries while the normal derivative $\partial u / \partial n$ vanishes on the other boundary segments. It therefore suffices to carry out integration along the internal interface, 2–1 in Fig. 2.8, so that

$$W(u + \theta h) = W(u) + \theta^2 W(h)$$
$$+ \theta \left(\int_{21} \epsilon_L h \frac{\partial u}{\partial n} \bigg|_L \, ds + \int_{12} \epsilon_R h \frac{\partial u}{\partial n} \bigg|_R \, ds \right), \quad (3.25)$$

or, since the path of integration is exactly the same in both integrals,

$$W(u + \theta h) = W(u) + \theta^2 W(h) + \theta \int_{21} h \left(\epsilon_L \frac{\partial u}{\partial n} \bigg|_L - \epsilon_R \frac{\partial u}{\partial n} \bigg|_R \right) ds.$$
$$(3.26)$$

The contour integral term is the only one involving an odd power of θ. In the finite element method, a true minimum is found, not merely a point of inflection; this implies that the odd-power terms in θ must vanish identically. (Note that the *integral* must vanish, not necessarily the *integrand*).

Were the integrand to vanish, then at every point of the interface the normal derivatives from the left and right would be related by

$$\epsilon_L \left. \frac{\partial u}{\partial n} \right|_L = \epsilon_R \left. \frac{\partial u}{\partial n} \right|_R. \tag{3.27}$$

The physical significance of this equality is easy to see. The normal derivative may be written as the projection of the gradient onto the surface normal, so that

$$\epsilon_L \left. \frac{\partial u}{\partial n} \right|_L = \epsilon_L \nabla u \cdot \mathbf{1}_L = -\epsilon_L E_L. \tag{3.28}$$

This, however, is nothing more than the normal component of electric flux density,

$$\epsilon_L \left. \frac{\partial u}{\partial n} \right|_L = -D_L = \mathbf{D}_L \cdot \mathbf{1}_n, \tag{3.29}$$

where $\mathbf{1}_n$ is the unit normal vector. Consequently, forcing the interface integral in (3.26) to vanish amounts to imposing the restriction

$$\int_{21} h(\mathbf{D}_R - \mathbf{D}_L) \cdot \mathbf{1}_n \, ds = 0 \tag{3.30}$$

for each and every error function h possible with the finite element mesh used. Thus the rule of normal flux continuity is satisfied approximately over the interface between unlike media, but it is not satisfied identically at all points. Such a boundary condition is usually called a *natural* condition of the finite element method, for it comes to be satisfied naturally in the process of minimization without needing to be imposed explicitly.

3.5 *Energy and capacitance evaluation*

To evaluate the stored energy per unit length of the transmission lines of Fig. 2.1 or 2.7 it is easiest to proceed element by element. That is, one calculates the energy for each element and sums over all the elements:

$$W = \sum_e W^{(e)} = \frac{1}{2} \sum_e \mathbf{U}^{(e)\mathrm{T}} \mathbf{S}^{(e)} \mathbf{U}^{(e)}. \tag{3.31}$$

This is an undemanding computation because each step only involves the matrix $\mathbf{S}^{(e)}$ of a single element. The capacitance of a two-conductor line is then found from the stored energy as

$$C = \frac{2W}{(\Delta u)^2}, \tag{3.32}$$

where Δu is the potential difference between the two conductors.

Quantities of major importance in the analysis of TEM-mode transmission lines include, first and foremost, the propagation velocity v in the line and the line characteristic impedance Z_0. In terms of the capacitance and inductance per unit length, these are given by

$$v = \frac{1}{\sqrt{LC}}, \tag{3.33}$$

$$Z_0 = \sqrt{\frac{L}{C}}. \tag{3.34}$$

Their determination requires solving two field problems associated with the transmission line: (1) the electrostatic problem dealt with in considerable detail above, and (2) the corresponding magnetostatic problem, whose solution yields the inductance L per unit length. Rather than set up the magnetostatic problem, however, most analysts resort to solving a second electrostatic problem. Suppose that the substrate material in Fig. 2.7 is replaced by air and the electrostatic problem is solved again, yielding this time a capacitance C_0. The propagation velocity in this case is known to be c, the velocity of light in free space. From (3.33), the inductance is then immediately

$$L = \frac{1}{c^2 C_0}, \tag{3.35}$$

and this same inductance value must hold for both the true dielectric-loaded line and the hypothetical air-filled line, for neither contains any magnetic material. Substituting into (3.33) and (3.34),

$$v = c\sqrt{\frac{C_0}{C}}, \tag{3.36}$$

$$Z_0 = \frac{1}{c\sqrt{C_0 C}}. \tag{3.37}$$

From the program designer's point of view this procedure is superior to solving the electrostatic and magnetostatic problems separately, because it only requires a single program and a single data set.

4.   Program construction issues

The mathematical and physical principles set out so far suffice to establish the basis of finite element programs for a wide variety of problems. The design of finite element programs, and their actual construction, raise a few additional issues which will now be dealt with.

4.1   *Handling of boundary conditions*

Setting up finite element problems so that they will fit the theoretical framework of Eqs. (2.31) and (3.17) requires the problem variables to be numbered in a special fashion. To arrange the equations in the manner shown, all potentials free to vary must be numbered first, all potentials with prescribed values must come last. In practice, it is not always convenient to renumber variables, nor to partition matrices in this way. Happily, renumbering and partitioning are only required for purposes of explanation; in practical computing they are never necessary.

Consider again the very simple two-element problem shown in Fig. 2.3(*b*). It is assumed that potentials 3 and 4 are prescribed, 1 and 2 are free to vary; in other words, the variable numbering is fully in accordance with the scheme used above. Following (2.32), the matrix equation to be solved reads in this case

$$\begin{bmatrix} S_{11} & S_{12} \\ S_{21} & S_{22} \end{bmatrix} \begin{bmatrix} U_1 \\ U_2 \end{bmatrix} = - \begin{bmatrix} S_{13} & S_{14} \\ S_{23} & S_{24} \end{bmatrix} \begin{bmatrix} U_3 \\ U_4 \end{bmatrix}. \tag{4.1}$$

Not much can be said regarding the high-numbered potentials, whose values are prescribed and therefore known from the outset. They are what they are — a statement that can be put in the form of a surprisingly useful identity,

$$\begin{bmatrix} D_{33} & \\ & D_{44} \end{bmatrix} \begin{bmatrix} U_3 \\ U_4 \end{bmatrix} = \begin{bmatrix} D_{33} & \\ & D_{44} \end{bmatrix} \begin{bmatrix} U_3 \\ U_4 \end{bmatrix}. \tag{4.2}$$

Here **D** may be chosen to be any definite diagonal matrix. Equation (4.2) may be combined with (4.1) to give

$$\begin{bmatrix} S_{11} & S_{12} & & \\ S_{21} & S_{22} & & \\ & & D_{33} & \\ & & & D_{44} \end{bmatrix} \begin{bmatrix} U_1 \\ U_2 \\ U_3 \\ U_4 \end{bmatrix} = \begin{bmatrix} -S_{13} & -S_{14} \\ -S_{23} & -S_{24} \\ D_{33} & \\ & D_{44} \end{bmatrix} \begin{bmatrix} U_3 \\ U_4 \end{bmatrix}. \tag{4.3}$$

This reformulation still keeps all known quantities on the right, as is conventional; but by reintroducing U_3 and U_4 on the left as well, it hints at a computational procedure in which the right-hand side is formed first, and all four potentials are then computed without regard to whether they belong to the prescribed or free subsets. That, however, implies their order is unimportant!

Let an arbitrary numbering be introduced for the potentials, one which does not take the free and prescribed potentials in any particular sequence. For example, let the vertices 1–2–3–4 be renumbered 2–4–1–3, so that the fixed-potential nodes are now numbered 1 and 3. The physical problem obviously does not change in any way as a result of the renumbering; only the rows and columns of the coefficient matrix in Eq. (4.3) are permuted to be consistent with the new numbering. Equation (4.3) becomes

$$\begin{bmatrix} D_{11} & & & \\ & S_{22} & & S_{24} \\ & & D_{33} & \\ & S_{42} & & S_{44} \end{bmatrix} \begin{bmatrix} U_1 \\ U_2 \\ U_3 \\ U_4 \end{bmatrix} = \begin{bmatrix} D_{11}U_1 \\ -S_{21}U_1 - S_{23}U_3 \\ D_{33}U_3 \\ -S_{41}U_1 - S_{43}U_3 \end{bmatrix}. \quad (4.4)$$

The coefficient matrix on the left is positive definite, so (4.4) can be solved for all four potentials just as if none had known values.

Equation (4.4) has more rows and columns than (4.1), so solving it is more costly; it requires more storage and more computer time. On the other hand, vertices and potentials may be numbered as desired. If computing methods are appropriately selected, advantage can be taken of the known fact that the coefficient matrix is sparse, and in particular that the rows corresponding to prescribed potentials only contain a single non-zero entry. The increase in both storage and computing time can then be kept quite small. Thus the increased cost of handling the matrix problem is more than compensated by the work saved in not renumbering and rearranging equations.

In practice, the diagonal matrix **D** is often taken to be the unit matrix, **D** = **I**. Occasionally, matrices **S** are encountered whose entries are very large or very small compared to unity, so that numerical convenience may suggest a different choice of **D**. Fortunately, such circumstances do not often arise; and when they do, it is usually quite sufficient to give D_{ii} a value of the same order as the other diagonal matrix entries, e.g., the average value of the S_{ii}. If taking **D** = **I** is acceptable, Eq. (4.4) becomes

$$\begin{bmatrix} 1 & & & \\ & S_{22} & & S_{24} \\ & & 1 & \\ & S_{42} & & S_{44} \end{bmatrix} \begin{bmatrix} U_1 \\ U_2 \\ U_3 \\ U_4 \end{bmatrix} = \begin{bmatrix} U_1 \\ -S_{21}U_1 - S_{23}U_3 \\ U_3 \\ -S_{41}U_1 - S_{43}U_3 \end{bmatrix}. \quad (4.5)$$

Equation (4.5) implies that setting up the finite element equations and imposing the boundary conditions can be done conveniently at the same time, on an element-by-element basis. As each element matrix is constructed, row and column numbers are scanned to determine whether they correspond to free or prescribed potentials. Matrix entries which correspond to free potentials are entered in the natural fashion. Prescribed potential values, on the other hand, are treated by substituting rows and columns of the unit matrix on the left, and by augmenting the right-hand side. The right-hand side of Eq. (4.5) is thus built up piece-meal, one element at a time.

4.2   *Interconnection of multiple elements*

Assembly of all individual element matrices to form the global matrix representation requires the connection transformation, Eqs. (2.23)–(2.24), to be executed. All the required topological information is contained in the connection matrix **C**. However, it would clearly be most inefficient to store the connection matrix in the explicit form of Eq. (2.25), and the disjoint global matrices in the form given by Eq. (2.22), for both matrices contain a high proportion of zero entries. An efficient way of assembling the global matrices proceeds recursively, building up the finite element representation one element at a time.

Suppose a triangular mesh and its accompanying matrices are partly built, and one more element is to be added. The kth step of such a process is illustrated in Fig. 2.9, where element 7–8–9 is to be added to an existing mesh of four elements. Prior to its addition, the mesh may be thought to possess a set of connected potentials representable by a potential vector $\mathbf{U}_{\text{con}}^{(k-1)}$. Introduction of the new element creates a partially disconnected potential vector $\mathbf{U}_{\text{dis}}^{(k)}$ which is composed of $\mathbf{U}_{\text{con}}^{(k-1)}$ and three additional entries. This vector must be collapsed into a new connected potential vector $\mathbf{U}_{\text{con}}^{(k)}$ by a connection transformation that enforces potential con-tinuity. In Fig. 2.9, vector $\mathbf{U}_{\text{con}}^{(k-1)}$ has six components. The new discon-nected potential vector $\mathbf{U}_{\text{dis}}^{(k)}$ has nine components, but when the new element 7–8–9 is connected to the already existing mesh, the new nodal potentials at 8 and 9 merge with already existing potentials at nodes 5 and

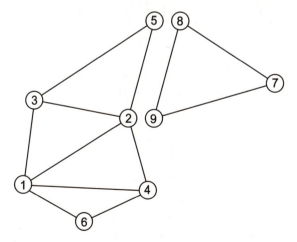

Fig 2.9   Recursive model buildup: a finite element mesh is extended by the addition of one more element.

2, so the connected potential vector $\mathbf{U}_{\text{con}}^{(k)}$ contains only seven components. Inspection shows that $\mathbf{U}_{\text{dis}}^{(k)}$ is related to $\mathbf{U}_{\text{con}}^{(k)}$ by a connection matrix $\mathbf{C}$,

$$
\begin{bmatrix} U_1 \\ U_2 \\ U_3 \\ U_4 \\ U_5 \\ U_6 \\ U_7 \\ U_8 \\ U_9 \end{bmatrix}_{\text{dis}} =
\begin{bmatrix} 1 & & & & & & \\ & 1 & & & & & \\ & & 1 & & & & \\ & & & 1 & & & \\ & & & & 1 & & \\ & & & & & 1 & \\ & & & & & & 1 \\ & & & & 1 & & \\ & 1 & & & & & \end{bmatrix}
\begin{bmatrix} U_1 \\ U_2 \\ U_3 \\ U_4 \\ U_5 \\ U_6 \\ U_7 \end{bmatrix}_{\text{con}} . \tag{4.6}
$$

Nodal potentials in the new vector $\mathbf{U}_{\text{dis}}^{(k)}$ may be classified into three categories, and the connection matrix may be partitioned accordingly:

(o) 'Old' potentials and nodes, which appear in both $\mathbf{U}_{\text{dis}}^{(k)}$ and $\mathbf{U}_{\text{dis}}^{(k-1)}$. In Fig. 2.9, nodes numbered 1 through 6 belong to this class.

(n) New modes whose potentials are independent, that will enlarge the node set. In Fig. 2.9, only node 7 belongs to this class.

(r) 'Removable' nodes whose potentials are dependent on existing ones, that will not increase the node set. In Fig. 2.9, nodes 8 and 9 belong to this class.

Partitioned by these classes, the connection relationship of (4.6) may be written

$$
\mathbf{U}_{\text{dis}}^{(k)} = \begin{bmatrix} \mathbf{U}_{\text{con}}^{(k-1)} \\ \mathbf{U}_{\text{n}} \\ \mathbf{U}_{\text{r}} \end{bmatrix} = \begin{bmatrix} \mathbf{I} & 0 \\ 0 & \mathbf{I} \\ \mathbf{C}_{\text{or}} & 0 \end{bmatrix} \begin{bmatrix} \mathbf{U}_{\text{con}}^{(k-1)} \\ \mathbf{U}_{\text{n}} \end{bmatrix} = \mathbf{C}\mathbf{U}_{\text{con}}^{(k)}. \tag{4.7}
$$

As elements are added to the mesh, the matrices $\mathbf{S}$ and $\mathbf{T}$ grow in a succession $\mathbf{S}^{(0)}, \ldots, \mathbf{S}^{(k-1)}, \mathbf{S}^{(k)}, \ldots$ To create $\mathbf{S}^{(k)}$, write first the disconnected matrix $\mathbf{S}_{\text{dis}}^{(k)}$ corresponding to the potential vector $\mathbf{U}_{\text{dis}}^{(k)}$, in the partitioned form

$$
\mathbf{S}_{\text{dis}}^{(k)} = \begin{bmatrix} \mathbf{S}_{\text{con}}^{(k-1)} & & \\ & \mathbf{S}_{\text{nn}} & \mathbf{S}_{\text{nr}} \\ & \mathbf{S}_{\text{rn}} & \mathbf{S}_{\text{rr}} \end{bmatrix} . \tag{4.8}
$$

Here the upper left submatrix is the matrix obtained at the previous recursion step; the remaining four submatrices contain the new element matrix, partitioned into its new and removable node rows and columns. Next, the removable rows and columns are removed by the connection transformation

$$\mathbf{S}^{(k)} = \mathbf{C}^{\mathrm{T}}\mathbf{S}_{\mathrm{dis}}^{(k)}\mathbf{C}. \tag{4.9}$$

Carrying out the matrix multiplication, the enlarged matrix $\mathbf{S}^{(k)}$ is

$$\mathbf{S}^{(k)} = \begin{bmatrix} \mathbf{S}_{\mathrm{con}}^{(k-1)} + \\ \mathbf{C}_{\mathrm{rn}}^{\mathrm{T}}\mathbf{S}_{\mathrm{rr}}\mathbf{C}_{\mathrm{rn}} & \mathbf{C}_{\mathrm{rn}}^{\mathrm{T}}\mathbf{S}_{\mathrm{rn}} \\ \mathbf{S}_{\mathrm{nr}}\mathbf{C}_{\mathrm{rn}} & \mathbf{S}_{\mathrm{nn}} \end{bmatrix}. \tag{4.10}$$

The matrix $\mathbf{T}$ is handled in precisely the same way. This procedure applies to any combining of partial meshes, not solely to individual elements. It is memory-effective; at each step the matrix is enlarged by only a border of new rows and columns. It is computationally rapid, for the submatrix $\mathbf{C}_{\mathrm{rn}}$ is both small and very sparse, so the matrix multiplications indicated in (4.10) can be carried out with high speed.

4.3 *Programming and data structures*

A major strength of the finite element method is its great geometric flexibility. Using triangular elements, any two-dimensional region may be treated whose boundary can be satisfactorily approximated by a series of straight-line segments. It should also be noted that the triangular element mesh in the interior of the problem region is regular neither geometrically nor topologically, the triangles being of varying sizes and shapes while their interconnection does not need to follow any fixed pattern.

Approximate solution of a given physical problem by means of finite elements may be regarded as comprising six distinct stages:

(i) Creation of finite element mesh, i.e., subdivision of the problem region into elements.

(ii) Definition of the sources and imposed boundary values of the problem.

(iii) Construction of the matrix representation of each element.

(iv) Assembly of all elements, by matrix transformations such as Eq. (4.10), and imposition of boundary conditions.

(v) Solution of the resulting simultaneous algebraic equations.

(vi) Display and evaluation of the results.

In essence, the geometrically and mathematically complicated boundary-value problem is decomposed into a disjoint set of elements in the first stage; all the subsequent stages merely reassemble the pieces in a systematic fashion and thereby produce the desired solution. The middle three stages are conveniently carried out one element at a time, for the matrix representation of each triangular element can be found if the vertex locations of that one triangle are known, without knowing anything about the remainder of the finite element mesh. Conversely, assembly and imposition of boundary conditions only require knowledge of the mesh topology, i.e., of the manner in which the triangles are interconnected, without regard to the nature of the material that the element contains, the source density, or indeed the differential equation that must be satisfied.

To proceed by way of example, consider the very simple mesh representation of Fig. 2.10 for the slot-conductor problem (Poisson's equation). The N triangles that constitute the problem region can be identified by an $N \times 3$ array giving the node numbers, and an $N \times 1$ array of the source densities in the elements. These arrays may be read from a file resident on some input device (e.g., a disk file). With one element per input line, data for this problem appear as follows:

$$
\begin{array}{ccc@{\qquad}c}
1 & 2 & 3 & 1.000 \\
2 & 4 & 3 & 1.000 \\
3 & 4 & 5 & 0.000 \\
4 & 7 & 6 & 0.000 \\
4 & 6 & 5 & 0.000 \\
\end{array}
$$

Matrix assembly proceeds as indicated by Eq. (4.10). The procedure starts by setting $\mathbf{S}_{con}^{(0)} = 0$ and continues by recursion, forming the element matrix for triangle 1–2–3, embedding this 3×3 matrix to form $\mathbf{S}_{con}^{(1)}$, then computing the element matrix for triangle 2–4–3, embedding it, ... until

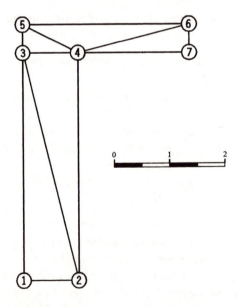

Fig 2.10 A very simple model for the slot problem, to illustrate data handling techniques.

the entire matrix **S** has been assembled. Computing the matrix representation of each individual element requires knowledge of its M vertex
coordinates. For this purpose, two $M \times 1$ arrays of the x and y coordinates are needed:

0.000	0.000
1.000	0.000
0.000	4.000
1.000	4.000
0.000	4.500
3.000	4.500
3.000	4.000

Finally, boundary conditions must be entered. For this purpose, it suffices
to specify boundary point numbers and corresponding potential values:

6	0.000
7	0.000

A program which implements this procedure in the Fortran 77 language is given at the end of this chapter. It will be noted that the program
itself contains no information specific to the particular physical problem
being solved. All geometric, topological, and other problem-dependent
information is contained in data arrays, which are prepared separately
and read in when required. Another quite different problem in Poisson's
equation can be solved using exactly the same program, by supplying a
different file of data.

5.   A sample program

The methods set out above are embodied in the Fortran program
shown at the end of this chapter. Only very simple methods are employed
in this program, in an attempt to keep it easy to read and easy to modify.
The overall program structure, on the other hand, is similar to most of
the very large, complex programs now in industrial and research use.

The SIMPLE2D package as given here comprises a main program
and various subroutines. In addition to these, it also calls upon subroutines contained in the general-purpose library package of Appendix
4. The main program contains no substantive executable statements other
than subroutine calls. It communicates with subroutines through
common blocks which reappear identically in all the subroutines. In
other words, the arrays and other data items are placed in a storage area
accessible to all program segments alike. The practical significance of
this data arrangement is important: if any alteration is required (for
example, if array dimensions are to be changed) it is only necessary to

restructure the common block appropriately, and then to replace the common block in every subroutine with an identical copy of the new version. No other reprogramming, and no piecemeal program testing, will be required.[1]

The main program shown here is purely a steering mechanism; it defines the data structure and sequences the subroutine calls, but it does not actually perform any computations. Such a program organization is common, for it allows individual subroutines to be replaced without affecting the operation of the others. For example, partial differential equations other than Poisson's equation could be solved by the program shown here, provided the element matrices used were appropriate to the new differential equation. But since the element matrices are generated by one subroutine, which carries out no other operations, the only program alteration needed to create the new program consists of removing that one subroutine and replacing it with a different one. All other program functions, such as data input, finite element assembly and equation solving, remain unaffected. For example, axisymmetric problems can be tackled with this program, provided an axisymmetric element subroutine is substituted for the *x–y* version shown.

The subroutines called by the main program, in the order of their appearance, are:

> meshin reads in and stores the problem data. It does only a very limited amount of validity checking. More sophisticated programs of this type often differ in the amount of data verification performed.

> elmatr computes the matrix representation of one first-order triangular element. It uses techniques discussed in detail in Chapter 4, slightly more general than those given above.

> elembd embeds the matrix contributions of one element into the global coefficient matrix and right-hand side.

> output prints out the solution once it has been obtained. This routine is very complicated in many finite element programs; it often incorporates graphic plotting, calculation of total stored energy, determination of maximal field values, and many other quantities of interest. Here it has been reduced to the most ele-

---

[1] Fortran 90 permits such repeated segments of code to be placed in a separate file and to be inserted where appropriate by means of the INCLUDE statement. A similar construct is available in C and C++. Various Fortran 77 compilers provide some variant of the INCLUDE mechanism, but only as a violation (usually referred to as an *extension*) of the language standard.

mentary form conceivable, a neatly formatted dump of all known values.

Although simple in concept, finite element methods can lead to relatively complex programs if reasonable flexibility in geometry and problem range is desired. On the other hand, even at the simplest level, finite element programs can (and should) be written to be highly problem-independent. It is important that the user of finite element methods acquires some understanding of the program structure involved, as well as an appreciation of the underlying mathematics. However, it is usually unwise to plunge into *ad hoc* program development for a specific problem — the development of efficient and error-free programs is a complex task often best left to specialists. The user primarily interested in solving particular problems will often find it more rewarding to modify already existing programs, rather than succumb to the temptation to start over again from scratch.

Like all programs given in this book, SIMPLE2D is truly simple-minded in its handling of errors. Checking for input data errors often makes up a significant part a finite element program because the number of possible mistakes is very large. Are there any overlapping triangles? Have all the triangle vertices been properly declared as nodes? Are they all geometrically distinct? Is the problem ill-posed (e.g., all potentials fixed)? Is the problem solvable (e.g., are there any fixed boundary values at all)? Are there are conflicts in boundary conditions (e.g., same node assigned two values)? SIMPLE2D hardly worries about such matters, and when it does discover an error, it makes no attempt at graceful recovery. Its response to an error consists of calling the error exception subroutine errexc included in the general utilities package. This error handler prints out the name of the routine where the error occurred, along with an error number which usually gives some further information about the error — and then simply stops the program run. While this procedure may seem inelegant, it does reduce managerial clutter in programs and thereby makes them much more readable than they other-wise might have been.

Input to SIMPLE2D is generally free-format, there is no need to worry much about which character position in a line any particular data item might occupy. This means, however, that blank lines are ambiguous — do they mean 'stop looking for data' or are they just white space to be ignored? Like the other programs in this book, SIMPLE2D assumes blank lines are generally of no significance, and that any line intended as a data terminator will contain the / character, which Fortran 77 interprets to mean 'end of data record'.

6.   Further reading

The finite element method using first-order elements has been applied to a large variety of electrical engineering problems in the past, and will no doubt continue to be applied. Although first-order elements do not produce solutions of particularly high accuracy, the method is simple to understand, simple to program, and above all simple to formulate where the fundamental physical equations are more complicated than those illustrated in this chapter.

The first application of triangular elements to the calculation of electric or other potential fields was probably that of Courant (1943). In his paper, piecewise approximation methods similar to finite elements were first developed. First-order triangular elements in essentially their present form were developed by Duffin (1959), who indicated the methods for solution and pointed out the availability not merely of approximate solutions, but of bounds for the stored field energy.

There are very few textbooks on finite elements as applied to electrical engineering, so it is not difficult to give a nearly exhaustive list of references. The material treated in this chapter is also dealt with by Jin (1993), approaching it a little differently but quite readably. Sabonnadière and Coulomb (1987) are less concerned than Jin with presenting an introduction to the beginner, but the introductory portions of their monograph may be helpful to readers anyway. Of course, there are many monographs and textbooks on finite element theory as applied to structural engineering, though unfortunately these can be quite difficult for the electrical engineer to read. Of the widely used civil engineers' finite element reference, the two-volume text by Zienkiewicz and Taylor (1989), only Chapter 10 is both useful and readily accessible to electrical engineers. Bickford (1990) can be recommended as an introduction to two-dimensional problems of the kind discussed in this chapter. His treatment may be excessively lengthy and detailed for some tastes, but in an introductory text it may be well to err on the side of prolixity. The text by Norrie and de Vries (1978) is written from the viewpoint of a mechanical engineer, but is sufficiently interdisciplinary to satisfy some electrical engineers as well. The little book by Owen and Hinton (1980) is easy to read, but many will find it not really sufficient, for it covers little more than the content of this chapter.

The paucity of textbook material in this field, however, is richly compensated by articles in periodicals and conference records. With the growing popularity of finite element methods for structural analysis in the 1960s, Zienkiewicz and Cheung (1965) attempted solution of practical potential problems and reported on their results. First-order triangular elements were applied to electrical problems shortly thereafter by Silvester (1969), as well as by Ahmed and Daly (1969).

The geometric flexibility of first-order triangles has endeared them to many analysts, so that they are at times used even where higher-order elements would probably yield better accuracy with lower computing costs. Andersen (1973) gives examples of well-working computer programs using first-order elements for daily design work.

In principle, circuits are mathematical abstractions of physically real fields; nevertheless, electrical engineers at times feel they understand circuit theory more clearly than fields. Carpenter (1975) has given a circuit interpretation of first-order elements.

Particularly if iterative equation-solving methods are employed, first-order finite element techniques sometimes resemble classical finite difference methods. The question is occasionally asked as to which method should then be preferred. This problem was investigated by Demerdash and Nehl (1976), who compared results for sample problems solved by both methods. Finite elements seem preferable even at the first-order level.

An interesting point to note about all finite element methods is that the approximate solution is uniquely defined at all points of interest, not merely on certain discretely chosen points as in finite difference methods. Further use of the solutions therefore often requires no further approximation. For example, Daly and Helps (1972) compute capacitances directly from the finite element approximation, without additional assumptions. Similarly, Bird (1973) estimates waveguide losses by reference to the computed solution.

7.   Bibliography

Ahmed, S. and Daly, P. (1969), 'Waveguide solutions by the finite-element method', *Radio and Electronic Engineer*, **38**, pp. 217–23.

Anderson, O.W. (1973), 'Transformer leakage flux program based on finite element method', *IEEE Transactions on Power Apparatus and Systems*, **PAS-92**, pp. 682–9.

Bickford, W.B. (1990), *A First Course in the Finite Element Method.* Homewood, Illinois: Irwin. xix + 649 pp.

Bird, T.S. (1973), 'Evaluation of attenuation from lossless triangular finite element solutions for inhonogemeously filled guiding structures', *Electronics Letters*, **9**, pp. 590–2.

Carpenter, C.J. (1975), 'Finite-element network models and their application to eddy-current problems', *Proceedings IEEE*, **122**, pp. 455–62.

Courant, R.L. (1943), 'Variational method for the solution of problems of equilibrium and vibration', *Bulletin of the American Mathematical Society*, **49**, pp. 1–23.

Daly, P. and Helps, J.D. (1972), 'Direct method of obtaining capacitance from finite-element matrices', *Electronics Letters*, **8**, pp. 132–3.

Demerdash, N.A. and Nehl, T.W. (1976), 'Flexibility and economics of the finite element and difference techniques in nonlinear magnetic fields of power devices', *IEEE Transactions on Magnetics*, **MAG-12**, pp. 1036–8.

Duffin, R.J. (1959), 'Distributed and lumped networks', *Journal of Mathematics and Mechanics*, **8**, pp. 793–826.

Jin, J.-M. (1993), *The Finite Element Method in Electromagnetics*. New York: John Wiley. xix + 442 pp.

Norrie, D.H. and de Vries, G. (1978), *An Introduction to Finite Element Analysis*. New York: Academic Press. xiii + 301 pp.

Owen, D.R.J. and Hinton, E. (1981), *A Simple Guide to Finite Elements*. Swansea: Pineridge Press. viii + 136 pp.

Sabonnadière, J.-C. and Coulomb, J.-L. (1987), *Finite Element Methods in CAD: Electrical and Magnetic Fields*. [S. Salon, transl.] New York: Springer-Verlag. viii + 194 pp.

Silvester, P. (1969), 'Finite element solution of homogeneous waveguide problems', *Alta Frequenza*. **38**, pp. 313–17.

Zienkiewicz, O.C. and Cheung, Y.K. (1965), 'Finite elements in the solution of field problems', *The Engineer*, Sept. 24, 1965, 507–10.

Zienkiewicz, O.C. and Taylor, R.L. (1989), *The Finite Element Method*. 4th edn. Vol. 1 of 2 vols. London: McGraw-Hill. xix + 648 pp.

8.   Programs

Program S IMPLE 2D, reproduced opposite, is closely related to other programs in this book and shares various utility subroutines with them. Apart from these, it is self-contained. The program listing is followed by a simple data set which should permit testing for correct compilation and execution.

```
C***************************************************************
C***********                                        ***********
C***********  First-order demonstration program  ***********
C***********                                        ***********
C***************************************************************
C        Copyright (c) 1995  P.P. Silvester and R.L. Ferrari
C***************************************************************
C
C        The subroutines  that make up this program  communicate
C        via named common blocks.  The variables in commons are:
C
C        Problem definition and solution
C        common /problm/
C          nodes   =  number of nodes used in problem
C          nelmts  =  number of elements in model
C          x, y    =  nodal coordinates
C          constr  =  logical, .true. for fixed potentials
C          potent  =  nodal potential array
C          nvtx    =  list of nodes for each element
C          source  =  source density in each element
C
C        Global matrix and right-hand side
C        common /matrix/
C          s       =  global s-matrix for whole problem
C          rthdsd  =  right-hand side of system of equations
C
C        Temporary working arrays and variables
C        common /workng/
C          sel     =  S for one element (working array)
C          tel     =  T for one element (working array)
C          intg    =  integer working array
C
C        Predefined problem size parameters (array dimensions):
C          MAXNOD  =  maximum number of nodes in problem
C          MAXELM  =  maximum number of elements in problem
C          HUGE    =  upper bound for any numeric value
C
C==============================================================
C        Global declarations -- same in all program segments
C==============================================================
        parameter (MAXNOD = 50, MAXELM = 75, HUGE = 1.E+35)
        logical constr
        common /problm/ nodes, nelmts, x(MAXNOD), y(MAXNOD),
     *                  constr(MAXNOD), potent(MAXNOD),
     *                  nvtx(3,MAXELM), source(MAXELM)
        common /matrix/ s(MAXNOD,MAXNOD), rthdsd(MAXNOD)
        common /workng/ sel(3,3), tel(3,3), intg(3)
C==============================================================
C
C        Fetch input data from input file
        call meshin
C
C        Set global s-matrix and right side to all zeros.
        call matini(s, MAXNOD, MAXNOD)
```

```
      call vecini(rthdsd, MAXNOD)
C
C     Assemble global matrix, element by element.
      do 40 i = 1,nelmts
C          Construct element s and t matrices
        ie = i
        call elmatr(ie)
C          Embed matrices in global s; augment right side:
        call elembd(ie)
   40   continue
C
C     Solve the assembled finite element equations
      call eqsolv(s, potent, rthdsd, nodes, MAXNOD)
C
C     Print out the resulting potential values
      call output
C
      stop
      end
C
C****************************************************************
C
      Subroutine meshin
C
C****************************************************************
C
C     Reads input data file in three parts:   nodes, elements,
C     fixed potentials.  Each part is concluded by a line that
C     contains only the / character at its leftmost position.
C
C     Nodes:              node number, x, y.
C     Elements:           node numbers, source density.
C     Potentials:         node number, fixed value.
C
C     All data are echoed as read  but little checking is done
C     for validity.
C
C================================================================
C     Global declarations -- same in all program segments
C================================================================
      parameter (MAXNOD = 50, MAXELM = 75, HUGE = 1.E+35)
      logical constr
      common /problm/ nodes, nelmts, x(MAXNOD), y(MAXNOD),
     *                constr(MAXNOD), potent(MAXNOD),
     *                nvtx(3,MAXELM), source(MAXELM)
      common /matrix/ s(MAXNOD,MAXNOD), rthdsd(MAXNOD)
      common /workng/ sel(3,3), tel(3,3), intg(3)
C================================================================
C
      Dimension nold(3), nnew(3)
C
C     Read in the node list and echo input lines.
C         Start by printing a heading for the node list.
      write (*, 1140)
```

```
1140 format (1x // 8x, 'Input node list' / 3x, 'n', 8x, 'x',
   1          11x, 'y' / 1x)
C
C        Read and echo nodes
     nodes = 0
     xold = HUGE
     yold = HUGE
  20 read (*,*, end=911) xnew, ynew
     if (xnew .ne. xold .or. ynew .ne. yold) then
       nodes = nodes + 1
       x(nodes) = xnew
       y(nodes) = ynew
       xold = xnew
       yold = ynew
       write (*, 1105) nodes, x(nodes), y(nodes)
       go to 20
     endif
1105 format (1x, i3, 2(2x, f10.5))
C
C    Read in the element list and echo all input as received.
C        Print heading to start.
     Write (*, 1160)
1160 format (1x // 6x, 'Input element list' / 3x, 'i', 5x,
   1          'j', 5x, 'k', 6x, 'Source' / 1x)
C
C        Read elements in turn.  Echo and count.
     nelmts = 0
     do 25 i = 1,3
  25   nold(i) = 0
  30 read (*,*, end=911) nnew, srcnew
     if (nnew(1) .ne. nold(1) .or.
   *     nnew(2) .ne. nold(2) .or.
   *     nnew(3) .ne. nold(3)) then
       nelmts = nelmts + 1
       do 35 i = 1,3
         nvtx(i,nelmts) = nnew(i)
  35     nold(i) = nnew(i)
       source(nelmts) = srcnew
       write (*, 1180) nnew, srcnew
       go to 30
     endif
1180 format (1x, i3, 2i6, 2x, g10.5)
C
C    Read list of fixed potential values and print.
C        Print header to start.
 120 write (*, 1200)
1200 format (1x // 5x, 'Input fixed potentials' / 6x,
   1          'node', 12x, 'value' / 1x)
C        Declare all nodes to start off unconstrained.
     do 40 m = 1,nodes
       constr(m) = .false.
  40   continue
     call vecini(potent, nodes)
C
```

```
C              Read and echo input.
       nconst = 0
       iold = 0
   60 read (*,*, end=911) inew, potnew
       if (inew .ne. iold) then
         nconst = nconst + 1
         constr(inew) = .true.
         potent(inew) = potnew
         iold = inew
         write (*, 1210) inew, potnew
         go to 60
       endif
 1210 format (6x, i3, 9x, f10.5)
C
C     Return to calling program.
       return
  911 call errexc('MESHIN', 3)
       end
C
C*************************************************************
C
       Subroutine elmatr(ie)
C
C*************************************************************
C
C     Constructs element matrices  S and T for a single first-
C     order triangular finite element.  ie = element number.
C
C===============================================================
C     Global declarations -- same in all program segments
C===============================================================
       parameter (MAXNOD = 50, MAXELM = 75, HUGE = 1.E+35)
       logical constr
       common /problm/ nodes, nelmts, x(MAXNOD), y(MAXNOD),
      *                constr(MAXNOD), potent(MAXNOD),
      *                nvtx(3,MAXELM), source(MAXELM)
       common /matrix/ s(MAXNOD,MAXNOD), rthdsd(MAXNOD)
       common /workng/ sel(3,3), tel(3,3), intg(3)
C===============================================================
C
C     Set up indices for triangle
       i = nvtx(1,ie)
       j = nvtx(2,ie)
       k = nvtx(3,ie)
C
C     Compute element T-matrix
       area = abs((x(j) - x(i)) * (y(k) - y(i)) -
      1           (x(k) - x(i)) * (y(j) - y(i))) / 2.
       do 20 l = 1,3
         do 10 m = 1,3
   10      tel(l,m) = area / 12.
   20    tel(l,l) = 2. * tel(l,l)
C
C     Compute element S-matrix
```

```
      i1 = 1
      i2 = 2
      i3 = 3
      call matini(sel, 3, 3)
      do 50 nvrtex = 1,3
        ctng = ((x(j) - x(i)) * (x(k) - x(i)) +
     1         (y(j) - y(i)) * (y(k) - y(i))) / (2. * area)
        ctng2 = ctng / 2.
C
        sel(i2,i2) = sel(i2,i2) + ctng2
        sel(i2,i3) = sel(i2,i3) - ctng2
        sel(i3,i2) = sel(i3,i2) - ctng2
        sel(i3,i3) = sel(i3,i3) + ctng2
C
C         Permute row and column indices once
        i4 = i1
        i1 = i2
        i2 = i3
        i3 = i4
        l = i
        i = j
        j = k
        k = l
   50   continue
C
      return
      end
C
C*************************************************************
C
      Subroutine elembd(ie)
C
C*************************************************************
C
C     Embeds single-element S and T matrices currently in sel
C     and tel (in common block "workng") in the global matrix
C     s.  Argument ie is the element number.
C
C=============================================================
C     Global declarations -- same in all program segments
C=============================================================
      parameter (MAXNOD = 50, MAXELM = 75, HUGE = 1.E+35)
      logical constr
      common /problm/ nodes, nelmts, x(MAXNOD), y(MAXNOD),
     *                constr(MAXNOD), potent(MAXNOD),
     *                nvtx(3,MAXELM), source(MAXELM)
      common /matrix/ s(MAXNOD,MAXNOD), rthdsd(MAXNOD)
      common /workng/ sel(3,3), tel(3,3), intg(3)
C=============================================================
C
C     Run through element S and T matrices (sel and tel),
C     augmenting the global S and the right-hand side.
      do 60 i = 1,3
        irow = nvtx(i,ie)
```

```
C
C             Does row correspond to a fixed potential?
          if (constr(irow)) then
C                Constrained row number.  Set global s and rthdsd.
             s(irow,irow) = 1.
             rthdsd(irow) = potent(irow)
          else
C                No, potential is free to vary.  Do all 3 columns.
             do 40 j = 1,3
                icol = nvtx(j,ie)
C
C                Does column correspond to a fixed potential?
                if (constr(icol)) then
C                   Yes; so augment right side only:
                   rthdsd(irow) = rthdsd(irow) + tel(i,j)
     1                    * source(ie) - sel(i,j) * potent(icol)
                else
C                   No; so augment s and rthdsd.
                   s(irow,icol) = s(irow,icol) + sel(i,j)
                   rthdsd(irow) = rthdsd(irow)
     1                       + tel(i,j) * source(ie)
                endif
   40        continue
          endif
   60    continue
C
C    All done -- return to calling program.
       return
       end
C
C
C*************************************************************
C
       Subroutine output
C
C*************************************************************
C
C    Prints the results on the standard output stream.
C
C=============================================================
C    Global declarations -- same in all program segments
C=============================================================
       parameter (MAXNOD = 50, MAXELM = 75, HUGE = 1.E+35)
       logical constr
       common /problm/ nodes, nelmts, x(MAXNOD), y(MAXNOD),
     *                 constr(MAXNOD), potent(MAXNOD),
     *                 nvtx(3,MAXELM), source(MAXELM)
       common /matrix/ s(MAXNOD,MAXNOD), rthdsd(MAXNOD)
       common /workng/ sel(3,3), tel(3,3), intg(3)
C=============================================================
C
C    Print the nodes and the output potential values.
C
       write (*,1000) (i, x(i), y(i), potent(i),
```

```
        1                                       i = 1, nodes)
   1000 format (1x // 12x, 'Final solution' / 3x, 'i', 8x,
      1             'x', 9x, 'y', 7x, 'potential' // (1x, i3, 2x,
      2             f10.5, F10.5, 3X, f10.5))
c
        return
        end
c


     0.010      0.020
     0.010      0.000
     0.030      0.000
     0.000      0.030
     0.000      0.000
/
     1  4  5          0.
     1  5  2          0.
     1  2  3      10000.
/
     4      0.000
     5      0.000
/
```

3

Electromagnetics of finite elements

1. Introduction

Finite element methods have been successful in electromagnetics largely because the conventional field equations permit numerous different reformulations. These bring the electromagnetic field within the scope of numerical methods that rely on high-order local approximations while permitting comparative laxity with respect to boundary and interface conditions. After a very brief review of electromagnetic theory as it is relevant to finite elements, this chapter describes the projective and variational reformulations that lead directly to finite element methods, the distinctions between various types of boundary conditions, and the sometimes confusing terminology attached to them.

2. Maxwell's equations

Electromagnetic field problems occupy a relatively favourable position in engineering and physics in that their governing laws can be expressed very concisely by a single set of four equations. These evolved through the efforts of several well-known scientists, mainly during the nineteenth century, and were cast in their now accepted form as differential equations by Maxwell. There also exists an equivalent integral form.

2.1 Differential relations

The variables that the Maxwell equations relate are the following set of five vectors and one scalar:

electric field intensity   **E**   volt/metre
magnetic field intensity   **H**   ampere/metre
electric flux density   **D**   coulomb/metre^2
magnetic flux density   **B**   tesla
electric current density   **J**   ampere/metre^2
electric charge density ρ coulomb/metre^3.

Each of these may be a function of three space coordinates x, y, z and the time t. The four Maxwell equations in differential form are usually written as follows:

$$\nabla \times \mathbf{E} = -\frac{\partial \mathbf{B}}{\partial t}, \tag{2.1}$$

$$\nabla \times \mathbf{H} = \mathbf{J} + \frac{\partial \mathbf{D}}{\partial t}, \tag{2.2}$$

$$\nabla \cdot \mathbf{D} = \rho, \tag{2.3}$$

$$\nabla \cdot \mathbf{B} = 0. \tag{2.4}$$

To these differential relations are added the constitutive relations

$$\mathbf{D} = \epsilon \mathbf{E}, \tag{2.5}$$

$$\mathbf{B} = \mu \mathbf{H}, \tag{2.6}$$

$$\mathbf{J} = \sigma \mathbf{E}, \tag{2.7}$$

describing the macroscopic properties of the medium being dealt with in terms of its permittivity ϵ, permeability μ and conductivity σ. The quantities ϵ, μ and σ are not necessarily simple constants, a notable exception being the case of ferromagnetic materials for which the **B–H** relationship may be a highly complicated nonlinear law. Furthermore, ϵ and μ may represent anisotropic materials, with flux densities differing in direction from their corresponding field intensities. In such cases the constitutive constants have to be written as tensors. For free space, $\epsilon_0 = 8.854 \cdots \times 10^{-12}$ farad/metre and $\mu_0 = 4\pi \times 10^{-7}$ henry/metre. The curiously exact value of μ_0 reflects the basic fact that the units of electromagnetism are adjustable within any framework which preserves the velocity of light c in free space, $c = 1/\sqrt{\mu_0 \epsilon_0} = 2.998 \cdots \times 10^8$ m/s, so either the value of ϵ_0 or of μ_0 may be assigned arbitrarily.

A current density **J** may exist under conditions where σ is considered to be zero, as for example in a vacuum, but with free electric charge ρ moving with velocity **v**, so $\mathbf{J} = \rho\mathbf{v}$. On other occasions materials are dealt with which have infinite conductivity (superconductors) or very high conductivity (copper) in which the current density **J** is effectively controlled by external current generators. Equation (2.2) does not describe such situations in a very convenient way. A useful alternative is obtained on observing that $\nabla \cdot \nabla \times \mathbf{H} = 0$ for any sufficiently differentiable vector $\mathbf{H}$. On taking divergences, Eq. (2.2) therefore becomes

$$\nabla \cdot \mathbf{J} + \frac{\partial}{\partial t} \nabla \cdot \mathbf{D} = 0. \tag{2.8}$$

When combined with Eq. (2.3) it expresses the indestructibility of charge:

$$\nabla \cdot \mathbf{J} + \frac{\partial \rho}{\partial t} = 0. \tag{2.9}$$

In the truly steady state, of course, Eq. (2.9) becomes $\nabla \cdot \mathbf{J} = 0$. In time-varying situations within good conductors it is also often a good approximation to assume a solenoidal current density. A substantial time-changing flow may be accounted for by a high density of drifting electrons neutralized by immobile positive ions locked to the lattice of the solid conductor.

2.2   *Integral relations*

A mathematically equivalent *integral* form of Eqs. (2.1)–(2.4) exists. Integrals are taken over an open surface S or its boundary contour C as illustrated in Fig. 3.1. The first two of Maxwell's equations become

$$\oint_C \mathbf{E} \cdot \mathbf{ds} = -\frac{\partial \phi}{\partial t} \tag{2.10}$$

and

$$\oint_C \mathbf{H} \cdot \mathbf{ds} = i + \frac{\partial \psi}{\partial t} \tag{2.11}$$

where ϕ and ψ are the total magnetic and electric fluxes encompassed by the contour C. Equation (2.10) is an expression of Faraday's law, and (2.11) is Ampère's circuital rule with Maxwell's addition of *displacement current* $\partial \psi / \partial t$. The divergence equations, Eqs. (2.3) and (2.4), correspond to Gauss's flux laws

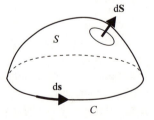

Fig 3.1   The surface S and contour C for the integral form of Maxwell's equations.

$$\oint_{\partial\Omega} \mathbf{D} \cdot d\mathbf{S} = q \tag{2.12}$$

and

$$\oint_{\partial\Omega} \mathbf{B} \cdot d\mathbf{S} = 0 \tag{2.13}$$

which state that the flux of **B** over any closed surface $\partial\Omega$ is zero whilst the corresponding integral of **D** equals the total charge q within the volume Ω enclosed by the surface $\partial\Omega$,

$$\int_{\Omega} \rho \, d\Omega = q. \tag{2.14}$$

As integral equations, these require no further boundary or interface conditions; they can be solved as they stand. However, they may need to be accompanied by additional constraints where these are an additional and independent part of the physical problem itself.

2.3   *Complex phasor notation*
     In many circumstances the problem being dealt with concerns the steady state reached when a system is excited sinusoidally. In such cases it is convenient to represent the variables in a complex phasor form. This means that each and any of the electromagnetic variables is represented by a phasor quantity, say in the case of electric field $\mathbf{E}_p = \mathbf{E}_r + j\mathbf{E}_i$. The corresponding physical, time-varying field is recovered as the real part of $\mathbf{E}_p \exp(j\omega t)$, where ω is the angular frequency of the sinusoidal excitation. Then when the partial differential operator $\partial/\partial t$ acts upon one of the electromagnetic variables it may be replaced by the simple arithmetic product factor $j\omega$. Thus Eq. (2.1) becomes

$$\nabla \times \mathbf{E} = -j\omega\mathbf{B}, \tag{2.15}$$

and so forth. Somewhere within the system being investigated an arbitrary reference of phase will have been established, corresponding to at least one of the spatial components of $\mathbf{E}_p$, say E_{p1}, being purely real. The physical, time-varying quantity here is $E_{p1} \cos(\omega t)$. Elsewhere, **B** will generally assume complex values corresponding to a phase shift from the reference of amount θ_1, so that

$$E_{p1} = |E_{p1}| \exp(j\theta_1) \tag{2.16}$$

for this particular component. In this case the physical field is $|E_{p1}| \cos(\omega t + \theta_1)$. This procedure is an extension of the familiar one-dimensional phasor circuit analysis.

2.4　　*Boundary rules*

Like any other differential equations, problems in Maxwell's equations can only be solved if accompanied by boundary conditions applicable at the limits of the geometric region covered by the problem at hand, and by interface conditions where distinct materials join. The laws governing the field vectors at discontinuities in material properties follow from the necessity that the integral forms of the Maxwell equations must be valid for surfaces and contours straddling such interfaces. If *I* and *II* represent two materials with differing properties then Eqs. (2.10)–(2.13) are equivalent to a set of rules at material interfaces as follows:

(i)　Tangential **E** is always continuous,

$$\mathbf{1}_n \times (\mathbf{E}_I - \mathbf{E}_{II}) = 0. \tag{2.17}$$

(ii)　Tangential **H** is discontinuous by an amount corresponding to any surface current $\mathbf{J}_s$ which may flow,

$$\mathbf{1}_n \times (\mathbf{H}_I - \mathbf{H}_{II}) = \mathbf{J}_s. \tag{2.18}$$

(iii)　Normal **B** is always continuous,

$$\mathbf{1}_n \cdot (\mathbf{B}_I - \mathbf{B}_{II}) = 0. \tag{2.19}$$

(iv)　Normal **D** is discontinuous by an amount corresponding to any surface charge ρ_s which may be present,

$$\mathbf{1}_n \cdot (\mathbf{D}_I - \mathbf{D}_{II}) = \rho_s. \tag{2.20}$$

The surface current $\mathbf{J}_s$ and surface charge ρ_s are most often encountered when one of the materials is a good conductor. At high frequencies there is a well-known effect which confines current largely to surface regions. The so-called skin depth in common situations is often sufficiently small for the surface phenomenon to be an accurate representation. Thus the familiar rules for the behaviour of time-varying fields at a boundary defined by good conductors follow directly from consideration of the limit when the conductor is perfect. No time-changing field can exist in a perfect conductor, so it must be that:

(i)　Electric field is entirely normal to the conductor and is supported by a surface charge $\rho_s = D_n$.

(ii)　Magnetic field is entirely tangential to the conductor and is supported by a surface current $J_s = H_t$.

True boundary conditions, i.e., conditions at the extremities of the boundary-value problem, are obtained by extending the interface conditions. The finite problem region is embedded in fictitious materials with infinite (or zero) permeabilities, permittivities or conductivities. The fictitious material, say material *II*, is chosen to force the relevant field vector (e.g., $\mathbf{H}_{II}$) to zero, and thereby to provide a true boundary condition (e.g., $\mathbf{1}_n \times \mathbf{E}_I = 0$) in the remaining material. Choosing the boundary shape and the exterior fictitious property value (zero or infinity) belongs to the art of engineering, rather than to engineering science.

3.   Waves and potentials

Multiple coupled differential equations in many variables can be very difficult to solve. Analysts often prefer to confront a single, more complicated, differential equation in a single variable, or at most a few variables. One common device for reducing mathematical complexity is therefore to formulate problems in terms of wave equations rather than the Maxwell equations themselves. This approach will now be reviewed briefly.

3.1   *Wave equations*

Wave equations in their conventional form are easily derived from the Maxwell curl equations, by differentiation and substitution. Consider, for example, the magnetic curl equation (2.2). Taking curls on both sides leads to

$$\nabla \times \nabla \times \mathbf{H} = \nabla \times \mathbf{J} + \frac{\partial}{\partial t} \nabla \times \mathbf{D}. \tag{3.1}$$

To eliminate $\mathbf{J}$ and $\mathbf{D}$ in favour of $\mathbf{E}$, the material property equations $\mathbf{J} = \sigma\mathbf{E}$ and $\mathbf{D} = \epsilon\mathbf{E}$ may be used. Assuming uniform scalar material properties, there results

$$\nabla \times \nabla \times \mathbf{H} = \sigma\nabla \times \mathbf{E} + \epsilon \frac{\partial}{\partial t} \nabla \times \mathbf{E}. \tag{3.2}$$

But $\nabla \times \mathbf{E}$ appears in the Maxwell electric curl equation (2.1), and according to this equation the rate of change of magnetic flux density may be substituted for it:

$$\nabla \times \nabla \times \mathbf{H} = -\sigma \frac{\partial \mathbf{B}}{\partial t} - \epsilon \frac{\partial^2 \mathbf{B}}{\partial t^2}. \tag{3.3}$$

If the magnetic material is simple enough to be described by the scalar relationship $\mathbf{B} = \mu\mathbf{H}$, then there finally results the wave equation

$$\nabla \times \nabla \times \mathbf{H} = -\mu\sigma \frac{\partial \mathbf{H}}{\partial t} - \mu\epsilon \frac{\partial^2 \mathbf{H}}{\partial t^2}. \tag{3.4}$$

Starting from the Maxwell electric curl equation (2.1) and following exactly the same process, a similar equation in the electric field **E** results:

$$\nabla \times \nabla \times \mathbf{E} = -\mu\sigma\,\frac{\partial \mathbf{E}}{\partial t} - \mu\epsilon\,\frac{\partial^2 \mathbf{E}}{\partial t^2}. \tag{3.5}$$

Some field analysts appear to have aesthetic (and possibly also mathematical) objections to the double curl operator $\nabla \times \nabla \times \mathbf{E}$. They use the standard vector identity, which is valid for any vector **E**,

$$\nabla \times \nabla \times \mathbf{E} \equiv \nabla\nabla \cdot \mathbf{E} - \nabla^2 \mathbf{E} \tag{3.6}$$

to recast the wave equations. The result is

$$\nabla^2 \mathbf{H} - \mu\sigma\,\frac{\partial \mathbf{H}}{\partial t} - \mu\epsilon\,\frac{\partial^2 \mathbf{H}}{\partial t^2} = 0, \tag{3.7}$$

$$\nabla^2 \mathbf{E} - \mu\sigma\,\frac{\partial \mathbf{E}}{\partial t} - \mu\epsilon\,\frac{\partial^2 \mathbf{E}}{\partial t^2} = -\nabla\left(\frac{\rho}{\epsilon}\right). \tag{3.8}$$

While mathematically indistinguishable from (3.4) and (3.5), this form of the wave equations is considered conventional. Note particularly that the magnetic wave equation (3.7) is homogeneous, while its electric counterpart (3.8) is not. This implies that all electromagnetic phenomena arise from electric charges ρ.

The electric and magnetic wave equations (3.7)–(3.8), valid for uniform regions, represent an enormous degree of mathematical simplification, as compared to the Maxwell equations. Many practical problems can be tackled by solving one wave equation without reference to the other, so that only a single boundary-value problem in a single (albeit three-component) vector variable remains. For example, Fig. 3.2 shows a hollow resonant cavity with perfectly conductive walls. The assumed infinite conductivity ensures that the magnetic field can only have a tangential component at wall surfaces, and that the electric field can only have a normal component. The two wave equations thus lead to twin formulations. One is framed solely in terms of the magnetic field,

$$\nabla^2 \mathbf{H} = \mu\epsilon\,\frac{\partial^2 \mathbf{H}}{\partial t^2}, \tag{3.9}$$

$$\mathbf{1}_n \cdot \mathbf{H} = 0 \qquad \text{on surfaces,} \tag{3.10}$$

the other solely in terms of the electric field,

$$\nabla^2 \mathbf{E} = \mu\epsilon\,\frac{\partial^2 \mathbf{E}}{\partial t^2}, \tag{3.11}$$

$$\mathbf{1}_n \times \mathbf{E} = 0 \qquad \text{on surfaces,} \tag{3.12}$$

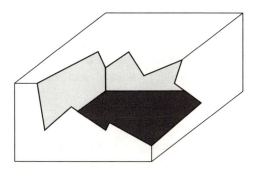

Fig 3.2   Resonant cavity. The interior is hollow, the walls are assumed to be perfectly conductive.

Either formulation taken by itself yields a computable problem, solvable by finite element methods.

3.2   *The classical potentials*

The mathematical treatment of fields is often simplified by working with potential functions instead of fields. The classical potentials follow directly from the field equations themselves. To introduce them into this discussion, begin with the well-known vector identity

$$\nabla \cdot \nabla \times \mathbf{A} \equiv 0 \tag{3.13}$$

which holds for any twice differentiable vector $\mathbf{A}$. It guarantees that the Maxwell magnetic divergence equation $\nabla \cdot \mathbf{B} = 0$ will always be satisfied if the flux density $\mathbf{B}$ is to be expressed in terms of an auxiliary vector $\mathbf{A}$ as

$$\mathbf{B} = \nabla \times \mathbf{A}. \tag{3.14}$$

With this substitution, the Maxwell electric curl equation (2.1) becomes

$$\nabla \times \left(\mathbf{E} + \frac{\partial \mathbf{A}}{\partial t} \right) = 0. \tag{3.15}$$

A second useful identity of vector calculus asserts that the curl of the gradient of any sufficiently differentiable scalar U always vanishes,

$$\nabla \times \nabla U \equiv 0, \qquad \text{any } U. \tag{3.16}$$

Now the bracketed quantity in Eq. (3.15) is an irrotational vector, so Eq. (3.16) ensures that it may always be represented as the gradient of some scalar $-V$. Thus

$$\mathbf{E} = -\frac{\partial \mathbf{A}}{\partial t} - \nabla V \tag{3.17}$$

results. The variables **A** and V are usually referred to as the *magnetic vector potential* and the *electric scalar potential*. Where no risk of confusion exists, these terms are often abbreviated by dropping the qualifiers 'magnetic' and 'electric', thus referring to **A** simply as the vector potential and V as the scalar potential.

Most electrical engineers feel quite at home with scalar potentials, but the notion of a magnetic vector potential seems mysterious to some. To help interpret its physical significance, Fig. 3.3 shows a wire loop fed by a source of electric current. The current creates a magnetic flux ϕ, of which some part ϕ_c links the open-circuited wire C. The flux ϕ_c may be calculated by integrating the local flux density **B** over the surface bounded by the open-circuited wire C:

$$\phi_c = \int \mathbf{B} \cdot d\mathbf{S}. \tag{3.18}$$

Substituting the vector potential **A**, Eq. (3.14),

$$\phi_c = \int \text{curl } \mathbf{A} \cdot d\mathbf{S} \tag{3.19}$$

The surface integral may be converted to a contour integral by Stokes' theorem. The flux threading C is therefore given by

$$\phi_c = \oint_C \mathbf{A} \cdot d\mathbf{s}. \tag{3.20}$$

To find the voltage $e(t)$ appearing at the break in the wire C, the linking flux ϕ_c is differentiated with respect to time,

$$e(t) = -\frac{\partial}{\partial t} \phi_c = -\oint_C \frac{\partial \mathbf{A}}{\partial t} \cdot d\mathbf{s}. \tag{3.21}$$

The time derivative $-\partial \mathbf{A}/\partial t$ clearly must have units of volts per meter, so it may be interpreted as the voltage induced in wire C per unit wire length. The vector potential may thus be viewed as a measure of the voltage-inducing ability of a magnetic field.

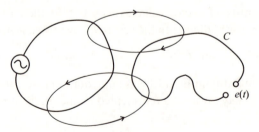

Fig 3.3 Voltage induced in contour C by currents flowing in a wire loop. The vector potential determines the terminal voltage.

Finding the potentials still implies solving boundary-value problems, so the potential equations must be accompanied by appropriate interface and boundary conditions. To determine these, a process is followed much like that applicable to the field vectors. Figure 3.4 shows an interface between materials *I* and *II*. A closed contour parallels the interface within material *I* and returns in material *II*, very nearly paralleling its path in material *I*. The magnetic flux that threads through the contour is given by the line integral of the vector potential **A**, as in Eq. (3.20). The surface spanned by the contour can be made arbitrarily small by keeping the two long sides of the contour arbitrarily close to each other while remaining at opposite sides of the material interface. The flux enclosed by the contour can therefore be made arbitrarily small as well, so that it vanishes in the limit of vanishing area. This can only happen if the tangential component of **A** has the same value on both sides of the interface. The boundary condition applicable to **A** is therefore

$$\mathbf{1}_n \times (\mathbf{A}_I - \mathbf{A}_{II}) = 0. \tag{3.22}$$

A corresponding condition for the scalar potential may be obtained from Faraday's law, which relates the line integral of electric field to the time derivative of magnetic flux. Because the flux vanishes for the contour of Fig. 3.4, the contour integral of the electric field must vanish too:

$$\oint \mathbf{E} \cdot d\mathbf{s} = -\frac{\partial \phi}{\partial t} = 0. \tag{3.23}$$

Substituting the expression for **E** in terms of potentials, Eq. (3.17),

$$\oint \mathbf{E} \cdot d\mathbf{s} = -\oint \nabla V \cdot d\mathbf{s} - \frac{\partial}{\partial t} \oint \mathbf{A} \cdot d\mathbf{s}. \tag{3.24}$$

Because the contour integrals of both **E** and **A** vanish, this equation can only be satisfied if

$$\oint \nabla V \cdot d\mathbf{s} = 0. \tag{3.25}$$

This, however, requires equality of potentials at adjacent points on the material interface,

$$V_I = V_{II}, \tag{3.26}$$

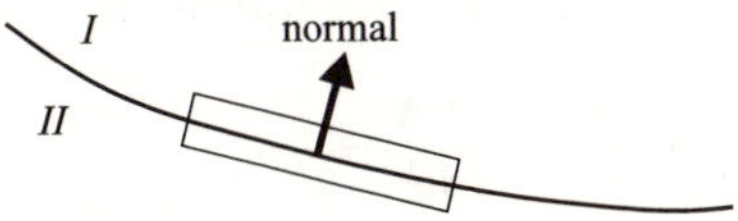

Fig 3.4 Interface conditions at a surface separating two dissimilar materials.

In plain words: the scalar potential must always be continuous across the material interface.

3.3   *Potential wave equations*

The classical potentials satisfy differential equations much like the wave equations that the fields must satisfy. These equations may be developed beginning with the definition of $\mathbf{A}$,

$$\nabla \times \mathbf{A} = \mathbf{B}. \tag{3.14}$$

Suppose for the moment that all materials have single-valued scalar (though not necessarily linear) magnetic properties, $\mathbf{B} = \mu\mathbf{H}$. Then

$$\frac{1}{\mu}\nabla \times \mathbf{A} = \mathbf{H}. \tag{3.27}$$

Wave-like equations involving only the potentials can be derived by means of a procedure that resembles the development applied to the field vectors. Taking curls on both sides,

$$\nabla \times \left(\frac{1}{\mu}\nabla \times \mathbf{A}\right) = \nabla \times \mathbf{H}, \tag{3.28}$$

then substituting the simple material property descriptions $\mathbf{D} = \epsilon\mathbf{E}$ and $\mathbf{J} = \sigma\mathbf{E}$, there results

$$\nabla \times \left(\frac{1}{\mu}\nabla \times \mathbf{A}\right) = \sigma\mathbf{E} + \epsilon\,\frac{\partial \mathbf{E}}{\partial t}. \tag{3.29}$$

The electric field $\mathbf{E}$ may be eliminated, substituting its value as given by the potentials in Eq. (3.17):

$$\nabla \times \frac{1}{\mu}\nabla \times \mathbf{A} + \sigma\,\frac{\partial \mathbf{A}}{\partial t} + \epsilon\,\frac{\partial^2 \mathbf{A}}{\partial t^2} = -\sigma\nabla V - \epsilon\nabla\,\frac{\partial V}{\partial t}. \tag{3.30}$$

This equation has a generally wave-like appearance. If desired, the curl-curl operator may be changed to the equivalent ∇^2 form by means of the general vector identity, valid for any sufficiently differentiable vector $\mathbf{A}$ and scalar ν,

$$\nabla \times \nu\nabla \times \mathbf{A} \equiv -\nu\nabla^2\mathbf{A} + \nu\nabla\nabla \cdot \mathbf{A} + \nabla\nu \times \nabla \times \mathbf{A}. \tag{3.31}$$

The result, obtained on setting $\nu = 1/\mu$ and substituting, is

$$\frac{1}{\mu}\nabla^2\mathbf{A} - \nabla\frac{1}{\mu} \times \nabla \times \mathbf{A} - \sigma\,\frac{\partial \mathbf{A}}{\partial t} - \epsilon\,\frac{\partial^2 \mathbf{A}}{\partial t^2} =$$
$$\nabla\left(\nabla \cdot \mathbf{A} + \sigma V + \epsilon\,\frac{\partial V}{\partial t}\right). \tag{3.32}$$

The left-hand side includes an extra term, as compared to the wave equations previously encountered; it permits material inhomogeneity, through the appearance of $\nabla(1/\mu)$. In regions with homogeneous and uniform magnetic material properties, Eq. (3.32) reduces precisely to the form of a standard wave equation on the left,

$$\frac{1}{\mu}\nabla^2 \mathbf{A} - \sigma \frac{\partial \mathbf{A}}{\partial t} - \epsilon \frac{\partial^2 \mathbf{A}}{\partial t^2} = \frac{1}{\mu}\nabla\left(\nabla \cdot \mathbf{A} + \sigma V + \epsilon \frac{\partial V}{\partial t}\right). \qquad (3.33)$$

Its right-hand side, on the other hand, involves both electric and magnetic potentials.

3.4 *Coulomb gauge*

Although the potentials $\mathbf{A}$ and V may appear to be fully defined by the wave equations, they are actually incompletely determined. A fundamental theorem due to Helmholtz states that a vector field can be uniquely specified only by specifying both its curl and its divergence. The divergence of $\mathbf{A}$ is as yet undetermined, though Eq. (3.14) specifies its curl as exactly equal to the magnetic flux density $\mathbf{B}$. Consequently, $\mathbf{A}$ and V remain ambiguous. They are fixed, and the potential equations made solvable, by assigning a value to the divergence of $\mathbf{A}$. This value may be chosen at will without affecting the physical problem, for all possible values still yield the same magnetic flux density $\mathbf{B}$. On the other hand, the various possible choices affect both ease of computation and ease of result visualization.

Different choices for the divergence of $\mathbf{A}$ are referred to as choices of *gauge*. A single physical problem may thus be described by several different potentials fields, one for each choice of gauge; these are said to be equivalent to within a *gauge transformation* and each particular choice of $\nabla \cdot \mathbf{A}$ is called a *gauge condition*. The best-known gauge condition used in electromagnetics is extremely simple,

$$\nabla \cdot \mathbf{A} = 0. \qquad (3.34)$$

With this choice, boundary conditions applicable to $\mathbf{A}$ can be further sharpened. I :t the thin box-shaped volume of Fig. 3.5 enclose a section of the interface between media I and II. Equation (3.34) requires that

$$\oint \mathbf{A} \cdot d\mathbf{S} = 0. \qquad (3.35)$$

The box volume can be made arbitrarily small by moving together its flat faces, one located each side of the material interface. Any contributions to integral (3.35) from the four box surfaces that intersect the interface can thereby be made arbitrarily small. There remains

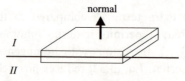

Fig 3.5 An interface between two different media. The vector potential is fully continuous under the Coulomb gauge.

$$\mathbf{1}_n \cdot (\mathbf{A}_I - \mathbf{A}_{II}) = 0. \tag{3.36}$$

Clearly, the normal components of **A** must now be continuous under this choice of gauge, as well as the tangential components which must be continuous under all circumstances. This choice of $\nabla \cdot \mathbf{A}$, usually called the *Coulomb gauge*, is most widely employed in magnetostatic problems. If there is no time variation, Eq. (3.33) becomes

$$\frac{1}{\mu}\nabla^2 \mathbf{A} = \nabla(\sigma V). \tag{3.37}$$

But in the static case, $-\nabla V$ is readily identified as the electric field. Consequently, $-\sigma \nabla V$ represents the electric current density **J**. Provided the conductivity σ is uniform, Eq. (3.37) may be written

$$\frac{1}{\mu}\nabla^2 \mathbf{A} = -\mathbf{J}. \tag{3.38}$$

A major part of computational magnetostatics is based on this equation.

3.5   *Diffusion gauge*

A second choice of gauge, though quite popular, appears not to be associated with any one person's name. It will be referred to as the *diffusion gauge* in this book. The divergence of **A** is selected to be

$$\nabla \cdot \mathbf{A} = -\epsilon \frac{\partial V}{\partial t}. \tag{3.39}$$

This choice annihilates the time-varying term on the right of Eq. (3.33), thereby reducing even time-varying problems to a simplicity comparable to magnetostatics as expressed by (3.37):

$$\frac{1}{\mu}\nabla^2 \mathbf{A} - \sigma \frac{\partial \mathbf{A}}{\partial t} - \epsilon \frac{\partial^2 \mathbf{A}}{\partial t^2} = \frac{1}{\mu}\nabla(\sigma V). \tag{3.40}$$

The boundary conditions that govern **A** must now be reexamined, however. The divergence of **A** no longer vanishes and

$$\oint \mathbf{A} \cdot d\mathbf{S} = \int \nabla \cdot \mathbf{A} \, d\Omega = \int (\epsilon_I - \epsilon_{II}) \frac{\partial V}{\partial t} \, d\Omega \tag{3.41}$$

for the small box volume shown in Fig. 3.5. Note that only a single value V of the scalar potential appears in Eq. (3.41), because the scalar potential must be continuous regardless of the choice of gauge. The difference in permittivity $(\epsilon_I - \epsilon_{II})$ is finite and independent of the box volume. Reducing the box volume by making its two flat faces approach each other then reduces the volume of integration in Eq. (3.41) while leaving the integrand unaltered. As the volume approaches zero, the integral approaches zero also, and the conclusion follows:

$$\mathbf{1}_n \cdot (\mathbf{A}_I - \mathbf{A}_{II}) = 0. \tag{3.42}$$

Thus the normal components of $\mathbf{A}$ must still be continuous, just as they were under the Coulomb choice of gauge. The quasi-wave equation that results,

$$\frac{1}{\mu}\nabla^2\mathbf{A} - \sigma\frac{\partial\mathbf{A}}{\partial t} - \epsilon\frac{\partial^2\mathbf{A}}{\partial t^2} = \nabla(\sigma V) \tag{3.43}$$

now possesses a right-hand side which may again be interpreted in a manner similar to Eq. (3.37). Its right side may be written as $-\mu\mathbf{J}_{static}$, where the current density $\mathbf{J}_{static}$ physically signifies the current density that *would* flow in the absence of any induced electromotive forces, i.e., that would flow if all time variations were very slow. A large part of the published literature on eddy-current and skin-effect problems relies on this interpretation. To illustrate, Fig. 3.6 shows a pair of long current-carrying conductors. Here a time-varying current is kept flowing in the conductors by a voltage source. The right-hand factor $\nabla(\sigma V)$ in Eq. (3.43) is viewed as the current density distribution that would result from very slow variations in the source voltage — in other words, the resistance-limited current distribution that would result at dc. In addition, it is conventional in low-frequency problems (though it is not

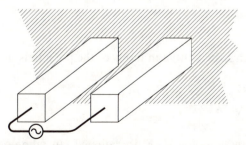

Fig 3.6   Skin effect in a long conductor pair. The driving term in the relevant wave equation is the current that would flow if there were no time variation of potential.

strictly necessary), to neglect displacement currents so that the potential equation (3.43) becomes

$$\frac{1}{\mu}\nabla^2 \mathbf{A} - \sigma \frac{\partial \mathbf{A}}{\partial t} = -\mathbf{J}_{\text{static}}. \tag{3.44}$$

With this viewpoint, finding the time-varying current density distribution in conductors becomes a three-step process: (1) drop all time-dependent terms and solve for the dc current distribution, (2) use the dc distribution as the source term on the right and solve Eq. (3.44) for $\mathbf{A}$, (3) determine the time-varying current distribution from $\mathbf{A}$. Note that the vector potential in this problem has only a single component A_z, because the current density is wholly z-directed. The resulting computational economy has made this formulation very popular in a large class of problems, both time-varying and static, where a suitable choice of coordinates results in $\mathbf{J}$ possessing only one component.

3.6   *Lorentz gauge*

A third time-honoured choice of gauge is justly popular as well as useful in wave-propagation problems. It seeks to eliminate the troublesome right-hand terms in Eq. (3.33) entirely, by setting

$$\nabla \cdot \mathbf{A} = -\mu\sigma V - \mu\epsilon \frac{\partial V}{\partial t}. \tag{3.45}$$

The wave equation that results for $\mathbf{A}$ is homogeneous, so that its solution is entirely driven by the boundary conditions. The scalar potential V then turns out to be governed by a similar but separate wave equation. This choice for the divergence of $\mathbf{A}$ is commonly known as the *Lorentz gauge*. It has been widely used in wave-propagation problems.

The wave equation for $\mathbf{A}$ under the Lorentz gauge follows directly from the Maxwell electric divergence equation (2.3), which may be written in the form

$$\nabla \cdot \mathbf{E} = -\frac{\rho}{\epsilon}. \tag{3.46}$$

The divergence of $\mathbf{E}$ may be written in terms of the potentials, following (3.17), as

$$\nabla \cdot \mathbf{E} = -\frac{\partial}{\partial t} \nabla \cdot \mathbf{A} - \nabla^2 V. \tag{3.47}$$

When these two expressions for $\nabla \cdot \mathbf{E}$ are equated, an inhomogeneous wave equation in V results. The problem to be solved thus consists of the two wave equations

$$\nabla^2 \mathbf{A} - \mu\sigma \frac{\partial \mathbf{A}}{\partial t} - \mu\epsilon \frac{\partial^2 \mathbf{A}}{\partial t^2} = 0, \tag{3.48}$$

$$\nabla^2 V - \mu\sigma \frac{\partial V}{\partial t} - \mu\epsilon \frac{\partial^2 V}{\partial t^2} = -\frac{\rho}{\epsilon}. \tag{3.49}$$

The boundary conditions that this choice of gauge imposes on the normal component of **A** are no more complicated than their counterparts in the diffusion gauge — in fact they are the same, Eq. (3.42). The mathematical argument that leads to them exactly parallels that of Eq. (3.41), so it will not be repeated here.

Quite a few articles and papers on electromagnetics, particularly in the area of low-frequency electromagnetics, assert that in certain cases it is 'not necessary to gauge the vector potential'. Since a gauge is defined by a *gauge condition*, a particular choice of $\nabla \cdot \mathbf{A}$, the assertion that 'there is no gauge' must be understood to mean that there is no need to choose any particular gauge, any choice will work equally well. It is clearly impossible to arrive at numerically computed answers unless some choice is made. If the analyst enforces no particular choice consciously, a choice is in reality made anyway — either hidden in the problem formulation, or supplied by the computer, through the mechanism of numerical roundoff error!

Leaving the gauge condition to chance is probably not a good idea. It may reduce precision unpredictably, if the irrotational part of **A** (the part fixed by the gauge condition) is numerically large. It can lead to numerical instability in iterative solutions, with successive estimates of the solution giving similar values of $\nabla \times \mathbf{A}$ but widely differing $\nabla \cdot \mathbf{A}$. Even stable solutions may not be reproducible from one computation to the next, and almost certainly from one computer to the next, for the roundoff behaviour may depend on such variables as machine word length.

3.7 *Separation of variables*

Solutions to the full wave equations are avoided wherever possible in applied electromagnetics. Many practical problems are first simplified by a process of *separation of variables*, which separates the wave equation into two or more differential equations in distinct independent variables. This method appears in most textbooks on differential equatons as a restricted but useful method for solving boundary-value problems. Here, however, it is viewed not as a method for seeking solutions, but of partitioning the problem into manageable pieces. To apply it to the wave equation in **A**, Eq. (3.48), it will be supposed that there exists a solution of the form

$$\mathbf{A}(x, y, z, t) = \mathbf{S}(x, y, z)T(t) \tag{3.50}$$

where $S(x, y, z)$ is a vector-valued function of the space coordinates, while $T(t)$ is scalar and depends only on time. This supposition can be tested easily by substituting it into the wave equation; should the supposition prove impossible, a conflict will become evident at once. On substitution,

$$T\nabla^2 S - \mu\sigma S T' - \mu\epsilon S T'' = 0 \tag{3.51}$$

which becomes, on dividing by T,

$$\nabla^2 S - \left(\frac{\mu\sigma T' - \mu\epsilon T''}{T}\right)S = 0. \tag{3.52}$$

The bracketed term involves only the time variable t, not the space variables; and time t appears nowhere except in this term. If the equation is to remain an equation for all values of t, this quantity cannot change with time! It must therefore be a constant, say $-k^2$,

$$\frac{\mu\sigma T' - \mu\epsilon T''}{T} = -k^2. \tag{3.53}$$

Rewriting, (3.52) and (3.53) thus become

$$\mu\epsilon \frac{d^2 T}{dt^2} + \mu\sigma \frac{dT}{dt} + k^2 T = 0, \tag{3.54}$$

$$(\nabla^2 + k^2)S = 0. \tag{3.55}$$

Equation (3.54) is an ordinary differential equation whose solution presents no difficulty. The partial differential equation (3.55) is commonly referred to as the *Helmholtz equation*. It is easier to solve than the wave equation, for it involves only the space variables x, y, z, but not t — hence one independent variable less than the wave equation.

The evident success of variable separation leaves an important question still open: are there any circumstances when separation does not work? The answer lies in the accompanying boundary conditions. Since the Helmholtz equation is a partial differential equation, it can only be solved if sufficient boundary conditions are known — and these must involve exactly the same variables, no more and no fewer, then the differential equation. Now the separated space equation (3.55) involves x, y, z but not t. To make the separation process useful, the boundary conditions must therefore not involve t either; if they do, the separation achieved by rewriting the differential equation is lost because they reintroduce the eliminated variable t. Acceptable boundary conditions must then relate space variables without involving time,

$$h_i(x, y, z) = 0. \tag{3.56}$$

A particular boundary value problem may involve numerous boundary conditions, $i = 1, \ldots, N$, but they must all have this form. Should there be any that involve time,

$$h_i(x, y, z, t) = 0, \tag{3.57}$$

separation is immediately lost. Although the wave equation separates, the Helmholtz equation and its boundary conditions no longer constitute a time-independent problem — the equation does not involve time, but its boundary conditions do.

Space-time interdependence in boundary conditions probably occurs most commonly in problems with moving boundaries or (in the inhomogeneous case) moving sources. As a general rule, then, problems of electrostatics or magnetostatics are amenable to this treatment. So are those wave problems in which the boundary motion is slow compared to the speed of wave propagation. For example, radar echoes of aircraft are invariably computed by assuming that the aircraft is stationary, an assumption convenient electromagnetically even if it is aerodynamically indefensible.

3.8 *Retarded potentials*

The wave equations can be solved completely in a small number of special cases. These are of practical interest because they generate valuable test cases for program testing. They are of greater value, however, in allowing the differential equations that describe waves to be reformulated as integral equations, thereby allowing the analyst to develop a computational method based on either the differential or the integral equation form.

To develop the procedure for inverting the wave differential operator — in other words, for formally solving the wave equation — it is easiest to begin with a specific concrete case. Consider again the scalar potential wave equation

$$\nabla^2 V - \mu\sigma \frac{\partial V}{\partial t} - \mu\epsilon \frac{\partial^2 V}{\partial t^2} = -\frac{\rho(t)}{\epsilon}, \tag{3.58}$$

under the assumption that a solution is desired in unbounded space, and that the charge density $\rho(t)$, which may vary with time, differs from zero only in an infinitesimal space $d\Omega$ surrounding the origin. In other words, the right-hand side is taken to represent a time-varying point charge. The solution may be presumed rotationally symmetric, for there are neither boundaries nor shaped source distributions to break its symmetry. The equation is therefore most easily stated in spherical coordinates. Normalizing the time variable, by setting $\tau = t/\sqrt{(\mu\epsilon)}$, it reads

$$\frac{1}{r^2}\frac{\partial}{\partial r}\left\{r^2\frac{\partial V(r,\tau)}{\partial r}\right\} - \sqrt{\frac{\mu}{\epsilon}}\,\sigma\,\frac{\partial V(r,\tau)}{\partial \tau} - \frac{\partial^2 V(r,\tau)}{\partial r^2} = 0, \qquad (3.59)$$

for all $r > 0$. This equation can be formally solved by a simple artifice. The substitution

$$V = \frac{U}{r} \qquad (3.60)$$

brings it into the form

$$\frac{\partial^2 U}{\partial r^2} - \sqrt{\frac{\mu}{\epsilon}}\,\sigma\,\frac{\partial U}{\partial \tau} - \frac{\partial^2 U}{\partial \tau^2} = 0. \qquad (3.61)$$

Clearly, Eq. (3.61) is nothing more than the conventional description of a one-dimensional propagating wave, as encountered in transmission line theory. It is easily solved in two particular circumstances: (1) sinusoidal steady-state waves, or (2) waves of any kind if the medium is lossless, $\sigma = 0$. The latter case is of particularly great interest for antenna analysis, keeping in mind that the region of solution is taken as unbounded. For this special but important case,

$$\frac{\partial^2 U}{\partial r^2} - \frac{\partial^2 U}{\partial \tau^2} = 0. \qquad (3.62)$$

This differential equation is satisfied, for $r > 0$, by any sufficiently differentiable function $U = f(\tau - r)$. In terms of the original problem variables, then

$$V = \frac{f(t - r\sqrt{\mu\epsilon})}{r}. \qquad (3.63)$$

The central point here is that V depends neither on r nor on t by itself, rather it is a function of their difference. This is true for any time-dependence whatever of the point charge; but which particular function $f(\tau - r)$ holds in any given case naturally depends on how $\rho(t)$ varies. To deduce this dependence, take r to be very small, so that $(\tau - r) \simeq \tau$. Surround the point charge with a very small sphere whose radius is r. The total electric flux $\rho(t)d\Omega$ emanating from the point charge covers a spherical surface $4\pi r^2$, so the flux density at that surface is $p(t)d\Omega/4\pi r^2$, correspondingly the potential $V = \rho(t)d\Omega/4\pi\epsilon r$. Since this is valid for very small r only, it represents the limiting value of V,

$$\lim_{r \to 0} V(t - r\sqrt{\mu\epsilon}) = \frac{\rho(t)d\Omega}{4\pi\epsilon r}. \qquad (3.64)$$

Comparison of the last two equations then indicates that

$$V = \frac{\rho(t - r\sqrt{\mu\epsilon})}{4\pi\epsilon r}\,d\Omega. \tag{3.65}$$

Physically, this equation says that the potential behaves with distance exactly as an electrostatic potential would, but with a difference. Instead of being proportional to the present value of the charge $\rho(t)d\Omega$, the potential responds to the value that the charge had a time $r\sqrt{\mu\epsilon}$ earlier. Thus the potential still corresponds to the charge causing it, but with a time retardation which equals the time taken for light to propagate from the charge to the point of potential measurement. The electrostatic potential may be viewed simply as the special case of very small retardation, i.e., a simplification valid at close distances.

The point charge is very special and quite artificial case. But Eq. (3.65) may be applied to the general case of time-varying charges distributed over some finite volume of space Ω, by dividing the volume into small portions and treating the charge in each as a point charge at Q. Summing over all the elementary volumes, the potential at some point P is then

$$V_P = \int_\Omega \frac{\rho(t - r_{PQ}\sqrt{\mu\epsilon})}{4\pi\epsilon r_{PQ}}\,d\Omega_Q. \tag{3.66}$$

This result is often called a *retarded potential*, and written in the abbreviated form

$$V_P = \int_\Omega \frac{[\rho]}{4\pi\epsilon r_{PQ}}\,d\Omega_Q, \tag{3.67}$$

with the square brackets denoting that the time-retarded value of the integrand is meant.

The vector potential may be given a similar treatment. In much of antenna theory, the conduction current $-\mu\sigma\partial A/\partial t$ is taken to be prescribed by the sources $\mathbf{J}_s$ that feed the antenna, so that the vector potential wave equation becomes

$$\nabla^2 \mathbf{A} - \mu\epsilon\frac{\partial^2 \mathbf{A}}{\partial t^2} = \mathbf{J}_s. \tag{3.68}$$

The formal solution is surprisingly easy to accomplish. It is only necessary to separate this vector equation into its Cartesian components. The result is a trio of equations identical in form to the scalar potential equation. When the components are recombined, there results

$$\mathbf{A}_P = \int_\Omega \frac{[\mathbf{J}_s]}{4\pi\epsilon r_{PQ}}\,d\Omega_Q. \tag{3.69}$$

The square brackets again signify time retardation.

4.   Functionals over potentials and fields

The first two chapters of this book have demonstrated by example how the actual field solution of Laplace's equation in an electrostatics problem corresponds to an electrical stored energy which is smaller than that for all other possible field distributions fitting the Dirichlet boundary conditions of the problem. Possible functions for the potential U were considered — and it was asserted from consideration of the minimum stored energy principle that the functional[1] $\mathfrak{F}(U)$,

$$\mathfrak{F}(U) = \frac{1}{2} \int_\Omega \nabla U \cdot \nabla U \, d\Omega, \tag{4.1}$$

integrated over the whole two-dimensional problem region Ω, had to be stationary about the true solution $U = u$. This variational principle was exploited to assemble a patchwork of piecewise-planar functions of the space variables in two dimensions which, within the limitations of the linear approximation, was closest to the correct electrostatic solution. Such a procedure is typical of the methods commonly used for the finite element solution of electromagnetics problems. In general, it may be said that corresponding to any boundary value problem arising in electromagnetics it will be possible to find at least one functional $\mathfrak{F}$ that has the property of being stationary, either as a maximum, minimum or saddle point, at the correct solution. The functional may be expressible in terms of the basic field variables **E**, **H**, **D** and **B**, or the potentials **A** and V, and will be computed by an integration over the volume concerned in the problem.

4.1   *The general scalar Helmholtz equation*

It is useful to examine a differential equation that may be termed the *inhomogeneous scalar Helmholtz equation*,

$$\nabla \cdot (p\nabla u) + k^2 qu = g \tag{4.2}$$

where u is the scalar variable that constitutes the solution required. The problem is defined in some space region Ω, filled with some materials

---

[1] The term *functional* denotes a scalar quantity whose value depends on the shape of a function. It differs from a *function*; the argument of a function is a *value* while the argument of a functional is a *function*. For example, the *root-mean-square value* $E(v(t))$ of a periodic voltage waveform $v(t)$ is a functional: it has a scalar value which depends on the amplitude and shape of the waveform throughout its entire cycle.

whose properties are supposed to be represented by $p(x, y, z)$ and $q(x, y, z)$. The quantity k^2 is a constant, invariant with position, and may or may not be known; $g(x, y, z)$ is a driving function, assumed to be given. Boundary conditions are specified on the surface $\partial\Omega$ that encloses Ω, and may be of either Dirichlet or homogeneous Neumann type, with u prescribed on some part ∂D of the boundary (Dirichlet), or $\partial u/\partial n = 0$ on ∂N (homogeneous Neumann); $\partial D \cup \partial N = \partial\Omega$.

The general Helmholtz equation is important mainly because it encompasses many specific problems as special cases. For example, Laplace's equation is obtained on setting $p = 1$, $k = 0$, $g = 0$:

$$\nabla^2 u = 0. \tag{4.3}$$

This represents the simple electrostatics problem already discussed in some detail. Allowing the material property to vary with position, $\epsilon = p(x, y, z)$ and permitting sources $g = -\rho(x, y, z)$ to exist, a general form of Poisson's equation results:

$$\nabla \cdot (\epsilon \nabla u) = -\rho. \tag{4.4}$$

Here the permittivity ϵ may vary with position in any specified manner and the space-charge distribution ρ is regarded as given.

Dropping the source terms and again requiring the right-hand side to vanish, a scalar Helmholtz equation is obtained. In two dimensions, it reads in one particular case

$$\nabla_T^2 H_z + k_c^2 H_z = 0. \tag{4.5}$$

This equation models the behaviour of transverse electric (TE) modes in a waveguide. Here ∇_T^2 is the transverse Laplacian operator, in Cartesian coordinates $\partial^2/\partial x^2 + \partial^2/\partial y^2$, whilst k_c is a cut-off wavenumber, determined as part of the solution process. Boundary conditions in this problem are of the homogeneous Neumann type at the waveguide boundary. The TM modes of the same guide may be found by solving a similar equation, but with Dirichlet boundary conditions and E_z as the dependent variable.

4.2 A functional for the Helmholtz equation

Instead of solving the general Helmholtz equation (4.2), it suffices to render stationary the functional

$$\mathfrak{F}(U) = \frac{1}{2} \int_\Omega (p\nabla U \cdot \nabla U - k^2 q U^2 + 2gU) d\Omega, \tag{4.6}$$

the region of integration Ω being the entire geometric domain over which the equation is to be solved; p and q may be local scalar-valued material properties. This as yet unsupported assertion requires proof; indeed it also requires further clarification, for it does not even state over what admissible space of functions U the functional is to be stationary.

The admissible functions U can be found, and the assertion justified at the same time. Let the (incorrect) trial solution U be written as the sum of the correct solution u and an error θh,

$$U = u + \theta h. \tag{4.7}$$

Here the error is represented as the product of a numeric parameter θ and a function $h(x, y, z)$. The reason for this apparently curious representation is simple: it allows the amount of error to be varied, without altering the relative distribution of error, by varying the numeric multiplier θ. With the trial solution U written in the form (4.7), the functional may be expanded into

$$\mathfrak{F}(U) = \mathfrak{F}(u) + \theta \delta \mathfrak{F}(u) + \frac{\theta^2}{2} \delta^2 \mathfrak{F}(u) \tag{4.8}$$

where all terms involving θ have been collected into $\delta\mathfrak{F}$, the so-called *first variation* of $\mathfrak{F}$,

$$\delta\mathfrak{F}(u) = \int_\Omega p \nabla u \cdot \nabla h \, d\Omega - \int_\Omega h(k^2 q u - g) \, d\Omega, \tag{4.9}$$

while the terms in θ^2 make up $\delta^2 \mathfrak{F}$, the *second variation* of $\mathfrak{F}$,

$$\delta^2 \mathfrak{F}(u) = \int_\Omega (p \nabla h \cdot \nabla h - k^2 q h^2) \, d\Omega. \tag{4.10}$$

To determine conditions for stationarity, the first and second variations must now be examined. Taking the simpler expression first: the second variation is clearly a number characteristic of the error function h. Its value depends on the size of k^2, for the factors p, q, h^2 and $\nabla h \cdot \nabla h$ are individually all positive. If indeed $\mathfrak{F}(U)$ can be made stationary at $U = u$, the stationary point may be a maximum, a minimum, or a saddle point; which one it is, depends on the ratio of integrals $\int k^2 q h^2 d\Omega / \int p \nabla h \cdot \nabla h \, d\Omega$.

The expression (4.9) that defines the first variation $\delta\mathfrak{F}(u)$ can be simplified considerably. Start with the general rule of vector calculus

$$\nabla \cdot (h\mathbf{\Phi}) = h\nabla \cdot \mathbf{\Phi} + \mathbf{\Phi} \cdot \nabla h, \tag{4.11}$$

valid for any sufficiently differentiable h and $\mathbf{\Phi}$. On setting $\mathbf{\Phi} = p\nabla u$ and integrating over the region Ω, Eq. (4.11) becomes

$$\int_\Omega \nabla \cdot (hp\nabla u) \, d\Omega = \int_\Omega h\nabla \cdot (p\nabla u) \, d\Omega + \int_\Omega p\nabla u \cdot \nabla h \, d\Omega. \tag{4.12}$$

The rightmost term appears in $\delta\mathfrak{F}$, but the other two may not be immediately recognizable. The divergence theorem of vector calculus allows changing the left side of Eq. (4.12) into a surface integral. Rearranging terms, Eq. (4.12) thus becomes

$$\int_\Omega p\nabla u \cdot \nabla h \, d\Omega = \oint_{\delta\Omega} ph\nabla u \cdot dS - \int_\Omega h\nabla \cdot (p\nabla u) \, d\Omega. \qquad (4.13)$$

On substituting (4.13), the first variation assumes the form

$$\delta\mathfrak{F}(u) = \oint_{\partial\Omega} ph\nabla u \cdot dS - \int_\Omega h\{\nabla \cdot (p\nabla u) \, d\Omega + k^2 qu - g\} \, d\Omega. \qquad (4.14)$$

The term in braces is merely the difference between the two sides of the Helmholtz equation and therefore vanishes identically. Consequently, the volume integral on the right is always zero. The final form of $\delta\mathfrak{F}$ is therefore quite simple,

$$\delta\mathfrak{F}(u) = \oint_{\partial\Omega} ph\nabla u \cdot dS. \qquad (4.15)$$

For a stationary point to exist, the first variation must vanish. The sufficient and necessary condition for this to happen is

$$\theta\oint_{\partial\Omega} ph\nabla u \cdot dS = 0 \qquad (4.16)$$

for all possible h. Remembering that $h = U - u$, this condition may be written as

$$\int_{\delta\Omega} p(U - u)\nabla u \cdot dS = 0. \qquad (4.17)$$

Note that $\delta\mathfrak{F}(u)$ involves no integral over the region Ω, only an integral around its boundary. The functions U are therefore unrestricted, except at the boundaries where they must be constrained in such a way that the boundary integral vanishes. Over any homogeneous Neumann boundary segment, the integrand in Eq. (4.17) vanishes identically, for ∇u has no component parallel to the surface normal dS; no restriction need be placed on U there. Over any Dirichlet boundary segments, values of u are known in advance because they are prescribed as part of the problem. It is therefore possible to make $U = u$, forcing the error function h to vanish so the integrand of Eq. (4.17) again vanishes. To conclude: the functional $\mathfrak{F}(U)$ is stationary at $U = u$, provided that (1) all boundary conditions are either of the Dirichlet type (u prescribed) or of the homogeneous Neumann type; (2) U satisfies all Dirichlet boundary conditions of the problem. The space of admissible functions U is therefore composed of all functions that satisfy the Dirichlet boundary conditions of the problem, and that possess first derivatives in the region Ω (otherwise the functional $\mathfrak{F}$ cannot be evaluated).

It should be evident that the stationary value of $\mathfrak{F}(U)$ deviates from the true $\mathfrak{F}(u)$ only by a term in θ^2. For small θ, it is therefore much more precise than the potential $U = u + \theta h$, which differs from u by a term proportional to θ. The functional $\mathfrak{F}$ itself is often closely related to global quantities of practical interest, such as stored energy or total power dissipation.

4.3   *General boundary conditions*

The foregoing development tacitly assumed that only very simple boundary conditions were of interest. This is, of course, not a realistic assumption. The argument must therefore be broadened to allow a wider range of conditions to apply.

Let the boundary $\partial\Omega$ of a problem domain Ω be composed of three distinct boundary segments ∂D, ∂N and ∂C, each of which may consist of multiple nonoverlapping patches,

$$\partial D \cup \partial N \cup \partial C = \partial\Omega, \tag{4.18}$$

$$\partial D \cap \partial N = \partial D \cap \partial C = \partial C \cap \partial N = 0. \tag{4.19}$$

Let the solution u satisfy the following conditions on the three segments:

$$u = u_0 \quad \text{on } \partial D \quad \text{(Dirichlet)}, \tag{4.20}$$

$$\frac{\partial u}{\partial n} = v_0 \quad \text{on } \partial N \quad \text{(Neumann)}, \tag{4.21}$$

$$\frac{\partial u}{\partial n} + au = v_0 \quad \text{on } \partial C \quad \text{(Cauchy)}. \tag{4.22}$$

The Dirichlet condition, where it applies, is satisfied by restricting the admissible approximating functions U, by setting $U = u$. The homogeneous Neumann boundary conditions

$$\nabla u \cdot \mathbf{1}_n = \frac{\partial u}{\partial n} = 0 \tag{4.23}$$

are satisfied approximately, without any restriction on the functions U; they are termed *natural boundary conditions* because they 'come naturally' to the functional (4.6). The Dirichlet conditions, on the other hand, are enforced by restricting the functions U; they are called *essential* or *principal* boundary conditions. These terms are generally used to describe variational functionals: boundary conditions implicit in the functional expression are termed *natural*, boundary conditions implicit in the admissible set of functions U are called *essential*.

4.4   *Inhomogeneous Neumann boundaries*

In general, the Neumann boundary segment ∂N may include prescribed values of normal derivative other than zero, say

$$\frac{\partial u}{\partial n} = \nabla u \cdot \mathbf{1}_n = v_0. \tag{4.24}$$

These conditions cannot be imposed by enforcing any particular values of the error function h, for the correct solution u is unknown on ∂N. In other words, inhomogeneous Neumann conditions cannot be made into essential boundary conditions. However, it is possible to modify the functional so as to make them into natural conditions. To do so, rewrite the contribution of the Neumann boundary ∂N to the first variation $\delta\mathfrak{F}(u)$, using $\theta h = U - u$ as in Eq. (4.17):

$$\theta\delta\mathfrak{F}(u) = \int_{\partial N} pUv_0 \, dS - \int_{\partial N} puv_0 \, dS. \tag{4.25}$$

The interesting point here is that the integrals on the right do not include θ or h explicitly. The first variation can therefore be made to vanish by adding its negative to the functional. However, even that is not really necessary, for the second integral in Eq. (4.25) is a fixed number because on ∂N, the value v_0 of the normal derivative is known. Although the value of this second integral term is not known in advance, it is a constant, has no effect on stationarity, does not depend on θ or h, and can be included or excluded in the functional at will. It is therefore sufficient to annihilate the variable integral by adding its negative to the functional:

$$\mathfrak{F}(U) = \frac{1}{2}\int_\Omega (p\nabla U \cdot \nabla U - k^2qU^2 + 2gU)d\Omega$$

$$- \int_{\partial N} pUv_0\mathbf{1}_S \cdot dS. \tag{4.26}$$

The unit vector $\mathbf{1}_S$ is collinear with dS, i.e., normal to the boundary segment ∂N. This functional form caters to both Dirichlet conditions (they are still essential conditions) and to Neumann conditions, by making them natural.

4.5   *Inhomogeneous Cauchy boundaries*

Cauchy boundary conditions appear in electromagnetics most often in the form of surface impedances. They too are convertible into natural conditions, by modifying the functional $\mathfrak{F}(U)$ to read

$$\mathfrak{F}(U) = \frac{1}{2}\int_{\Omega}(p\nabla U \cdot \nabla U - k^2 q U^2 + 2gU)\,d\Omega$$
$$+ \frac{1}{2}\int_{\partial C} ap U^2 \,dS - \int_{\partial C} pUv_0 \,dS. \tag{4.27}$$

To prove this assertion, the now traditional procedure may be followed of substituting $U = u + \theta h$. Then, after expanding and collecting terms,

$$\mathfrak{F}(U) = \mathfrak{F}(u) + \theta\left\{\int_{\partial C} ph\left(\frac{\partial u}{\partial n} + au - v_0\right)\,dS\right\}$$
$$+ \frac{\theta^2}{2}\left\{\int_{\Omega}(p\nabla h \cdot \nabla h - k^2 q h^2)\,d\Omega + \int_{\partial C} aph^2 \,dS\right\}. \tag{4.28}$$

The first term $\mathfrak{F}(u)$ clearly has a constant value; it is in general an unknown value, but fixed nevertheless for any given problem. The last term in braces (the second variation) has a fixed value for any given h, it does not vary with θ; further, it depends on the error quadratically. Whether its value is positive, negative, or zero depends on the problem constants. Thus the stationary value, if there is one, may be a minimum, maximum, or saddle point. Most importantly, the first variation of $\mathfrak{F}(U)$, i.e., the first term in braces, vanishes identically because the term in parentheses vanishes everywhere — it is the definition of the Cauchy boundary condition! The proposed form of functional therefore incorporates the Cauchy condition on ∂C as a natural boundary condition.

To solve the fully general Helmholtz equation problem incorporating all three common types of boundary condition, it suffices to find a stationary point of the functional

$$\mathfrak{F}(U) = \frac{1}{2}\int_{\Omega}(p\nabla U \cdot \nabla U - k^2 q U^2 + 2gU)\,d\Omega$$
$$+ \frac{1}{2}\int_{\partial C} ap U^2 \,dS - \int_{\partial C} pUv_0 \,dS - \int_{\partial N} pUv_0 \,dS \tag{4.29}$$

over all *admissible* functions U. To be admissible, a function must satisfy two conditions: it must be once differentiable and it must satisfy all Dirichlet boundary conditions. Differentiability is necessary in order to make it possible to evaluate the functional $\mathfrak{F}(U)$ at all, and essential boundary conditions must be satisfied to ensure that a stationary point exists

4.6   Natural continuity conditions

Finite element methods invariably evaluate functionals on an element-by-element basis. If integral expressions such as (4.29) above are to be applied to each element separately, the boundary terms clearly relate to the boundaries of an element, not to boundaries of the whole problem region. Suppose, for example, that the whole problem region Ω is subdivided into two finite elements I and II, separated by the element interface ∂E, as in Fig. 3.7. Along the element interface, neither the potential U nor its gradient are known. The unknown value of gradient ∇U must therefore be viewed as a Neumann condition at the boundary of an element when integration over that element is performed. Let $\mathfrak{F}(E, U)$ represent the functional value of Eq. (4.29), evaluated element by element:

$$
\begin{aligned}
\mathfrak{F}(E, U) = \frac{1}{2}\int_{\Omega} & (p\nabla U \cdot \nabla U - k^2 q U^2 + 2gU)\, d\Omega \\
& + \frac{1}{2}\int_{\partial C} apU^2\, dS - \int_{\partial C} pUv_0\, dS \\
& - \int_{\partial N} pUv_0\, dS \\
& - \int_{\partial E} p_I U_I (\nabla U|_{II} \cdot \mathbf{1}_I)\, dS - \int_{\partial E} p_{II} U_{II}(\nabla U|_I \cdot \mathbf{1}_{II})\, dS.
\end{aligned}
$$

$$(4.30)$$

Here p_I and p_{II} represent the material property values on the two sides of the interface, U_I and U_{II} are the potential values. The first four terms of the functional refer to external boundaries and to bulk integrals, so they are unchanged from the original functional formulation. The unit vectors $\mathbf{1}_I$ and $\mathbf{1}_{II}$ are outward normals from the two elements; since there is only one element interface ∂E, they are simply opposites, $\mathbf{1}_I = -\mathbf{1}_{II} = \mathbf{1}_E$. Because the potential itself must be continuous across the interface, $U_I = U_{II} = U_E$, Eq. (4.30) may be simplified to read

$$
\mathfrak{F}(E, U) = \mathfrak{F}(U) + \int_{\partial E} U_E(p_{II}\nabla U|_I \cdot \mathbf{1}_E - p_I \nabla U|_{II} \cdot \mathbf{1}_E)\, dS.
$$

$$(4.31)$$

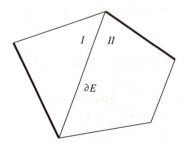

Fig 3.7   The region Ω is subdivided into two finite elements I and II by the element interface ∂E.

Note that the gradients ∇U are specified as being evaluated at one or other side of the element interface ∂E. Although the potential is continuous across the interface, its derivatives are not necessarily so. The distinction between gradients must therefore be retained.

Although finite elements are usually set up so that element edges follow the material interfaces of the physical problem, any element subdivision is acceptable in principle, even one that crosses physical boundaries between materials or source regions. The element interface ∂E might therefore be located anywhere, indeed there might be any number of such interfaces. Since the functional $\mathfrak{F}$ must have a unique value, the additional interface integral terms must vanish identically.

$$\int_{\partial E} U_E p_{II} \nabla U|_I \cdot \mathbf{1}_E \mathrm{d}S = \int_{\partial E} U_E p_I \nabla U|_{II} \cdot \mathbf{1}_E \,\mathrm{d}S. \tag{4.32}$$

Their omission thus implies that there is a *natural continuity condition* between elements, analogous to the natural boundary condition at region edges. Equation (4.32) implies this condition must be

$$\frac{\mathbf{1}_E \cdot \nabla U|_I}{p_I} = \frac{\mathbf{1}_E \cdot \nabla U|_{II}}{p_{II}}. \tag{4.33}$$

If U is the electric scalar potential, this may be interpreted as the condition of continuity of electric flux, $\nabla \cdot (\mathbf{E}/\epsilon) = 0$. The physical problem satisfies this equation strictly at every interface point. In contrast, the finite element solution satisfies Eq. (4.32), which may be viewed as the weak equivalent of Eq. (4.33).

4.7   *The vector Helmholtz equation*

Many electromagnetics problems cannot conveniently be framed in terms of scalar potentials and must be attacked by a full vector field treatment. These are treated much like their scalar counterparts, though the mathematical details are a little different. To develop a stationary functional for the vector Helmholtz equation, analogous to the scalar case, let

$$\mathfrak{F}(U) = \frac{1}{2} \int_\Omega (\nabla \times \mathbf{U} \cdot p \nabla \times \mathbf{U} - k^2 q \mathbf{U} \cdot \mathbf{U}) \,\mathrm{d}\Omega, \tag{4.34}$$

where p and q represent scalar-valued local material properties. As in the scalar case, it will be supposed that $\mathbf{U}$ is an approximation to the desired vector quantity, $\mathbf{u}$ is its exact value. The approximate vector may be written as

$$\mathbf{U} = \mathbf{u} + \theta \mathbf{h}, \tag{4.35}$$

where **h** is a vector-valued error function, θ is a scalar parameter. With this substitution, the functional $\mathfrak{F}(U)$ becomes

$$\mathfrak{F}(\mathbf{U}) = \mathfrak{F}(\mathbf{u}) + \theta \delta \mathfrak{F}(\mathbf{u}) + \frac{\theta^2}{2} \delta^2 \mathfrak{F}(\mathbf{u}) \tag{4.36}$$

where the second variation has a value that characterizes the function **h**, as in the scalar case,

$$\delta^2 \mathfrak{F}(\mathbf{u}) = \mathfrak{F}(\mathbf{h}), \tag{4.37}$$

and

$$\delta \mathfrak{F}(u) = \int_\Omega (\nabla \times \mathbf{u} \cdot p \nabla \times \mathbf{h} - k^2 q \mathbf{u} \cdot \mathbf{h}) \, d\Omega. \tag{4.38}$$

Now $\mathfrak{F}(\mathbf{U})$ is stationary if the term in θ vanishes, so the issue at hand is to determine the conditions for that to happen. Expression (4.38) for $\delta \mathfrak{F}(\mathbf{u})$ may be rewritten, using the vector differentiation rule

$$\nabla \cdot (\mathbf{a} \times \mathbf{h}) = (\nabla \times \mathbf{a}) \cdot \mathbf{h} - \mathbf{a} \cdot (\nabla \times \mathbf{h}). \tag{4.39}$$

Setting $\mathbf{a} = \nabla \times \mathbf{u}$ and integrating over region Ω,

$$\int_\Omega \nabla \times \mathbf{u} \cdot p \nabla \times \mathbf{h} \, d\Omega = \int_\Omega \mathbf{h} \cdot \nabla \times (p \nabla \times \mathbf{u}) \, d\Omega$$
$$- \oint_{\partial \Omega} p \nabla \times \mathbf{u} \times \mathbf{h} \cdot d\mathbf{S}. \tag{4.40}$$

With this substitution, the first variation $\delta \mathfrak{F}(\mathbf{u})$ becomes

$$\delta \mathfrak{F}(\mathbf{u}) = \int_\Omega (\nabla \times p \nabla \times \mathbf{u} - k^2 q \mathbf{u}) \cdot \mathbf{h} \, d\Omega$$
$$- \oint_{\partial \Omega} p(\nabla \times \mathbf{u}) \times \mathbf{h} \cdot d\mathbf{S}. \tag{4.41}$$

The integral over Ω vanishes identically because the field **u** satisfies (by definition!) the homogeneous Helmholtz equation

$$\nabla \times p \nabla \times \mathbf{u} - k^2 q \mathbf{u} = 0. \tag{4.42}$$

To make sense of the boundary integral term, rearrange the vector triple product $(\nabla \times \mathbf{u}) \times \mathbf{h} \cdot d\mathbf{S}$:

$$(\nabla \times \mathbf{u}) \times \mathbf{h} \cdot d\mathbf{S} = (\nabla \times \mathbf{u}) \times \mathbf{h} \cdot \mathbf{1}_n dS = \mathbf{h} \times \mathbf{1}_n \cdot \nabla \times \mathbf{u} \, dS \tag{4.43}$$

so that, reintroducing the definition $\theta \mathbf{h} = \mathbf{U} - \mathbf{u}$ from Eq. (4.35),

$$\theta \oint_{\partial \Omega} p \nabla \times \mathbf{u} \times \mathbf{h} \cdot d\mathbf{S} = \oint_{\partial \Omega} (\mathbf{U} - \mathbf{u}) \times \mathbf{1}_n \cdot (p \nabla \times \mathbf{u}) \, dS. \tag{4.44}$$

The triple product on the right may have its terms interchanged cyclically, yielding

$$\theta \oint_{\partial\Omega} p\nabla \times \mathbf{u} \times \mathbf{h} \cdot d\mathbf{S} = \oint_{\partial\Omega} \mathbf{1}_n \times (p\nabla \times \mathbf{u}) \cdot (\mathbf{U} - \mathbf{u})\, dS. \qquad (4.45)$$

The integrand of this boundary integral clearly vanishes if either of two conditions holds. As in the scalar case, these are termed *natural* and *essential* boundary conditions, respectively. The first arises when the curl of the correct solution $\mathbf{u}$ is purely normal to the boundary,

$$\mathbf{1}_n \cdot \nabla \times \mathbf{u} = 0, \qquad (4.46)$$

and this is the condition that comes naturally if no boundary conditions of any sort are imposed on the finite element solution. The essential boundary condition is

$$\mathbf{U} \times \mathbf{1}_n = \mathbf{u} \times \mathbf{1}_n, \qquad (4.47)$$

and this must be imposed by appropriately constraining the vector fields. The interesting point to note is that both the boundary conditions refer to selected vector components, not to all components. In particular, the solution of a vector Helmholtz equation turns out to be fully determinate if the tangential vector field is prescribed, as in Eq. (4.47); the normal components take care of themselves. To summarize, the conclusion is that the boundary value problem

$$\left.\begin{array}{ll} \nabla \times p\nabla \times \mathbf{u} - k^2 q\mathbf{u} = 0 & \text{in } \Omega, \\ \mathbf{u} \times \mathbf{1}_n = \mathbf{u}_t & \text{on } \partial D \text{ (where } \partial D \cup \partial N = \partial\Omega), \end{array}\right\} \quad (4.48)$$

is solved by finding a stationary point of the functional (4.34) over all $\mathbf{U}$ that satisfy the boundary condition on ∂D. Clearly, the functions $\mathbf{U}$ must be differentiable once, so the general characteristics of the vector functions $\mathbf{U}$ are broadly similar to the function U of the scalar case: they must be differentiable, and must satisfy the essential boundary conditions.

4.8   *Electric and magnetic walls*

Two situations of particular importance to microwave engineering arise when $\mathbf{u} = \mathbf{H}$ and $\mathbf{u} = \mathbf{E}$. The natural boundary conditions in these two formulations have interesting physical interpretations and are worth investigating briefly.

If a field problem is formulated in terms of the electric field vector $\mathbf{E}$, the natural boundary condition is

$$\mathbf{1}_n \cdot \nabla \times \mathbf{E} = 0, \qquad (4.49)$$

which may be written, in accordance with Maxwell's electric curl equation (2.1), as

$$\mathbf{1}_n \cdot \frac{\partial \mathbf{B}}{\partial t} = \frac{\partial}{\partial t}(\mathbf{1}_n \mathbf{B}) = 0. \tag{4.50}$$

Time-varying fields can meet this requirement in only one way: the magnetic flux density vector **B** must lie parallel to the bounding surface. Such boundaries correspond to perfect conductors and are also sometimes referred to as *electric walls*; they are surfaces through which no magnetic flux can penetrate, a condition satisfied by a perfect electric conductor.

If a field problem is formulated in terms of the magnetic field vector **H**, on the other hand, the natural boundary condition becomes

$$\mathbf{1}_n \cdot \nabla \times \mathbf{H} = 0. \tag{4.51}$$

Maxwell's magnetic curl equation (2.2) suggests rewriting this as

$$\mathbf{1}_n \cdot \frac{\partial}{\partial t}\left(\mathbf{J} + \frac{\partial \mathbf{D}}{\partial t}\right) = 0. \tag{4.52}$$

Inside a waveguide, cavity, or any other dielectric medium, there cannot exist any conduction current density **J**. Consequently,

$$\frac{\partial^2}{\partial t^2}(\mathbf{1}_n \cdot \mathbf{D}) = 0. \tag{4.53}$$

Here the electric flux density vector **D**, hence also the electric field vector **E**, must lie parallel to the bounding surface; there cannot be any component of **D** normal to the boundary. Such boundaries are sometimes referred to as *magnetic walls*; no electric flux can penetrate them, and that would be the case were the material perfectly permeable magnetically. They arise frequently as the consequence of geometric symmetry.

In the engineering literature, the terms *Dirichlet condition* and *Neumann condition* are sometimes used in connection with vector variables, extending the terms by analogy with scalar fields. Given that the terms *natural* and *essential* are much more fully descriptive, it is probably best to avoid the alternative names, which in any case have no historical justification.

5.   Projective solutions

The finite element method as used in electromagnetics may be regarded as one particular form of the general class of mathematical methods called *projective approximation*. This mathematical viewpoint applies equally to integral equations or boundary-value problems of differential equations, and it has the added advantage of sometimes allowing finite element methods to be created where no underlying variational principle is known. Projective solution techniques are backed by an extensive mathematical theory. Because this specialist field of mathematics is well beyond the scope of this book, the principles of projective methods will be outlined informally, with no attempt at a rigorously valid mathematical development.

5.1   *Functions and spaces*

Projective methods are most easily developed and understood in the mathematical language of linear spaces, whose terminology will be briefly reviewed in the next few pages. In general, the electromagnetic field problems considered in this book may all be stated in symbolic form as

$$\mathscr{P}u = v. \tag{5.1}$$

Here $\mathscr{P}$ is a symbolic operator, u and v are symbolic representations of fields; they may stand for vectors, scalars, or combinations of both. The operator $\mathscr{P}$ may be viewed as a transformation rule between sets of functions, because it associates at least one solution u with every possible source function v — it *maps* every u into some v. To take a concrete example, $\mathscr{P}$ might represent the two-dimensional Poisson's equation problem of Fig. 3.8, where

$$\left. \begin{aligned} \nabla^2 u &= v &&\text{in } \Omega, \\ u &= 0 &&\text{on boundary } \partial D, \\ \frac{\partial u}{\partial n} &= 0 &&\text{on boundary } \partial N. \end{aligned} \right\} \tag{5.2}$$

Although no single mathematical expression specifies $\mathscr{P}$, Eq. (5.2) does state precisely which function v corresponds to any given function u. It therefore describes $\mathscr{P}$ exactly, if perhaps a little clumsily.

Suppose $R_{\mathscr{P}}$ represents the *range* of the operator $\mathscr{P}$, i.e., the set of all functions v that might conceivably appear as source functions in this

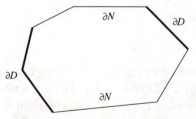

Fig 3.8 Two-dimensional region Ω where Poisson's equation holds. Boundary conditions are partly of Dirichlet, partly of Neumann type.

problem. These functions form a *linear space* (sometimes called a *vector space*) $\mathbb{V}$, a set of functions with three fundamental properties:

(1) If ψ_1 and ψ_2 belong to $\mathbb{V}$ (one usually writes $\psi_1 \in \mathbb{V}$), then so do all their linear combinations, such as $k_1\psi_1 + k_2\psi_2$.
(2) Addition and subtraction are commutative and associative for all members of $\mathbb{V}$.
(3) $\mathbb{V}$ contains a null element 0 such that $\psi + 0 = \psi$.

The source functions v of an electromagnetic field problem obviously possess many other special properties. For example, they are defined over the geometric region Ω of the physical problem and they are square integrable on Ω. Such mathematical characteristics often arise from physical conservation laws; square integrability usually corresponds to finiteness of stored energy or power. For example, current density $\mathbf{J}$ must be square integrable in problems of electric current flow, otherwise power density $\mathbf{J} \cdot \mathbf{E} = \mathbf{J} \cdot \mathbf{J}/\sigma$ is not defined and power cannot be calculated.

In a similar fashion, suppose $D_{\mathscr{P}}$ is the *domain* of the operator $\mathscr{P}$, the set of all possible solutions u of the problem (5.1). These functions also form a linear space, usually one different from $R_{\mathscr{P}}$. For example, in the problem represented by Eqs. (5.2) all $u \in D_{\mathscr{P}}$ must be twice differentiable while the $v \in R_{\mathscr{P}}$ need not be differentiable at all.

5.2   Inner product spaces

The solutions of most problems arising in computational electromagnetics belong to *inner product spaces*, entities much more narrowly defined than linear spaces. The linear space $\mathbb{V}$ is an inner product space if it can be endowed with an *inner product*, a form $\langle \psi_1, \psi_2 \rangle$ defined for any two functions ψ_1 and ψ_2 in $\mathbb{V}$, with the following properties:

$$\langle \psi_1, \psi_2 \rangle > 0 \quad \text{if } \psi_1 \neq 0, \quad \langle \psi_1, \psi_1 \rangle = 0 \quad \text{if } \psi_1 = 0, \tag{5.3}$$

$$\langle \psi_1, \psi_2 \rangle = \langle \psi_2, \psi_1 \rangle \tag{5.4}$$

$$\langle a\psi_1, \psi_2 \rangle = \langle \psi_1, a\psi_2 \rangle = a\langle \psi_1, \psi_2 \rangle, \quad \text{with } a = \text{any number.} \tag{5.5}$$

One possible inner product that may be defined on the domain $D_{\mathscr{P}}$ of the Poisson's equation problem (5.2) is

$$\langle \psi_1, \psi_2 \rangle = \int_{\Omega} \psi_1 \psi_2 \, d\Omega. \tag{5.6}$$

This product is not the only one possible. Another, quite independent, valid inner product for functions $\psi_i \in D_{\mathscr{P}}$ would be

$$\langle \psi_1, \psi_2 \rangle = \int_\Omega \nabla \psi_1 \cdot \nabla \psi_2 \, d\Omega. \tag{5.7}$$

It may seem at first glance that (5.7) fails to distinguish between constant functions and therefore violates (5.3). However, this objection is not valid. Equation (5.2) fixes any possible member function of $D_\mathscr{P}$ to be zero everywhere on the boundary segment ∂D so the only possible constant function belonging to $D_\mathscr{P}$ is *zero*. The same does not hold true for $R_\mathscr{P}$, because its members may have any values anywhere along the region boundary. This example illustrates that an inner product can only be defined with respect to some specific set of functions; the mathematical form of the product and the restrictions placed on the functions go hand in hand. Merely writing Eq. (5.7), without saying what rules the functions ψ_i must obey, is not enough.

The term *inner product* alludes to the fact that the dot-product of two ordinary vectors in three-space, $\mathbf{a} \cdot \mathbf{b} = a_x b_x + a_y b_y + a_z b_z$, satisfies properties (5.3)–(5.5), so an inner product of two functions may be viewed as its generalization. Pursuing this geometric analogy, $\langle \psi_1, \psi_2 \rangle$ is also called the *projection* of ψ_1 onto ψ_2. The projection of a function onto itself say $\langle \psi, \psi \rangle$, is said to yield the *inner product norm* or *natural norm* of ψ,

$$\parallel \psi \parallel = \sqrt{\langle \psi, \psi \rangle}. \tag{5.8}$$

The inner product norm may be thought to measure the size of a function, and in that sense it corresponds to the idea of vector length, which is also a size measure. Many different norms can be defined for functions in addition to the inner product norm; the conventional L_1 and L_∞ norms are examples. When no confusion can arise, the inner product norm of Eq. (5.8) is simply called the *norm*. As the words *space* and *projection* suggest, both geometric intuition and geometric terminology can be extended from Euclidean vectors to functions. Thus one speaks of the *angle* between functions as computed by their inner product, and says ψ_1 and ψ_2 are *orthogonal* if $\langle \psi_1, \psi_2 \rangle = 0$. In the same way, $\parallel \psi_1 - \psi_2 \parallel$ is referred to as the *distance* between functions ψ_1 and ψ_2.

An inner product space $\mathbb{V}$ is said to be *complete* if it includes all functions whose near neighbours are also members of $\mathbb{V}$. This very informal definition can be sharpened to greater precision. Suppose ψ_1, $\psi_2, \psi_3, \ldots$ form a sequence with decreasing distance between functions,

$$\lim_{m,n \to \infty} \parallel \psi_m - \psi_n \parallel = 0. \tag{5.9}$$

With increasing m and n the functions ψ_m and ψ_n come closer to each other, so they approach some ultimate function ψ in the limit — provided such a function ψ exists in the space $\mathbb{V}$. (If $\psi \notin \mathbb{V}$, distance is undefined and the word 'approach' means nothing.) A sequence with the property (5.9) is called a *Cauchy sequence*. It *converges* to ψ if $\psi \in \mathbb{V}$. If all possible limits of Cauchy sequences are included in $\mathbb{V}$, every Cauchy sequence converges and $\mathbb{V}$ is said to be complete. If it is complete with respect to its natural norm, $\mathbb{V}$ is called a *Hilbert space*. Most of the function spaces of interest in numerical electromagnetics are Hilbert spaces.

It may seem far-fetched to suppose that the limit ψ of a sequence might not be included in the set of functions that make up the sequence, but this is actually a very common occurrence. To illustrate why and how, consider the linear space formed by all differentiable periodic time functions $p(t)$ of period T. Any such function can be described by writing its Fourier series, and may be approximated by a truncated Fourier series. The sequence of truncated Fourier series $s_n(t)$,

$$s_n(t) = \sum_{k=1}^{n} \frac{4(k-1)^{-1}}{(2k+1)\pi} \cos \frac{(2k+1)\pi t}{T}, \qquad (5.10)$$

approaches a unit-amplitude square wave as n increases. Although every term $s_n(t)$ in the sequence is infinitely differentiable, the sequence does not converge to a differentiable function — for its limit is the square wave, and the square wave is not differentiable! Thus the differentiable periodic functions constitute an inner product space under the inner product definition (5.6), but they do not form a Hilbert space. However, the square-integrable periodic functions do. This follows from a simple observation: every function $s_n(t)$ is square integrable over the period T, and so is every function that can be approached by a sequence of the functions $s_n(t)$.

5.3 *Projective approximation*

Much like variational methods, projective methods avoid the original boundary-value problem and seek alternative representations. The function u is the correct solution of Eq. (5.1) under the condition, which is both necessary and sufficient, that

$$\langle w, \mathscr{P}u \rangle = \langle w, v \rangle, \qquad \forall\, w \in R_{\mathscr{P}} \qquad (5.11)$$

where $R_{\mathscr{P}}$ is the range of the operator $\mathscr{P}$. This statement has a straightforward geometric interpretation. In Euclidean three-space, two vectors are equal if they have equal projections on three distinct, independent vectors (e.g., on the unit coordinate vectors). Note that their projections onto *any* vector are equal if their projections onto the three independent vectors are. Similarly, in Eq. (5.11) equality is guaranteed for all functions w if equality holds for all w that belong to a spanning set[2] of $R_{\mathscr{P}}$. Now $R_{\mathscr{P}}$ is an inner product space but not necessarily a Hilbert space. The usual procedure is to broaden the problem somewhat, by requiring equal projections onto a spanning set of the Hilbert space W that includes the range $R_{\mathscr{P}}$ of the operator $\mathscr{P}$. A function $\bar{u}$ is said to satisfy Eq. (5.1) in a weak sense if the left and right sides of Eq. (5.11) have equal projections onto any function w in a Hilbert space $\mathsf{W} \supseteq R_{\mathscr{P}}$,

$$\langle w, \mathscr{P}\bar{u} \rangle = \langle w, v \rangle, \qquad \forall\, w \in \mathsf{W}. \tag{5.12}$$

It is said that u satisfies (5.1) in a strong sense if it satisfies (5.11). The reason for this approach is pragmetic: weak solutions are usually much easier to construct than strong ones.

The general principle of projective methods is to satisfy Eq. (5.12) only with respect to an M-dimensional subspace W_{RM} of W, rather than the entire operator range $R_{\mathscr{P}}$. The resulting solution should therefore really be described as an *approximate weak solution with respect to* W_{RM}. Suppose that the set of functions $\{\beta_k | k = 1, \ldots, M\}$ spans the space W_{RM}, i.e., that any function s in W_{RM} may be expressed exactly as a linear combination of the β_k,

$$s = \sum_{i=1}^{M} s_i \beta_i, \qquad \forall\, s \in \mathsf{W}_{RM}. \tag{5.13}$$

(Note that s_i here denotes a numerical coefficient, not a function.) A weak solution of Eq. (5.1) with respect to W_{RM} is then the function $\bar{u}_{(M)}$ that satisfies

$$\langle \beta_j, \mathscr{P}\bar{u}_{(M)} \rangle = \langle \beta_j, v \rangle, \qquad j = 1, \ldots, M. \tag{5.14}$$

This equation does not yet provide a method for finding $\bar{u}_{(M)}$, which for the moment remains an abstract symbol. It can be made computable by a further approximation. Let $\bar{u}_{(M)}$ be approximated by $\bar{u}_{(MN)}$, a function constrained to lie within an N-dimensional linear subspace of W, say $\mathsf{W}_{DN} \subset \mathsf{W}$. Suppose $\{\alpha_i | i = 1, \ldots, N\}$ is a spanning set for W_{DN} so that the approximate solution $\bar{u}_{(MN)} \in \mathsf{W}_{DN}$ is given by

$$\bar{u}_{(MN)} = \sum_{i=1}^{N} u_{(M)i} \alpha_i \tag{5.15}$$

where the $u_{(M)i}$ are numerical coefficients (not functions), $i = 1, \ldots, N$. Then (5.14) reads

----

[2] $\mathbb{B}$ is a spanning set of some space $\mathbb{A}$ if (1) its members $b_i \in \mathbb{B}$ are linearly independent, (2) every function $a \in \mathbb{A}$ can be represented as a linear combination of the members of $\mathbb{B}$, $a = \sum \bar{a}_i b_i$.

$$\left\langle \beta_j, \mathscr{P} \sum_{i=1}^{N} u_{(M)i}\alpha_i \right\rangle = \langle \beta_j, v \rangle, \quad j = 1, \dots, M. \tag{5.16}$$

For a linear operator $\mathscr{P}$, Eq. (5.16) reduces to the special form

$$\sum_{i=1}^{N} \langle \beta_j, \mathscr{P}\alpha_i \rangle u_{(M)i} = \langle \beta_j, v \rangle, \quad j = 1, \dots, M. \tag{5.17}$$

This may be identified as a matrix equation, whose solution provides the set of N coefficients $u_{(M)i}$. These define an approximation $\bar{u}_{(MN)}$ to the weak solution $\bar{u}$. This approximation is generally useful only if it is *convergent*, in the sense that an enlargement of the spaces $\mathbb{W}_{DN}$ and $\mathbb{W}_{RM}$ will bring the approximate weak solution arbitrarily close to the strong solution,

$$\lim_{M,N\to\infty} \| \bar{u}_{(MN)} - u \| = 0. \tag{5.18}$$

Just how the spaces $\mathbb{W}_{DN}$ and $\mathbb{W}_{RM}$ should be defined, how they can be enlarged to ensure convergence, and how rapid the convergence may be, are questions of major theoretical importance to which a great deal of work has been devoted. These issues, however, fall within the area of finite element mathematics rather than electromagnetics, so they will not be treated in this book.

5.4   *Geometric interpretation*

The argument presented above may perhaps be illustrated in geometric or pictorial terms. If the functions that make up a function space are thought to correspond to points, then the space itself must correspond to a cloud of points, and may be shown in a picture as a portion of the drawing space. In Fig. 3.9, $\mathbb{W}$ is a Hilbert space which encompasses both the range $R_{\mathscr{P}}$ and domain $D_{\mathscr{P}}$ of the operator $\mathscr{P}$; in other words, $\mathscr{P}$ maps every domain point u into the corresponding range point v. An approximate solution is sought by selecting two subspaces within $\mathbb{W}$: $\mathbb{W}_{DN}$ to serve as a model for the operator domain $D_{\mathscr{P}}$ and $\mathbb{W}_{RM}$ as a model for the range $R_{\mathscr{P}}$. Note that $\mathbb{W}_{DN}$ must lie within the domain, $D_{\mathscr{P}} \supseteq \mathbb{W}_{DN}$, otherwise the term $\mathscr{P}\alpha_i$ of Eq. (5.17) means nothing. The same is not true for the range, however; the inner product $\langle \beta_j, v \rangle$ remains well-defined so long as both functions belong to the same inner product space $\mathbb{W}$. The range $R_{\mathscr{P}}$ and the range model $\mathbb{W}_{RM}$ need not even overlap! If they do not, or if v does not lie in the model space, $v \notin W_{RM}$, then Eq. (5.18) will find the best available approximation: that whose projection onto W_{RM} satisfies it.

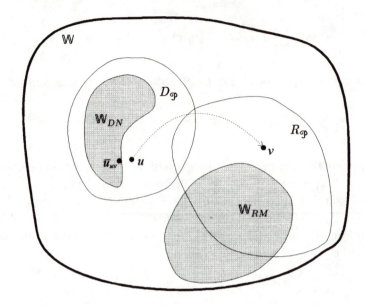

Fig 3.9   Approximate solution may be sought in a Hilbert space
which includes both the domain $D_{\mathscr{P}}$ and range $R_{\mathscr{P}}$ of the operator $\mathscr{P}$.

What then is the difference between strong and weak solutions? It lies
in the sorts of approximating functions that may be chosen to make up
W_{DN} and W_{RM}. The Hilbert space W is generally larger than the range
$R_{\mathscr{P}}$ of the operator $\mathscr{P}$, so it is perfectly possible to construct an approximate solution out of functions possessing weaker continuity properties
and fewer boundary constraints than the strong solution. This indeed
forms the very essence of many finite element methods, where approximations to smooth, continuous solutions are built up from element functions which are only piecewise smooth.

5.5   *Galerkin's method for the Poisson equation*

The most common projective technique in finite element analysis
is undoubtedly the Galerkin method. It is closely related to the variational formulations already discussed at some length, as well as to the
method of weighted residuals used in various other engineering disciplines.
This section presents Galerkin's method from a projective viewpoint,
illustrating it by Poisson's equation, then relates it to the other common
techniques.

Consider again the Poisson equation problem (5.2). Here all the possible solutions u, and all possible source distributions v, lie within the
linear space of square-integrable functions, which form a Hilbert space
$\mathsf{W}^{(0)}$ under the inner product definition

$$\langle \psi_i, \psi_j \rangle = \int_\Omega \psi_i \psi_j \, d\Omega, \qquad \forall \; \psi_i, \psi_j \in \mathbf{W}^{(0)}. \tag{5.19}$$

Hilbert spaces with this or similar inner products over square-integrable functions are common in electromagnetics and mechanics, because the squared integrands often bear a close relationship to energy, power, momentum, or other densities subject to conservation laws. Under this inner product definition, the weak approximation (5.17) to Eq. (5.2) is

$$\sum_{i=1}^{N} \left(\int_\Omega \beta_j \nabla^2 \alpha_i \, d\Omega \right) u_{(M)i} = \int_\Omega \beta_j v \, d\Omega, \qquad j = 1, \ldots, M. \tag{5.20}$$

This form of expression, however, is not applicable if the α_i merely belong to the Hilbert space $\mathbf{W}^{(0)}$ of square-integrable functions, for the integral in parentheses is meaningful only if α_i is sufficiently differentiable for $\nabla^2 \alpha_i$ to exist. In other words, $\mathscr{P}$ as described by Eq. (5.2) applies to twice-differentiable functions but not to any others. To develop a weak solution, the definition of $\mathscr{P}$ must be broadened, creating another operator $\mathscr{P}'$ which applies to a larger collection of functions but coincides exactly with $\mathscr{P}$ for all those functions within the latter's domain $D_\mathscr{P}$. The traditional way to do so is to integrate by parts, i.e., to apply Green's second identity once again,

$$\sum_{i=1}^{N} \left(\int_\Omega \nabla \beta_j \cdot \nabla \alpha_i \, d\Omega - \oint_{\partial\Omega} \beta_j \frac{\partial \alpha_i}{\partial n} \, dS \right) u_{(M)i} = \int_\Omega \beta_j v \, d\Omega. \tag{5.21}$$

The boundary integral here vanishes in consequence of (5.2), so there remains

$$\sum_{i=1}^{N} \left(\int_\Omega \nabla \beta_j \cdot \nabla \alpha_i \, d\Omega \right) u_{(M)i} = \int_\Omega \beta_j v \, d\Omega. \tag{5.22}$$

This equation amounts to an implicit definition of $\mathscr{P}'$. It does not say exactly how to construct $\mathscr{P}'u$ for any u, but it does give a prescription for calculating $\langle w, \mathscr{P}'u \rangle$ for any function u which need be differentiable only *once*. Hence the approximating function space $\mathbf{W}_{DN}$ is a proper subspace of $\mathbf{W}^{(1)}$, the space of once-differentiable functions that satisfy the essential boundary conditions. It is a proper subspace of $\mathbf{W}^{(0)}$, for $\mathbf{W}_{DN} \subset \mathbf{W}^{(1)} \subseteq \mathbf{W}^{(0)}$. Note, however, that with this broader definition of $\mathscr{P}'$ the approximate range $\mathbf{W}_{RM}$ also changes. Under the narrow definition (5.2) of the operator $\mathscr{P}$, it was only necessary that $\beta_j \in \mathbf{W}^{(0)}$. Under the broadened definition of $\mathscr{P}'$, its domain has been enlarged but the possible choice of $\mathbf{W}_{RM}$ has shrunk, for Eq. (5.22) requires every β_j to be once differentiable, i.e., $\mathbf{W}_{RM} \subset \mathbf{W}^{(1)} \subseteq \mathbf{W}^{(0)}$.

The extended definition of $\mathscr{P}'$ makes the differentiability requirements on members of W_{DN} and W_{RM} similar. The essence of the Galerkin approach is to make these spaces symmetric, through integration by parts, and then to use the same approximation space for both domain and range, $W_{DN} = W_{RM}$. This does not necessarily mean $\alpha_i = \beta_i$, though many analysts find this choice convenient. True symmetric (self-adjoint) operators, for which

$$\langle u, \mathscr{P}v \rangle = \langle u, \mathscr{P}v \rangle, \tag{5.23}$$

then invariably yield symmetric matrix representations.

A geometric interpretation of the Galerkin process appears in Fig. 3.10, where $W_{DN} = W_{RM}$.

5.6   Weighted residual methods

One time-honoured group of engineering techniques for approximate solution of operator equations is known as the *method of weighted residuals*. These may be viewed as a very general approach, of which Galerkin's method is an important special case. Their basic principle is very simple. The operator equation (5.1) is first rewritten as

$$\mathscr{P}u - v = r, \tag{5.24}$$

to form the residual r, which is then minimized. To do so, the function u is given an approximate representation, much as in Eq. (5.13),

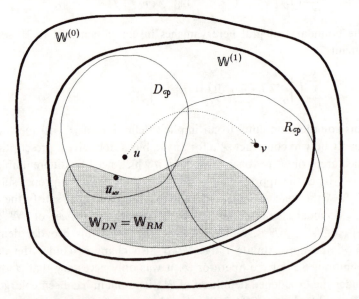

Fig 3.10   Galerkin's method: approximating spaces for range and domain of $\mathscr{P}'$ are made the same by symmetrizing $\mathscr{P}'$.

$$u(N) = \sum_{i=1}^{N} u_{(N)i}\alpha_i \qquad (5.25)$$

so that Eq. (5.24) is replaced by the approximation

$$\sum_{i=1}^{N} u_{(N)i}\mathscr{P}\alpha_i - v = r. \qquad (5.26)$$

The word 'minimize' can be taken to mean many things. Accordingly, there are many varieties of the weighted residual method, corresponding to the various interpretations of minimization.

5.7 *Projective weighted residuals*

One reasonable approach to minimizing the residual in Eq. (5.26) is to minimize it in a projective sense. That is to say, r is required to have exactly zero projection onto each of a set of M functions β_j,

$$\sum_{i=1}^{N} \langle\beta_j, \mathscr{P}\alpha_i\rangle u_{(MN)i} - \langle\beta_j, v\rangle = 0, \quad j = 1,\ldots M. \qquad (5.27)$$

This equation is similar to Eq. (5.14), save for one major difference: the operator $\mathscr{P}$ still has its original meaning, so all the functions α_i must belong to the operator domain $D_{\mathscr{P}}$. In this form, the weighted residual method coincides with the general projective method. Since the method is the same, its difficulties are the same: the functions α_i must be within the domain of $\mathscr{P}$. For the Helmholtz or Poisson equations, this means they must be twice differentiable and not at all easy to construct. The differentiability requirement may be reduced by the same means as in the development of Galerkin's method; an application of Green's second identity. The operator $\mathscr{P}$ is thereby generalized to the new and symmetric form $\mathscr{P}'$, exactly as for the Galerkin method:

$$\sum_{i=1}^{N} \langle\beta_j, \mathscr{P}'\alpha_i\rangle u_{(MN)i} - \langle\beta_j, v\rangle = 0, \quad j = 1,\ldots M. \qquad (5.28)$$

This weak form, however, remains more general than Galerkin's method, for the requirement of identical approximate domain and range spaces, $W_{DN} = W_{RM}$, has not been imposed. As in Galerkin's method, the approximating functions α_i must belong to W_{DN}; they must be once differentiable, $W_{DN} \subseteq W^{(1)}$, and must satisfy the essential boundary conditions of the boundary-value problem. As sketched in Fig. 3.10, however, there is no need for the domain $D_{\mathscr{P}}$ to be included in W_{DN}; even $D_{\mathscr{P}} \cap W_{DN} = 0$ is permissible! Because $\mathscr{P}'$ is created by an integration by parts that transfers differentiations, the functions β_j must be once differentiable also, $\beta_j \in W_{RM} \subseteq W^{(1)}$. However, W_{DN} and W_{RM} need not be the same, indeed they need not even intersect. The choice of β_j, which are frequently called *weighting functions*, is sometimes said to be arbitrary. In fact, it is anything but arbitrary; the requirement $\beta_j \in W^{(1)}$ is essential for the method to make sense. Beyond that, however, the choice is open. Figure 3.11 sketches the relationship of the square-integrable functions $W^{(0)}$ to the space $W^{(1)}$ (once-differentiable functions that satisfy the essential boundary conditions) and $D_{\mathscr{P}} = W^{(2)}$ (twice-differentiable functions that satisfy all the boundary conditions). The strong solution u lies in the space $W^{(2)}$ whose members satisfy both the differentiability requirements and all of the boundary conditions, i.e. in the domain of $\mathscr{P}$. The approximate weak solution $\bar{u}_{MN}$, on the other hand, is sought among the members of the larger space $W^{(1)} \supset W^{(2)}$ that also satisfy the essential boundary conditions but are only once differentiable. The finite approximation in turn must choose from among a limited subset W_{DN} of the functions in $W^{(1)}$.

On triangular elements of the kind already considered, the projective weighted residual method comes quite naturally. The approximate solution U is expressed in terms of the element functions, with continuity enforced at the element interfaces. Thus the space of *trial functions* W_{DN} is made up of 'pyramid' functions as sketched in Fig. 3.12. The weighting functions, sometimes also called the *testing functions*, need not be the

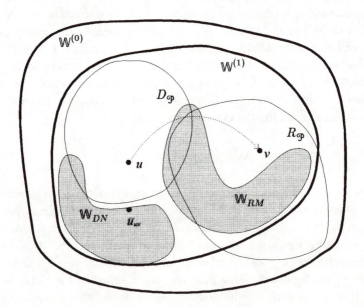

Fig 3.11   The method of weighted residuals: W_{DN} and W_{RM} may, but need not, intersect.

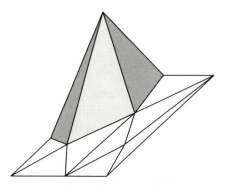

Fig 3.12   One of the continuous basis functions on first-order triangular finite elements.

same, but it is usually convenient to choose them so. As a result, the projective weighted residual formulation becomes indistinguishable from the Galerkin process.

5.8   *Collocation methods*

An altogether different family of weighted residual methods is obtained by abandoning the idea of projecting r onto the functions β_j and demanding instead that r vanish exactly at a set of geometric points P_j in the problem domain Ω. Such a *pointwise* weighted residual technique leads to the matrix equation

$$\sum_{i=1}^{N} (\mathscr{P}\alpha_i)|P_j \bar{\bar{u}}_{(MN)i} = v(P_j). \tag{5.29}$$

The solution $\bar{\bar{u}}_{(MN)}$ is of course not the same as the projective solution $\bar{u}_{(MN)}$. This technique has the advantage of simplicity, because it avoids computing any projections at all; it only requires evaluating the functions on the left and right at a set of points P_j. It is usually termed *collocation*, for it collocates ('places together') at a set of discrete points the values on the left and right sides of the equation. It has two disadvantages. First, the functions α_i must be within the domain of $\mathscr{P}$ in its strong form. This causes more difficulties with differential than integral operators, so collocation is more popularly used for solving integral equations. A second problem is uncertain accuracy, for nothing at all can be known about error behaviour at geometric domain points other than P_j.

Weighted residual methods, both projective and collocative, have been widely used in various branches in engineering since the early 1900s, and have been popular in electromagnetics since the late 1960s. For historical reasons they have been referred to, especially by the antennas community, as the *method of moments*, without distinguishing between projection and collocation. In fact, some writers have suggested that collocation is merely a special case of projection, with $\beta_j = \delta(P_j)$, the Dirac delta function. This assertion is untenable; the Dirac delta functions are not square integrable and do not belong to a Hilbert space with inner product of the form (5.17), so notions of projection, distance, and convergence borrowed from Hilbert spaces are not applicable. Both the theory and practice of collocation methods must therefore be developed separately from projective methods, for the two families of methods really are quite different. A few collocative finite element methods have appeared in the literature, but they are rare and will not be treated in this book. Projective weighted residual methods, on the other hand, will be found in several chapters.

The projective formulations, and particularly the weighted residual formulation, provide an interesting possibility for self-checking. The coefficient matrices obtained by projections have M rows and N columns. Analysts commonly take $N = M$, but this all too frequently reflects haste rather than deliberate choice. Taking $M > N$ allows least-squares solutions to be constructed which can give a useful guide to the solution precision.

6.   Two-dimensional field problems

The general techniques sketched in the foregoing can be applied to fully three-dimensional field problems, or to problems that have been simplified to two-dimensional form. This section outlines the methods that lead to such forms. The development is straightforward in principle, but nevertheless harbours some surprises in the potential functions and their formulation.

6.1   *Translational uniformity*

Many practical problems can be idealized by assuming that the problem boundaries can be described by equations of the form $f(x, y) = 0$, i.e., boundaries with a prescribed shape in the x–y plane but infinite and uniform in z. The field description in such cases can best be deduced from the general case through a separation of variables. Consider again the vector Helmholtz equation (3.55),

$$(\nabla^2 + k^2)\mathbf{S} = 0, \tag{3.55}$$

and assume that the field $\mathbf{S}$ can be represented in the separated form

$$\mathbf{S} = \mathbf{P}(x, y)Z(z). \tag{6.1}$$

On substitution,

$$(\nabla^2 + k^2)\mathbf{S} = Z\nabla_T^2\mathbf{P} + \mathbf{P}Z'' + k^2\mathbf{P}Z = 0, \tag{6.2}$$

where $\nabla_T^2 = \partial^2/\partial x^2 + \partial^2/\partial y^2$ is the transverse Laplacian. Dividing by Z,

$$(\nabla_T^2 + k^2)\mathbf{P} + \frac{Z''}{Z}\,\mathbf{P} = 0. \tag{6.3}$$

This equation contains only one expression involving z, the ratio Z''/Z. The classical separation argument now applies: if Eq. (6.3) is to hold true for all values of z, this ratio cannot depend on z; and since $Z(z)$ is a function of nothing else, the ratio Z''/Z must in fact be a constant, say $-m^2$. Thus the vector Helmholtz equation separates into two parts,

$$\frac{d^2}{dz^2} + m^2 Z = 0, \tag{6.4}$$

$$(\nabla_T^2 + k^2 - m^2)\mathbf{P} = 0. \tag{6.5}$$

The former may be recognized immediately as having sinusoidal solutions $Z = \cos m(z - z_0)$, which may, if desired, be written in an alternative exponential form. The latter, Eq. (6.5), is a two-dimensional vector Helmholtz equation which is generally much simpler to solve than its three-dimensional counterpart.

Several special cases of the Helmholtz equation (6.5) are of importance in practical applications. One particularly interesting group of situations arises when Eq. (6.5) degenerates into Laplace's equation. This can happen in two ways: either $k^2 - m^2 = 0$ with $m \neq 0$, or else $k = m = 0$. The latter is the classical static longitudinally uniform case, exhaustively mined for examples by almost every field theory textbook. The former is of equally great interest. To keep $k^2 = m^2$, the time equation (3.54) then must read

$$\mu\epsilon\frac{d^2 T(t)}{dt^2} + \mu\sigma\frac{dT(t)}{dt} + m^2 T(t) = 0. \tag{6.6}$$

Note if the conductivity σ is not taken as zero, both k^2 and m^2 are complex. In that case, Eq. (6.4) has exponentially decaying sinusoidal solutions, and so does (6.6). The product $\mathbf{S}(x,y)Z(z)T(t)$ therefore represents a pair of travelling waves,

$$\mathbf{S}ZT = \mathbf{S}(x,y)\cos mz \cos\frac{m}{\sqrt{(\mu\epsilon)}}t$$

$$= \frac{\mathbf{S}(x,y)}{2}\left(\cos m\left\{z + \frac{t}{\sqrt{(\mu\epsilon)}}\right\}\cos m\left\{z - \frac{t}{\sqrt{(\mu\epsilon)}}\right\}\right). \tag{6.7}$$

These represent TEM waves, propagating with velocity $1/\sqrt{\mu\epsilon}$ in the $\pm z$ directions.

Waves of quite different sorts arise from $\lambda^2 = k^2 - m^2 \neq 0$. The z and t functions still combine to form travelling waves; but the wave function **S** must now satisfy an eigenvalue problem, Eq. (6.5). Thus there can exist many different waves travelling at different velocities, one corresponding to each of the eigenvalues λ^2. This precisely describes the classical hollow waveguide problem of microwave technology, so the waveguide problem is often used by finite element analysts as a test case and demonstration example.

6.2   *Axisymmetric regularity or uniformity*

Many problems are best set up in cylindrical coordinates, because truly axisymmetric geometric shapes are involved; the electromagnetic sources in the problem need not necessarily be axisymmetric. The general method of variable separation applies here as it does in translationally uniform problems, but the results are somewhat different. To develop the methods, consider the scalar Helmholtz equation, this time in cylindrical coordinates,

$$\frac{1}{r}\frac{\partial}{\partial r}\left(r\frac{\partial \mathscr{U}}{\partial r}\right) + \frac{1}{r^2}\frac{\partial^2 \mathscr{U}}{\partial \varphi^2} + \frac{\partial^2 \mathscr{U}}{\partial z^2} + k^2 \mathscr{U} = 0. \tag{6.8}$$

To separate variables, let

$$\mathscr{U}(r,\varphi,z) = U(r,z)\Phi(\varphi). \tag{6.9}$$

Substituting into the Helmholtz equations and dividing by Φ, there results

$$\left\{\frac{1}{r}\frac{\partial}{\partial r}\left(r\frac{\partial U}{\partial r}\right) + \frac{\partial^2 U}{\partial z^2} + k^2 U\right\} + \frac{1}{r^2}U\frac{\Phi''}{\Phi} = 0. \tag{6.10}$$

Now the only term that involves the angular variable φ in any way is the ratio Φ''/Φ. The usual separation argument thus applies: if the equation is to hold for any and all φ, only one term in it cannot vary with φ. Thus, although the ratio Φ''/Φ involves the symbol φ, its value must be independent of φ; and since it does not involve any of the other variables, it must be a constant. One therefore sets $\Phi''/\Phi = -m^2$, and separation is achieved:

$$\frac{\mathrm{d}^2 \Phi}{\mathrm{d}\varphi^2} + m^2 \Phi = 0, \tag{6.11}$$

$$\frac{1}{r}\frac{\partial}{\partial r}\left(r\frac{\partial U}{\partial r}\right) + \frac{\partial^2 U}{\partial z^2} + \left(k^2 - \frac{m^2}{r^2}\right)U = 0. \tag{6.12}$$

In contrast to the Cartesian case, m may only take on integer values here, for the variation of solutions with φ must have a period of exactly 2π or its integer multiples.

The first of the separated equations, (6.11), is formally identical to its counterpart in the translationally uniform case, and clearly has similar solutions: sines and cosines of the angle φ. However, Eq. (6.12) is different from the separation equation obtained in the Cartesian case, for it contains singularities of the form $1/r$ and $1/r^2$. This equation may be regarded as a generalized Bessel equation, for it reduces exactly to the familiar Bessel differential equation if the term $\partial^2 U/\partial z^2$ is removed:

$$\frac{1}{r}\frac{\partial}{\partial r}\left(r\frac{\partial U}{\partial r}\right) + \left(k^2 - \frac{m^2}{r^2}\right)U = 0. \tag{6.13}$$

Were there no z-variation, the solutions would thus be the well-known and fully tabulated Bessel functions. In the present case, they are not; but given that the radial variation, and the radial singularities, are identical in the two equations, it is reasonable to expect that the solutions will behave near their singular points in a manner not similar to the Bessel functions. This behaviour depends mainly on the separation constant m. Near $r = 0$ the Bessel functions are bounded and smooth for $m = 0$, but they grow without bound for all $m \geq 1$. Thus numerical methods may be expected to work well for $m = 0$, but may give trouble for larger m. Of course, $m = 0$ corresponds to true axial symmetry (no variation of sources or solutions with φ) so it is an enormously important special case; but other values of m must not be dismissed too lightly, as will quickly become evident.

The vector Helmholtz equation may be treated in a like fashion to the scalar, but the mathematical results are not exactly similar because the detailed appearance of the vector Laplacian differs from its scalar counterpart. The case of particular importance in applications is the solenoidal potential or field

$$\mathbf{A} = \mathbf{1}_\varphi A_\varphi, \tag{6.14}$$

for which the Helmholtz equation assumes the relatively simple form

$$\frac{\partial^2 A_\varphi}{\partial r^2} + \frac{1}{r}\frac{\partial A_\varphi}{\partial r} + \left(k^2 - \frac{1}{r^2}\right)A_\varphi + \frac{\partial^2 A_\varphi}{\partial z^2} = 0. \tag{6.15}$$

Comparison with Eq. (6.12) quickly shows that this equation is identical to the scalar case with $m = 1$; in other words, the completely axisymmetric vector equation has the same form and the same solution as the scalar problem with periodicity 1. Clearly, there is a singularity of form $1/r^2$ at the axis of symmetry and numerical troubles may be expected in methods that attempt to model A_φ with straightforward polynomials.

6.3   *Modified axisymmetric potentials*

The difficulties that arise from solution singularities in axisymmetric or axi-periodic problems can be avoided by introducing a new dependent variable, a *modified potential*. In the scalar problem, set

$$U = r^q V \tag{6.16}$$

where q is some number chosen to eliminate the singularity. With this substitution, the modified Bessel equation (6.12) becomes

$$r^q \left(\frac{\partial^2 V}{\partial r^2} + \frac{\partial^2 V}{\partial z^2} \right) + (2q + 1)r^{q-1} \frac{\partial V}{\partial r}$$
$$+ (k^2 r^2 - m^2 + q^2)r^{q-2} V = 0. \tag{6.17}$$

The lowest power of r in this equation is $q - 2$, so there is certain to be no singularity if q is assigned a value of at least 2. Of course, any larger value $q \geq 2$ will eliminate the singularity also. (Note that q does not depend on the value of m, even though the existence of a singularity does.) The approach to numerical solution is to solve the modified equation (6.17) in V, a smooth and well-behaved function unlikely to raise any numerical difficulties; and subsequently to find U by reintroducing the factor r^q.

Finite element methods, in which either a variational or a projective formulation leads to a solution in weak form, can make do with multipliers r^q much less draconian than $q = 2$. To see why, consider the scalar functional

$$\mathfrak{F}(U) = \frac{1}{2} \int_\Omega (\nabla U \cdot \nabla U - k^2 U + 2gU) \, d\Omega \tag{6.18}$$

or its vector analogue. To evaluate it in detail it is first necessary to express the gradient in cylindrical coordinates:

$$\nabla U = \nabla(r^q V)$$
$$= \mathbf{1}_r \left(r^q \frac{\partial V}{\partial r} + qr^{q-1} V \right) + \mathbf{1}_\theta r^{q-1} \frac{\partial V}{\partial \varphi} + \mathbf{1}_z r^q \frac{\partial V}{\partial z}. \tag{6.19}$$

The key term in the integrand of Eq. (6.18) then reads

$$\nabla U \cdot \nabla U = r^{2q}\left\{ \left(\frac{\partial V}{\partial r}\right)^2 + \left(\frac{\partial V}{\partial z}\right)^2 \right\}$$

$$+ r^{2q-1}2qV\frac{\partial V}{\partial r} + r^{2q-2}\left(\frac{\partial V}{\partial \varphi}\right)^2 \qquad (6.20)$$

where the lowest power of r is $2q - 2$, so that regularization only requires $q = 1$! In fact, even this value is excessive, for what needs to be evaluated is not the squared gradient, but its integral. Expressed in cylindrical coordinates as an iterated integral, the functional $\mathfrak{F}(U)$ reads

$$\mathfrak{F}(U) = \frac{1}{2}\int\int\int (\nabla U \cdot \nabla U - k^2 U + 2gU)r\,dr\,d\varphi\,dz. \qquad (6.21)$$

Note that the integration with respect to the azimuthal variable φ contributes an extra r! Retaining only the critical term for clarity's sake,

$$\mathfrak{F}(U) = \frac{1}{2}\sum\sum\sum\left\{ \ldots + r^{2q-2}\left(\frac{\partial V}{\partial \varphi}\right)^2 \ldots \right\}r\,dr\,d\varphi\,dz \qquad (6.22)$$

whence it is obvious that the lowest power of r in integrand is $2q - 1$. The functional therefore involves no singularity, if $q \geq \frac{1}{2}$, i.e., if the potential is modified so that $U = V\sqrt{r}$.

For reasons that probably have a basis in aesthetics rather than arithmetic, some analysts have traditionally eschewed the factor $\sqrt{r}$, preferring an integer power such as $q = 1$. As a result, some practical programs use $U = V\sqrt{r}$. some others use $U = Vr$, while still others ignore the singularity, set $U = V$, and hope for the best. Their hopes are usually answered well enough when solving problems where the axis of symmetry does not enter the problem region at all. If it does, not only does the solution accuracy suffer, but the element construction usually requires *ad hoc* program patches to avoid evaluating the integrand at any singular points.

7.   Magnetic scalar potentials

The Maxwell equations possess an appealing symmetry. There are two curl equations and two divergence equations — two electric equations and two magnetic — two in flux densities and two in field intensities. This symmetry suggests that a set of symmetrically related potentials can also be defined: a magnetic scalar potential and an electric vector potential. Of these, the magnetic scalar potential has found extensive use, while applications of the electric vector potential have been more restricted, though still useful. The major characteristics of the magnetic scalar potential, and some extensions of it, will be outlined in the following.

7.1   *The scalar potential*

In the very special case of a current-free region, $\mathbf{J} + \partial \mathbf{D}/\partial t = 0$, the Maxwell magnetic curl equation degenerates to a statement of irrotationality,

$$\nabla \times \mathbf{H} = 0. \tag{7.1}$$

This property permits $\mathbf{H}$ to be derived from a scalar potential $\mathscr{P}$,

$$\mathbf{H} = -\nabla \mathscr{P}. \tag{7.2}$$

Many useful magnetics problems can be phrased conveniently in terms of this potential. The Maxwell magnetic divergence equation

$$\nabla \cdot \mathbf{B} = 0 \tag{7.3}$$

may be rewritten, setting $\mathbf{B} = \mu \mathbf{H}$, as

$$\nabla \cdot (\mu \nabla \mathscr{P}) = 0 \tag{7.4}$$

and this may be viewed as a nonlinear generalization of Laplace's equation. It is valid only in regions containing no currents, so the range of problems that can be solved is somewhat restricted. On the other hand, it permits magnetic field problems to be solved in general three-dimensional cases without involving a vector potential, so it is inherently attractive for reasons of computational economy as well as ease of representation of the results.

To solve boundary-value problems governed by Eq. (7.4), the equation itself must be augmented by boundary conditions. Consider first a boundary on which the tangential component of $\mathbf{H}$ is prescribed. The boundary condition on such a surface would be an essential boundary condition were the problem to be framed variationally in terms of $\mathbf{H}$. By definition of $\nabla \mathscr{P}$, the potential difference between any two surface points P and Q is given by

$$\mathscr{P}_P - \mathscr{P}_Q = \int_Q^P \nabla \cdot \mathbf{ds} = -\int_Q^P \mathbf{H} \cdot \mathbf{ds}. \tag{7.5}$$

The distance increment vector $\mathbf{ds}$ always lies tangential to the surface, so only the tangential components of $\mathbf{H}$ enter into the integration. Consequently, the integral is determined once the tangential value of $\mathbf{H}$ is specified. Surfaces with prescribed tangential $\mathbf{H}$ thus correspond to surfaces with prescribed $\mathscr{P}$, i.e., Dirichlet boundaries. Thus the essential boundary conditions of the vector problem map into essential boundary conditions of the scalar problem. Correspondingly, surfaces with prescribed normal component of **H**, which enter the vector field problem as natural boundary conditions, map into natural boundary conditions (Neumann conditions) in the scalar potential. Consider a surface where the normal component $\mathbf{1}_n \cdot \mathbf{H}$ is given. This component may be written

$$\mathbf{1}_n \cdot \mathbf{H} = -\mathbf{1}_n \cdot \nabla \mathscr{P} = -\frac{\partial \mathscr{P}}{\partial n}. \tag{7.6}$$

Prescribing the normal vector component is therefore equivalent to prescribing the normally directed derivative of the scalar potential. Boundary value problems in the scalar magnetic potential are therefore amenable to exactly the same sort of treatment as boundary value problems in the electric scalar potential.

7.2   *Carpenter–Noether potentials*

In current-carrying regions the magnetic scalar potential cannot be defined, for it relies on irrotationality of the magnetic field for its existence. However, a change of variable can serve to define a closely related potential valid even where currents flow. The idea seems to be traceable to Maxwell, though its currently accepted form is due to Carpenter and some earlier work by Noether. Given the current distribution $\mathbf{J} + \partial \mathbf{D}/\partial t$ in the problem region, Carpenter begins by defining a vector quantity $\mathbf{H}_c$ such that

$$\nabla \times \mathbf{H}_c = \mathbf{J} + \frac{\partial \mathbf{D}}{\partial t}. \tag{7.7}$$

The quantity $\mathbf{H}_c$ is not the magnetic field sought, for it is not subject to any boundary conditions at all; nor is it necessary for $\mathbf{H}_c$ to be solenoidal. In fact, $\mathbf{H}_c$ is not unique; many different definitions are possible because $\nabla \times \mathbf{H}_c = \nabla \times (\mathbf{H}_c + \nabla \psi)$ for any scalar function ψ. Finding an auxiliary function $\mathbf{H}_c$ is therefore much easier than finding the correct magnetic field. To solve the original problem, it remains to observe that the difference between $\mathbf{H}_c$ and the desired field **H** is irrotational,

$$\nabla \times (\mathbf{H} - \mathbf{H}_c) = 0, \tag{7.8}$$

because the curls of both equal the same current density. However, their difference is obviously irrotational, so it can be represented by a scalar potential $\mathscr{P}_c$,

$$\mathbf{H} - \mathbf{H}_c = \nabla \mathscr{P}_c. \tag{7.9}$$

Solving a magnetic field problem is thus done in three stages. The auxiliary function $\mathbf{H}_c$ is determined first, then a potential problem is solved to find $\mathscr{P}_c$, and finally its gradient is combined with $\mathbf{H}_c$ to form the total solution.

To find the differential equation $\mathscr{P}_c$ must satisfy, multiply both sides of (7.9) by the local permeability μ, then take divergences of both sides:

$$\nabla \cdot \{\mu(\mathbf{H} - \mathbf{H}_c)\} = \nabla(\mu\nabla\mathscr{P}_c). \tag{7.10}$$

Since $\mu\mathbf{H} = \mathbf{B}$, the true and exact flux density that is the solution of the magnetic field problem, $\nabla(\mu\mathbf{H}_c)$ must vanish. There remains

$$\nabla \cdot (\mu\nabla\mathscr{P}_c) = -\nabla \cdot (\mu\mathbf{H}_c), \tag{7.11}$$

a nonlinear version of the scalar Poisson equation. Boundary conditions to accompany it are obtainable following exactly the same arguments as given for the plain scalar potential $\mathscr{P}$. The resulting boundary-value problem is amenable to the same solution techniques as any other Poisson-like equation.

The Carpenter potential is often referred to in the literature as the *reduced* magnetic scalar potential, in contrast to the *total* magnetic scalar potential $\mathscr{P}$. It is not often used in wave propagation problems where $\partial\mathbf{D}/\partial t$ is nonzero but unknown, for the simple reason that finding a suitable $\mathbf{H}_c$ is then difficult. It is extensively used in magnetostatics and in diffusion problems, however. The extra work required to find $\mathbf{H}_c$ is usually modest compared to the computational labour involved in the alternative: solution of a full three-component vector field problem.

7.3   Tozoni–Mayergoyz potentials

The Carpenter–Noether reduced scalar potential can be quite difficult to incorporate in computer programs, at least in the form described above, because no systematic way is prescribed for finding the reducing function $\mathbf{H}_c$. A method for doing so, and thereby rendering $\mathbf{H}_c$ unique, was proposed by Tozoni and Mayergoyz, following suggestions in the work of Sommerfeld and Noether. In this special case of the Carpenter approach, the reducing field is derived from the Biot–Savart law in its volume integral form,

$$\mathbf{H}_c(P) = \frac{1}{4\pi} \int \frac{\mathbf{J}(Q) \times \mathbf{r}}{|r|^3} \, d\Omega_Q, \tag{7.12}$$

where P is the point of observation, Q is a source point, and $\mathbf{r}$ is the distance vector from Q to P, as in Fig. 3.13. The Biot–Savart integral in Eq. (7.12) is the formal inverse of the curl operator. Hence $\mathbf{H}_c$ obtained in this fashion is precisely the magnetic field that would be caused by the current distribution $\mathbf{J}$, if the problem had no boundaries and there were no magnetic materials anywhere, a situation sometimes described as 'wooden magnets in open space'.

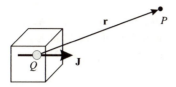

Fig 3.13 The Tozoni–Mayergoyz choice of H_c is obtained by integration over all current-carrying regions.

An obvious benefit of the Tozoni–Mayergoyz approach is that $\mathbf{H}_c$ is exactly solenoidal because it is a physically realizable magnetic field. It is readily verified that

$$\nabla \cdot \mathbf{H}_c(P) = \frac{1}{4\pi} \nabla_P \cdot \int \frac{\mathbf{J}(Q) \times \mathbf{r}}{|r|^3} \, d\Omega_Q \equiv 0. \tag{7.13}$$

Note that the divergence operator here operates in the coordinates of the observation point P; the coordinates of Q are dummy variables that do not survive the integration.

To solve the actual magnetic field problem at hand, the real material permeabilities must be reintroduced. To do so, rewrite the potential equation in the form

$$\nabla \cdot (\mu \nabla \mathscr{P}_c) = \nabla \cdot (\mu \mathbf{H}_c) \tag{7.14}$$

and convert, using the usual vector identities, to

$$\nabla \cdot (\mu \nabla \mathscr{P}_c) = \mu \nabla \cdot \mathbf{H}_c + \mathbf{H}_c \cdot \nabla \mu. \tag{7.15}$$

By the nature of $\mathbf{H}_c$, the first term on the right vanishes identically, leaving

$$\nabla \cdot (\mu \nabla \mathscr{P}_c) = \mathbf{H}_c \cdot \nabla \mu \tag{7.16}$$

as the inhomogeneous potential for solution. Boundary conditions to accompany it are defined in precisely the same way as for the more general Carpenter–Noether formulation.

Various workers have reported computational difficulties with the Tozoni–Mayergoyz potential formulation. Their root cause is not hard to find. In most computer programs the volume integral of (7.12) is evaluated by some method of numerical integration, which involves some error. Conventional integration schemes (e.g., Gaussian quadrature) implicitly approximate the integrand by a polynomial, then integrate the polynomial exactly. Now the best degree-N polynomial approximation to a given vector function may be quite good, but it is not likely to be solenoidal; yet solenoidality is firmly assumed in Eq. (7.16). Thus quadrature error may cause the computed $\mathbf{H}_c$ to have a finite divergence, which is neglected in (7.16). Solenoidality, however, is a convenience, not a necessity. Error propagation of this kind is largely eliminated by regarding the computed $\mathbf{H}_c$ as a Carpenter–Noether potential and solving Eq. (7.15), not (7.16).

7.4   *Two scalar potentials*

The reduced potential formulations, including the Tozoni–Mayergoyz version, sometimes suffer from a numerical instability that arises from purely arithmetic sources. In problems with large material property contrasts, such as are often encountered at air–iron interfaces in magnetic devices, recovery of the true magnetic field by

$$\mathbf{H} = \mathbf{H}_c - \nabla P_c \tag{7.17}$$

may encounter roundoff error difficulties because $\mathbf{H}_c$ and $\nabla \mathscr{P}_c$ may approach equality. Although each may be computed with good accuracy, their difference may be gravely in error. One conventional cure, especially popular in the early 1980s, avoids subtraction by dividing the problem region into two subregions, one encompassing all current-carrying conductors, the other all remaining space. The reduced potential (in its Tozoni–Mayergoyz version) is employed in the current-carrying region, the total scalar potential elsewhere.

It will be evident on brief reflection that this approach is indistinguishable from the original Carpenter formulation. At the region interface, the field $\mathbf{H}_c$ is discontinuous — as indeed were nearly all the reducing fields shown in the early Carpenter work — and surface distributions of sources appear. These may be interpreted physically as surface currents or sheets of magnetic poles. Although such discontinuities may be clumsy to incorporate into computer programs, there is no fundamental difficulty in doing so and various practical problems have been solved by this technique.

8.   **Further reading**

Not surprisingly, electromagnetic field theory relevant to finite elements overlaps quite extensively with engineering electromagnetics in general. However, it also involves some peculiarities of its own. Finite element analysts are often more concerned about variational properties than most authors of general textbooks consider necessary. They employ a wider selection of potentials than have been common in purely pencil-and-paper methods, and they pay more heed to gauge problems than engineers have traditionally been wont to. The references given for this chapter reflect these biases.

The subject of field theory is long established, so any suggested readings can be but a selection of those available. The authors themselves, Silvester (1968) and Ferrari (1975), have each written shortish works on electromagnetism, biased toward numerical methods in their selection of theoretical material. These are at a fairly introductory level, but they have the merit of being consistent in style and notation with the present book. At a slightly more advanced level, Ramo, Whinnery and Van Duzer (1994) outlines the main points of engineering electromagnetics in a time-proven fashion — the book is now in its third edition, each edition having lasted for about ten years. For communication electromagnetics in particular, Balanis (1989) details the major points in almost 1000 pages of text. The classical text by Collin (1991), which originally appeared in 1960 and was long out of print, has fortunately appeared in a second edition; for travelling-wave engineering, it is hard to find a better book. Of the more general classically physics-oriented books on electromagnetics, Stratton (1941) has served well for over half a century and continues to do so. Jones (1986) gives a superb review of modern work, particularly as it concerns electromagnetic travelling waves; but his book is hardly easy reading.

Variational methods and their relationship to electromagnetic fields have traditionally not been treated nearly so thoroughly as one might wish; but the number of available and accessible books has been increasing in recent years. Dudley (1994) has outlined the basic ideas clearly though briefly, with emphasis on wave problems. Beltzer (1990) gives a much lighter mathematical treatment but implements much of the theory immediately in the MACSYMA algebraic manipulation language. The underlying variational and projective mathematical principles, and their relationship to linear spaces, are well set out in the now classic two-volume textbook by Stakgold (1967–1968). For an introduction to linear spaces themselves, it is hard to find a better book than Reza's (1971) which develops the material entirely in terms of electrical engineering problems. These notions are applied to finite element analysis in a clear introductory fashion by Mori (1986); they are developed more extensively, though still very readably, in a finite element context by several authors. Strang and Fix (1973) set out the theory mainly in classical terms with one-dimensional applications; a much shorter, though useful alternative is the chapter in Szabo and Babuška (1991). Jin's textbook (1993) also includes an eminently readable chapter of perhaps slightly more classical material, set entirely in terms of electromagnetics.

Delving further into the mathematical aspects of finite elements, Oden and Reddy (1976) develop the theory much further; their book can probably be considered a fundamental contribution to the literature, though hardly easy to read. A broader, more readable, and certainly more introductory treatment will be found in Mitchell and Wait (1978). Johnson (1987) adopts a more didactic, but also more comprehensive, approach and deals well with two-dimensional problems of the sorts likely to find application in electromagnetics; his book is noteworthy for being one of the easiest to read. Marti (1986) is a serious work but treats the fundamental mathematics well and remains accessible (as a *second* book on the subject); much the same goes for Carey and Oden (1983). The monograph by Ciarlet (1978) is detailed, but mathematically rather harder going; so is the very complete section on finite element mathematics in the encyclopedic handbook edited by Kardestuncer (1987). Unfortunately, even the mathematically oriented books on finite elements often draw applications from structural engineering rather than electromagnetics. That may at times render them difficult to read for the electrical engineer.

9.   Bibliography

Balanis, C.A. (1989), *Advanced Engineering Electromagnetics*, New York: Wiley. xx + 981 pp.

Beltzer, A. (1990), *Variational and Finite Element Methods: a Symbolic Computation Approach*. Berlin: Springer-Verlag. xi + 254 pp.

Carey, G.F. and Oden, J.T. (1983), *Finite Elements: Mathematical Aspects*. (Vol. 4 of the Texas Finite Element Series.) Englewood Cliffs: Prentice-Hall. viii + 195 pp.

Ciarlet, P.G. (1978), *The Finite Element Method for Elliptic Problems*. Amsterdam: North Holland. xvii + 530 pp.

Collin, R.E. (1991), *Field Theory of Guided Waves*, 2nd edn. New York: IEEE Press. xii + 851 pp.

Dudley, D.G. (1994), *Mathematical Foundations for Electromagnetic Theory*. New York: IEEE Press. xiii + 250 pp.

Ferrari, R.L. (1975), *An Introduction to Electromagnetic Fields*. London: Van Nostrand Reinhold. xi + 202 pp.

Jin, J.-M. (1993), *The Finite Element Method in Electromagnetics*. New York: John Wiley. xix + 442 pp.

Johnson, C. (1987), *Numerical Solution of Partial Differential Equations by the Finite Element Method*. Cambridge: Cambridge University Press. vi + 279 pp.

Jones, D.S. (1986), *Acoustic and Electromagnetic Waves*. Oxford: Clarendon. xix + 745 pp.

Kardestuncer, H. [ed.] (1987), *Finite Element Handbook*. New York: McGraw-Hill. xxiv + 1380 pp.

Marti, J.T. (1986), *Introduction to Sobolev Spaces and Finite Element Solution of Elliptic Boundary Value Problems*. [J.R. Whiteman, transl.] London: Academic Press. viii + 211 pp.

Mitchell, A.R. and Wait, R. (1978), *The Finite Element Method in Partial Differential Equations*. New York: Wiley. x + 198 pp.

Mori, M. (1986), *The Finite Element Method and its Applications*. New York: Macmillan. xi + 188 pp.

Oden, J.T. and Reddy, J.N. (1976), *An Introduction to the Mathematical Theory of Finite Elements*. New York: Wiley. xii + 429 pp.

Ramo, S., Whinnery, J.R. and Van Duzer, T. (1994), *Fields and Waves in Communication Electronics*, 3rd edn. New York: Wiley. xix + 844 pp.

Reza, F. (1971), *Linear Spaces in Engineering*. Waltham, Massachusetts: Ginn. xiii + 416 pp.

Silvester, P. (1968), *Modern Electromagnetic Fields*. Englewood Cliffs: Prentice-Hall. xiv + 332 pp.

Stakgold, I. (1967–1968), *Boundary Value Problems of Mathematical Physics*. New York: Macmillan. Vol. I (1967) viii + 340 pp.; Vol. II (1968) viii + 408 pp.

Strang, G. and Fix, G.J. (1973), *An Analysis of the Finite Element Method*. Englewood Cliffs: Prentice-Hall. xiv + 306 pp.

Stratton, J.A. (1941), *Electromagnetic Theory*. New York: McGraw-Hill. xv + 615 pp.

Szabo, B.A. and Babuška, I. (1991), *Finite Element Analysis*. New York: Wiley-Interscience. xv + 368 pp.

4

Simplex elements for the scalar Helmholtz equation

1.   Introduction

Any polygon, no matter how irregular, can be represented exactly as a union of triangles, and any polyhedron can be represented as a union of tetradehra. It is thus reasonable to employ the triangle as the fundamental element shape in two dimensions, and to extend a similar treatment to three-dimensional problems by using tetrahedra.

The solution accuracy obtained with simple elements may be satisfactory in some problems, but it can be raised markedly by using piecewise polynomials instead of piecewise-linear functions on each element. If desired, this increased accuracy can be traded for computing cost, by using high-order approximations on each element but choosing much larger elements than in the first-order method. Indeed, both theory and experience indicate that for many two-dimensional problems, it is best to subdivide the problem region into the smallest possible number of large triangles, and to achieve the desired solution accuracy by the use of high-order polynomial approximants on this very coarse mesh.

In the following, details will be given for the construction of simplicial elements — triangles and tetrahedra — for the inhomogeneous scalar Helmholtz equation. This equation is particularly valuable because of its generality; a formulation valid for the inhomogeneous Helmholtz equation allows problems in Laplace's equation, Poisson's equation, or the homogeneous Helmholtz equation to be solved by merely dropping terms from this general equation. Scalar and quasi-scalar problems will be considered throughout, while all materials will be assumed locally linear and isotropic.

2.   Interpolation functions in simplex coordinates

Two fundamental properties of the approximating functions used with first-order triangular finite elements make the method simple and attractive. First, these functions can always be chosen to guarantee continuity of the desired potential or wave function across all boundaries between triangles, provided only that continuity is imposed at triangle vertices. Second, the approximations obtained are in no way dependent on the placement of the triangles with respect to the global x–y coordinate system. The latter point is probably geometrically obvious: the solution surface is locally defined by the three vertex values of potential and is therefore unaltered by any redefinition of the x- and y-axes, even though the algebraic expressions for the approximating functions may change. This section develops local basis functions for triangles so as to retain both of these desirable properties.

2.1   *Simplex coordinates*

Basis polynomials suited to simplicial finite elements are most easily developed in terms of the so-called simplex coordinates. In general, a simplex in N-space is defined as the minimal possible nontrivial geometric figure in that space; it is always a figure defined by $N + 1$ vertices. Thus a one-dimensional simplex is a line segment, a two-simplex is a triangle, a simplex in three dimensions is a tetrahedron. Its vertex locations suffice to define any simplex uniquely. The size $\sigma(S)$ of a simplex S may be defined by

$$\sigma(S) = \frac{1}{N!} \begin{vmatrix} 1 & x_1^{(1)} & x_1^{(2)} & \cdots & x_1^{(N)} \\ 1 & x_2^{(1)} & & & \\ \vdots & & & & \\ 1 & x_{N+1}^{(1)} & x_{N+1}^{(2)} & \cdots & x_{N+1}^{(N)} \end{vmatrix}. \tag{2.1}$$

The elements of this determinant are the vertex coordinates of the simplex. Their subscripts identify the vertices, while the superscripts in parentheses denote space directions. Under this definition, the size of a triangle is its area, the size of a tetrahedron is its volume, and so on.

Let some point P be located inside the simplex S. Then P uniquely defines a subdivision of S into $N + 1$ subsimplexes, each of which has the point P as one vertex, the remaining N vertices being any N taken from the set of $N + 1$ vertices of S. For example, a line segment is partitioned into two line segments by a point P placed somewhere along its length; similarly, a triangle is partitioned into three subtriangles by any interior point P, as shown in Fig. 4.1(a). Each of the $N+1$ subsimplexes S_1, $S_2, \ldots, S_{N+1}$ is entirely contained within the simplex S. The size of S must be the sum of the sizes of the subsimplexes,

$$\sigma(S) = \sum_{k=1}^{N+1} \sigma(S_k). \tag{2.2}$$

This relationship may be obtained geometrically by inspection, or it may be derived algebraically from the defining equation (2.2).

The splitting of S into subsimplexes is uniquely defined by the choice of point P. Conversely, the sizes of the $N+1$ subsimplexes (relative to the size of S) may be regarded as defining the location of point P within the simplex S. Thus, one may define $N+1$ numbers,

$$\zeta_m = \sigma(S_m)/\sigma(S), \tag{2.3}$$

which specify the point P uniquely within S. These numbers are usually known as the *simplex coordinates* (or the homogeneous coordinates, or the barycentric coordinates) of P. Geometrically, they measure the perpendicular distance between one vertex of S and the opposite side or face, as indicated in Fig. 4.1(b). This interpretation will be clear, at least for triangles, if one observes that the ratio of areas of triangles 1–2–3 and P–2–3 in Fig. 4.1(b) is identical to the ratio of triangle heights, since both triangles have the line 2–3 as base. Similarly, it is geometrically evident, and easy to deduce from Eqs. (2.2) and (2.3), that

$$\sum_{k=1}^{N+1} \zeta_k = 1. \tag{2.4}$$

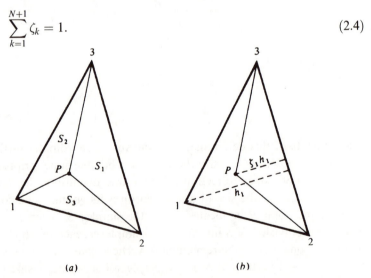

(a)   (b)

Fig 4.1 (a) Any interior point P defines a splitting of the triangle into subtriangles. (b) Simplex coordinates measure relative distance towards each vertex from the opposite side.

This linear dependence of the simplex coordinates should not be surprising, since it takes $N + 1$ simplex coordinates to define a point in N-space.

2.2 *Cartesian–simplex conversion*

Conversion from global Cartesian coordinates to simplex coordinates locally is readily accomplished by Eq. (2.3). For example, let the point (x, y) lie within a triangle in the x–y plane. The first simplex coordinate ζ_1 of this point is then given by

$$\zeta_1 = \frac{\begin{vmatrix} 1 & x & y \\ 1 & x_2 & y_2 \\ 1 & x_3 & y_3 \end{vmatrix}}{\begin{vmatrix} 1 & x_1 & y_1 \\ 1 & x_2 & y_2 \\ 1 & x_3 & y_3 \end{vmatrix}}. \tag{2.5}$$

The denominator represents twice the triangle area A. Expanding the numerator by minors of its first column, (2.5) may be restated as

$$\zeta_1 = \frac{1}{2A} \left(\begin{vmatrix} x_2 & y_2 \\ x_3 & y_3 \end{vmatrix} - \begin{vmatrix} x & y \\ x_3 & y_3 \end{vmatrix} + \begin{vmatrix} x & y \\ x_2 & y_2 \end{vmatrix} \right). \tag{2.6}$$

Rearranging and collecting terms in x and y, there results

$$\zeta_1 = \frac{x_3 y_3 - x_2 y_2}{2A} + \frac{y_2 - y_3}{2A} x + \frac{x_2 - x_3}{2A} y. \tag{2.7}$$

This equation has the expected linear form $\zeta_1 = a + bx + cy$. A similar calculation for the remaining two simplex coordinates yields the conversion equation

$$\begin{bmatrix} \zeta_1 \\ \zeta_2 \\ \zeta_3 \end{bmatrix} = \frac{1}{2A} \begin{bmatrix} x_2 y_3 - x_3 y_2 & y_2 - y_3 & x_3 - x_2 \\ x_3 y_1 - x_1 y_3 & y_3 - y_1 & x_1 - x_3 \\ x_1 y_2 - x_2 y_1 & y_1 - y_2 & x_2 - x_1 \end{bmatrix} \begin{bmatrix} 1 \\ x \\ y \end{bmatrix}. \tag{2.8}$$

It should be noted that the simplex coordinates are strictly local; Eq. (2.3) always yields the same simplex coordinate values for a given point, regardless of any translation or rotation of the global coordinate system. This property is of great value. It allows most of the mathematical formulation of element equations to be accomplished once and for all, for a prototypal triangle. The results can be used subsequently for any triangle whatever, by applying a few simple coordinate transformation rules.

2.3 *Interpolation on 1-simplexes*

To construct finite elements for potential problems, suitable interpolation functions are next required. Such functions are readily defined, beginning with a certain family of auxiliary polynomials $R_m(n, \zeta)$ of degree n. Member m of this family is defined by

$$
\left.
\begin{aligned}
R_m(n, \zeta) &= \prod_{k=0}^{m-1} \frac{\zeta - k/n}{m/n - k/n} = \frac{1}{m!} \prod_{k=0}^{m-1} (n\zeta - k), \quad m > 0, \\
R_0(n, \zeta) &= 1.
\end{aligned}
\right\} \quad (2.9)
$$

The zeros of $R_m(n, \zeta)$ are perhaps most easily found by rewriting it, for $m > 0$, in the form

$$
\begin{aligned}
R_m(n, \zeta) &= \prod_{k=0}^{m-1} \left(\frac{\zeta - \dfrac{k}{n}}{\dfrac{m}{n} - \dfrac{k}{n}} \right) \\
&= \frac{(\zeta)\left(\zeta - \dfrac{1}{n}\right)\left(\zeta - \dfrac{2}{n}\right)\cdots\left(\zeta - \dfrac{m-1}{n}\right)}{\left(\dfrac{m}{n}\right)\left(\dfrac{m}{n} - \dfrac{1}{n}\right)\left(\dfrac{m}{n} - \dfrac{2}{n}\right)\cdots\left(\dfrac{m}{n} - \dfrac{m-1}{n}\right)}.
\end{aligned} \quad (2.10)
$$

This polynomial clearly has zeros at $\zeta = 0, \ 1/n, \ldots, (m-1)/n$, and reaches unity value at $\zeta = m/n$. In other words, it has exactly m equi-spaced zeros to the left of $\zeta = m/n$, and none to the right, as illustrated in Fig. 4.2 (*a*).

On a 1-simplex (a line segment of some given length), a family of polynomials $\alpha_{ij}(\zeta_1, \zeta_2)$ of degree n, with exactly n zeros each, may be defined by

$$
\alpha_{ij}(\zeta_1, \zeta_2) = R_i(n, \zeta_1) R_j(n, \zeta_2), \quad (2.11)
$$

where it is understood that $i + j = n$. The construction of the functions α_{ij} is indicated in Fig. 4.2. They may be recognized as being nothing more than the classical Lagrange interpolation polynomials, though expressed in an unusual notation. Each interpolation polynomial, and each inter-polation node, is identified here by a double index, indicating its place-ment in the two simplex coordinates. This manner of labelling emphasizes the relationship between interpolation functions and their corresponding interpolation points, and at the same time exhibits the symmetry inherent in a simplex. Figure 4.2(*b*) shows the complete cubic family of interpola-tion polynomials constructed in this manner.

2.4   *Interpolation functions on n-simplexes*

The technique for developing interpolation functions on a 1-simplex (line segment) is easily extended to triangles, tetrahedra, and indeed to simplexes of any dimensionality. Again, each interpolation function is constructed as a product of separate factors, one in each of

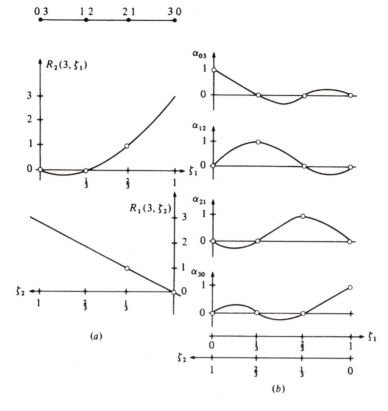

Fig 4.2 (*a*) A line segment, and the polynomials $R_1(3,\zeta_2)$ and $R_2(3,\zeta_1)$, for which $i+j=3$, but which do not vanish at node 21. (*b*) The four cubic interpolation polynomials for the line segment.

the simplex coordinates. Interpolation functions for triangles are thus defined by

$$\alpha_{ijk} = R_i(n,\zeta_1)R_j(n,\zeta_2)R_k(n,\zeta_3), \qquad i+j+k = n. \qquad (2.12)$$

Again, the resulting polynomials are interpolatory on regularly spaced point sets, as shown in Fig. 4.3. That is to say, α_{ijk} vanishes on all the points in the set, except for *ijk* where it has unity value. It is easily seen that there are $(n+1)(n+2)/2$ such points on the triangle.

The natural numbering of interpolation nodes and polynomials on triangles involves three indices. Sometimes such multi-index numbering becomes inconvenient because the index strings grow long, and a single-index numbering scheme is preferred even though it does not clearly exhibit the geometric symmetry inherent in a simplex. Corresponding single-index and triple-index numbering schemes are shown in Fig. 4.3. Here α_{300} is given the alternative name α_1, function α_{210} is referred to as

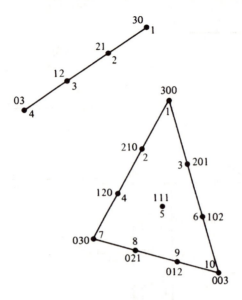

Fig 4.3   Interpolation node sets on line segments and triangles, showing alternative (single and multiple) numbering schemes.

α_2, and so on. No confusion can arise because the natural multi-index numbering never leads to single indices.

Any smooth function ψ defined over a finite element may now be approximated in terms of the functions α_m (of degree N) by writing

$$\psi \simeq \sum_{m=1}^{M} \psi_m \alpha_m \tag{2.13}$$

where the numbers ψ_m are the values taken on by the function ψ at the nodal points of the functions α_m, and M is the number of nodal points, i.e., $M = (N+1)(N+2)/2!$ for triangles, $M = (N+1)(N+2)(N+3)/3!$ for tetradehra. This is said to be an *interpolative* approximation, for it matches the function values exactly at the M nodal points and approximates all intermediate values by means of the interpolation polynomials α_m. Of course, if ψ happens to be a polynomial of the same degree N as the family of α_m, or of any lower degree, Eq. (2.13) is exact. It is then appropriate to speak of a *representation* of ψ, rather than an approximation.

The polynomials defined by Eqs. (2.11)–(2.12) possess two properties important in finite element construction. First, they are defined in terms of the simplex coordinates, and are therefore invariant under all rotations and translations of the global coordinate system into which the finite element is to be fitted. Second, they are always defined over the variable range $0 \leq \zeta_i \leq 1$, for the simplex coordinates ζ_i span this range of values in any simplex whatever. These two properties allow much of the lengthy algebraic work in creating finite element equations to be done just once, the results being applicable to any simplex.

2.5   *Differentiation in simplex coordinates*

Polynomial approximations to functions can be differentiated and integrated by purely algebraic operations expressible in terms of matrices. This fact is fundamental to the finite element method, for it allows differential equations to be replaced by matrix equations that computers can rapidly solve.

To see how matrix operators can replace differential operators, suppose $\psi(x, y)$ is a polynomial of degree N (in x and y jointly). Any directional derivative of ψ is also a polynomial, and always of degree $N - 1$. It can therefore be represented exactly in terms of the interpolation polynomials of degree $N - 1$. Any higher degree would serve equally well, so there is some room for choice. Usually it is most convenient to write the derivative in terms of interpolation polynomials of the same degree N as the original polynomial. For example, if the polynomial ψ is representable as

$$\psi = \sum_{m=1}^{M} \psi_m \alpha_m \tag{2.14}$$

then its x-directed derivative $\partial \psi / \partial x$ can be represented without approximation as

$$\frac{\partial \psi}{\partial x} = \sum_{m=1}^{M} \psi_{(x)m} \alpha_m \tag{2.15}$$

where $\psi_{(x)m}$ is the value of $\partial \psi / \partial x$ at interpolation node m. Clearly, the set of nodal values of $\psi_{(x)m}$ is uniquely related to the set of nodal values ψ_m. Differentiation can therefore be replaced by matrix multiplication, the differentiation operator $\partial / \partial x$ being replaced by a matrix $\mathbf{D}_{(x)}$. If the coefficient vector $\Psi = [\psi_1, \psi_2 \ldots \psi_M]^\mathrm{T}$ defines the polynomial ψ and the coefficient vector $\Psi_{(x)} = [\psi_{(x)1} \psi_{(x)2} \ldots \psi_{(x)M}]^\mathrm{T}$ similarly defines its x-derivative, then (2.15) can be replaced by

$$\Psi_{(x)} = \mathbf{D}_{(x)} \Psi. \tag{2.16}$$

To find the matrix $\mathbf{D}_{(x)}$, first perform the formal differentiation,

$$\frac{\partial \psi}{\partial x} = \sum_{m=1}^{M} \psi_m \frac{\partial \alpha_m}{\partial x}.$$ (2.17)

The chain rule of differentiation permits rewriting the derivative on the right in terms of the simplex coordinates,

$$\frac{\partial \alpha_m}{\partial x} = \sum_{i=1}^{3} \frac{\partial \alpha_m}{\partial \zeta_i} \frac{\partial \zeta_i}{\partial x}.$$ (2.18)

But Eq. (2.8) above implies that the typical x-directed derivative $\partial \zeta_i / \partial x$ in (2.18) is just a geometric constant $b_i/2A$,

$$\frac{\partial \zeta_i}{\partial x} = \frac{y_{i+1} - y_{i-1}}{2A} = \frac{b_i}{2A}.$$ (2.19)

In a similar way, the y-directed derivative may be rewritten as

$$\frac{\partial \zeta_i}{\partial y} = \frac{x_{i-1} - x_{i+1}}{2A} = \frac{c_i}{2A}.$$ (2.20)

The derivative $\partial \psi / \partial x$ in Eq. (2.17) is therefore easily rewritten in terms of derivatives with respect to the simplex coordinates. It remains to find a set of numbers $D_{mk}^{(i)}$ that allows expressing these derivatives in terms of the interpolation polynomials, as

$$\frac{\partial \alpha_m}{\partial \zeta_i} = \sum_k D_{mk}^{(i)} \alpha_k.$$ (2.21)

These numbers are easily found by evaluating both sides of Eq. (2.21) at each interpolation node P_m of the set of α_k. At the typical node P_m all the interpolation functions vanish except α_m, which has the value $\alpha_m(P_m) = 1$. Formal differentiation of the α_m, followed by evaluation at each node P_k in turn, thus turns Eq. (2.21) into

$$D_{mk}^{(i)} = \frac{\partial \alpha_m}{\partial \zeta_i}\bigg|_{P_k}.$$ (2.22)

The resulting matrices $\mathbf{D}^{(i)}$ are square, possessing as many rows and columns as there are functions α_m of degree N. They are universal, i.e., their element values are independent of simplex size and orientation. For triangles of second order, for example

$$\mathbf{D}^{(1)} = \begin{bmatrix} 3 & 1 & 1 & -1 & -1 & -1 \\ 0 & 2 & 0 & 4 & 2 & 0 \\ 0 & 0 & 2 & 0 & 2 & 4 \\ 0 & 0 & 0 & 0 & 0 & 0 \\ 0 & 0 & 0 & 0 & 0 & 0 \\ 0 & 0 & 0 & 0 & 0 & 0 \end{bmatrix}.$$ (2.23)

The formal process of differentiation maps second-order polynomials, of which there are six in the plane, onto first-order polynomials, of which there are only three. The differentiation matrix $\mathbf{D}^{(1)}$ can therefore be expected to possess a three-dimensional nullspace. Inspection of the matrix in (2.23) bears out this expectation. Its leading three rows are clearly linearly independent (because the leading 3×3 submatrix is triangular!) so its range is at least three-dimensional; and there are three null rows, so the nullspace is also at least three-dimensional.

The matrices $\mathbf{D}^{(i)}$ are sometimes called *differentiation matrices* or *finite differentiation operators*, because they perform differentiation on interpolation polynomials in much the same formal way that the operator $\partial/\partial x$ does on general functions. To evaluate an x-directed derivative, for example, it remains to combine the above steps, yielding

$$
\begin{aligned}
\frac{\partial \psi}{\partial x} &= \sum_{m=1}^{M} \psi_m \frac{\partial \alpha_m}{\partial x} \\
&= \sum_{m=1}^{M} \psi_m \sum_{i=1}^{3} \frac{b_i}{2A} \frac{\partial \alpha_m}{\partial \zeta_i} \\
&= \sum_{m=1}^{M} \psi_m \sum_{i=1}^{3} \frac{b_i}{2A} \sum_{k} D_{mk}^{(i)} \alpha_k.
\end{aligned}
\tag{2.24}
$$

If this expression seems a bit involved, it can be simplified by putting it in the matrix form

$$
\mathbf{\Psi}_{(x)} = \sum_{i=1}^{3} \left(\frac{b_i}{2A} \mathbf{D}^{(i)\mathrm{T}} \right) \mathbf{\Psi}
\tag{2.25}
$$

which shows how the nodal values $\mathbf{\Psi}_{(x)}$ of the x-directed derivative are related to the nodal values $\mathbf{\Psi}$ of the function being differentiated. (Note the matrix transposition implied by summation over the *first* index of $\mathbf{D}^{(i)}$.) A similar argument applies to differentiation in the y-direction, or indeed to any other directional derivative.

Although the argument has been developed here in terms of triangles, analogous results hold for the one-dimensional and three-dimensional cases, i.e., for line segments and tetrahedra. The differentiation matrices are of course different for the different numbers of dimensions. Appendix 3 includes tables of values for the most commonly needed differentiation matrices, i.e., those applicable to the first few orders of interpolation polynomial. Only one differentiation matrix, say $\mathbf{D}^{(1)}$, actually needs be tabulated, the remaining ones being derived by the row and column interchanges that correspond to cyclic renumbering of coordinate indices.

3.   General simplex elements

Numerous problems in electromagnetics may be regarded as special cases of the inhomogeneous Helmholtz equation

$$(\nabla^2 + k^2)u = g, \tag{3.1}$$

subject to various boundary conditions, of which the homogeneous Neumann condition (vanishing normal derivative) or Dirichlet conditions (fixed boundary value) are probably the most common in applications. An approximate solution to this equation, using high-order elements will now be constructed, by means of a procedure much like that followed with first-order triangles.

3.1   *Approximation of the variational functional*

As the last chapter has shown in detail, solutions of Eq. (3.1) render stationary the functional

$$F(U) = \frac{1}{2}\int \nabla U \cdot \nabla U \, d\Omega - \frac{1}{2}k^2 \int U^2 \, d\Omega + \int Ug \, d\Omega, \tag{3.2}$$

provided all candidate solution functions U are continuous within the problem region, and provided they all satisfy the Dirichlet boundary conditions of the problem. The problem region is approximated by a union of simplicial elements, and on each element the approximate potential function is represented as a linear combination of the approximating functions developed above,

$$U = \sum_i U_i \alpha_i. \tag{3.3}$$

Each approximating function α_m vanishes at every nodal point, except at its associated node P_m where α_m has unity value. Hence the coefficients U_i in (3.3) will be potential values at the interpolation nodes P_i. Furthermore, interelement continuity will be guaranteed, as shown by the following argument, which is strictly applicable to triangles but can be generalized easily to tetrahedra. If $U(x, y)$ is a polynomial function of degree at most N in x and y jointly, then along any straight line in the x–y plane the function $U = U(s)$ must be a polynomial of at most degree N in the distance s measured along the straight line (such as a triangle edge). A triangular finite element of order N has exactly $N + 1$ nodes along each edge. Since U must be a polynomial function of degree N along each triangle edge, these $N + 1$ nodal values fix the values of U everywhere along the edge. The edge nodes are invariably shared between every pair of adjacent triangles. Therefore the potential values on the two sides of the interelement boundary are determined by the same $N + 1$ nodal values, and it follows that U must be continuous across triangle edges. A similar argument can be made for simplex element of any other dimensionality.

A single element will now be developed in detail, interconnection of elements to form the entire problem region being deferred until later. Substituting the approximation (3.3) into (3.2), the functional $F(U)$ becomes

$$F(U) = \frac{1}{2}\sum_i\sum_j U_i U_j \int \nabla\alpha_i \cdot \nabla\alpha_j \, d\Omega$$
$$-\frac{k^2}{2}\sum_i\sum_j U_i U_j \int \alpha_i \alpha_j \, d\Omega + \sum_i U_i \int \alpha_i g \, d\Omega. \qquad (3.4)$$

Although this is not in principle necessary, the forcing function g is commonly approximated over each triangle by the same polynomials as the response U,

$$g = \sum_j g_j \alpha_j. \qquad (3.5)$$

There are various practical ways of determining the coefficients in this approximation. For example, interpolative approximation may be used; the coefficients are then simply the values of g at the interpolation nodes. Using an approximation of the form (3.5), no matter what the method for finding coefficients, the functional $F(U)$ may be expressed as the matrix form

$$F(U) + \frac{1}{2}\mathbf{U}^{\mathrm{T}}\mathbf{S}\mathbf{U} - \frac{k^2}{2}\mathbf{U}^{\mathrm{T}}\mathbf{T}\mathbf{U} + \mathbf{U}^{\mathrm{T}}\mathbf{T}\mathbf{G}. \qquad (3.6)$$

Here $\mathbf{U}$ is the vector of coefficients in Eq. (3.3), while $\mathbf{G}$ represents the vector of coefficients g_j in Eq. (3.5); the square matrices $\mathbf{S}$ and $\mathbf{T}$ are given by

$$S_{mn} = \int \nabla\alpha_m \cdot \nabla\alpha_n \, d\Omega \qquad (3.7)$$

and

$$T_{mn} = \int \alpha_m \alpha_n \, d\Omega. \qquad (3.8)$$

The matrices $\mathbf{S}$ and $\mathbf{T}$ are sometimes termed the *Dirichlet matrix*, and the *metric*, of the particular set of approximating functions $\{\alpha_m | m = 1, 2, \ldots\}$. Somewhat similar matrices that arise in the theory of elasticity are often referred to as the *stiffness matrix* and the *mass matrix*, respectively, and these terms are sometimes encountered in the electromagnetics literature even though the words are not appropriate there.

The fundamental requirement is for $F(U)$ to be stationary. Since $F(U)$ in Eq. (3.6) is an ordinary function of a finite number of variables (the components of the array $\mathbf{U}$), the stationarity requirement amounts to demanding that

$$\frac{\partial F}{\partial U_m} = 0 \qquad (3.9)$$

for all m whose corresponding components U_m of $\mathbf{U}$ are free to vary. Substituting Eq. (3.6) into (3.9) and carrying out the indicated differentiations, there is obtained the matrix equation

$$\mathbf{SU} - k^2\mathbf{TU} = -\mathbf{TG}. \qquad (3.10)$$

Solution of this equation for U determines the approximate value of U everywhere in the region of interest, and thus solves the problem.

3.2   *High-order triangular-element matrices*

Several different element types will now be developed in detail, as specializations of the general formulation given above. The first, and probably the most important, is the planar (two-dimensional) triangular element. Its metric (the matrix $\mathbf{T}$) requires little work; it is only necessary to write it in a normalized form independent of the element size. Multiplying and dividing by the element area A,

$$T_{mn} = A \int \alpha_m \alpha_m \mathrm{d}\left(\frac{\Omega}{A}\right). \qquad (3.11)$$

The integral on the right is dimensionless and universal; it is independent of triangle size and shape, for it involves only quantities expressed in the simplex coordinates. For example, for a triangular element of order 2,

$$\mathbf{T} = \frac{1}{180}\begin{bmatrix} 6 & 0 & 0 & -1 & -4 & -1 \\ 0 & 32 & 16 & 0 & 16 & -4 \\ 0 & 16 & 32 & -4 & 16 & 0 \\ -1 & 0 & -4 & 6 & 0 & -1 \\ -4 & 16 & 16 & 0 & 32 & 0 \\ -1 & -4 & 0 & -1 & 0 & 6 \end{bmatrix}. \qquad (3.12)$$

Such matrices have been evaluated and tabulated for polynomials up to order 6, and symbolic algebra programs exist to compute still higher-order values should they be required. For the first few orders, tables of element matrices are given in Appendix 3.

To obtain the Dirichlet matrix in a convenient form, the first step is to write out the integrand of Eq. (3.7) in Cartesian coordinates, as

$$S_{mn} = \int \left(\frac{\partial \alpha_m}{\partial x} \frac{\partial \alpha_n}{\partial x} + \frac{\partial \alpha_m}{\partial y} \frac{\partial \alpha_n}{\partial y} \right) d\Omega. \tag{3.13}$$

The chain rule of differentiation permits recasting this equation as

$$\frac{\partial \alpha_m}{\partial x} = \sum_{i=1}^{3} \frac{\partial \alpha_m}{\partial \zeta_i} \frac{\partial \zeta_i}{\partial x}. \tag{3.14}$$

But from Eq. (2.8) above, the typical derivative in (3.14) may be restated as

$$\frac{\partial \zeta_i}{\partial x} = \frac{y_{i+1} - y_{i-1}}{2A} = \frac{b_i}{2A}. \tag{3.15}$$

On combining Eqs. (3.14) and (3.15), and writing corresponding expressions for derivatives with respect to y, Eq. (3.13) can be stated entirely in terms of simplex coordinates for the triangle,

$$S_{mn} = \frac{1}{4A^2} \sum_i \sum_j (b_i b_j + c_i c_j) \int \frac{\partial \alpha_m}{\partial \zeta_i} \frac{\partial \alpha_n}{\partial \zeta_j} d\Omega. \tag{3.16}$$

Here A again represents the triangle area, while the geometric coefficients b and c are given by

$$b_i = y_{i+1} - y_{i-1}, \tag{3.17}$$
$$c_i = x_{i-1} - x_{i+1}. \tag{3.18}$$

The integral in Eq. (3.16) can be evaluated without actually performing any formal differentiations or integrations, by using the differentiation matrices developed earlier. Substituting the matrix forms,

$$\int \frac{\partial \alpha_m}{\partial \zeta_i} \frac{\partial \alpha_n}{\partial \zeta_j} d\Omega = \sum_k \sum_l D_{mk}^{(i)} \left(\int \alpha_k \alpha_l \, d\Omega \right) D_{nl}^{(j)}. \tag{3.19}$$

But the bracketed quantity is immediately recognized as the **T** matrix already developed and evaluated. Because **T** was earlier developed for a triangle of unit area, A appears as a factor, so that

$$\mathbf{S} = \frac{1}{4A} \sum_i \sum_j (b_i b_j + c_i c_j) \mathbf{D}^{(i)} \mathbf{T} \mathbf{D}^{(j)\mathrm{T}} \tag{3.20}$$

with **T** as tabulated (i.e., correct for unit area).

Throughout the above discussion the subscripts always progress modulo **3**, i.e., cyclically around the three vertices of the triangle. The double summations of Eq. (3.20) can be reduced to a single summation, if it is recognized that

$$\left.\begin{array}{l} b_ib_j + c_ic_j = -2A\cot\theta_k \quad i \neq j \\ b_i^2 + c_i^2 = 2A(\cot\theta_j + \cot\theta_k), \end{array}\right\} \quad (3.21)$$

where i, j, k denote the three vertices of the triangle. (This trigonometric identity is probably far from obvious; a proof is given in Appendix 1.) Substituting Eqs. (3.21) into Eq. (3.16), expanding, and collecting terms, the double summation may be replaced by a single summation,

$$S_{mn} = \sum_{k=1}^{3}\cot\theta_k \int \left(\frac{\partial\alpha_m}{\partial\zeta_{k+1}} - \frac{\partial\alpha_m}{\partial\zeta_{k-1}}\right)\left(\frac{\partial\alpha_n}{\partial\zeta_{k+1}} - \frac{\partial\alpha_n}{\partial\zeta_{k-1}}\right)\frac{d\Omega}{2A}. \quad (3.22)$$

The integral remaining on the right-hand side of this equation is dimensionless, and only involves quantities expressed in terms of the simplex coordinates. Indeed, Eq. (3.22) may be written

$$S_{mn} = \sum_{k=1}^{3} Q_{mn}^{(k)}\cot\theta_k \quad (3.23)$$

to exhibit this fact more clearly. The three matrices $\mathbf{Q}^{(k)}$ are purely numeric and have exactly the same values for a triangle of any size or shape. They too may therefore be calculated and tabulated for permanent reference, along with the matrices $\mathbf{T}$, $\mathbf{D}$ and $\mathbf{C}$. However, there is little need to carry out this work explicitly; it suffices to substitute the algebraic differentiation operators ($\mathbf{D}$ matrices) in the defining expression for $Q_{mn}^{(k)}$ to obtain

$$Q_{mn}^{(k)} = \frac{1}{2}\sum_{i}\sum_{j}\left(D_{mi}^{(k+1)} - D_{mi}^{(k-1)}\right)T_{ij}\left(D_{nj}^{(k+1)} - D_{nj}^{(k-1)}\right) \quad (3.24)$$

or in matrix form

$$\mathbf{Q}^{(k)} = \frac{1}{2}\left(\mathbf{D}^{(k+1)} - \mathbf{D}^{(k-1)}\right)\mathbf{T}\left(\mathbf{D}^{(k+1)} - \mathbf{D}^{(k-1)}\right)^{\mathrm{T}}. \quad (3.25)$$

Note that the matrices that pre- and postmultiply $\mathbf{T}$ are transposes of each other. Since $\mathbf{T}$ is obviously symmetric, the resulting matrix $\mathbf{Q}^{(k)}$ is symmetric also. Further, only one of the matrices $\mathbf{Q}^{(k)}$ need be tabulated, the remaining two being easily derivable by row and column permutations.

3.3   Interconnection and continuity

The interconnection of elements to form a connected whole follows precisely the same methodology as for first-order triangles, though the details are a little more complicated. To proceed by example, Fig. 4.4 shows the interconnection of two second-order triangular elements. If viewed as a disjoint pair whose potentials are not related in any way, these two elements possess twelve independent nodal potentials, while the connected pair has only nine. To force the potentials of the two representations to coincide, the disjoint and connected potential vectors must be related by the equation

$$\mathbf{U}_{\text{dis}} = \mathbf{C}\mathbf{U}_{\text{con}}, \tag{3.26}$$

where $\mathbf{C}$ is the connection matrix which expresses the constraints placed on the set of twelve potentials. For the case shown in Fig. 4.4,

$$\mathbf{C} = \begin{bmatrix} 1 & & & & & & & & & \\ & 1 & & & & & & & & \\ & & 1 & & & & & & & \\ & & & 1 & & & & & & \\ & & & & 1 & & & & & \\ & & & & & 1 & & & & \\ 1 & & & & & & & & & \\ & & 1 & & & & & & & \\ & & & & & & 1 & & & \\ & & & & & 1 & & & & \\ & & & & & & & 1 & & \\ & & & & & & & & 1 & \end{bmatrix}. \tag{3.27}$$

Following precisely the same arguments as for first-order elements, the matrix $\mathbf{S}$ for the connected system is

$$\mathbf{S} = \mathbf{C}^{\text{T}}\mathbf{S}_{\text{dis}}\mathbf{C} \tag{3.28}$$

and the matrix $\mathbf{T}$ for the connected system is

$$\mathbf{T} = \mathbf{C}^{\text{T}}\mathbf{T}_{\text{dis}}\mathbf{C}. \tag{3.29}$$

In practical computing, the connection matrix $\mathbf{C}$ is not stored explicitly. By its very nature, $\mathbf{C}$ contains in each row only a single nonzero element equal to unity. One suitable method of storing $\mathbf{C}$ compactly is to specify an addressing string whose kth entry indicates the column number of the nonzero element in the kth row of $\mathbf{C}$. In the present example, $\mathbf{C}$ of Eq. (3.27) would be represented in this way as an array of 12 numeric entries, i.e., as the string [1, 2, 3, 4, 5, 6, 1, 3, 7, 6, 8, 9]. Because $\mathbf{C}$ is a large matrix, such compact storage arrangements lead to very considerable computational economies.

3.4 *Rotations and column permutations*

The three matrices $\mathbf{Q}^{(i)}$ are row and column permutations of each other so it is unnecessary, and may be wasteful, to compute and tabulate all three. To see how these matrices can be derived from each other, let the triple-index numbering of interpolation nodes and func-

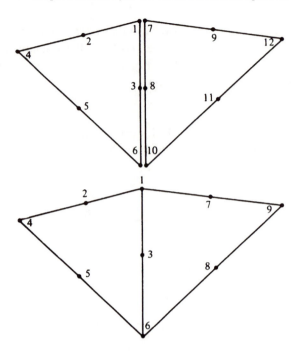

Fig 4.4 Continuity of potential between adjacent elements is guaranteed by continuity of potential at the shared edge nodes.

tions, and correspondingly the triple-index numbering of matrix elements, be restored. From the symmetry inherent in the triple-index numbering, see Fig. 4.5, it may be seen that

$$\left.\frac{\partial\alpha_{ijk}}{\partial\zeta_1}\right|_{rst} = \left.\frac{\partial\alpha_{kij}}{\partial\zeta_2}\right|_{trs} = \left.\frac{\partial\alpha_{jki}}{\partial\zeta_3}\right|_{str}. \tag{3.30}$$

In other words: all three directional derivatives, and therefore also all three matrices $\mathbf{Q}$, can be produced from a single one, by permuting rows and columns in a manner which corresponds to advancing the triple indices cyclically. Each of the three matrices is related to the other two by transformations of the form

$$\mathbf{Q}^{(2)} = \mathbf{P}^{\mathrm{T}}\mathbf{Q}^{(1)}\mathbf{P}, \tag{3.31}$$

$$\mathbf{Q}^{(3)} = \mathbf{P}^{\mathrm{T}}\mathbf{Q}^{(2)}\mathbf{P} = \mathbf{P}^{\mathrm{T}}\mathbf{P}^{\mathrm{T}}\mathbf{Q}^{(1)}\mathbf{P}\mathbf{P}. \tag{3.32}$$

Here $\mathbf{P}$ represents the relevant permutation matrix.

The permutation transformations on the matrices $\mathbf{Q}$ have a straightforward geometric significance, as may be seen from Fig. 4.5. The permutation matrix $\mathbf{P}$ is a matrix whose columns are the columns of the unit matrix, rearranged to correspond to rotation of the triangle

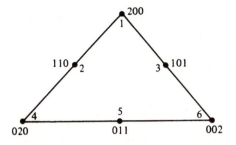

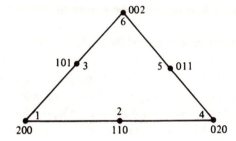

Fig 4.5 A cyclic permutation of the triangle coordinate labels corresponds to a cyclic permutation of the finite element node labels.

about its centroid in a counterclockwise direction. In Fig. 4.5, such a rotation carries point 6 into the location formerly occupied by point 1, point 3 into point 2, point 5 into point 3, etc. Correspondingly, the permutation matrix **P** is

$$
\mathbf{P} =
\begin{bmatrix}
 & & 1 & & & \\
 & & & 1 & & \\
 & 1 & & & & \\
 & & & & 1 & \\
 & & & 1 & & \\
1 & & & & &
\end{bmatrix}.
\tag{3.33}
$$

For brevity, it is often convenient to denote the permutation transformations by an operator rot which symbolizes the subscript mappings required:

$$
\mathbf{Q}^{(2)} = \text{rot } \mathbf{Q}^{(1)},
\tag{3.34}
$$

$$
\mathbf{Q}^{(3)} = \text{rot rot } \mathbf{Q}^{(1)}.
\tag{3.35}
$$

Because the permutations represent rotations of the simplex coordinate numbering, it is hardly surprising that

$$\mathbf{Q}^{(1)} = \text{rot rot rot } \mathbf{Q}^{(1)}, \tag{3.36}$$

for three rotations of a triangle should map every point of the triangle into itself.

Any permutation matrix $\mathbf{P}$ consists of the columns of the unit matrix, rearranged in an altered sequence. Consequently, only the location of the nonzero element in each column needs to be known to convey full information about it. Thus, the indexing string IS = [6, 3, 5, 1, 2, 4] represents $\mathbf{P}$ of Eq. (3.33) and provides all the information required to perform matrix transformations of the form (3.34) on second-order elements. Indeed, the matrix transformation itself is best computationally executed not as a multiplication, but as a lookup operation in an indexing string, which substitutes one set of subscripts for another. All popular high-level languages (e.g., Fortran and C) will accept nested subscripts in the form S(IS(M), IS(N)) to represent one rotation, and S(IS(IS(M))), IS(IS(N)) to represent two, so coding is relatively easy. Alternatively, many Pascal or C programmers prefer to store the indexing strings as linked lists. Pointers to pointers then take the place of nested subscripts.

The permutation operators as represented by indexing strings have an interesting and valuable property: the indexing string for elements of order $N - 1$ is an ordered proper subset of the indexing string for order N. This property permits indexing strings for low orders to be derived from high-order operators with little effort, by simply casting out unwanted indices. For example, the correct indexing string IS = [3, 1, 2] or order 1 is easily obtained from IS = [6, 3, 5, 1, 2, 4], valid for $N = 2$, by simply throwing out all node numbers greater than 3, but keeping the remaining nodes in their original order. Appendix 3 includes the relevant indexing information for elements of high order, from which indexing strings for all lower orders can be derived easily.

The use of simplex coordinates permits carrying out all differentiations and integrations in a universal manner, valid for any triangle. The triangle area ultimately appears as a multiplicative factor in the matrix $\mathbf{T}$, and nowhere else. The triangle shape, expressed by the cotangents of included angles at the three vertices, enters the matrix $\mathbf{S}$ as three weighting factors. Practical computer programs therefore store the universal matrices $\mathbf{Q}^{(1)}$ and $\mathbf{T}$ for a triangle of unit area, then assemble the matrix representation of each element by weighting the matrices by the included-angle cotangents and the triangle area. Tables of these two matrices appear in Appendix 3, for the first few orders of triangular element.

4.   A high-order finite element computer program

A computer program HITRIA2D suitable for experiments with high-order triangular elements is given at the end of this chapter. This section briefly describes its structure and use.

4.1   *Program description*

Program HITRIA2D is a generalization of the earlier program SIMPLE2D. The two are similar in outline, indeed they share some basic utility routines, but the high-order program is considerably more ambitious.

Data input to HITRIA2D follows the same general pattern as for SIMPLE2D. A list of nodal points is first read in, each as an *x–y* coordinate pair. The node list is followed by a list of elements. In contrast to the simpler program, HITRIA2D uses several different orders of triangular element, so the first item in each input line that describes elements is the number of nodes on that element: 3, 6, 10, etc. The source density appropriate to the element appears next, followed by the list of node numbers. It is not necessary to specify the order of element, since the order is implied by the total number of vertices. The final portion of input data is a list of constrained (fixed-value) nodes and the potential values to be imposed on them. Fixed-value nodes are expected to be numbered with the highest numbers. An illustrative data set follows the program listing at the end of this chapter.

Element formation in HITRIA2D follows precisely the techniques discussed in the foregoing section. A block data segment of substantial size houses the element matrices in compacted form. It is accompanied by short indexing routines able to expand the elementary matrices as required to form the full element matrices up to fourth order. The elements are treated one by one, creating the matrices **S** and **T** appropriate to that matrix alone. Each one in turn is then fitted into the already existing element assembly. Edited for brevity, the key segment of the program reads

```
c     assemble global matrices, element by element.
      do 40 ie = 1,nelmts
         If (nve(ie) .eq. 3 .or. nve(ie) .eq. 6 .or.
     *       nve(ie) .eq. 10 .or. nve(ie) .eq. 15)
     *       call trielm(ie)
         call elembd(ie)
40    continue
```

Subroutine trielm constructs the matrices for a single element, elembd embeds them in the global element assembly. This program structure lends itself to easy modification by the inclusion of other types of element. Extending the program to, say, four-noded square elements requires only one additional line in the main program, such as

```
if (nve(ie) .eq. 4) call sqrelm\rm (ie)
```

Of course, a subroutine `sqrelm` must be provided to return the **S** and **T** matrices for that element type. In fact, this is precisely the technique used in later chapters to extend the program.

HITRIA2D is a test bed program suitable for experimentation, designed for easy modification and legibility but almost without regard to computational economy. All matrices are stored in full form — even where they are known to possess special properties such as symmetry. No computational short-cuts are taken, even where they might effect significant time savings. Few provisions are included for error detection; there is no streamlined method for easy data input; nor is output available in any form but nodal potential values. While all these considerations would weigh heavily in the design of industrial-grade software, they are not very important in constructing experimental programs; indeed they are often a hindrance to rapid program alteration.

The maximum numbers of nodes, elements, nodes per element, and the like are included in HITRIA2D via Fortran `parameter` statements. As listed in this chapter, the data space taken by the program amounts to some 32K–35K bytes and the code space is similar. Even on quite small computers, there should be ample room to enlarge data areas, to accommodate considerably larger meshes than the listed limit of 50 nodes.

4.2   *High-order elements in waveguide analysis*

As discussed earlier, the possible propagating modes of a perfectly conducting waveguide homogeneously filled with perfect dielectric are described by the wave functions u which satisfy

$$(\nabla^2 + k_c^2)u = 0, \tag{4.1}$$

where k_c^2 is a cut-off wavenumber. This homogeneous Helmholtz equation is exactly of the form discussed above, so it lends itself to investigation of error behaviour and program running characteristics.

The finite element version of the problem becomes the solution of the matrix equation Eq. (3.10), but with zero right-hand side. The boundary conditions are of homogeneous Dirichlet or Neumann type for the transverse electric (TE) and transverse magnetic (TM) modes respectively. The object here is to determine the unknown values k_c^2 which occur as an infinite set of *eigenvalues* when Eq. (4.1) itself is solved, and the *eigenfunctions* to go with them. Using triangular element functions, the Helmholtz equation is discretized and the discrete (matrix) form is treated by standard techniques of linear algebra, to yield a finite set of eigenvalues k^2 which are approximations to the truncated set k_c^2. The eigensolutions U then correspond to waveguide field distributions.

High-order triangular elements may be expected to perform particularly well where the guide shape is a convex polygon, for the polygonal shape can be expressed exactly as a union of triangles. Hence the accuracy obtained from a finite element solution may be assessed by comparing it with analytic exact solutions, which are available for rectangular waveguides and for isosceles right-triangular guides. The rectangular shape may be decomposed into two or more triangles, while for the triangular guide, quite accurate answers may be obtained even from a single element of high order!

Since triangular elements of various different orders have been calculated and tabulated, the practical analyst will in each case need to decide whether to employ just a few high-order elements, or to use many elements of lower order. To gain an idea of error behaviour over a spectrum of modes, Table 4.1 shows the error encountered in calculating the cut-off wavelengths of the first eight TM modes of an isosceles right-triangular waveguide. Four different orders of element were used. However, all subdivisions were so chosen as to keep the matrix equation exactly of order 55 each time, making the required memory size and computing time practically the same in each case. Obviously, the use of few high-order

Table 4.1 Percentage error in calculated cut-off frequencies – right-triangular waveguide

Cut-off frequency	$N = 1$ $E = 144$	$N = 2$ $E = 36$	$N = 3$ $E = 16$	$N = 4$ $E = 9$
1.00000	1.46	0.122	0.017	0.002
1.41421	3.38	0.560	0.125	0.030
1.61245	3.81	0.73	0.24	0.045
1.84391	5.38	1.33	0.68	0.160
2.00000	7.25	2.13	0.95	0.63
2.23607	6.84	2.08	1.41	0.42
2.28035	8.04	2.70	1.34	0.70
2.40832	10.2	3.71	2.21	1.04

Note: N = order of elements, E = number of elements used in model.

elements produces much better results than a correspondingly larger number of low-order triangles.

Convex guide shapes do not have any reentrant corners at which field singularities might occur. For such shapes, it can be shown on theoretical grounds that the error ϵ behaves according to

$$\epsilon = Kh^{-2N}, \tag{4.2}$$

where h is the maximum dimension of any element in the model, and N is the polynomial order. K represents a proportionality constant, which is not in general known. It is evident from Eq. (4.2), as well as from Table 4.1, that the use of high-order polynomials is advantageous; the gain from an increase in the exponent will always outweigh the loss from a corresponding increase in h.

The approximating functions used in this study are polynomial in each element, but only have continuity of the function sought (not of its derivatives) across element interfaces. The resulting approximate solutions characteristically will show 'creases' at the element edges, if the capabilities of the approximating functions are being stretched hard. For example, Fig. 4.6 shows two guided wave modes in a rectangular guide, calculated with two fifth-order elements. The fundamental TM mode of Fig. 4.6(a) is clearly very well approximated. The modal pattern

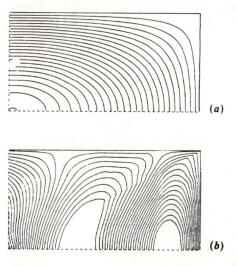

Fig 4.6 Two modes of a rectangular waveguide, modelled by two fifth-order triangular elements. (a) The fundamental H (TM) mode. Although some error may be detected near the centre of the guide, the plot closely resembles the analytic solution. (b) The $(5,1)$ H (TM) mode. In contrast to the fundamental mode, the element subdivision is quite clearly visible.

in Fig. 4.6(*b*) should resemble a sequence of patterns similar to Fig. 4.6(*a*), but clearly the fifth-order polynomials cannot quite manage to model so intricate a pattern accurately. Further, the diagonal inter-element joint is clearly visible in Fig. 4.6(*b*).

For nonconvex guide shapes, where reentrant sharp corners are encountered, no simple analytically solvable test case is available. However, comparisons with experimental measurements on real guides are possible. While the field singularities encountered in nonconvex shapes do produce a loss in accuracy, the use of high-order elements still remains attractive.

5. Embeddings, projections, and their uses

The planar problem of Helmholtz's equation can be solved accurately and conveniently if means exist for differentiating and integrating sets of interpolation functions — in other words, if matrices $\mathbf{T}$ and $\mathbf{D}^{(i)}$ are known. These do not suffice, however, for solving various other two-dimensional problems, including the very common case of axial symmetry. Satisfactory treatment of axisymmetric problems requires additional operators which do not have any immediately recognizable counterparts in continuum analysis: embeddings and projections. These are developed in this section, and immediately applied to the construction of axisymmetric finite elements.

5.1 *Axisymmetric scalar fields*

Many scalar fields in isotropic but inhomogeneous media are described by the inhomogeneous Helmholtz equation

$$\nabla \cdot (p\nabla u) + k^2 u = g, \tag{5.1}$$

where p is the local value of some material property — permittivity, reluctivity, conductivity, etc. In many problems both the excitation function g and all boundary conditions possess rotational symmetry; some examples are coaxial cable junctions, high-voltage insulators, and resonant cavities. In such cases the true mathematical problem is two-dimensional, since finding the potential distribution over only one azimuthal plane suffices. The two-dimensional differential equation and the corresponding functional, however, differ from their translationally uniform counterparts. New finite elements must therefore be devised. Such boundary value problems are solved by minimizing the two-dimensional (r–z plane) functional

$$F(U) = \int\int 2\pi r(p\nabla U \cdot \nabla U - k^2 U^2 + 2Ug)\, dr\, dz, \tag{5.2}$$

where the differential and integral operations refer to the coordinates r and z only. This functional differs from its planar counterpart in having the integrand weighted by the radial variable r.

Just as in the two-dimensional planar case, the solution region in the r–z plane is subdivided into triangular elements, and the integral (5.2) is evaluated over each triangle in turn. As before, U and g are approximated on each triangle, using the triangle interpolation polynomials. Thus

$$U = \sum_m U_m \alpha_m,$$ (5.3)

$$g = \sum_m g_m \alpha_m.$$ (5.4)

Element-independent integration is not immediately possible, because the factor r in the integrand is dependent on the position of the triangular element with respect to the axis of symmetry. The key to further progress lies in observing that r is trivially a linear function of position, i.e., of the coordinates r and z, so that over any one element it may be expressed exactly as the linear interpolate of its three vertex values:

$$r = \sum_{i=1}^{3} r_i \zeta_i.$$ (5.5)

Hence the functional $F(U)$ may be written, substituting Eq. (5.5) into (5.2), as

$$F(U) = 2\pi \sum_{i=1}^{3} r_i \int\int \zeta_i (p\nabla U \cdot \nabla U - k^2 U^2 + 2Ug)\, dr\, dz.$$ (5.6)

The derivatives with respect to r and z that the gradient operator in Eq. (5.6) implies may be converted to derivatives with respect to the simplex coordinates, using exactly the same arguments as already advanced in the planar case. Thus

$$\frac{\partial U}{\partial r} = \frac{1}{2A} \sum_{j=1}^{3} b_j \frac{\partial U}{\partial \zeta_j},$$ (5.7)

$$\frac{\partial U}{\partial z} = \frac{1}{2A} \sum_{j=1}^{3} c_j \frac{\partial U}{\partial \zeta_j},$$ (5.8)

where

$$b_j = z_{i+1} - z_{i-1},$$ (5.9)

$$c_j = r_{i-1} - r_{i+1},$$ (5.10)

the subscript i being understood to progress cyclically around the triangle vertices; A represents the triangle area as before. Substituting, the functional $F(U)$ on each element takes on the rather lengthy form

$$
\begin{aligned}
F(U) = {} & \frac{2\pi}{4A^2} \sum_m \sum_n p U_m U_n \sum_{i=1}^{3} r_i \sum_{k=1}^{3} \sum_{l=1}^{3} \\
& (b_k b_l + c_k c_l) \iint \zeta_i \left(\frac{\partial \alpha_m}{\partial \zeta_k} \frac{\partial \alpha_n}{\partial \zeta_l} \right) dr\, dz \\
& - \frac{2\pi}{4A^2} \sum_m \sum_n k^2 U_m U_n \sum_{i=1}^{3} r_i \iint \zeta_i \alpha_m \alpha_n \, dr\, dz \\
& + \frac{\pi}{A^2} \sum_m \sum_n U_m g_n \sum_{i=1}^{3} r_i \iint \zeta_i \alpha_m \alpha_n \, dr\, dz.
\end{aligned}
\tag{5.11}
$$

Here the material property p has been taken outside the integral sign, on the assumption that it will have a constant value within each element. Now in contrast to all the matrices and integrals encountered so far, all three integrals in Eq. (5.11) are lopsided, containing only a single index i. Balance is restored by writing

$$
r_i = r_i \cdot 1 = r_i \sum_{j}^{3} \zeta_j
\tag{5.12}
$$

so that the functional $F(U)$ becomes

$$
\begin{aligned}
F(U) = {} & \frac{\pi}{2A^2} \sum_m \sum_n p U_m U_n \sum_{i=1}^{3} \sum_{j=1}^{3} r_i \sum_{k=1}^{3} \sum_{l=1}^{3} \\
& (b_k b_l + c_k c_l) \iint \zeta_i \zeta_j \left(\frac{\partial \alpha_m}{\partial \zeta_k} \frac{\partial \alpha_n}{\partial \zeta_l} \right) dr\, dz \\
& - \frac{\pi}{2A^2} \sum_m \sum_n (k^2 U_m U_n - 2 U_m g_n) \sum_{i=1}^{3} \sum_{j=1}^{3} \\
& r_i \iint \zeta_i \zeta_j \alpha_m \alpha_n \, dr\, dz.
\end{aligned}
\tag{5.13}
$$

Its stationary value is obtained, as usual, for

$$
\frac{\partial F(U)}{\partial U_i} = 0, \quad \text{for all } i \text{ where } U_i \text{ is free to vary,}
\tag{5.14}
$$

or, in detail,

$$\sum_m \left[p \sum_{i=1}^{3} r_i \sum_{j=1}^{3} \sum_{k=1}^{3} \sum_{l=1}^{3} (b_k b_l + c_k c_l) q_{ijmknl} \right] U_m$$
$$- \sum_m \left[\sum_{i=1}^{3} \sum_{j=1}^{3} r_i Y_{mn}^{(ij)} \right] (k^2 U_m - g_m) = 0 \qquad (5.15)$$

where

$$q_{ijmknl} = \int\!\!\int \left(\zeta_i \frac{\partial \alpha_m}{\partial \zeta_k} \right) \left(\zeta_j \frac{\partial \alpha_n}{\partial \zeta_l} \right) dr \, dz \qquad (5.16)$$

and

$$Y_{mn}^{(ij)} = \int\!\!\int (\zeta_i \alpha_m)(\zeta_j \alpha_n) \, dr \, dz. \qquad (5.17)$$

Now (5.15) is a matrix equation, but the matrices it contains are both new and complicated. They certainly do not fit the patterns of matrices encountered so far, for the multipliers ζ_i imply that the integrands are of order $2N + 2$ if the interpolation functions α_m are of order N. Nevertheless, the integrands are clearly polynomials, so the matrix elements can still be evaluated using an extension of the techniques that apply to the planar case. This extension relies on *embedding* operators, which will be dealt with in the next section.

5.2   *Embedding operators*

Interpolative representations of polynomials are not unique, in the sense that an N-dimensional polynomial p of degree n can be represented exactly in terms of the interpolation polynomials of that degree, or of any higher degree. Such switching of polynomial bases is of value and importance when complicated expressions, such as (5.16)–(5.17), include terms of differing polynomial orders. In simplex coordinates, unlike Cartesians, polynomials even lack a unique representation because the simplex coordinates themselves are linearly dependent, Eq. (2.4). It is always permissible to multiply p by unity; but (2.4) says that 1 may be written as the sum of simplex coordinates, so that

$$p = 1 \cdot p = \sum_{j=1}^{N+1} \zeta_j p. \qquad (5.18)$$

Here the $N + 1$ members of the right-hand sum are individually terms one degree higher (in the several ζ_j jointly) than p itself. To illustrate, suppose that p is the one-dimensional polynomial $p = x + 1$ in the domain $0 \le x \le 1$. Defining simplex coordinates ζ_1 and ζ_2 over this line segment, p may be written in the form $p = 2\zeta_1 + \zeta_2$. But because $\zeta_1 + \zeta_2 = 1$, alternative forms are

$$
\begin{aligned}
p &= 2\zeta_1 + \zeta_2 \\
&= (2\zeta_1 + \zeta_2)(\zeta_1 + \zeta_2) = 2\zeta_1^2 + 3\zeta_1\zeta_2 + \zeta_2^2 \\
&= (2\zeta_1 + \zeta_2)(\zeta_1 + \zeta_2)^2 = 2\zeta_1^3 + 5\zeta_1^2\zeta_2 + 4\zeta_1\zeta_2^2 + \zeta_2^3
\end{aligned}
\tag{5.19}
$$

and so on. This sequence of operations is sometimes called *embedding*, for the low-order polynomials are embedded within a higher-order framework.

A particularly useful embedding operation arises if p is itself one of the simplex interpolation polynomials of order n, say $p = \alpha_i$. Then

$$
\alpha_i = \sum_{j=1}^{N+1} \zeta_j \alpha_i
\tag{5.20}
$$

where each of the $N + 1$ terms on the right is clearly a polynomial of degree $n + 1$. Each of these can be represented exactly in terms of the interpolation polynomials of degree $n + 1$, as

$$
\zeta_j \alpha_i^{(n)} = \sum_k C_{ik}^{(j)} \alpha_k^{(n+1)}
\tag{5.21}
$$

where the summation is taken over all interpolation polynomials of degree $n + 1$. Note that superscripts have been attached to the $\alpha_i^{(n)}$ to distinguish polynomials of differing degree.

The coefficients $C_{ik}^{(j)}$ in Eq. (5.21) are pure numbers whose values can be found by evaluating Eq. (5.21) at each interpolation node associated with the polynomials of degree $n + 1$. At the interpolation node $P_m^{(n+1)}$, all $\alpha_k^{(n+1)}$ vanish except $\alpha_m^{(n+1)}$ which assumes the value 1. The summation in (5.21) there collapses into a single term, leaving

$$
C_{ik}^{(j)} = \zeta_j \alpha_i^{(n)} \Big|_{P_k^{(n+1)}}.
\tag{5.22}
$$

The numbers $C_{ik}^{(j)}$ form a family of three rectangular matrices, possessing as many rows as there are polynomials of degree n, and as many columns as there are polynomials of degree $n + 1$. (Note that the emebedding matrix $\mathbf{C}^{(j)}$ has nothing to do with the connection matrices $\mathbf{C}$ discussed elsewhere in this chapter.) For example, the matrix $\mathbf{C}^{(1)}$ that embeds polynomials of order 1 into an element of order 2 reads

$$
\mathbf{C}^{(1)} = \frac{1}{4}
\begin{bmatrix}
4 & 0 & 0 & 0 & 0 & 0 \\
1 & 1 & 0 & 0 & 0 & 0 \\
1 & 0 & 1 & 0 & 0 & 0
\end{bmatrix}.
\tag{5.23}
$$

Like **T** and $\mathbf{D}^{(1)}$, these matrices are universal, in the sense that they are independent of simplex size or shape. They can therefore be computed once and tabulated, there is no need ever to recompute them. In fact, there is not even any need to tabulate all the matrices for $j = 1, \ldots, N + 1$; just one suffices, for all the rest can be derived from it by the row and column interchanges that correspond to cyclic renumbering of simplex coordinates.

The embedding matrices $\mathbf{C}^{(j)}$ are immediately useful for finding higher-order representations of arbitrary polynomials. If ψ is a polynomial of degree n, then it can be represented in at least two forms,

$$\psi = \sum_m \psi_m^{(n)} \alpha_m^{(n)} = \sum_m \psi_m^{(n+1)} \alpha_m^{(n+1)} \tag{5.24}$$

where $\psi_m^{(n)}$ and $\psi_m^{(n+1)}$ represent function values at the interpolation nodes. There is no need to perform function evaluations to determine the higher-order coefficients $\psi_m^{(n+1)}$; these are readily derived from those of lower order. Write, in accordance with Eqs. (5.20) and (5.21),

$$\alpha_i^{(n)} = \sum_j \zeta_j \alpha_i^{(n)} = \sum_k \left(\sum_j C_{ik}^{(j)} \right) \alpha_k^{(n+1)} \tag{5.25}$$

and substitute into Eq. (5.24) to obtain

$$\sum_k \psi_k^{(n+1)} \alpha_k^{(n+1)} = \sum_k \sum_i \psi_i^{(n)} \left(\sum_j C_{ik}^{(j)} \right) \alpha_k^{(n+1)}. \tag{5.26}$$

The summations must be equal termwise because the interpolation polynomials are linearly independent. Hence

$$\psi_k^{(m+1)} = \sum_i \left(\sum_j C_{ik}^{(j)} \right) \psi_i^{(n)}. \tag{5.27}$$

This equation is more conveniently written in the matrix form

$$\Psi^{(n+1)} = \left(\sum_j \mathbf{C}^{(j)\mathrm{T}} \right) \Psi^{(n)}, \tag{5.28}$$

where $\Psi^{(n)}$ is the column vector of function values $\psi_j^{(n)}$ at the nodes of the interpolation nodes of degree n, while $\Psi^{(n+1)}$ represents the column vector of function values $\psi_j^{(n+1)}$ at interpolation nodes of the next higher order. Note the matrix transposition that corresponds to summing over the *first* index in Eq. (5.27).

The embedding matrices $\mathbf{C}^{(1)}$ are useful in this and several other applications. Tables of $\mathbf{C}^{(1)}$ for the most frequently encountered element types are therefore given in Appendix 3.

5.3   *Axisymmetric scalar matrices*

The embedding matrices derived in the previous section are immediately useful for forming the matrices needed in axisymmetric analysis. Using Eq. (5.21), a typical element of the matrix $\mathbf{Y}^{(ij)}$ as given by Eq. (5.17) may be written

$$Y_{mn}^{(ij)} = \sum_u \sum_v C_{mu}^{(i)} C_{nv}^{(j)} \iint \alpha_u^{(N+1)} \alpha_v^{(N+1)} \, dr \, dz. \tag{5.29}$$

But the remaining integral is clearly the uv component of the matrix $\mathbf{T}^{(N+1)}$ of order $N+1$. Thus

$$\mathbf{Y}^{(ij)} = \mathbf{C}^{(i)} \mathbf{T}^{(N+1)} \mathbf{C}^{(j)\mathrm{T}}. \tag{5.30}$$

The matrix $\mathbf{q}_{ij}$ is a little more complicated because it involves derivatives. These can be dealt with by using the finite differentiation operator, so that

$$q_{ijmknl} = \sum_u \sum_v D_{mv}^{(k)} D_{nu}^{(l)} \iint \zeta_i \alpha_v^{(N)} \zeta_j \alpha_u^{(N)} \, dr \, dz. \tag{5.31}$$

But the integral in Eq. (5.31) is precisely the same as the definition of $Y_{vu}^{(ij)}$, which has already been dealt with. Hence

$$q_{ijmknl} = \sum_u \sum_v D_{mv}^{(k)} Y_{vu}^{(ij)} D_{nu}^{(l)}, \tag{5.32}$$

which may be reorganized as a matrix equation,

$$\mathbf{q}_{ijkl} = \mathbf{D}^{(k)} \mathbf{Y}^{(ij)} \mathbf{D}^{(l)\mathrm{T}}. \tag{5.33}$$

If desired, this may be recast solely in terms of the tabulated matrices as

$$\mathbf{q}_{ijkl} = \mathbf{D}^{(k)} \mathbf{C}^{(i)} \mathbf{T}^{(N+1)} \mathbf{C}^{(j)\mathrm{T}} \mathbf{D}^{(l)\mathrm{T}}. \tag{5.34}$$

Much as in the planar case, the cotangent identity of Eq. (3.21) can be exploited to reduce the number of independent matrices actually needed. Instead of summing over k and l, it then suffices to sum over just one index, say, k, writing

$$\left(p \sum_{i=1}^{3} r_i \sum_{j=1}^{3} \sum_{k=1}^{3} \cot \theta_k \mathbf{Q}^{(ijk)} \right) \mathbf{U}$$

$$- \left(\sum_{i=1}^{3} \sum_{j=1}^{3} r_i \mathbf{Y}^{(ij)} \right) (k^2 \mathbf{U} - \mathbf{G}) = 0 \tag{5.35}$$

where

$$\mathbf{Q}^{(ijk)} = \left(\mathbf{D}^{(k+1)} - \mathbf{D}^{(k-1)}\right)\mathbf{C}^{(i)}\mathbf{T}^{(N+1)}\mathbf{C}^{(j)\mathrm{T}}\left(\mathbf{D}^{(k+1)} - \mathbf{D}^{(k-1)}\right)^{\mathrm{T}}.$$

(5.36)

Clearly, three matrices suffice to carry out any operations required: $\mathbf{D}^{(1)}$, $\mathbf{C}^{(1)}$, and $\mathbf{T}$. This conclusion is actually quite widely applicable, so there is little point in tabulating the $\mathbf{Q}^{(ijk)}$ and other specialized matrices; tabulation of the three fundamental universals suffices for most purposes. Within any one computer program, on the other hand, it may be useful to store intermediate sums or products, so as to avoid their recomputation. For example, in the axisymmetric case it may be useful to store $\sum_j \mathbf{Q}^{(ijk)}$ and $\sum_j \mathbf{Y}^{(ij)}$, because only the summed quantities are of immediate use in problem solving.

5.4   *Solution of coaxial-line problems*

The axisymmetric scalar elements are clearly ideally suited to the analysis of coaxial-line discontinuity problems. At the same time, coaxial-line problems with known solutions can provide some indication of the performance of these elements. The simplest possible problem of this class is that of an infinite coaxial line, readily modelled as a rectangular region in the r–z plane with Neumann boundary conditions imposed on two sides parallel to the r-axis, Dirichlet boundary conditions on the remaining two. The rectangular region is easily subdivided into various different numbers of triangles, permitting a rough assessment to be obtained of the relative performance of the various orders of elements. The analytic solution in this case is of course known, so that no difficulty attaches to comparing computed and exact results. Table 4.2 shows the radial variation of potential values obtained for seven different element subdivisions and orders N.

As may be seen from Table 4.2, the first-order solutions shown do not quite achieve two significant figure accuracy locally, while the second-order solution deviates from the analytic result only in the third digit. The third-order solution provides full three-figure accuracy, while the fourth-order solution differs from the exact results only slightly in the fourth figure. Using elements of fifth order, a solution accuracy similar to fourth-order elements is observed, but the error behaviour has become erratic. With sixth-order elements, this erratic behaviour is still more pronounced, showing that discretization error has entirely vanished from the solution, while roundoff error due to the finite word length used in computation has taken over as the governing factor in limiting accuracy. In the cases shown, matrix orders were roughly similar, ranging

Table 4.2 Nodal potential values for a long coaxial structure

r	N = 1	N = 2	N = 3	N = 4	N = 5	N = 6	exact
1.0	1.0000	1.0000	1.0000	1.0000	1.0000	1.0000	1.0000
2.0	0.7363	0.7334	0.7305	0.7283		0.7293	0.7298
2.2	0.7076				0.6922		0.6926
3.0	0.5782	0.5737	0.5716	0.5718		0.5717	0.5717
4.0	0.4652	0.4611	0.4593	0.4606		0.4593	0.4595
4.6	0.3948				0.4052		0.4050
5.0	0.3773	0.3736	0.3727	0.3723		0.3723	0.3723
5.8	0.3153				0.3144		0.3147
6.0	0.3053	0.3023	0.3017	0.3015		0.3015	0.3014
7.0	0.2438				0.2411		0.2413
8.0	0.1918	0.1898	0.1894	0.1893		0.1892	0.1893
8.2	0.1820				0.1797		0.1797
9.0	0.1453	0.1438	0.1435	0.1434		0.1433	0.1434
9.4	0.1282				0.1264		0.1264
10.0	0.1037	0.1026	0.1024	0.1023		0.1022	0.1023
10.6	0.0807				0.0796		0.0796
11.0	0.0660	0.0653	0.0652	0.0651		0.0651	0.0651
11.9	0.0383				0.0378		0.0378
12.0	0.0316	0.0313	0.0312	0.0312		0.0312	0.0312
13.0	0.0000	0.0000	0.0000	0.0000	0.0000	0.0000	0.0000
Number of elements:							
	288	72	32	18	8	8	∞

from 150 to 175. For this numerical experiment, a computer with 24-bit floating-point mantissa was used, corresponding to approximately seven decimal digits. In matrix problems of this general character, four-figure accuracy may often be expected from seven-digit calculations. There are two principal sources for the roundoff error: error accumulation in the numerous additive processes involved in setting up the matrices, and error accumulation in the solution of simultaneous equations. Roundoff error accruing from the latter can be practically eliminated by means of iterative refinement in the equation solving process. However, little can be done about error arising in the process of creating the equations, aside from performing all calculations in double precision.

5.5   *Axisymmetric vector fields*

Many field problems are not conveniently described by scalar potentials. In these cases, it becomes necessary to solve for a vector quantity, which may be one of the field vectors $\mathbf{E}$, $\mathbf{H}$, or the flux densities $\mathbf{D}$, $\mathbf{B}$; or it may be a vector potential. In any case, it is necessary then to solve a vector Helmholtz equation of the form

$$(\nabla^2 + k^2)\mathbf{A} = \mathbf{G}, \tag{5.37}$$

where $\mathbf{A}$ denotes the vector to be determined, while $\mathbf{G}$ is some given source vector. It should be noted that the Laplacian operator in Eq. (5.37) is the vector Laplacian,

$$\nabla^2\mathbf{A} = \nabla(\nabla \cdot \mathbf{A}) - \nabla \times \nabla \times \mathbf{A}, \tag{5.38}$$

which is in general different from its scalar counterpart. The functional

$$F(\mathbf{U}) = \int (\nabla \times \mathbf{U} \cdot \nabla \times \mathbf{U} - \mathbf{U} \cdot \nabla\nabla \cdot \mathbf{U} - k^2\mathbf{U} + 2\mathbf{G} \cdot \mathbf{U})\, \mathrm{d}\Omega$$
$$- \oint \mathbf{U} \times \nabla \times \mathbf{U} \cdot \mathrm{d}\mathbf{S} \tag{5.39}$$

is then minimized by the vector $\mathbf{U}$ nearest the correct solution $\mathbf{A}$ to Eq. (5.37), provided that

$$\nabla(\nabla \cdot \mathbf{U}) = 0. \tag{5.40}$$

As in the scalar case, the most common types of boundary condition that occur in practice cause the surface-integral term to vanish,

$$\oint \mathbf{U} \times \nabla \times \mathbf{U} \cdot \mathrm{d}\mathbf{S} = 0. \tag{5.41}$$

For the moment, only axisymmetric situations will be considered in which both $\mathbf{G}$ and $\mathbf{A}$ have a single component, and that in the azimuthal direction. That is to say, all possible approximate solution vectors must be of the form

$$\mathbf{U} = U_\phi \mathbf{1}_\phi, \tag{5.42}$$

where $\mathbf{1}_\phi$ is the unit vector in the azimuthal direction. Under this restriction, the boundary integral of (5.41) becomes

$$\oint U_\phi \left(\frac{\partial U_\phi}{\partial n} + \frac{U_\phi}{r} \frac{\partial r}{\partial n} \right) \mathrm{d}S = 0. \tag{5.43}$$

It will be seen that the homogeneous Dirichlet condition of vanishing $\mathbf{U}$ at surfaces satisfies the surface integral requirement; so do various other commonly encountered situations.

Under the assumptions and restrictions outlined, the functional of Eq. (5.39) becomes

$$F(\mathbf{U}) = 2\pi \int \left(r \left\{ \left(\frac{\partial U_\phi}{\partial x} \right)^2 + \left(\frac{\partial U_\phi}{\partial r} \right)^2 - k^2 U_\phi^2 + 2G_\phi U_\phi \right\} \right.$$
$$\left. + 2U_\phi \frac{\partial U_\phi}{\partial r} + \frac{U_\phi^2}{r} \right) d\Omega, \tag{5.44}$$

where the differentiations and integrations refer to the r–z plane only. This functional is not only different from its scalar axisymmetric counterpart; it is particularly inconvenient because it contains the singular term $1/r$, which does not admit a polynomial expansion. However, $F(\mathbf{U})$ can be brought into the general finite element framework by a change in variable: instead of solving for the vector originally desired, one solves for

$$U' = U_\phi / \sqrt{r}. \tag{5.45}$$

Correspondingly, it is convenient to write the right-hand term as

$$G' = G_\phi / \sqrt{r} \tag{5.46}$$

so that the functional of (5.39) finally becomes

$$F(\mathbf{U}) = \int r^2 \left\{ \left(\frac{\partial U'}{\partial z} \right)^2 + \left(\frac{\partial U'}{\partial r} \right)^2 \right\} d\Omega + \frac{9}{4} \int (U')^2 \, d\Omega$$
$$- k^2 \int f^2 (U')^2 \, d\Omega + 2 \int r^2 G' U' \, d\Omega + 2 \int r U' \frac{\partial U'}{\partial r} \, d\Omega, \tag{5.47}$$

where the region Ω of integration is again the relevant portion of the r–z plane.

To develop finite elements for the axisymmetric vector problem, it remains to substitute the usual polynomial expansion for the trial function,

$$U' = \sum_j U'_j \alpha_j(\zeta_1, \zeta_2, \zeta_3). \tag{5.48}$$

The expansion is lengthy because the functional (5.47) contains many terms, so it will not be carried out in detail here. However, it should be evident that all terms in (5.47) are now polynomials, and therefore directly integrable without approximation. Furthermore, the integrals are all expressible in terms of the simplex coordinates, so that once again the integrals can be calculated for a reference triangle whose shape, size and placement eventually appear as weighting coefficients. The detailed generation of elements, as well as subsequent solution techniques, therefore follow exactly the same pattern as for the scalar field case.

5.6   *Projection operators*

The embedding operation changes the representation of a function from a polynomial basis of order n to a polynomial basis of order $n + 1$. Its inverse process, in which the order of representation is reduced, can be defined; but this operation cannot be exact. Nevertheless, a *projection* operator can be defined, with the properties that (1) it will invert the embedding process where exact inversion is possible, and (2) it will compute the nearest possible approximation where it is not. To find a representation of this operator, subtract an approximation $\sum_j u_j^{(n)} \alpha_j^{(n)}$ of degree n from the true polynomial u represented in terms of the functions $\alpha_j^{(n+1)}$ of degree $n + 1$, and minimize their mean squared difference over the element:

$$\frac{\partial}{\partial u_m^{(n)}} \int_\Omega \left(\sum_k u_k^{(n+1)} \alpha_k^{(n+1)} - \sum_j u_j^{(n)} \alpha_j^{(n)} \right)^2 d\Omega = 0. \qquad (5.49)$$

Rewriting in detail, there results

$$\frac{\partial}{\partial u_m^{(n)}} \int_\Omega \left(\sum_k u_k^{(n+1)} \alpha_k^{(n+1)} \right)^2 d\Omega$$

$$- 2 \frac{\partial}{\partial u_m^{(n)}} \int_\Omega \left(\sum_k u_k^{(n+1)} \alpha_k^{(n+1)} \sum_j u_j^{(n)} \alpha_j^{(n)} \right) d\Omega$$

$$+ \frac{\partial}{\partial u_m^{(n)}} \int_\Omega \left(\sum_j u_j^{(n)} \alpha_j^{(n)} \right)^2 d\Omega = 0. \qquad (5.50)$$

On differentiating and compacting the expressions,

$$\sum_j \left(\int_\Omega \alpha_j^{(n)} \alpha_m^{(n)} \, d\Omega \right) u_j^{(n)} = \sum_k \left(\int_\Omega \alpha_k^{(n+1)} \alpha_m^{(n)} \, d\Omega \right) u_k^{(n+1)}. \qquad (5.51)$$

The polynomial $\alpha_m^{(n)}$ in the right-hand integral is of degree n. It can be embedded into the family of polynomials of degree $n + 1$, by writing

$$\alpha_m^{(n)} = \sum_l \zeta_l \alpha_m^{(n)} = \sum_l \sum_i C_{mi}^{(l)} \alpha_i^{(n+1)}. \qquad (5.52)$$

There then results

$$\sum_j \left(\int_\Omega \alpha_j^{(n)} \alpha_m^{(n)} \, d\Omega \right) u_j^{(n)} =$$
$$\sum_k \sum_i \left(\sum_l C_{mi}^{(l)} \right) \left(\int_\Omega \alpha_k^{(n+1)} \alpha_i^{(n+1)} \, d\Omega \right) u_k^{(n+1)}. \tag{5.53}$$

The integral that occurs on both sides of this equation is immediately recognized as an element of the metric $\mathbf{T}$. Consequently, Eq. (5.53) may be put in the matrix form

$$\mathbf{T}^{(n)} \mathbf{U}^{(n)} = \left(\sum_l \mathbf{C}^{(l)} \right) \mathbf{T}^{(n+1)} \mathbf{U}^{(n+1)} \tag{5.54}$$

where $\mathbf{U}$ is the column of u_k, of order n or $n+1$, as indicated by the bracketed superscript. This equation can immediately be solved for the column vector $\mathbf{U}^{(n)}$ that contains the approximation coefficients $u_j^{(n)}$. The result is

$$\mathbf{U}^{(n)} = \mathbf{P}\mathbf{U}^{(n+1)}, \tag{5.55}$$

where

$$\mathbf{P} = \mathbf{T}^{-1} \left(\sum_j \mathbf{C}^{(j)} \right) \mathbf{T}^{(n+1)}. \tag{5.56}$$

Linear independence of the polynomials α_m, guarantees that $\mathbf{T}$ is invertible, so this solution is always possible. The matrix $\mathbf{P}$ is usually called the *projection matrix* or *projector* onto the linear manifold of polynomials α_m. Because the projection is always *downward* in polynomial degree, $\mathbf{P}$ has as many rows as there are interpolation polynomials of degree n, and as many columns as there are interpolation polynomials of degree $n+1$. For example, the projector from degree 2 to degree 1 on triangles reads

$$\mathbf{P} = \frac{1}{5} \begin{bmatrix} 2 & 3 & 3 & -1 & -1 & -1 \\ -1 & 3 & -1 & 2 & 3 & -1 \\ -1 & -1 & 3 & -1 & 3 & 2 \end{bmatrix}. \tag{5.57}$$

Like most of the matrices encountered so far, this matrix too is universal; it is the same for any simplex element of a given order and spatial dimensionality. The element size does appear as a multiplier in the matrix $\mathbf{T}^{(n+1)}$, but it also enters into $\mathbf{T}^{-1}$ as a divisor. The entries in $\mathbf{P}$ are therefore independent of both the size and shape of the element.

An interesting check on this development can be based on the following observation. The embedding

$$\Psi^{(n+1)} = \left(\sum_j \mathbf{C}^{(j)\mathrm{T}} \right) \Psi^{(n)} \qquad (5.58)$$

replaces a polynomial representation on one basis by another, without changing the function itself in any way; the original polynomial of degree n,

$$\psi = \sum_l \psi_l \alpha_l^{(n)}, \qquad (5.59)$$

retains its previous values everywhere. The projection

$$\Psi^{(n)} = \mathbf{P}\Psi^{(n+1)} \qquad (5.60)$$

is exact on this polynomial, since the polynomial itself is only of degree n. Embedding, then projecting, should therefore leave the polynomial unaltered — the projection operation exactly undoes the embedding. Therefore

$$\mathbf{P}\left(\sum_j \mathbf{C}^{(j)\mathrm{T}} \right) = \mathbf{I}, \qquad (5.61)$$

the identity matrix. This assertion is readily verified by multiplying out the two rectangular matrices, indeed it provides a useful check in programming work.

6.   Tetrahedral elements

A discussion of simplex elements cannot be complete until tetrahedral elements have been dealt with fully since software for three-dimensional problems relies on simplex elements even more heavily than its two-dimensional counterpart. The main reason lies in the algebraic topology of mesh generation rather than in any algebraic properties of tetrahedra: failure-proof algorithms exist for subdividing a polyhedron or polygon into simplexes, but not into any other standard element shape.

6.1   *Interpolation and the metric* **T**

The interpolation theory and matrix construction relevant to tetrahedra closely follow the analogous development for triangles. Indeed the definitions of the **S** and **T** matrices given by Eqs. (3.7) and (3.8) are valid for simplexes of any dimensionality, provided the differentiations and integrations are properly carried out. It thus remains to evaluate **S** and **T** in the three-dimensional case, with the integrations taken over the tetrahedron volume Ω. The metric **T** is easily evaluated; the algebraic operations required to form

$$T_{mn} = \int \alpha_m \alpha_n \, \mathrm{d}\Omega \tag{6.1}$$

are precisely the same as in two dimensions, with the sole difference that the interpolation polynomials are now functions of four simplex coordinates, not three,

$$\alpha_{ijkl}(\zeta_1, \zeta_2, \zeta_3, \zeta_4) = R_i(n, \zeta_1) R_j(n, \zeta_2) R_k(n, \zeta_3) R_l(n, \zeta_4); \tag{6.2}$$

and $\mathrm{d}\Omega$ refers to a volume rather than surface element. The association between single indices α_m and multiple indices α_{ijkl} is analogous to its two-dimensional counterpart, as illustrated in Fig. 4.7. Like the metrics of a triangle, numerical values of the T_{mn} are best computed using a symbolic computation language, in an obvious and straightforward fashion: the polynomials are explicitly written out, expanded, and integrated term by term.

Just as for the line segment and the triangle, the metric of a tetrahedron consists of the tetrahedron volume, multiplied by a matrix of numerical values independent of tetrahedron size or shape. For example, **T** for a first-order tetrahedral element reads

$$\mathbf{T} = \frac{1}{20} \begin{bmatrix} 2 & 1 & 1 & 1 \\ 1 & 2 & 1 & 1 \\ 1 & 1 & 2 & 1 \\ 1 & 1 & 1 & 2 \end{bmatrix}. \tag{6.3}$$

Tables of the first few orders of tetrahedra appear in Appendix 3. The tabulation extends to somewhat lower orders than for triangles, because the number of nodes on a tetrahedral element rises rather quickly with element order.

6.2   *Dirichlet matrices for tetrahedra*

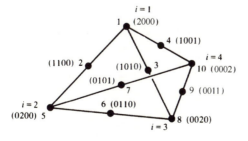

Fig 4.7   Tetrahedron showing modal numbering schemes for a second-order polynomial approximation.

The Dirichlet matrix **S** for a tetrahedron is treated similarly, except that the gradient operators must be reexpressed in simplex coordinates. By definition, the tetrahedron coordinates are

$$
\zeta_1 = \frac{\begin{vmatrix} 1 & x & y & z \\ 1 & x_2 & y_2 & z_2 \\ 1 & x_3 & y_3 & z_3 \\ 1 & x_4 & y_4 & z_4 \end{vmatrix}}{\begin{vmatrix} 1 & x_1 & y_1 & z_1 \\ 1 & x_2 & y_2 & z_2 \\ 1 & x_3 & y_3 & z_3 \\ 1 & x_4 & y_4 & z_4 \end{vmatrix}},
\tag{6.4}
$$

and similarly for the remaining three ζ. Expanding the numerator by minors of its first column, and noting that the denominator above represents six times the tetrahedron volume V, Eq. (6.4) may be restated as

$$
\zeta_1 = \frac{1}{6V}\left(\begin{vmatrix} x_2 & y_2 & z_2 \\ x_3 & y_3 & z_3 \\ x_4 & y_4 & z_4 \end{vmatrix} - \begin{vmatrix} x & y & z \\ x_3 & y_3 & z_3 \\ x_4 & y_4 & z_4 \end{vmatrix} \right.
$$
$$
\left. + \begin{vmatrix} x & y & z \\ x_2 & y_2 & z_2 \\ x_4 & y_4 & z_4 \end{vmatrix} - \begin{vmatrix} x & y & z \\ x_2 & y_2 & z_2 \\ x_3 & y_3 & z_3 \end{vmatrix} \right).
\tag{6.5}
$$

The last three determinants in this equation are next expanded by minors of their first rows, so as to extract the multipliers of x, y and z. Rearranging and collecting terms, there results the conversion equation

$$
\begin{bmatrix} \zeta_1 \\ \zeta_2 \\ \zeta_3 \\ \zeta_4 \end{bmatrix} = \frac{1}{6V} \begin{bmatrix} a_1 & b_1 & c_1 & d_1 \\ a_2 & b_2 & c_2 & d_2 \\ a_3 & b_3 & c_3 & d_3 \\ a_4 & b_4 & c_4 & d_4 \end{bmatrix} \begin{bmatrix} 1 \\ x \\ y \\ z \end{bmatrix}
\tag{6.6}
$$

where the sixteen matrix entries depend only on the vertex locations of the tetrahedron,

$$
a_1 = \begin{vmatrix} x_2 & y_2 & z_2 \\ x_3 & y_3 & z_3 \\ x_4 & y_4 & z_4 \end{vmatrix},
\tag{6.7}
$$

$$
b_1 = -\begin{vmatrix} y_3 & z_3 \\ y_4 & z_4 \end{vmatrix} + \begin{vmatrix} y_2 & z_2 \\ y_4 & z_4 \end{vmatrix} - \begin{vmatrix} y_2 & z_2 \\ y_3 & z_3 \end{vmatrix},
\tag{6.8}
$$

and similarly for the rest. The chain rule of differentiation is now easily applied to form the Dirichlet matrix **S**. It says

$$\frac{\partial \alpha_m}{\partial x} = \sum_{i=1}^{4} \frac{\partial \alpha_m}{\partial \zeta_i} \frac{\partial \zeta_i}{\partial x} = \sum_{i=1}^{4} \frac{\partial \alpha_m}{\partial \zeta_i} \frac{b_i}{6V}. \tag{6.9}$$

Using this, and the analogous expressions in y and z, the general expression

$$S_{mn} = \int \nabla \alpha_m \cdot \nabla \alpha_n \, d\Omega$$

$$= \int \left(\frac{\partial \alpha_m}{\partial x} \frac{\partial \alpha_n}{\partial x} + \frac{\partial \alpha_m}{\partial y} \frac{\partial \alpha_n}{\partial y} + \frac{\partial \alpha_m}{\partial z} \frac{\partial \alpha_n}{\partial z} \right) d\Omega \tag{6.10}$$

is modified to read

$$S_{mn} = \frac{1}{6^2 V^2} \sum_{i=1}^{4} \sum_{j=1}^{4} (b_i b_j + c_i c_j + d_i d_j) \int \frac{\partial \alpha_m}{\partial \zeta_i} \frac{\partial \alpha_n}{\partial \zeta_j} \, d\Omega. \tag{6.11}$$

This expression fits readily into the matrix formalism already developed. In three dimensions, exactly as in two, the integral may be rewritten in terms of differentiation matrices as

$$\int \frac{\partial \alpha_m}{\partial \zeta_i} \frac{\partial \alpha_n}{\partial \zeta_j} \, d\Omega = \sum_k \sum_l D_{mk}^{(i)} \left(\int \alpha_k \alpha_l \, d\Omega \right) D_{nl}^{(j)}. \tag{6.12}$$

The final result for tetrahedra is therefore

$$\mathbf{S} = \frac{1}{36V} \sum_{i=1}^{4} \sum_{j=1}^{4} (b_i b_j + c_i c_j + d_i d_j) \mathbf{D}^{(i)} \mathbf{T} \mathbf{D}^{(j)\mathrm{T}}. \tag{6.13}$$

The sixteen-term summation of Eq. (6.13) can be condensed, just as the corresponding nine-term summation for triangles was contracted to only three terms. The key to finding the condensed form is the observation

$$\sum_i b_i = \sum_v c_i = \sum_i d_i = 0 \tag{6.14}$$

whose proof is straightforward, though tedious — and clearly another useful application of computer algebra! It implies that the terms for $i = j$ in Eqs. (6.11) and (6.13) sum to zero, so that the latter equation may be rewritten as

$$\mathbf{S} = -\sum_{(ij)} \frac{b_i b_j + c_i c_j + d_i d_j}{36V^2} \mathbf{Q}^{(ij)} \tag{6.15}$$

where the summation ranges over the six index pairs $(ij) = (12), (13), (14), (23), (24), (34)$, and the matrix $\mathbf{Q}^{(ij)}$ is given by exactly the same equation as for the two-dimensional case,

$$\mathbf{Q}^{(ij)} = \frac{1}{2}\left(\mathbf{D}^{(i)} - \mathbf{D}^{(j)}\right)\mathbf{T}\left(\mathbf{D}^{(i)} - \mathbf{D}^{(j)}\right)^{\mathrm{T}}. \tag{6.16}$$

There are as many matrices $\mathbf{Q}^{(ij)}$ as there are tetrahedron edges, one corresponding to each index pair (ij) that identified an edge as joining vertices i and j.

6.3   *Permutations and rotations of tetrahedra*

Tetrahedral elements, like any other simplexes, have no privileged coordinate directions; vertex numbering could start with any one of the four vertices and the homogeneous coordinates would be assigned numbers to suit. It follows that only one of the six matrices $\mathbf{Q}^{(ij)}$ is independent, the other five are obtainable by suitable row and column permutations.

The permutations, and the method for discovering them, can be demonstrated by taking as an example the second-order tetrahedral element shown in Fig. 4.7. Here each tetrahedron edge is represented by one of the six possible index pairs (ij). The $N = 10$ nodal points involved are shown with their multi-index labelling $(pqrs)$, with $p + q + r + s = 2$. Also shown is an arbitrarily chosen single-index scheme defining m (or n). The fact that a systematic single indexing is followed is immaterial, although the mapping rule between multiple and single indices does need to be known when employing matrices presented in the single-index form.

Suppose that the set of numbers $Q^{(12)}_{mn}$ has been worked out. It is clear on comparing the diagrams (a) and (b) of Fig. 4.8 that

$$Q^{(13)}_{37} = Q^{(12)}_{29} \tag{6.17}$$

holds, since the numbers in the $\mathbf{Q}$-matrix are determined only by the relative positions of points m, n and the edge (ij) and not by the particular labels assigned to these. In the case here, the same number $Q^{(13)}_{37}$ repre-

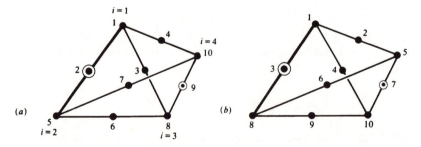

Fig 4.8 (a) Second-order tetrahedron. (b) The same, viewed after a rotation about the ζ_1 axis.

sents the edge shown heavily lined (edge *i–j*) and the two circled nodal points *m, n*, no matter how these may be identified. It may be confirmed that Eq. (6.17) is a particular equation taken from the complete matrix element permutation

$$\mathbf{Q}^{(13)} = \mathbf{P}_{1342}\mathbf{Q}^{(12)}\mathbf{P}_{1342}^{\mathrm{T}},\tag{6.18}$$

where **P** is the permutation matrix

$$\mathbf{P}_{1342} = \begin{bmatrix} 1 & & & & & & & & \\ & & 1 & & & & & & \\ & 1 & & & & & & & \\ & & & 1 & & & & & \\ & & & & & & & 1 & \\ & & & & & 1 & & & \\ & & & & & & 1 & & \\ & & & & 1 & & & & \\ & & & & & & & & 1 \\ & & & & & 1 & & & \end{bmatrix},\tag{6.19}$$

and $\mathbf{P}_{1342}^{\mathrm{T}}$ is its transpose. This permutation represents a relabelling of vertex numbers corresponding to a right-handed rotation about the ζ_1 axis; that is to say, vertex 1 retains its label while vertices 2, 3 and 4 have their numbers advanced cyclically so they become 3, 4 and 2 respectively. Extending this argument, all five $\mathbf{Q}^{(ij)}$ matrices are obtained by

$$\begin{aligned} \mathbf{Q}^{(13)} &= \mathbf{P}_{1342}\mathbf{Q}^{(12)}\mathbf{P}_{1342}^{\mathrm{T}}, \\ \mathbf{Q}^{(14)} &= \mathbf{P}_{1342}\mathbf{Q}^{(13)}\mathbf{P}_{1342}^{\mathrm{T}}, \\ \mathbf{Q}^{(24)} &= \mathbf{P}_{4213}\mathbf{Q}^{(12)}\mathbf{P}_{4213}^{\mathrm{T}}, \\ \mathbf{Q}^{(23)} &= \mathbf{P}_{4213}\mathbf{Q}^{(24)}\mathbf{P}_{4213}^{\mathrm{T}}, \\ \mathbf{Q}^{(34)} &= \mathbf{P}_{1342}\mathbf{Q}^{(23)}\mathbf{P}_{1342}^{\mathrm{T}}. \end{aligned}\tag{6.20}$$

Note that two, and only two, permutation matrices are needed, $\mathbf{P}_{1342}$ and $\mathbf{P}_{4213}$. Taking the component equations of (6.20) in order, all six required matrices are thus generated from $\mathbf{Q}^{(12)}$ alone. Clearly, there is no need to tabulate more than this one matrix.

Values of the various matrices have been computed and tabulated for tetrahedral matrices as well as their lower-dimensionality counterparts. Tables appear in Appendix 3, in the form of integer quotients. It will be noted that the matrix size grows rapidly for tetrahedra, so that in practical computing elements of second order are commonly used, but fourth or higher orders are rarely if ever encountered.

7.   Further reading

High-order finite elements are in widespread use in almost all branches of engineering analysis, having first been introduced in elasticity problems. A good source for background material in this area is the textbook by Desai and Abel (1972), or perhaps the more recent work of Rao (1989).

Electrical engineering applications of high-order simplex elements have been very varied. An interesting overview of electromagnetic problems solvable by one particular commercial program is given by Brauer, in a chapter of about a hundred pages, replete with examples, contributed to a multi-authored volume (Brauer 1993). A different and smaller selection of sample problems, using another commercial package, appears in the book of Sabonnadière and Coulomb (1987). Historically, early application of high-order elements was made to the solution of waveguide problems (Silvester 1969b). Of considerable interest is their use in diffusion problems by Kaper, Leaf and Lindeman (1972), who discuss high-order triangles extensively. A large-scale program using elements up to sixth order, with extensive application-oriented further processing of the field results, is reported by Kisak and Silvester (1975). The latter deal with complex diffusion. McAulay (1975) extends the technique to wave propagation in lossy media, while Stone (1973) and others have constructed matrices for the acoustic surface wave problem. The mathematical aspects of waveguide and resonator problems are set out by Jin (1993) and more briefly by Mori (1986). Straightforward potential problem solutions have been reported, among others, by Andersen (1973), who describes a large-scale applications package with extensive input and output data management facilities.

A large part of the work on simplex elements has led to the creation of tabulated universal matrices. Systematic methods for generating interpolation functions on triangles were proposed at a relatively early date by Silvester (1969a), with immediate application to the solution of potential problems. In addition to the planar triangles already referred to, matrices for the various related problems have been recorded in the literature, for example the axisymmetric matrices of Konrad and Silvester (1973). The general approach, construction of any 'universal' matrices from a few primitives, was given by Silvester (1978) later — no doubt another case of clear scientific hindsight! The theory has been extended since, and the presentation appearing in this chapter is probably the most complete available in the literature.

8. Bibliography

Andersen, O.W. (1973), 'Transformer leakage flux program based on finite element method', *IEEE Transactions on Power Apparatus and Systems*, **PAS-92**, 682–9.

Brauer, J.R. [ed.] (1993), *What Every Engineer Should Know about Finite Element Analysis*. 2nd edn. New York: Marcel Dekker. xiii + 321 pp.

Desai, C.S. and Abel, J.F. (1972), *Introduction to the finite element method*. New York: Van Nostrand. xvii + 477 pp.

Jin, J.-M. (1993), *The Finite Element Method in Electromagnetics*. New York: John Wiley. xix + 442 pp.

Kaper, H.G., Leaf, G.K. and Lindeman, A.J. (1972), 'The use of interpolatory polynomials for a finite element solution of the multigroup diffusion equation', in *The mathematical foundations of the finite element method with applications to partial differential equations*, pp. 785–9, ed. A.K. Aziz. New York: Academic Press. (See also *Transactions of the American Nuclear Society*, **15**, 298).

Kisak, E. and Silvester, P. (1975), 'A finite-element program package for magnetotelluric modelling', *Computer Physics Communications*, **10**, 421–33.

Konrad, A. and Silvester, P. (1973), 'Triangular finite elements for the generalized Bessel equation of order m', *International Journal of Numerical Methods in Engineering*, **7**, 43–55.

McAulay, A.D. (1975), 'Track-guided radar for rapid transit headway control', *Journal of Aircraft*, **12**, 676–81.

Mori, M. (1986), *The Finite Element Method and its Applications*. New York: Macmillan. xi + 188 pp.

Rao, S.S. (1989), *The Finite Element Method in Engineering*. 2nd edn. Oxford: Pergamon. xxvi + 643 pp.

Sabonnadière, J.-C. and Coulomb, J.-L. (1987), *Finite Element Methods in CAD: Electrical and Magnetic Fields* [S. Salon, transl.] New York: Springer-Verlag. viii + 194 pp.

Silvester, P. (1969a), 'High-order polynomial triangular finite elements for potential problems', *International Journal of Engineering Science*, **7**, 849–61.

Silvester, P. (1968b), 'A general high-order finite-element waveguide analysis program', *IEEE Transactions on Microwave Theory and Technology*, **MTT-17**, 204–10.

Silvester, P. (1978), 'Construction of triangular finite element universal matrices', *International Journal of Numerical Methods in Engineering*, **12**, pp. 237–44.

Stone, G.O. (1973), 'High-order finite elements for inhomogeneous acoustic guiding structures', *IEEE Transactions on Microwave Theory and Technology*, **MTT-21**, 538–42.

9. Programs

Program HITRIA2D is listed in the following. Like other programs in this book, it is written in Fortran 77. The main program listing is followed by an element-construction routine trielm and a data-reading routine gettri which fetches triangular element data from file. There follows a set of element data sufficient for the first two element orders, in the format expected by `gettri`, and a small data set which should suffice for checking compilation and correct execution. It should be noted that `gettri` as listed expects to find eight data files (Q and T matrices for orders 1 through 4). To run the program with only the data listed here, it will be necessary to remove the four segments of program code in `gettri` that attach and read the missing files (matrices of orders 3 and 4). Tables of higher-order elements will be found in Appendix 3.

```
C****************************************************************
C***********                              ***********
C*********** High-order finite element program  ***********
C***********                              ***********
C****************************************************************
C      Copyright (c) 1995  P.P. Silvester and R.L. Ferrari
C****************************************************************
C
C      The subroutines  that make up this program  communicate
C      via named common blocks.  The variables in commons are:
C
C      Problem definition and solution
C      common /problm/
C        nodes  = number of nodes used in problem
C        nelmts = number of elements in model
C        x, y   = nodal coordinates
C        constr = logical, .true. for fixed potentials
C        potent = nodal potential array
C        nvtx   = list of nodes for each element
C        source = source density in each element
C        nve    = number of nodes in each element
C
C      Global matrices and right-hand side
C      common /matrix/
C        s      = global S-matrix for whole problem
C        t      = global T-matrix for whole problem
C        rthdsd = right-hand side of system of equations
C
C      Temporary working arrays and variables
C      common /workng/
C        sel    = S for one element (working array)
C        tel    = T for one element (working array)
C        intg   = integer working array
C        intg0  = integer working array
C
C      Predefined problem size parameters (array dimensions):
C        MAXNOD = maximum number of nodes in problem
C        MAXELM = maximum number of elements in problem
C        MAXNVE = maximum number of vertices in any element
C
C================================================================
C     Global declarations -- same in all program segments
C================================================================
      parameter (MAXNOD = 50, MAXELM = 75, MAXNVE = 15)
      logical constr
      common /problm/ nodes, nelmts, x(MAXNOD), y(MAXNOD),
     *        constr(MAXNOD), potent(MAXNOD),
     *        nvtx(MAXNVE,MAXELM), source(MAXELM), nve(MAXELM)
      common /matrix/ s(MAXNOD,MAXNOD), t(MAXNOD,MAXNOD),
     *        rthdsd(MAXNOD)
      common /workng/ sel(MAXNVE,MAXNVE), tel(MAXNVE,MAXNVE),
     *        intg(MAXNVE), intg0(MAXNVE)
C================================================================
C
```

```
C       Set up triangular finite element matrices.
        call gettri
C
C       Fetch input data from input file.
        call meshin
C
C          Set global matrices and right side to all zeros.
        call matini(s, MAXNOD, MAXNOD)
        call matini(t, MAXNOD, MAXNOD)
        call vecini(rthdsd, MAXNOD)
C
C       Assemble element matrix.  Proceed element by element.
        do 40 ie = 1,nelmts
C
C          Construct element s and t matrices
          if (nve(ie) .eq.  3  .or.  nve(ie) .eq.  6  .or.
     *        nve(ie) .eq. 10  .or.  nve(ie) .eq. 15)
     *        call trielm(ie)
C
C          Embed matrices in global s; augment right side:
          call elembd(ie)
   40   continue
C
C       Solve the assembled finite element equations
        call eqsolv (s, potent, rthdsd, nodes, MAXNOD)
C
C       Print out the resulting potential values
        call output
C
        stop
        end
C
C**************************************************************
C
        Subroutine meshin
C
C**************************************************************
C
C       Reads input data file in  three parts:  nodes, elements,
C       fixed potentials.   Each part is followed by a line con-
C       taining nothing but a slant character /  which serves as
C       a data terminator.   Data lines are in free format, with
C       comma or blank  as acceptable  separators between items.
C       character data  (element types!) Must be  encased  in
C       apostrophes.
C
C       nodes:                                      x, y   values.
C       elements:        type, total nodes, source, node numbers.
C       potentials:                    node number, fixed value.
C
C       All data are echoed as read but little checking is done
C       for validity.
C==============================================================
C       Global declarations -- same in all program segments
```

```
C==============================================================
      parameter (MAXNOD = 50, MAXELM = 75, MAXNVE = 15)
      logical constr
      common /problm/ nodes, nelmts, x(MAXNOD), y(MAXNOD),
     *          constr(MAXNOD), potent(MAXNOD),
     *          nvtx(MAXNVE,MAXELM), source(MAXELM), nve(MAXELM)
      common /matrix/ s(MAXNOD,MAXNOD), t(MAXNOD,MAXNOD),
     *          rthdsd(MAXNOD)
      common /workng/ sel(MAXNVE,MAXNVE), tel(MAXNVE,MAXNVE),
     *          intg(MAXNVE), intg0(MAXNVE)
C==============================================================
C
      Parameter (HUGE = 1.e+30)
      logical done
C
C     Read in the node list and echo input lines.
C             Start by printing the heading for nodes.
      write (*, 200)
C
C             Read and echo until null list (line with / only)
      nodes = 0
      xi0 = HUGE
      yi0 = HUGE
   20 read (*,*) xi, yi
      if (xi .ne. xi0  .or.  yi .ne. yi0) then
        nodes = nodes + 1
        x(nodes) = xi
        y(nodes) = yi
        xi0 = xi
        yi0 = yi
        write (*, 210) nodes, xi, yi
        go to 20
      endif
C
C     Read in the element list and echo all input as received.
C           Print element heading to start.
      write (*, 220)
C
C             Read elements in turn.
      nelmts = 0
      do 30 i = 1,MAXNVE
   30   intg0(i) = 0
   40 read (*,*) nods, sourci,
     *                          (intg(j), j = 1,min0(nods,MAXNVE))
      done = .true.
      do 50 i = 1,nods
   50   done = done .and. (intg(i) .eq. intg0(i))
      if (.not. done) then
        nelmts = nelmts + 1
        source(nelmts) = sourci
        nve(nelmts) = nods
        write (*,230) nelmts, nods, sourci,
     *                          (intg(i), i = 1,nods)
        do 60 i = 1,nods
```

```
              nvtx(i,nelmts) = intg(i)
              intg0(i) = intg(i)
   60         continue
           go to 40
         endif
C
C      Read list of fixed potential values and print.
C            Print header first, then declare all nodes free.
         write (*, 240)
         call vecini(potent, MAXNOD)
         do 70 m = 1,nodes
   70    constr(m) = .false.
C
C            Read and echo input.
         last = 0
   80 read (*,*, end=90) i, xi
         if (i .ne. last) then
           constr(i) = .true.
           potent(i) = xi
           last = i
           write (*, 250) i, xi
           go to 80
         endif
C
   90 return
C
  200 format (1x // 8x, 'Input node list' // 2x, 'Node', 6x, 'x',
      *          11x, 'y' / 1x)
  210 format (1x, i3, 2(2x, f10.5))
  220 format (1x // 9x, 'Input element list' //
      * ' No   Nodes  Source', ' Element node numbers' / 1x)
  230 format (1x, i3, 2x, i4, f10.5, 1x, 15i3)
  240 format (1x // 4x, 'Input fixed potentials' // 6x,
      *          'Node', 12x, 'Value' / 1x)
  250 format (6x, i3, 9x, f10.5)
C
         end
C
C***********************************************************
C
         Subroutine elembd(ie)
C
C***********************************************************
C
C      Embeds single-element  s and t matrices currently in sel
C      and tel (in common block /workng/) in the global  matrix
C      s. Argument ie is the element number.
C
C============================================================
C      Global declarations -- same in all program segments
C============================================================
         parameter (MAXNOD = 50, MAXELM = 75, MAXNVE = 15)
         logical constr
         common /problm/ nodes, nelmts, x(MAXNOD), y(MAXNOD),
```

```
     *           constr(MAXNOD), potent(MAXNOD),
     *           nvtx(MAXNVE,MAXELM), source(MAXELM), nve(MAXELM)
      common /matrix/ s(MAXNOD,MAXNOD), t(MAXNOD,MAXNOD),
     *           rthdsd(MAXNOD)
      common /workng/ sel(MAXNVE,MAXNVE), tel(MAXNVE,MAXNVE),
     *           intg(MAXNVE), intg0(MAXNVE)
C================================================================
C
C     Run through element  s and t matrices  sel and tel, aug-
C     menting s, t, and the right-hand side as appropriate.
C
      nvertc = nve(ie)
      do 60 i = 1,nvertc
         irow = nvtx(i,ie)
C
C           Does row correspond to a fixed potential?
         if (constr(irow)) go to 50
C
C           No, potential may vary.  Do all nvertc columns.
         do 40 j = 1,nvertc
         icol = nvtx(j,ie)
C
C           Does column correspond to a fixed potential?
         if (constr(icol)) go to 30
C
C           No; so augment s, t and rthdsd.
         s(irow,icol) = s(irow,icol) + sel(i,j)
         t(irow,icol) = t(irow,icol) + tel(i,j)
         rthdsd(irow) = rthdsd(irow) + tel(i,j) * source(ie)
         go to 40
C
C           Yes; so augment right side only:
   30       continue
            rthdsd(irow) = rthdsd(irow) + tel(i,j) * source(ie)
    1                               - sel(i,j) * potent(icol)
   40       continue
         go to 60
C
C           Constrained row number.  Set global s and rthdsd.
   50    continue
         s(irow,irow) = 1.
         rthdsd(irow) = potent(irow)
C
   60    continue
C
C     All done -- return to calling program.
      return
      end
C
C****************************************************************
C
      Subroutine output
C
C****************************************************************
```

```
C
C      Outputs problem and results to standard output stream.
C
C================================================================
C      Global declarations -- same in all program segments
C================================================================
       parameter (MAXNOD = 50, MAXELM = 75, MAXNVE = 15)
       logical constr
       common /problm/ nodes, nelmts, x(MAXNOD), y(maxnod),
     *         constr(MAXNOD), potent(MAXNOD),
     *         nvtx(MAXNVE,MAXELM), source(MAXELM), nve(MAXELM)
       common /matrix/ s(MAXNOD,MAXNOD), t(MAXNOD,MAXNOD),
     *         rthdsd(MAXNOD)
       common /workng/ sel(MAXNVE,MAXNVE), tel(MAXNVE,MAXNVE),
     *         intg(MAXNVE), intg0(MAXNVE)
C================================================================
C
C      Print the nodes and the output potential values.
C
       write (*, 1000) (i, x(i), y(i), potent(i),
     1                                 i = 1, nodes)
 1000 format (1x /// 12x, 'Final solution' // '    i', 8x, 'x',
     1         9x, 'y', 7x, 'Potential' // (1x, i3, 2x, f10.5,
     2         F10.5, 3x, f10.5))
C
       return
       end


C***************************************************************
C
       Subroutine trielm(ie)
C
C***************************************************************
C      Copyright (c) 1994 P.P. Silvester and R.L. Ferrari
C***************************************************************
C
C      Constructs the element matrices s, t  for a triangular
C      element of order 1, 2, 3, or 4.   ie = element number.
C================================================================
C      Global declarations -- same in all program segments
C================================================================
       parameter (MAXNOD = 50, MAXELM = 75, MAXNVE = 15)
       logical constr
       common /problm/ nodes, nelmts, x(MAXNOD), y(MAXNOD),
     *         constr(MAXNOD), potent(MAXNOD),
     *         nvtx(MAXNVE,MAXELM), source(MAXELM), nve(MAXELM)
       common /matrix/ s(MAXNOD,MAXNOD), t(MAXNOD,MAXNOD),
     *         rthdsd(MAXNOD)
       common /workng/ sel(MAXNVE,MAXNVE), tel(MAXNVE,MAXNVE),
     *         intg(MAXNVE), intg0(MAXNVE)
C================================================================
C      High-order triangular finite element common blocks --
C         shared only by subroutines TRIELM and GETTRI
```

```
C================================================================
      common /qtblok/ t1(6), t2(21), t3(55), t4(120),
     1                q1(6), q2(21), q3(55), q4(120)
C================================================================
C
C       Set up corner indices for triangle;
      if      (nve(ie) .eq.  3) then
        norder = 1
      else if (nve(ie) .eq.  6) then
        norder = 2
      else if (nve(ie) .eq. 10) then
        norder = 3
      else if (nve(ie) .eq. 15) then
        norder = 4
      else
        call errexc('TRIELM', 1000 + ie)
      endif
      i = nvtx(1,ie)
      j = nvtx(nve(ie)-norder,ie)
      k = nvtx(nve(ie),ie)
C
C       Find element area and cotangents of included angles
      area = abs((x(j) - x(i)) * (y(k) - y(i)) -
     1           (x(k) - x(i)) * (y(j) - y(i))) / 2.
      if (area .le. 0.) then
        call errexc('TRIELM', 2000 + ie)
      endif
      ctng1 = ((x(j) - x(i)) * (x(k) - x(i)) +
     1         (y(j) - y(i)) * (y(k) - y(i))) / (2. * Area)
      ctng2 = ((x(k) - x(j)) * (x(i) - x(j)) +
     1         (y(k) - y(j)) * (y(i) - y(j))) / (2. * Area)
      ctng3 = ((x(i) - x(k)) * (x(j) - x(k)) +
     1         (y(i) - y(k)) * (y(j) - y(k))) / (2. * Area)
C
C     Compute element s and t matrix entries
      do 30 l = 1,nve(ie)
        do 20 m = 1,l
          go to (1, 2, 3, 4), norder
    1     tel(l,m) = t1(locate(l,m)) * area
          sel(l,m) = q1(locate(l,m)) * ctng1
     *         + q1(locate(indxp(l,1,1),indxp(m,1,1))) * ctng2
     *         + q1(locate(indxp(l,1,2),indxp(m,1,2))) * ctng3
          go to 10
    2     tel(l,m) = t2(locate(l,m)) * area
          sel(l,m) = q2(locate(l,m)) * ctng1
     *         + q2(locate(indxp(l,2,1),indxp(m,2,1))) * ctng2
     *         + q2(locate(indxp(l,2,2),indxp(m,2,2))) * ctng3
          go to 10
    3     tel(l,m) = t3(locate(l,m)) * area
          sel(l,m) = q3(locate(l,m)) * ctng1
     *         + q3(locate(indxp(l,3,1),indxp(m,3,1))) * ctng2
     *         + q3(locate(indxp(l,3,2),indxp(m,3,2))) * ctng3
          go to 10
    4     tel(l,m) = t4(locate(l,m)) * area
```

```fortran
        sel(l,m) = q4(locate(l,m)) * ctng1
     *            + q4(locate(indxp(l,4,1),indxp(m,4,1))) * ctng2
     *            + q4(locate(indxp(l,4,2),indxp(m,4,2))) * ctng3
10      tel(m,l) = tel(l,m)
        sel(m,l) = sel(l,m)
20      continue
30    continue
50  return
    end
C
C ************************************************************
C
      Function indxp (i, n, k)
C
C ************************************************************
C
C    Returns the  permuted row or column  index corresponding
C    to  row or column  i of a  triangular element matrix of
C    order n, with the rotation permutation applied k times.
C
      dimension ntwist(83), jump(6)
C
      data ntwist/  3,   1,   2,   6,   3,   5,   1,   2,   4,  10,   6,
     *  9,   3,   5,   8,   1,   2,   4,   7,  15,  10,  14,   6,   9,  13,
     *  3,   5,   8,  12,   1,   2,   4,   7,  11,  21,  15,  20,  10,  14,
     * 19,   6,   9,  13,  18,   3,   5,   8,  12,  17,   1,   2,   4,   7,
     * 11,  16,  28,  21,  27,  15,  20,  26,  10,  14,  19,  25,   6,   9,
     * 13,  18,  24,   3,   5,   8,  12,  17,  23,   1,   2,   4,   7,  11,
     * 16,  22/
      data jump/0,   3,   9,  19,  34,  55/
C
C    "ntwist" contains row and column mappings  for the first
C    four orders of triangular element. "jump" contains poin-
C    ters to starting points in "ntwist" (minus 1).
C
      km3 = mod(k, 3)
      indxp = i
      if (km3 .eq. 0) go to 90
      do 70 l = 1, km3
        indxp = ntwist(jump(n)+indxp)
70      continue
90    continue
      return
      end
C
C ************************************************************
C
      Subroutine gettri
C
C ************************************************************
C
C    Sets up the universal element matrices qe and te, single
C    precision,  by fetching data  from files.  q1 and t1 are
C    the first order matrices,  q2 and t2 second order,  etc.
```

```
C       Common data storage area /qtblok/ is shared with TRIELM.
C================================================================
C       High-order triangular finite element common blocks --
C          shared only by subroutines TRIELM and GETTRI
C================================================================
        common /qtblok/ t1(6), t2(21), t3(55), t4(120),
     1                  q1(6), q2(21), q3(55), q4(120)
C================================================================
        Double precision value
C
C            Attach and read Q of order 1.
        open (Unit=1, err=911, file='QTRIA1.DAT')
        read (1,*) i, n
        do 10 k = 1,n
          read (1,*) i, j, junk, value
          q1(locate(i,j)) = value
   10   continue
        close(0)
C
C            Attach and read T of order 1.
        open (Unit=1, err=911, file='TTRIA1.DAT')
        read (1,*) i, n
        do 11 k = 1,n
          read (1,*) i, j, junk, value
          t1(locate(i,j)) = value
   11   continue
        close(1)
C
C            Attach and read Q of order 2.
        open (Unit=1, err=911, file='QTRIA2.DAT')
        read (1,*) i, n
        do 20 k = 1,n
          read (1,*) i, j, junk, value
          q2(locate(i,j)) = value
   20   continue
        close(1)
C
C            Attach and read T of order 2
        open (Unit=1, err=911, file='TTRIA2.DAT')
        read (1,*) i, n
        do 21 k = 1,n
          read (1,*) i, j, junk, value
          t2(locate(i,j)) = value
   21   continue
        close(1)
C
C            Attach and read Q of order 3.
        open (Unit=1, err=911, file='QTRIA3.DAT')
        read (1,*) i, n
        do 30 k = 1,n
          read (1,*) i, j, junk, value
          q3(locate(i,j)) = value
   30   continue
        close(1)
```

```
C
C              Attach and read T of order 3.
       open (Unit=1, err=911, file='TTRIA3.DAT')
       read (1,*) i, n
       do 31 k = 1,n
         read (1,*) i, j, junk, value
         t3(locate(i,j)) = value
   31    continue
       close(1)
C
C              Attach and read Q of order 4.
       open (Unit=1, err=911, file='QTRIA4.DAT')
       read (1,*) i, n
       do 40 k = 1,n
         read (1,*) i, j, junk, value
         q4(locate(i,j)) = value
   40    continue
       close(1)
C
C              Attach and read T of order 4
       open (Unit=1, err=911, file='TTRIA4.DAT')
       read (1,*) i, n
       do 41 k = 1,n
         read (1,*) i, j, junk, value
         t4(locate(i,j)) = value
   41    continue
       close(1)
  911  continue
       end
```

1	6	2	QTRIA1.DAT
1	1	0	0.00000000000000e+000
1	2	0	0.00000000000000e+000
1	3	0	0.00000000000000e+000
2	2	1	5.00000000000000e-001
2	3	-1	-5.00000000000000e-001
3	3	1	5.00000000000000e-001
1	6	12	TTRIA1.DAT
1	1	2	1.66666666666667e-001
1	2	1	8.33333333333333e-002
1	3	1	8.33333333333333e-002
2	2	2	1.66666666666667e-001
2	3	1	8.33333333333333e-002
3	3	2	1.66666666666667e-001
2	21	6	QTRIA2.DAT
1	1	0	0.00000000000000e+000
1	2	0	0.00000000000000e+000
1	3	0	0.00000000000000e+000
1	4	0	0.00000000000000e+000
1	5	0	0.00000000000000e+000
1	6	0	0.00000000000000e+000
2	2	8	1.33333333333333e+000

2	3	-8	-1.33333333333333e+000
2	4	0	0.00000000000000e+000
2	5	0	0.00000000000000e+000
2	6	0	0.00000000000000e+000
3	3	8	1.33333333333333e+000
3	4	0	0.00000000000000e+000
3	5	0	0.00000000000000e+000
3	6	0	0.00000000000000e+000
4	4	3	5.00000000000000e-001
4	5	-4	-6.66666666666667e-001
4	6	1	1.66666666666667e-001
5	5	8	1.33333333333333e+000
5	6	-4	-6.66666666666667e-001
6	6	3	5.00000000000000e-001
2	21	180	TTRIA2.DAT
1	1	6	3.33333333333333e-002
1	2	0	0.00000000000000e+000
1	3	0	0.00000000000000e+000
1	4	-1	-5.55555555555556e-003
1	5	-4	-2.22222222222222e-002
1	6	-1	-5.55555555555556e-003
2	2	32	1.77777777777778e-001
2	3	16	8.88888888888889e-002
2	4	0	0.00000000000000e+000
2	5	16	8.88888888888889e-002
2	6	-4	-2.22222222222222e-002
3	3	32	1.77777777777778e-001
3	4	-4	-2.22222222222222e-002
3	5	16	8.88888888888889e-002
3	6	0	0.00000000000000e+000
4	4	6	3.33333333333333e-002
4	5	0	0.00000000000000e+000
4	6	-1	-5.55555555555556e-003
5	5	32	1.77777777777778e-001
5	6	0	0.00000000000000e+000
6	6	6	3.33333333333333e-002

0.000	0.000
1.000	0.000
0.000	4.000
1.000	4.000
0.000	4.500
3.000	4.500
3.000	4.000

/

```
     3    1.000    1  2  3
     3    1.000    2  4  3
     3    0.000    3  4  5
     3    0.000    4  7  6
     3    0.000    4  6  5
 /
     6       0.000
     7       0.000
 /
```

5

Differential operators in ferromagnetic materials

1.   Introduction

Many of the problems of classical electrophysics that interest electrical engineers are inherently nonlinear. In fact it may be said that nonlinearity is an essential ingredient in almost all practical device analysis; nonlinearities determine the values over which field variables may range, and thereby fix the limits of device performance. Many useful devices could not function at all were they linear; mixers, modulators, and microwave circulators spring to mind as examples, but others can be found easily at any power level or in any part of the frequency spectrum.

Nonlinear finite element methods found early use in the analysis of power-frequency magnetic devices, most of which were not easily amenable to any other analytic technique, and a large part of the finite element literature in electrical engineering still deals with nonlinear magnetics. The nonlinearities encountered in such problems can often be assumed single-valued and monotonic so that comparatively simple methods can be made computationally efficient and reliable. For this reason, the present chapter will outline the major nonlinear methods used in magnetic field analysis. Other areas, such as problems of plasma electrochemistry or semiconductor devices, are generally amenable to similar mathematical techniques, though the electromagnetic formulations are often much more complicated.

2.   Functional minimization for magnetic fields

The basic principles of linear finite element analysis carry over to nonlinear problems almost without modification. As always, a stationary functional is constructed and discretized over finite elements. Its stationary point — or its maximum or minimum, if a suitable extremum principle can be found — is established by solving a system of simultaneous algebraic equations whose behaviour mimics that of the underlying continuum problem. As might be expected, the algebraic equations that result from nonlinear boundary-value problems are nonlinear too. They can be solved by several different methods. Thus the field of nonlinear finite element analysis contains an extra dimension beyond that of linear problems: there is a choice of nonlinear solution method in addition to all the other choices the analyst must face.

2.1   *Energy functionals*

As set out in Chapter 3, magnetic field problems are commonly described either in terms of a scalar potential P, so defined that the magnetic field $\mathbf{H}$ is its gradient

$$\mathbf{H} = -\nabla P, \tag{2.1}$$

or in terms of a vector potential $\mathbf{A}$, whose curl gives the magnetic flux density $\mathbf{B}$:

$$\mathbf{B} = \nabla \times \mathbf{A}. \tag{2.2}$$

The scalar potential can be defined readily only in regions where $\mathbf{H}$ is irrotational, i.e., in current-free regions. There, P is governed by the nonlinear differential equation

$$\nabla \cdot (\mu \nabla P) = 0, \tag{2.3}$$

which is nonlinear because the material permeability is a function of the magnetic field itself. Correspondingly, the vector potential is governed by

$$\nabla \times (\nu \nabla \times \mathbf{A}) = \mathbf{J}, \tag{2.4}$$

which may be regarded as more general because the problem region may include nonzero current densities $\mathbf{J}$. These may of course represent imposed current densities as well as current densities which result from time variations of the fields themselves (eddy currents). The material reluctivity ν (the inverse of material permeability μ) is again field-dependent, hence Eq. (2.4) is nonlinear.

Equations (2.3)–(2.4) are valid for general three-dimensional fields. Although there is considerable practical interest in three-dimensional problems, a great majority of practical cases tends to be handled by two-dimensional approximation — not only because computing times are long for three-dimensional cases, but also because the representation of results and of geometric input data give rise to serious problems of visualization and interpretation. Two-dimensional problems often arise quite naturally; for example, electric machines are often analysed on the assumption that they have infinite axial length. In such cases, Eq. (2.3) still applies, subject to the understanding that the divergence and gradient operators refer to two dimensions. In Eq. (2.4), the vectors may be assumed to be entirely z-directed, so this equation reduces to the two-dimensional form

$$\frac{\partial}{\partial x}\left(\nu\,\frac{\partial A}{\partial x}\right) + \frac{\partial}{\partial y}\left(\nu\,\frac{\partial A}{\partial y}\right) = -J, \tag{2.5}$$

which is formally identical to the two-dimensional version of (2.3), except for the obvious changes of variable. Indeed, (2.5) has precisely the same general appearance as a linear problem in Poisson's equation; the difference is that the material reluctivity ν is not constant, but depends on the field represented by A.

It is important to note that the permeabilities of most common materials may be assumed to be scalar, single-valued, and monotonic; increasing fields invariably yield reduced permeability and augmented reluctivity. This fact is of considerable theoretical importance, for it guarantees solvability and uniqueness for the methods presented in this chapter.

To attempt solution of the nonlinear problem in any of these forms by finite element methods, a suitable functional must first be defined. It would be tempting to assume that, in view of the similarity of (2.5) to the corresponding linear differential equations, a suitable functional would be

$$F(\mathbf{U}) = \int_{\Omega} \frac{\mathbf{B} \cdot \mathbf{H}}{2}\,d\Omega - \int_{\Omega} \mathbf{J} \cdot \mathbf{U}\,d\Omega, \tag{2.6}$$

where $\mathbf{U}$ is the trial $\mathbf{A}$ exactly as for the linear case; but minimization of $F(\mathbf{U})$ does not prove this suspicion to be well founded. However, the integrand in the first term of Eq. (2.6) is readily spotted as proportional to the stored energy density in the linear problem (in fact, is just twice the stored energy density) so that (2.6) may be written in the form

$$F(\mathbf{U}) = \int_{\Omega} W(\mathbf{U})\,d\Omega - \int_{\Omega} \mathbf{J} \cdot \mathbf{U}\,d\Omega, \tag{2.7}$$

where $W(\mathbf{U})$ denotes the energy density associated with the trial solution $\mathbf{U}$. Happily, the energy density is not intrinsically tied to linear materials, so that Eq. (2.7) turns out to be valid for nonlinear materials as well — provided the energy density is correctly taken for magnetic materials as

$$W = \int_{\Omega} \mathbf{H} \cdot d\mathbf{B}. \tag{2.8}$$

In the vector potential problem, Eq. (2.4), one then has

$$\mathbf{B} = \nabla \times \mathbf{U} \tag{2.9}$$

and

$$\mathbf{H} = \nu\mathbf{B}, \tag{2.10}$$

with ν dependent on $\mathbf{B}$ itself.

It can be shown that, since the leading term in Eq. (2.7) actually represents energy, the conditions for existence and uniqueness of a mathematical solution are precisely those which are required for the existence of a unique stable state in a physical magnetic system: the magnetization curves of all materials in the problem must be monotonic increasing, while their first derivatives are monotonic decreasing. Thus, for the purposes of this chapter, it will suffice to consider methods for minimizing $F(\mathbf{U})$ as given by Eq. (2.7). To keep the examples simple, only two-dimensional problems will be dealt with, with $\mathbf{U}$ an entirely z-directed quantity U.

2.2   *Minimization over finite elements*

The discretization of functional (2.7) for the nonlinear case follows much the same techniques as were employed for linear problems. Let the problem region Ω be discretized into a set of nonoverlapping finite elements in the usual manner and let attention be concentrated on a single element. The potential U will be represented by the interpolative approximation

$$U = \sum_i U_i \alpha_i(x, y) \tag{2.11}$$

within a single element. With this substitution, $F(U)$ becomes an ordinary function of the finite number of finite element variables. The minimization process required is thus simply

$$\frac{\partial F}{\partial U_i} = 0, \tag{2.12}$$

where the index i ranges over all unconstrained variables. Differentiating Eq. (2.7) in order to minimize,

$$\int \left(\frac{\partial W}{\partial U_i} - J\alpha_i \right) d\Omega = 0. \tag{2.13}$$

From Eqs. (2.8) and (2.10) it follows that

$$\frac{\partial W}{\partial U_i} = \frac{\partial}{\partial U_i} \int_0^B \nu b \, db, \tag{2.14}$$

where b is a dummy variable of integration. A similar result holds for the scalar potential problem Eq. (2.3). In the following, it will be convenient to regard the reluctivity as a function of the square of the flux density, so it is preferable to rewrite Eq. (2.14) as

$$\frac{\partial W}{\partial U_i} = \frac{1}{2} \frac{\partial}{\partial U_i} \int_0^{B^2} \nu(b^2) \mathrm{d}(b^2), \tag{2.15}$$

which, by the usual chain rule of differentiation, becomes

$$\frac{\partial W}{\partial U_i} = \frac{1}{2} \nu(B^2) \frac{\partial}{\partial U_i} B^2. \tag{2.16}$$

Returning now to Eq. (2.13), which expresses the minimality requirement, let (2.16) be substituted. Since Eq. (2.9) relates the flux density **B** to the vector potential, Eq. (2.13) is readily brought into the matrix form

$$\mathbf{SU} = \mathbf{J}, \tag{2.17}$$

where **U** is the vector of nodal potential values, while **J** is a vector with entries

$$J_k = \int J\alpha_k \, \mathrm{d}\Omega \tag{2.18}$$

and the element coefficient matrix **S** contains the entries

$$S_{ij} = \int \nu \left(\frac{\partial \alpha_i}{\partial x} \frac{\partial \alpha_j}{\partial x} + \frac{\partial \alpha_i}{\partial y} \frac{\partial \alpha_j}{\partial y} \right) \mathrm{d}\Omega. \tag{2.19}$$

Of course, the reluctivity under the integral sign in Eq. (2.19) is not only a position function but also field dependent. Element matrix formation is therefore rather more complicated in the nonlinear than in the corresponding linear case, particularly since the material property may vary locally within any one element.

3.   Solution by simple iteration

In principle, any of the standard methods applicable to nonlinear equations may be employed for solving the simultaneous nonlinear algebraic equations that result from functional discretization. In practice, only a few methods have found wide use. The most common of these are *simple iteration* and *Newton iteration*. The simple iteration process may be considered obsolete in magnetics, but occasionally finds use in other areas where the Newton process may be difficult to formulate. It is therefore useful to deal briefly with the simple iteration procedure, then to consider Newton iteration in some detail.

3.1   *Simple iteration*

A conceptually simple iterative method may be set up as follows. Let a set of reluctivity values be assumed, and let the resulting linear problem be solved. Barring uncommon luck, the potential vector **U** thus calculated will be wrong for the nonlinear problem. But **U** may be used to compute a new set of reluctivities, which can in turn be employed to calculate another approximation to **U**. The process can be begun by assuming initial (zero-field) reluctivity values everywhere, and can be terminated when two successive solutions agree within some prescribed tolerance. A suitable program flow chart is indicated in Fig. 5.1(*a*).

The simple algorithm of Fig. 5.1(*a*) is not always stably convergent, but it can be improved considerably by undercorrecting the reluctivities. That is to say, the reluctivities used during step $k + 1$ are not those calculated from the approximate potential values at step k, but rather a weighted combination of the potential values obtained at step k and step $k - 1$:

$$\nu^{(k+1)} = \nu\left\{ p\mathbf{U}^{(k)} + (1 - p)\mathbf{U}^{(k-1)} \right\}, \tag{3.1}$$

where the superscripts denote iteration steps. Here p is a fixed numerical parameter, $0 < p < 1$. Without undercorrection, the iterative process

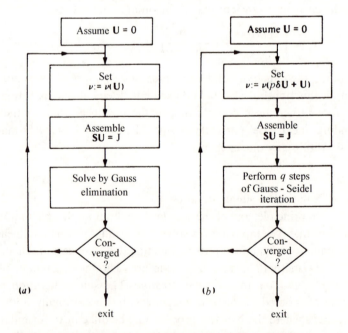

(*a*)   (*b*)

Fig 5.1   (*a*) A simple iterative algorithm. (*b*) Flow chart for a more reliable and possibly faster variant of the same method.

frequently oscillates over a wide range of reluctivity values. No extra memory is required by this change, since approximation $k - 1$ for **U** can be overwritten by approximation $k + 1$; the required storage still amounts to two full vectors.

At each step of the iterative algorithm, a full set of simultaneous linear equations must be solved, of the form (2.17). Solution by standard elimination methods may seem wasteful, for two reasons. First, a fairly good approximation to the solution is known at each step, possibly accurate to several significant figures; but no use is made of this prior information. Second, successive approximate solutions are all calculated to full precision even though the trailing figures are in most cases not significant at all. Rather than solve explicitly, it therefore seems reasonable to employ an iterative method, such as Gauss–Seidel iteration, for the simultaneous linear equations. Such methods can exploit the already existing approximate solution, and can also trade speed for accuracy by restricting the number of Gauss–Seidel iterations performed at each step of the iterative functional minimization.

A refined version of the iteration algorithm is shown in Fig. 5.1(*b*); numerous variants on it can be constructed by varying the number of steps q in the linear iteration between successive recalculations of reluctivity, and by choosing various different iterative methods for the linear equations. As a general rule, this class of methods tends to be memory economic since it is never required to assemble the entire matrix **S**. It is only necessary to form the product **SU** for given **U**, and this requirement can usually be met by accumulating products on an element-by-element basis. Of course, the labour involved in reconstructing **S** at each main iterative step is not reduced thereby, the element matrix must be calculated for every nonlinear element at every step. This is a formidable amount of work for complicated elements. However, in two-dimensional plane problems involving first-order triangular elements, the computations involved are relatively minor.

Consider a triangular element in the x–y plane and assume that all currents and vector potentials are purely z-directed. Let first-order triangular elements be used. Since the potential **A** varies linearly in each element, **B** must have a constant value throughout every element. To be consistent, the reluctivity must then also be constant in each element. Thus Eq. (2.19) becomes identical to its linear counterpart, so that it is very simple to evaluate. The only extra work required is the determination of the proper reluctivity value for each triangle.

Axisymmetric problems involving saturable materials are treated in much the same fashion as the corresponding problems with linear materials. It should be noted once again, however, that the vector and scalar

Laplacian operators in cylindrical coordinates are in fact two different operators; hence the functionals for the vector and scalar potential problems are not identical. The differences, however, are similar to those encountered in linear problems, and will not be treated further here.

3.2   *A lifting magnet*

To illustrate the manner of solving nonlinear problems, consider the simple lifting magnet shown in Fig. 5.2. It will be assumed that the magnet structure extends into the paper relatively far, so that the problem may be considered to be two-dimensional. Of course, the structure is in reality more complicated; not only is it three-dimensional, but it is also unbounded. However, the major interest usually centres on the local fields near the magnet poles. Hence, taking the problem region to be bounded at some relatively great distance, as well as treating it in two dimensions, are reasonable assumptions. In Fig. 5.2, the magnet is therefore encased in a rectangular box, whose sides are assumed to be at zero magnetic vector potential. Only half the problem need be analysed, because the vertical centre-line of Fig. 5.2 represents a symmetry plane.

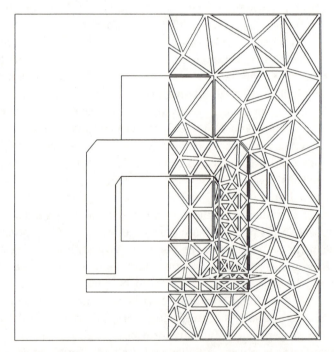

Fig 5.2   Outline drawing of a lifting magnet and the finite element mesh used for its analysis. Symmetry implies that only half the problem region need be modelled and analysed.

As in the various linear problems treated in Chapter 2, the problem region is first modelled by an assembly of finite elements. The choice of element size and shape should take account of the probable stored energy density in any given part of the region; the elements should be smallest where the energy density is high, and where it varies rapidly. However, the nonlinearity of the material prevents energy densities from rising quite so high as would be the case for linear materials, because the iron permeability will fall as the flux density rises. Thus a somewhat more even element subdivision is adequate for the nonlinear case. The right half of Fig. 5.2 shows a triangular decomposition of the problem region. Near the magnet pole, many quite small elements are employed; not all of these are distinctly visible in the figure. The mesh shown only includes some 200 elements, so it is far more crude than the 2000–20 000 elements that commercial CAD systems often use.

A flux plot of a solution is shown in Fig. 5.3. The flux distribution in the iron near the air gap and particularly at the corners of the magnet iron is relatively uniform, showing some, but not very strong, iron saturation. The iron plate being lifted is strongly saturated, being much thinner than the lifting magnet itself. A flux-line count shows that the leakage flux, which avoids the plate entirely, amounts to about 20 per cent. This

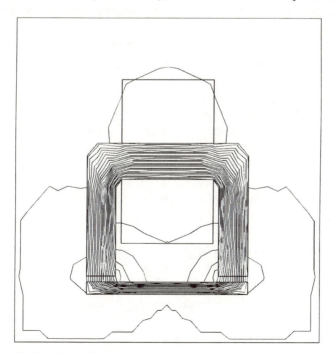

Fig 5.3   Flux distribution in the lifting magnet, with a current density of 1 megaampere/square metre in the coils.

figure is of interest to designers, for it is one of several useful indices of design effectiveness.

While it suffices to analyse and display only half the problem region, designers often draw full flux-line plots, as shown in Fig. 5.3. This drawing was constructed from a solution valid for half the region, so it cannot contain any additional quantitative information in its second half; but the human eye can often see features in a symmetric picture that remain hidden when only half is visible. For example, the full plot shows that homogeneous Neumann boundary conditions are satisfied at the symmetry plane, a point not obvious when only half the picture is plotted. To demonstrate this cover half the figure with a sheet of paper!

4.   Solution by Newton iteration

The simple iteration method given above converges rather slowly and consumes considerable quantities of computer time. Further, there is no known certain method for estimating the amount of reluctivity undercorrection that will guarantee stability without slowing the process unduly. It is therefore preferable to examine the Newton iteration process, which is unconditionally stable. It is very much more rapidly convergent than simple iteration, and surprisingly enough requires little if any additional computer memory. Its convergence rate is theoretically quadratic near the solution point, i.e., the number of significant digits in each iterative trial solution should be approximately double the number of significant digits in the preceding one. Thus if the first step of iteration achieves an approximate solution correct to within a factor of two (correct to one significant binary digit), the second step will produce two correct bits, the third one four, the next eight, and so on. Proceeding further, five or six quadratically convergent steps should yield the full precision available on a conventional 32-bit digital computer. Truly quadratic convergence is in fact only obtained very near the solution. Nevertheless, more than seven or eight Newton steps are not very often required, for precision exceeding the level physically justifiable.

4.1   *Formulation of Newton iteration*

The Newton process may be set up as follows. Let the energy functional (2.7) again be discretized by substituting the discrete approximation of Eq. (2.11). Let $\mathbf{A}$ be the correct solution to be found, while $\mathbf{U}$,

$$\mathbf{U} = \mathbf{A} - \delta\mathbf{U}, \tag{4.1}$$

represents an incorrect but reasonably close estimate of **A**. Let each component of the gradient of $F(\mathbf{U})$ be expanded in a multidimensional Taylor's series near **U**:

$$\frac{\partial F}{\partial U_i} = \frac{\partial F}{\partial U_i}\bigg|_U + \sum_j \frac{\partial^2 F}{\partial U_i \partial U_j}\bigg|_U \delta U_j + \ldots \tag{4.2}$$

But Eq. (2.12) requires that at $\mathbf{U} = \mathbf{A}$ all components of the gradient must vanish. Neglecting Taylor's series terms beyond the second, Eq. (4.2) therefore furnishes a prescription for calculating the deviation of **U** and **A**,

$$\delta \mathbf{U} = -\mathbf{P}^{-1}\mathbf{V}, \tag{4.3}$$

where **P** is the Jacobian matrix of the Newton iteration (the Hessian matrix of F),

$$P_{ij} = \frac{\partial^2 F}{\partial U_i \partial U_j}\bigg|_U, \tag{4.4}$$

while **V** represents the gradient of $F(\mathbf{U})$ at **U**:

$$V_i = \frac{\partial F}{\partial U_i}\bigg|_U. \tag{4.5}$$

An iterative method is now constructed by assuming some set of potentials **U** to start, and calculating its apparent difference from **A**, using Eq. (4.3). This computed difference (usually called the *step* or *displacement*) is then added to the initially estimated potentials. Since third- and higher-order derivatives of the Taylor's series (4.2) were discarded, the displacement as computed by Eq. (4.3) is only approximate. It is exact if $F(\mathbf{U})$ is exactly quadratic, so that no third- or higher-order derivatives exist; indeed this is the case for linear problems, and the iteration converges in one step. For nonlinear problems, the computed displacement is simply added to **U** to form a better estimate, and the entire process is repeated. Thus, the iterative prescription

$$\mathbf{U}^{(k+1)} = \mathbf{U}^{(k)} - (\mathbf{P}^{(k)})^{-1}\mathbf{V}^{(k)} \tag{4.6}$$

produces the sequence of Newton iterates converging to **A**.

Successful implementation of the Newton process (4.6) requires evaluation of the derivatives (4.4) and (4.5). Formal differentiation of $F(\mathbf{U})$ yields

$$\frac{\partial^2 F}{\partial U_i \partial U_j} = \int \frac{\partial^2 W}{\partial U_i \partial U_j}\, d\Omega. \tag{4.7}$$

On differentiating Eq. (2.16) the integrand in turn becomes

$$\frac{\partial^2 W}{\partial U_i \partial U_j} = \frac{\nu}{2} \frac{\partial^2 (B^2)}{\partial U_i \partial U_j} + \frac{1}{2} \frac{\mathrm{d}\nu}{\mathrm{d}B^2} \frac{\partial B^2}{\partial U_i} \frac{\partial B^2}{\partial U_j}. \tag{4.8}$$

In accordance with Eqs. (2.9) and (2.11),

$$B^2 = \sum_m \sum_n (\nabla \alpha_m \cdot \nabla \alpha_n) U_m U_n. \tag{4.9}$$

The derivatives in Eq. (4.8) may thus be explicitly evaluated, yielding

$$\frac{\partial^2 W}{\partial U_i \partial U_j} = \nu (\nabla \alpha_i \cdot \nabla \alpha_j)$$

$$+ 2 \frac{\partial \nu}{\mathrm{d}(B^2)} \sum_m \sum_n (\nabla \alpha_m \cdot \nabla \alpha_i)(\nabla \alpha_n \cdot \nabla \alpha_j) U_m U_n. \tag{4.10}$$

It is interesting to note that the first right-hand term in Eq. (4.10) is exactly what would have been expected for a linear problem. The second and more complicated term only appears in the nonlinear case. Thus somewhat more calculation will be required to construct the coefficient matrix **P** for a nonlinear problem than would have been required for the coefficient matrix **S** of a similar linear problem. The matrix structure, however, is exactly the same in both cases. The nonlinear problem requires a little more computation, but poses the same memory requirements as its linear counterpart.

4.2   *First-order triangular Newton elements*

In electric machine analysis, under the assumption of infinite axial machine length, first-order triangular elements have proved very popular. Their usefulness is due partly to their geometric flexibility which permits easy modelling of complicated electric machine cross-sections, and partly to their relative simplicity in programming.

To develop first-order triangular elements in detail, let the Jacobian matrix of Eq. (4.4) be written out in detail. It takes the somewhat long-winded form

$$P_{ij} = \int \nu (\nabla \alpha_i \cdot \nabla \alpha_j) \, \mathrm{d}\Omega$$

$$+ 2 \int \frac{\mathrm{d}\nu}{\mathrm{d}(B^2)} \sum_m \sum_n (\nabla \alpha_m \cdot \nabla \alpha_i)(\nabla \alpha_n \cdot \nabla \alpha_j) U_m U_n \, \mathrm{d}\Omega. \tag{4.11}$$

On a first-order triangular element, the flux density and therefore the reluctivity are constant everywhere. The first integral in Eq. (4.11) may thus be recognized as being simply the nonlinear matrix **S** of Eq. (2.19). Further, since the first-order element interpolation functions are linear, their gradients must be constants. A little algebra suffices to show that

$$P_{ij} = S_{ij} + \frac{2}{\Delta} \frac{\mathrm{d}\nu}{\mathrm{d}(B^2)} \sum_m \sum_n (\nabla \alpha_m \cdot \nabla \alpha_i)(\nabla \alpha_n \cdot \nabla \alpha_j) U_m U_n,$$

(4.12)

where Δ denotes the element area. To simplify notation, let **S**$'$ denote the matrix **S** obtained for a single element with unity reluctivity,

$$S_{ij} = \nu S'_{ij}$$

(4.13)

and let **E** be the auxiliary vector

$$E_k = \sum_m S'_{km} U_m.$$

(4.14)

With these notational alterations, Eq. (4.12) can easily be brought into the form

$$P_{ij} = \nu S'_{ij} + \frac{2}{\Delta} \frac{\mathrm{d}\nu}{\mathrm{d}(B^2)} E_i E_j.$$

(4.15)

The second term on the right, it may be noted, is merely the outer product of the auxiliary vectors **E**, hence it is computationally very cheap.

The remaining terms in the equations that describe the Newton process with first-order triangular elements are easily added, indeed they were largely worked out for the simple iterative process above. The residual vector (4.5) may be written

$$V_i = \int \frac{\partial W}{\partial U_i} \bigg|_U \mathrm{d}\Omega - J_i,$$

(4.16)

where **J** is the vector defined by Eq. (2.18). Differentiating in accordance with Eq. (2.16), and using the notations defined immediately above,

$$\int \frac{\partial W}{\partial U_i} \mathrm{d}\Omega = \nu \sum_m S'_{im} U_m = \nu E_i.$$

(4.17)

The entire iterative process thus runs as shown in Fig. 5.4.

An operation count shows that to form the Jacobian matrix of Eq. (4.15) and the residual vector of Eqs. (4.16)–(4.17) requires a total of twenty-eight multiplications per finite element, plus whatever arithmetic may be needed to determine the values of reluctivity and its derivative. This computing time is inherently short; furthermore, for many elements

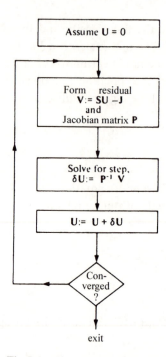

Fig 5.4 The Newton algorithm, stably convergent for magnetic field problems wherever the magnetization characteristics are monotonic.

the total time increases only linearly with element number. On the other hand, even the most efficient sparse-matrix equation-solving algorithms have solution times rising more rapidly than linearly with node number. In consequence, matrix setup time approaches the solution time for rather small numbers of elements (of the order of 50 or fewer) while in practical analysis problems involving many thousands of elements, matrix formation only requires about 1–5 per cent of total computing time. Thus total computing times are sensitive to the choice of equation-solving method for finding the Newton correction, Eq. (4.3).

4.3   *Material models*

It will be evident from the rough timing estimates given above that the computing time required for finding reluctivities from given potential values should be reasonably short, for the reluctivity and its first derivative must be determined for every finite element at every Newton step. Since the quantity that naturally flows from Eq. (4.9) is the square of flux density, not the flux density itself, it appears best to arrange material data so as to yield reluctivity values quickly, given the square of flux density. The classical *B–H* curve is decidedly not the best choice here, and curves relating reluctivity to the squared flux density directly are commonly used. For economy, it is best to avoid modelling the reluctivity curve by transcendental functions, which usually entail a substantial computing cost. For example, an exponential function typically entails a computing cost equivalent to ten multiply-and-add operations. Popular choices are cubic spline curves, or cubic Hermite interpolation polynomials. The difference lies in the nature of approximation: Hermite polynomial curves are continuous, with continuous first derivatives; splines have their second derivatives continuous also.

In the most common program designs, the $\nu = \nu(B^2)$ curve is taken to be piecewise cubic. In other words, the range $0 < B^2 < B_{max}^2$ is subdivided into segments and a separate polynomial approximation is written for each segment. For brevity in the following discussion, set $x = B^2$. Initial modelling of the magnetization curve beings with choosing an ascending sequence of values $0 = x_1 < \ldots < x_N$. A cubic polynomial is fully described by specifying four coefficients. The values of reluctivity at the interval endpoints, ν_{k-1} and ν_k, are invariably taken as two of these; the other two are usually the endpoint values of second derivatives ν_{k-1}'' and ν_k'' (common in spline models) or first derivatives ν_{k-1}' and ν_k'. In each interval $x_{k-1} < x < x_k$ the reluctivity may then be expressed as

$$\nu(x) = C\nu_{k-1} + D\nu_k + C(C^2 - 1)\nu_{k-1}'' + D(D^2 - 1)\nu_k'' \qquad (4.18)$$

and its first derivative can be evaluated as

$$\frac{d\nu}{dx} = \frac{\nu_k - \nu_{k-1}}{x_k - x_{k-1}} - \left\{(3C^2 - 1)\nu_{k-1}'' - 3(D^2 - 1)\nu_k''\right\}\frac{x_k - x_{k-1}}{6}$$

$$(4.19)$$

where C and D measure fractional distance from the ends of the segment (i.e., they are the homogeneous coordinates of the line segment),

$$\left.\begin{aligned}C &= \frac{x - x_{k-1}}{x_k - x_{k-1}}, \\ D &= \frac{x_k - x}{x_k - x_{k-1}}.\end{aligned}\right\} \qquad (4.20)$$

Since a cubic polynomial is a cubic polynomial no matter how it was obtained, the task of finding material property values is much the same whether the tabulated values of ν and ν'' (or ν') were derived from a spline or Hermite fit. There are two basic steps: (1) find the correct segment $x_{k-1} < x < x_k$, then (2) find the property values within the segment using the above two expressions. Approximately 12–15 multiplicative operations are required for the evaluation, the exact number depending on how the program code is structured. The correct segment is usually found by a binary search, which takes at worst $\log_2 N$ search steps for an N-segment curve model.

Piecewise-cubic curves are normally used because any lower order of approximation can conveniently provide continuity of function, but not continuity of its slope. Slope continuity is desirable because it allows the Newton process to converge smoothly, without the missteps which may occur if the first derivative of reluctivity, and hence the components of **P**, undergoes sharp changes. Although simple in principle, the fitting of twice differentiable polynomial models to experimentally obtained curves can be quite a complex task. It need only be done once for any given material, however, and perhaps in a perfect future world the material supplier will also supply the relevant tables of coefficients. Because it does not really fall within the craft of the finite element analyst, it will not be pursued further here.

5.   A magnetic analysis program

The principles outlined in the foregoing section have been incorporated in a demonstration program capable of solving small saturable-material problems. Like others given in this book, this program is primarily intended to be read by people, not computers; it therefore avoids any techniques that might obscure the flow of work. For example, it omits almost all error-checking and stores all matrices in full even where they are known to be symmetric. Major efficiency gains are possible by exploiting such matrix properties.

5.1   *Program structure*

Not surprisingly, the overall structure of program `SATURM` resembles that of the other finite element programs of this book. Input data are acquired from a data file by routine `meshin`, similar to its linear counterpart except in one major respect: every element must be described not only by giving the source density associated with it, but also specifying its material. The underlying assumption is that all meshes will be so constructed as to have material interfaces coincide with element edges; no element will ever straddle two materials. Once the data input phase is complete, the program runs a Newton iteration of precisely the sort already described. In each Newton step, the element matrices are generated by routine `elmatr` and embedded into the global matrix representation of the whole problem by `elembd`. The algebraic equations thus obtained are solved for the Newton step vector.

It is worth noting that the residual computation, Eqs. (4.16)–(4.17), is carried out on an element-by-element basis. Consequently, it is necessary to form the 3×3 matrix **S** for each element but there is no need to assemble these into a global matrix **S**. Thus the memory requirements of the nonlinear problem are almost exactly those of an equivalent linear problem: one large matrix **P** is all that need be stored.

The Newton process is iterative, so the user is informed of progress by printing out, at each Newton step, the L_∞ norms of the potential and displacement vectors:

Newton iteration

newt	damper	stepnorm	potnorm	convgc
1	5.00E-01	7.34E-02	7.34e-02	1.00e + 00
2	9.57E-01	1.08E-02	3.67E-02	2.94E-01
3	9.69E-01	6.62E-03	2.64E-02	2.51E-01
4	9.85E-01	3.45E-03	2.00E-02	1.73E-01
5	9.98E-01	1.16E-03	1.66E-02	6.99E-02
6	1.00E + 00	1.99E-04	1.54E-02	1.29E-02
7	1.00E + 00	2.08E-05	1.52E-02	1.37E-03
8	1.00E + 00	2.68E-07	1.52E-02	1.76E-05
9	1.00E + 00	5.56E-09	1.52E-02	3.65E-07

Here stepnorm and potnorm are the step and potential norms; convgc is the convergence criterion, here taken simply as the ratio of these two norms. (The possibly mysterious-looking quantity damper is a damping ratio, discussed in some detail below.) This example is not unusual, in that the Newton process takes about four or five iterations to establish the right magnitude of solution, and to achieve one or two correct binary digits in the mantissa; thereafter, about three or four more steps achieve all the precision possible with a 32-bit machine. By the eighth or ninth Newton step, no further improvement is possible because the correction steps have become smaller than the roundoff limit uncertainty of the machine.

5.2   *Material models*

The demonstration program includes a reluctivity calculation routine `reluc` based on cubic spline methods. It is wholly conventional: the correct segment is found by a binary search and the function values are interpolated as in Eqs. (4.18)–(4.20). This routine differs from a practical and commercial program only in one major respect: it incorporates one, and only one, saturable material, and that one is firmly built in. Commercial analysis programs would normally allow access to a large number of material descriptions stored in a library file. The interpolation data shown in the data statement of routine RELUC corresponds to a rather pedestrian old-fashioned electrical sheet steel, whose *B–H* curve is shown in Fig. 5.5.

While the *B–H* curve of Fig. 5.5 communicates well with magnetic device engineers, it is an inconvenient data structure for computing. As explained above, a curve of reluctivity vs. squared flux density is more appropriate. Such a curve, corresponding to Fig. 5.5, appears in Fig. 5.6. It is evident at once that the material is essentially linear up to a flux density of about 1.25 or 1.30 tesla, with its reluctivity rising thereafter, steadily from about 2.15 tesla.

The material behaviour shown in Fig. 5.6, with the reluctivity curve rising steadily, cannot be continued indefinitely to higher flux densities. At some high value of flux density, the relative reluctivity must approach 1 asymptotically, for all magnetic materials resemble vacuum at high enough saturation levels. For computational purposes, however, the steadily rising curve is both adequate and reasonable.

It will be evident from Fig. 5.5 that measurements beyond about $B = 2$ tesla will require very high currents and will be difficult to carry out. Despite that, the curve models must cover a much wider range. In the initial steps of any iterative process it is not unusual to encounter local flux densities well beyond the curves shown here. Even though the field solution, when it has converged, may not contain any flux densities higher than 1.5 or 1.8 tesla, it is not at all uncommon to encounter trial flux densities of 25 or 50 tesla in the initial one or two steps of a

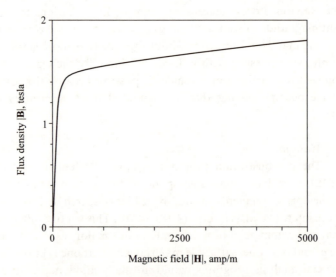

Fig 5.5 *B–H* curve plotted from the data incorporated in the example program. The material is a cheap sheet steel.

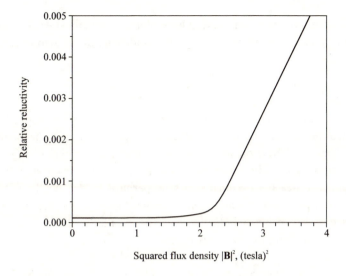

Fig 5.6   Reluctivity curve corresponding to Fig. 5.5. Note the flat region at the left, with constant low reluctivity where the material is essentially linear.

Newton iteration! The material model must provide for such events, by providing a curve that induces convergent behaviour — even though it may not be actually correct for any real material. One simple expedient is that shown in subroutine `reluc`, where the ν–B^2 curve is simply extrapolated linearly for all values beyond the range of available data.

5.3   *Practical Newton iteration*

Magnetic material properties — reluctivities or permeabilities — can easily range over three or four orders of magnitude. Semiconductor devices are even nastier in this respect, with swings of six or eight orders of magnitude not uncommon. The Newton iteration method converges marvellously once it is near the solution, but such very large changes can affect initial convergence drastically. Suppose, for example, than an analysis of the lifting magnet of Fig. 5.2 is begun with a simple starting field, setting $C = 0$ everywhere. Clearly, this implies $D = 0$, so the material reluctivity assigned to every element in the first iteration is the initial reluctivity. With so low a reluctivity, high values of **B** will result from even a small current density in the magnet coil; hence the vector potential will be overestimated for the second iteration. This overestimate of potentials and flux densities leads to high reluctivities (strongly saturated material) in the third iteration, which in turn leads to an underestimate of flux densities, and so on. The solution will converge, but in a very slow and strongly oscillatory fashion. The usual cure for this difficulty is to forbid the Newton program from taking large steps. This is most easily achieved by introducing a *damping factor q*, and modifying the Newton process to read

$$\mathbf{U}^{(k+1)} = \mathbf{U}^{(k)} - q[\mathbf{P}^{(k)}]^{-1}\mathbf{V}^{(k)}. \tag{5.1}$$

In other words, the full step is calculated but not actually applied.

Introduction of a damping factor q is certain to ruin the quadratic convergence of Newton iteration, so practical programs arrange for a nonstationary iterative process in which q is varied from step to step. The general idea is to reduce step size, perhaps so drastically as to half size, if the proposed displacement is large; but to allow the full step to be used if the proposed displacement is small. In the demonstration program, the damping factor is set as

$$q = 1 - \frac{1}{2}\frac{\| \delta\mathbf{U}^{(k)} \|}{\| \mathbf{U}^{(k)} \|} \tag{5.2}$$

with appropriate *ad hoc* measures taken to avoid pathological cases (e.g., problems with exactly null solution). Many other recipes are equally useful.

6.   Analysis of a DC machine

An example may illustrate the scale and manner of analysis now used in various industrial settings. Figure 5.7 shows an outline drawing and a finite element model for a small four-pole direct-current machine. It should be evident on inspection that this model is fairly minimal; very few portions of the machine are modelled by more finite elements than are actually necessary for correctly representing the geometric shapes. Only one pole pitch of the four-pole machine is modelled. The remaining portion of the machine cross-section will invariably be geometrically similar, similar in excitations, and therefore will have a similar magnetic field solution. This model comprises 772 elements and 404 node points. It can be shown on theoretical grounds that no first-order triangulation can contain more elements than twice the number of nodes. In practical analyses, ratios of about 1.8 are commonly encountered. This model is crude, though not absolutely minimal; the machine could be modelled by somewhat fewer elements, but the model would be very crude. In contemporary design practice, meshes of this quality are used for initial rough design studies. They are followed in due course by more refined and more detailed models, perhaps containing nearer 8000 rather than 800 elements.

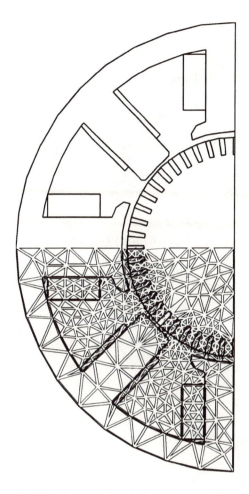

Fig 5.7 Cross-sectional view of a small four-pole direct-current machine and the finite element mesh used in its analysis.

6.1   *Boundary conditions*

A solution based on the model of Fig. 5.7, computed for full rated field current but no other winding currents, is shown in Fig. 5.8(*a*). The boundary conditions imposed to obtain this solution are fairly simple. The outer periphery of the machine may be taken to be a flux line, setting $A = 0$ all along it. When the machine is running unloaded, the rotor current and interpole current are absent, so that the field winding provides the only excitation. In these circumstances, the pole axis and the interpolar axis must be electromagnetic as well as geometric symmetry lines; a flux line must run along the pole axis, while the interpole axis must be orthogonal to all flux lines. The former condition is imposed by fixing the value of vector potential, the latter is a natural condition of the

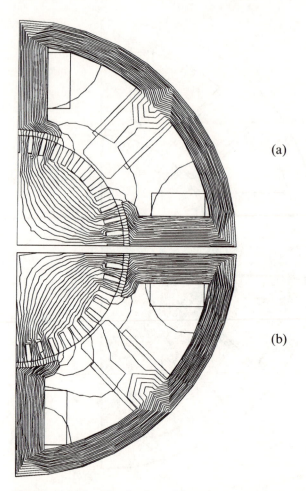

(a)

(b)

Fig 5.8   (*a*) Flux distribution with no load. (*b*) Lightly loaded machine: loading destroys magnetic symmetry lines.

finite element minimization process and does not need to be explicitly enforced.

It should be noted that the model of Fig. 5.7 actually contains a radial symmetry line (the interpolar axis, inclined at 45° to the horizontal) so it should only be necessary to build half the model. The no-load solution of Fig. 5.8(*a*) bears out this expectation: the flux distribution is symmetric, so the geometric symmetry is matched by magnetic symmetry. However, modelling only half a pole pitch of the machine would not suffice to analyse performance under load. This is evident from Fig. 5.8(*b*), which shows the machine running under a moderate load. It is evident at once that the interpole axis is no longer a plane of magnetic symmetry, despite being a plane of geometric symmetry; the geometric shapes are symmetric, but the electrical excitations are not. The resulting flux distribution has no clear symmetry plane within the pole pitch, so a full quadrant of the machine needs to be modelled if its performance under load is to be analysed.

6.2   *Performance evaluation*

The finite element model of Fig. 5.7 permits quite a large variety of machine performance characteristics to be calculated. The most important of these are probably the generated terminal voltages and perhaps the machine torque. These are most easily computed by finding the flux linked by each winding at any given instant.

The flux linking any given contour, such as for example the conductor in a given winding, may be calculated as the line integral of vector potential around that contour,

$$\varphi = \oint \mathbf{A} \cdot \mathbf{ds}. \tag{6.1}$$

Consider a contour such as might be spanned by an armature conductor. It lies in a rotor slot and extends in the machine axial direction, then runs along an essentially peripheral track to another slot, and returns again in a direction parallel to the machine axis. Since end effects are neglected here, only the longitudinally directed portions of the contour contribute to the integral. The flux linkages per unit length can be calculated by adding all the vector potentials at coil sides, being careful to multiply the potential values by -1 wherever the coil winding direction coincides with the negative z-direction, and by $+1$ where the winding direction is collinear with the z-axis. Thus

$$\varphi = \sum_{i=1}^{N} w_i(A_i). \tag{6.2}$$

Here the coefficients w_i take on values $_1$ or -1 according to winding direction, while N is the number of turns in the winding. The index i ranges over all the finite elements belonging to the winding for which the flux linkages are to be calculated, while the bracketed quantity on the right is the average vector potential in element i. If it is assumed that the winding is made of conductors much smaller in diameter than the element size, succesive winding turns will occupy all possible positions in each element, hence the need for averaging. If desired, an approximation to the generated voltage waveshape in the winding may also be found by this method without recomputing the field distribution. It suffices to recalculate the flux linkages, Eq. (6.2), for successive positions of the rotor, or what is equivalent, recalculating the summation (6.2) with all winding direction coefficients shifted one slot.

A quite minimal finite element model such as in Fig. 5.7 is usually sufficient for computing machine performance indices which depend on overall averages or weighted averages of local potential values. Flux linkages, generated voltages, stored energies, indeed nearly all terminal parameters belong to this class. Models of this level of sophistication may be expected to produce terminal voltages with an accuracy of the order of a few per cent at worst. This accuracy should not be surprising, for the classical simple magnetic circuit approach to generated voltage estimation may be expected to come within 5–6% of experiment; and it may itself be regarded as a very crude finite element method! For purposes of calculating terminal parameters, first-order triangular element solutions are quite good enough if only sufficient elements are used to model the machine shape accurately; very fine subdivision throughout the entire machine is usually quite unnecessary. On the other hand, if local details of flux distributions are of interest, locally refined meshes should be employed. For example, if tooth-tip flux density distributions are considered to be of importance, very considerably finer subdivision should be used for the tooth to be investigated and probably for its neighbours to either side as well.

6.3   *Periodicity conditions*

Analysis of the machine running under load is complicated by an additional consideration. Not only is the flux distribution in the loaded machine no longer symmetric about the interpolar axis; it is not symmetric about the polar axis either, so the centreline of a pole no longer coincides with a flux line. This is evident from Fig. 5.7(*b*), where flux lines do not follow the pole centreline; particularly near the machine axis, the asymmetry is quite obvious. It is still valid to hold the outside rim of the machine to be a flux line, $A = 0$, and to insist that this flux line must pass through the machine axis. However, while in the no-load case symmetry requires that this flux line follow the pole axis radially outward, Fig. 5.8(*b*) clearly shows that no radial line is a flux line and no position can be assigned *a priori* to the line corresponding to $A = 0$. It can be confidently asserted, on the other hand, that all phenomena in the machine must be periodic. All the vector potentials along any radial line must therefore repeat identically at an angular displacement of π; furthermore they must be exactly the negatives of the corresponding vector potentials along another radial line displaced by $\pi/2$,

$$A(r, \theta) = A(r, \theta + \pi) = -A\left(r, \theta + \frac{\pi}{2}\right). \tag{6.3}$$

For example, the vector potential values along the vertical pole axis in Fig. 5.8(*b*) must be the negatives of the vector potentials along the horizontal pole axis. This constraint is amenable to exactly the same treatment as all the usual boundary conditions. To keep matters simple, consider the mesh shown in Fig. 5.9(*a*) where potentials at nodes 5 and 6 will be required to equal the negatives of potentials at nodes 1 and 2. The corresponding disjoint element assemblage is shown in Fig. 5.9(*b*). Clearly in the assembled problem there will be only four independent potential values. Thus the vector of twelve disjoint potentials must be related to the connected potentials by the matrix transformation

$$\mathbf{A}_{\text{dis}} = \mathbf{C}\mathbf{A}_{\text{con}}, \tag{6.4}$$

where the connection matrix $\mathbf{C}$ is the transpose of

$$\mathbf{C}^{\mathrm{T}} = \begin{bmatrix} 1 & & & -1 & & & -1 & \\ & 1 & & -1 & 1 & & & \\ & & 1 & & & 1 & 1 & \\ & & & 1 & & & 1 & 1 \end{bmatrix}. \tag{6.5}$$

The periodicity requirement implies a process of matrix assembly which is in principle similar to any other encountered to date. The novelty is that it now involves subtraction of some entries as well as the addition of

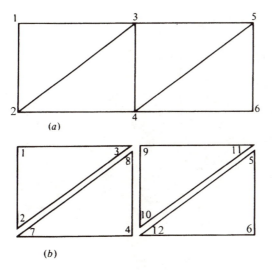

(a)

(b)

Fig 5.9 (*a*) Simple triangular mesh, constrained to have edge-to-edge periodicity. (*b*) Same mesh, disjoint representation.

others, for the connection matrix in this case involves negative as well as positive elements.

It is worth noting that Eq. (6.3) implies unexpectedly great freedom in mesh construction. There is no need to choose the quarter-machine to be bounded by radial lines; *any* radial path, straight or curved, will do equally well so long as it is matched by a similar boundary exactly $\pi/2$ away. This freedom can be used to simplify many otherwise messy geometric problems of mesh construction, such as rotor teeth in electric machines that happen to fall short of symmetry lines. Indeed periodicity conditions can be exploited to simulate relative motion, by joining meshes that represent physical parts of a device. By way of example, Fig. 5.10 shows a mesh equivalent in every respect to that of Fig. 5.7: unusual, perhaps, but perfectly valid and sometimes very convenient.

7.   Anisotropic materials

In many practical magnetic and electric devices, anisotropic materials are deliberately employed, as for example grain-oriented steels in transformer cores. Such materials have one reluctivity value in a preferred direction p, and a different value in the orthogonal direction q. The arguments earlier applied to general fields, Eqs. (2.5)–(2.10), remain valid, and the functional to be minimized remains $F(\mathbf{U})$ of Eq. (2.7), provided the reluctivity is treated as a tensor quantity. The difficulty in practical implementation lies in the fact that the reluctivity, and hence the

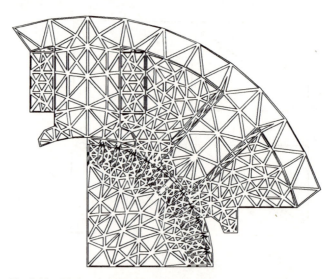

Fig 5.10 Finite element mesh equivalent to Fig. 5.7. Periodicity conditions apply to edge nodes $\pi/2$ removed.

stored energy density W, both depend not only on the magnitude of the
flux density vector but also on its direction. In practice, it may be
expected that curves of reluctivity against flux density are obtainable
for the two principal directions p and q but for few, if any, intermediate
directions. An assumption which usually lies fairly close to the truth (the
error rarely exceeds a few percentage points) is to suppose that loci of
constant reluctivity form ellipses in the B_p–B_q plane. In algebraic terms,
this statement is equivalent to assuming that

$$W = W(B_p^2 + r^2 B_q^2), \tag{7.1}$$

r being the ratio of semiaxes. The argument in (7.1) may be expressed as

$$B_p^2 + e^2 B_q^2 = \left(\frac{\partial A}{\partial q}\right)^2 + r^2 \left(\frac{\partial A}{\partial p}\right) 62. \tag{7.2}$$

The exchange of p and q stems from the rotation of directions in
$\mathbf{B} = \nabla \times \mathbf{A}$. Let it be desired to construct triangular elements. In terms
of the usual triangle area coordinates, Eq. (7.2) may be rewritten as

$$B_p^2 + r^2 B_q^2 = \sum_i \sum_j \frac{\partial A}{\partial \zeta_i} \frac{\partial A}{\partial \zeta_j} \left(\frac{\partial \zeta_i}{\partial q} \frac{\partial \zeta_j}{\partial q} + r^2 \frac{\partial \zeta_i}{\partial p} \frac{\partial \zeta_j}{\partial p}\right). \tag{7.3}$$

Just as in the case of linear elements, the derivatives of the homogeneous
coordinates with respect to the preferred coordinates p, q may be
expressed in terms of the preferred coordinate values at the triangle
vertices. There results

$$B_p^2 + r^2 B_q^2 = \sum_i \sum_j \frac{\partial A}{\partial \zeta_i} \frac{\partial A}{\partial \zeta_j} \{(q_{i-1} - q_{i+1})(q_{j-1} - q_{j+1})$$

$$+ r^2 (p_{i+1} - p_{i-1})(p_{j+1} - p_{j-1})\}. \tag{7.4}$$

To form element matrices, the minimization (2.12)–(2.13) next needs to
be carried out. Again using the chain rule of differentiation,

$$\frac{\partial W}{\partial U_i} = \frac{\partial W}{\partial b^2} \frac{\partial b^2}{\partial U_i}, \qquad b^2 = B_p^2 + r^2 B_q^2. \tag{7.5}$$

Here $\mathbf{U}$ is a trial solution, the correct solution being $\mathbf{U} = \mathbf{A}$. Using interpolatory approximating functions, as in Eq. (2.11), substitution yields

$$\frac{\partial W}{\partial U_k} = \frac{1}{2\Delta^2} \sum_i U_i \sum_j \frac{dW}{db^2} \frac{\partial \alpha_l}{\partial \zeta_i} \frac{\partial \alpha_k}{\partial \zeta_j} \{(q_{i-1} - q_{i+1})(q_{j-1} - q_{j+1})$$

$$+ r^2 (p_{i+1} - p_{i-1})(p_{j+1} - p_{j-1})\}. \tag{7.6}$$

The indices i and j are of course taken to be cyclic modulo 3.

The matrix **S** for the finite element method may be obtained by integration of Eq. (7.5). Analytic integration is only possible in relatively simple cases because W is in general some empirically known and complicated function of position. To construct a Newton method, the chain rule of differentiation may again be applied so as to shift differentiations from the potentials to the squared flux density. There results

$$\frac{\partial^2 W}{\partial U_k \partial U_m} = \frac{\mathrm{d}^2 W}{\mathrm{d}(b^2)^2} \frac{\partial b^2}{\partial U_k} \frac{\partial b^2}{\partial U_m} + \frac{\mathrm{d} W}{\mathrm{d}(b^2)} \frac{\partial^2 (b^2)}{\partial U_k \partial U_m}, \tag{7.7}$$

where the derivatives are evaluated exactly as for (7.5). Once more, analytic differentiation and integration are difficult for high-order elements but quite straightforward for first-order triangles.

Methods of this type are particularly easy to apply to laminated materials, where the energy expressions can be obtained analytically in terms of the properties of the component materials (usually iron and air). This approach has also been used for the study of flux distribution and rotational flux densities in polyphase multi-limbed transformer cores using grain-oriented steel, an inherently anisotropic material. In such cases, the preferred direction p is the direction of rolling of the steel sheet. The semiaxis ratio r above may sometimes rise as high as 10, so that an analysis based on assumed isotropy does not yield good results.

8.   Further reading
Nonlinear problems arise in many areas of continuum electrophysics, and finite element methods have been applied to a large variety of specific cases. Three main areas dominate the work to date: ferromagnetic nonlinear materials, semiconductors and plasmas. While details differ considerably, the basic theory remains essentially similar to that applicable elsewhere in mathematical physics. A simple introduction, at a level similar to this book, will be found in Sabonnadière and Coulomb (1987), a more general, shorter, and perhaps less satisfactory one in Mori (1986). The book by Oden (1972), followed by Oden and Reddy (1976), outlines the mathematical basis underlying the entire area and does so in considerably greater detail. A more recent survey of the techniques as applied to electrical engineering problems will be found in Chari and Silvester (1981).

The ferromagnetic materials problem was the first to be tackled by finite elements. To date, it remains the best researched and best established, with extensive published theory as well as substantial quantities of applicational examples available in the literature. The first papers dealing with this class of problems were those of Chari and Silvester (1971), which defined a methodology for static fields. The methods as set out in this chapter largely follow the approach of Silvester, Cabayan and Browne (1973). Numerous advances since that time have widened the scope of the method, e.g., the treatment of time-varying processes as exemplified by Hanalla and MacDonald (1980), or the use of curvilinear elements by Silvester and Rafinejad (1974).

Semiconductor modelling work follows techniques similar in kind, but perhaps a little more complicated in detail. A good survey of the earlier work in the area will be found in Barnes and Lomax (1977). In this area, as in magnetics, techniques are now sufficiently advanced to allow construction of commercially available, widely usable program packages, such as reported by Buturla, Cottrell, Grossman and Salsburg (1981).

The solution of Maxwell's equations in plasmas has applications in both the atomic energy (plasma containment) and radio propagation areas. Relatively less has been accomplished in that area by finite elements; the article by Tran and Troyon (1978) may provide a sample.

9. Bibliography

Barnes, J.J. and Lomax, R.J. (1977), 'Finite-element methods in semiconductor device simulation', *IEEE Transactions on Electron Devices*, **ED-24**, pp. 1082–9.

Buturla, E.M., Cottrell, P.E., Grossman, B.M. and Salsburg, K.A. (1981), 'Finite-element analysis of semiconductor devices: the FIELDAY program', *IBM Journal of Research and Development*, **25**, pp. 218–31.

Chari, M.V.K. and Silvester, P. (1971), 'Analysis of turboalternator magnetic fields by finite elements', *IEEE Transactions on Power Apparatus and Systems*, **PAS-90**, pp. 454–64.

Chari, M.V.K. and Silvester, P.P. [eds.] (1981), *Finite elements for electrical and magnetic field problems*. Chichester: John Wiley. xii + 219 pp.

Hanalla, A.Y. and MacDonald, D.C. (1980), 'Sudden 3-phase short-circuit characteristics of turbine generators from design data using electromagnetic field calculations', *Proceedings IEE*, **127**, part C, pp. 213–20.

Mori, M. (1986), *The Finite Element Method and its Applications*. New York: Macmillan. xi + 188 pp.

Oden, J.T. (1972), *Finite elements of nonlinear continua*. New York: McGraw-Hill.

Oden, J.T. and Reddy, J.N. (1976), *An introduction to the mathematical theory of finite elements*. New York: Wiley-Interscience. xiii + 429 pp.

Sabonnadièrre, J.-C and Coulomb, J.-L. (1987), *Finite Element Methods in CAD: Electrical and Magnetic Fields*. [S. Salon, transl.] New York: Springer-Verlag. viii + 194 pp.

Silvester, P., Cabayan, H. and Browne, B.T. (1973), 'Efficient techniques for finite-element analysis of electric machines', *IEEE Transactions on Power Apparatus and Systems*, **PAS-92**, pp. 1274–81.

Silvester, P. and Rafinejad, P. (1974), 'Curvilinear finite elements for two-dimensional saturable magnetic fields', *IEEE Transactions on Power Apparatus and Systems*, **PAS-93**, pp. 1861–70.

Tran, T.M. and Troyon, F. (1978), 'Finite element calculation of the anomalous skin effect in a homogeneous unmagnetized cylindrical plasma column', *Computer Physics Communications*, **16**, pp. 51–6.

10.   Programs

Program SATURMAT is listed in the following. It is modest in scale and adheres to much the same structure as other programs in this book. The program listing is followed by a very small data set which may be used for testing.

```
C**************************************************************
C***********                                      ***********
C***********           Saturable material program ***********
C***********                                      ***********
C**************************************************************
C       Copyright (c) 1995  P.P. Silvester and R.L. Ferrari
C**************************************************************
C
C       Solves the  problem in file "ninpt" and  writes results
C       to file "noutp"; file names are assigned interactively.
C
C       The subroutines  that make up this program  communicate
C       via named common blocks.  The variables in commons are:
C
C       Global parameter values needed by all program units
C       common /problm/
C          nodes   = number of nodes used in problem
C          nelmts  = number of elements in model
C          x, y    = nodal coordinates
C          constr  = logical, .true. for fixed potentials
C          potent  = nodal potential array
C          step    = Newton step array
C          nvtx    = list of nodes for each element
C          source  = source density in each element
C          materl  = material code for each element
C
C       Global matrix and right-hand side
C       common /matrix/
C          p       = global P-matrix for whole problem
C          resid   = residual vector (right-hand side)
C          step    = Newton step computed from resid
C          b2max   = largest B-squared in any element
C
C       Temporary working arrays and variables
C       common /workng/
C          pel     = P for one element (working array)
C          sel     = S for one element (working array)
C          tel     = T for one element (working array)
C
C       Predefined problem size parameters (array dimensions):
C          MAXNOD  = maximum number of nodes in problem
C          MAXELM  = maximum number of elements in problem
C
C=============================================================
C       Global declarations -- same in all program segments
C=============================================================
        parameter (MAXNOD = 50, MAXELM = 75, HUGE = 1.E+35)
        logical constr
        common /problm/ nodes, nelmts,
     *   x(MAXNOD), y(MAXNOD), constr(MAXNOD), potent(MAXNOD),
     *   nvtx(3,MAXELM), source(MAXELM), materl(MAXELM)
        common /matrix/ p(MAXNOD,MAXNOD), resid(MAXNOD),
     *                  step(MAXNOD), b2max
        common /workng/ pel(3,3), sel(3,3), tel(3,3), e(3)
```

```
C===============================================================
C
C     Fetch input data from input file
      call meshin
C
C     Newton iteration: set up to start
      Write (*,500)
  500 Format (1x/ 16x, 'Newton iteration' // 1x, 'newt', 4x,
     *  'B max', 3x, 'dampg', 3x, 'stepnorm', 4x, 'potnorm',
     *  4x,' convgc')
      convgc = 1.
      newt = 0
C
C         Do up to 12 steps while 'convgc' exceeds tolerance
    1 if (newt .lt. 12  .and.  convgc .gt. 1.e-5) then
      newt = newt + 1
      b2max = 0.
C         Set global p-matrix and residual to all zeros.
      call matini(p, MAXNOD, MAXNOD)
      call vecini(resid, MAXNOD)
C
C         Assemble global matrix, element by element.
      do 10 i = 1,nelmts
C           Construct element P, S and T matrices
        call elmatr(i)
C           Embed matrices in global P; augment right side:
        call elembd(i)
   10     continue
C
C         Solve the assembled finite element equations
      call eqsolv(p, step, resid, nodes, MAXNOD)
C         Fix up Newton step and check convergence.
C         First find vector norms
      ptnorm = 1.e-30
      stnorm = 1.e-30
      do 20 i = 1,nodes
        ptnorm = max(ptnorm, abs(potent(i)), abs(step(i)))
        stnorm = max(stnorm, abs(step(i)))
   20     continue
      convgc = stnorm/ptnorm
      damper = 1. - 0.5 * min(1.0, convgc**2)
      do 30 i = 1,nodes
        potent(i) = potent(i) + damper * step(i)
   30     continue
      Write (*,510) newt, sqrt(b2max), damper, stnorm,
     *                  ptnorm, convgc
      go to 1
      endif
  510   Format (1x, i3, 2x, 0p2f8.3, 1p5e11.2)
C
C     Print out the resulting potential values
      call output
C
      stop
```

```
      end
C
C****************************************************************
C
      Subroutine elmatr(ie)
C
C****************************************************************
C
C      Constructs element matrices  p and t for a single first-
C      order triangular finite element.  ie = element number.
C
C===============================================================
C      Global declarations -- same in all program segments
C===============================================================
      parameter (MAXNOD = 50, MAXELM = 75, HUGE = 1.E+35)
      logical constr
      common /problm/ nodes, nelmts,
     *   x(MAXNOD), y(MAXNOD), constr(MAXNOD), potent(MAXNOD),
     *   nvtx(3,MAXELM), source(MAXELM), materl(MAXELM)
      common /matrix/ p(MAXNOD,MAXNOD), resid(MAXNOD),
     *                step(MAXNOD), b2max
      common /workng/ pel(3,3), sel(3,3), tel(3,3), e(3)
C===============================================================
C
C      Set up indices for triangle
      i = nvtx(1,ie)
      j = nvtx(2,ie)
      k = nvtx(3,ie)
C
C      Compute element T-matrix
      area = abs((x(j) - x(i)) * (y(k) - y(i)) -
     1           (x(k) - x(i)) * (y(j) - y(i))) / 2.
      do 20 l = 1,3
        do 10 m = 1,3
          tel(l,m) = area / 12.
   10   continue
        tel(l,l) = 2. * tel(l,l)
   20 continue
C
C      Compute element geometric P-matrix
      i1 = 1
      i2 = 2
      i3 = 3
      do 30 l = 1,3
        do 30 m = 1,3
          sel(l,m) = 0.
   30   continue
      do 50 nvrtex = 1,3
        ctng = ((x(j) - x(i)) * (x(k) - x(i)) +
     1          (y(j) - y(i)) * (y(k) - y(i))) / (2. * area)
        ctng2 = ctng / 2.
        sel(i2,i2) = sel(i2,i2) + ctng2
        sel(i2,i3) = sel(i2,i3) - ctng2
        sel(i3,i2) = sel(i3,i2) - ctng2
```

```
            sel(i3,i3) = sel(i3,i3) + ctng2
C               Permute row and column indices once
            i4 = i1
            i1 = i2
            i2 = i3
            i3 = i4
            l = i
            i = j
            j = k
            k = l
   50    continue
C
C     Find squared flux density in element
      bsq = 0.
      do 70 i = 1,3
        e(i) = 0.
        do 60 j = 1,3
          e(i) = e(i) + sel(i,j)*potent(nvtx(j,ie))
   60     continue
        bsq = bsq + e(i) * potent(nvtx(i,ie))/area
   70    continue
      b2max = max(bsq, b2max)
C
C     Find reluctivity and derivative for material not 0
      if (materl(ie) .ne. 0) then
        call reluc(bsq, rnu, dnu)
      else
        rnu = 1.
        dnu = 0.
      endif
C
C     Create Jacobian P, and S matrix for element
      do 90 i = 1,3
        do 80 j = 1,3
          sel(i,j) = rnu * sel(i,j)
          pel(i,j) = sel(i,j) + 2.*dnu*e(i)*e(j)/area
   80     continue
   90    continue
C
      return
      end
C
C***********************************************************
C
      Subroutine reluc(b2, rval, rder)
C
C***********************************************************
C
C     Find the  relative reluctivity and  its derivative  with
C     respect to b2, the squared flux density. The material is
C     a rather ordinary sheet steel.
C
      Dimension b0(11), r0(11), r2(11)
      Data  N, b0, r0, r2        /     11, 0.0000000E+00,
```

```
     *          0.4000000E-01, 0.1600000E+00, 0.3600000E+00,
     *          0.6400000E+00, 0.1000000E+01, 0.1440000E+01,
     *          0.1690000E+01, 0.1960000E+01, 0.2250000E+01,
     *          0.2560000E+01, 0.1090000E-03, 0.1090000E-03,
     *          0.1090000E-03, 0.1100000E-03, 0.1110000E-03,
     *          0.1120000E-03, 0.1200000E-03, 0.1430000E-03,
     *          0.1960000E-03, 0.4170000E-03, 0.1250000E-02,
     *          0.0000000E+00, 0.0000000E+00, 0.0000000E+00,
     *          0.0000000E+00, 0.4393732E-05, 0.0000000E+00,
     *          0.2483705E-03, 0.4209161E-03, 0.4664167E-03,
     *          0.9512425E-02, 0.0000000E+00/
C
      if (b2 .le. b0(N)) then
        kl = 1
        kr = N
    1   if (kr-kl .gt. 1) then
          k = (kr+kl)/2
          if (b0(k) .gt. b2) then
            kr = k
          else
            kl = k
          endif
          go to 1
        endif
        dx = b0(kr)-b0(kl)
        du = (b0(kr)-b2)/dx
        dl = (b2-b0(kl))/dx
        du2 = du**2
        dl2 = dl**2
        rval = du*r0(kl)+dl*r0(kr) + ((du2-1.)*du*r2(kl)
     *         + (dl2-1.)*dl*r2(kr))*dx**2/6.
        rder = -(r0(kl)-r0(kr))/dx - ((3.*du2-1.)*r2(kl)
     *         - (3.*dl2-1.)*r2(kr))*dx/6.
      else
        dx = b0(N)-b0(N-1)
        rder = -(r0(N-1)-r0(N))/dx+(+r2(N-1)+2.*r2(N))*dx/6.
        rval = r0(N)+(b2-b0(N))*rder
      endif
      return
      end
C
C**********************************************************
C
      Subroutine elembd(ie)
C
C**********************************************************
C
C     Embeds single-element p and t matrices currently in sel
C     and tel (in general common block) in the global  matrix
C     p.  Argument ie is the element number.
C
C==========================================================
C     Global declarations -- same in all program segments
C==========================================================
```

```
      parameter (MAXNOD = 50, MAXELM = 75, HUGE = 1.E+35)
      logical constr
      common /problm/ nodes, nelmts,
     *  x(MAXNOD), y(MAXNOD), constr(MAXNOD), potent(MAXNOD),
     *  nvtx(3,MAXELM), source(MAXELM), materl(MAXELM)
      common /matrix/ p(MAXNOD,MAXNOD), resid(MAXNOD),
     *                step(MAXNOD), b2max
      common /workng/ pel(3,3), sel(3,3), tel(3,3), e(3)
C=================================================================
C
C     Run through element p and t matrices (pel, sel and tel),
C     augmenting the global p and the right-hand side as
C     appropriate.
C
      do 30 i = 1,3
        irow = nvtx(i,ie)
C
C         Does row correspond to a fixed potential?
        if (constr(irow)) then
          p(irow,irow) = 1.
          resid(irow) = 0.
        else
C         No, potential is free to vary.  Do all 3 columns.
          do 20 j = 1,3
            icol = nvtx(j,ie)
C
C           Column corresponds to fixed potential? Augment
C               residual only (otherwise also augment matrix)
            if (constr(icol)) then
              resid(irow) = resid(irow) + tel(i,j) *
     *                  source(ie) - sel(i,j) * potent(icol)
            else
              p(irow,icol) = p(irow,icol) + pel(i,j)
              resid(irow) = resid(irow) + tel(i,j) *
     *                  source(ie) - sel(i,j) * potent(icol)
            endif
   20     continue
        endif
   30 continue
C
C     All done -- return to calling program.
      return
      end
C
C****************************************************************
C
      Subroutine meshin
C
C****************************************************************
C
C     Reads input data file in three parts:   nodes, elements,
C     fixed potentials.  Each division  is followed by a line
C     containing nothing but the / character,  which serves as
C     a terminator.
```

```
c
c       nodes:                  node number, x, y.
c       elements:               node numbers, source.
c       potentials:             node number, fixed value.
c
c       All data are echoed as read but little or no checking is
c       done for validity.
c
c================================================================
c       Global declarations -- same in all program segments
c================================================================
        parameter (MAXNOD = 50, MAXELM = 75, HUGE = 1.E+35)
        logical constr
        common /problm/ nodes, nelmts,
     *   x(MAXNOD), y(MAXNOD), constr(MAXNOD), potent(MAXNOD),
     *   nvtx(3,MAXELM), source(MAXELM), materl(MAXELM)
        common /matrix/ p(MAXNOD,MAXNOD), resid(MAXNOD),
     *                  step(MAXNOD), b2max
        common /workng/ pel(3,3), sel(3,3), tel(3,3), e(3)
c================================================================
c
        Dimension nold(3), nnew(3)
c
c       Read in the node list and echo input lines.
c           Start by printing a heading for the node list.
        write (*, 1140)
 1140   format (1x // 8x, 'Input node list' // 3x, 'n', 8x, 'x',
     1            11x, 'y' / 1x)
c
c           Read and echo nodes
        nodes = 0
        xold = HUGE
        yold = HUGE
   20   read (*,*, end=911) xnew, ynew
        if (xnew .ne. xold  .or.  ynew .ne. yold) then
          nodes = nodes + 1
          x(nodes) = xnew
          y(nodes) = ynew
          xold = xnew
          yold = ynew
          write (*, 1105) nodes, x(nodes), y(nodes)
          go to 20
        endif
 1105   format (1x, i3, 2(2x, f10.5))
c
c       Read in the element list and echo all input as received.
c           Print heading to start.
        Write (*, 1160)
 1160   format (1x // 6x, 'Input element list' // 3x, 'i', 5x,
     1            'j', 5x, 'k', 6x, 'Source', 3x, 'Material' / 1x)
c           Read elements in turn.  Echo and count.
        nelmts = 0
        do 25 i = 1,3
   25   nold(i) = 0
```

```
   30 read (*,*, end=911) nnew, srcnew, matnew
      if (nnew(1) .ne. nold(1) .or.
    *    nnew(2) .ne. nold(2) .or.
    *    nnew(3) .ne. nold(3)) then
         nelmts = nelmts + 1
         do 35 i = 1,3
            nvtx(i,nelmts) = nnew(i)
   35       nold(i) = nnew(i)
         source(nelmts) = srcnew
         materl(nelmts) = matnew
         write (*, 1180) nnew, srcnew, matnew
         go to 30
      endif
 1180 format (1x, i3, 2i6, 2x, f10.5, i7)
C
C     Read list of fixed potential values and print.
C           Print header to start.
  120 write (*, 1200)
 1200 format (1x // 4x, 'Input fixed potentials' // 6x,
    1            'node', 12x, 'value' / 1x)
C           Declare all nodes to start off unconstrained.
      do 40 m = 1,nodes
         potent(m) = 0.
         constr(m) = .false.
   40    continue
C           Read and echo input.
      nconst = 0
      iold = 0
   60 read (*,*, end=911) inew, potnew
      if (inew .ne. iold) then
         nconst = nconst + 1
         constr(inew) = .true.
         potent(inew) = potnew
         iold = inew
         write (*, 1210) inew, potnew
         go to 60
      endif
 1210 format (6x, i3, 9x, f10.5)
C
C     Return to calling program.
  900 return
  911 call errexc('MESHIN', 1)
      end
C
C****************************************************************
C
      Subroutine output
C
C****************************************************************
C
C     Prints the results on the standard output device.
C
C================================================================
C     Global declarations -- same in all program segments
```

```
C=================================================================
      parameter (MAXNOD = 50, MAXELM = 75, HUGE = 1.E+35)
      logical constr
      common /problm/ nodes, nelmts,
     *  x(MAXNOD), y(MAXNOD), constr(MAXNOD), potent(MAXNOD),
     *  nvtx(3,MAXELM), source(MAXELM), materl(MAXELM)
      common /matrix/ p(MAXNOD,MAXNOD), resid(MAXNOD),
     *                step(MAXNOD), b2max
      common /workng/ pel(3,3), sel(3,3), tel(3,3), e(3)
C=================================================================
C
C     Print the nodes and the output potential values.
      write (*,1000) (i, x(i), y(i), potent(i),
     1                                        i = 1, nodes)
 1000 format (1x //// 12x, 'Final solution' // 3x, 'i', 8x,
     1          'x', 9x, 'y', 7x, 'potential' // (1x, i3, 2x,
     2          f10.5, F10.5, 3X, f10.5))
C
      return
      end



   0.010      0.020
   0.010      0.000
   0.030      0.000
   0.000      0.030
   0.000      0.000
/
   1  4  5      0.000    1
   1  5  2      0.000    1
   1  2  3      0.100    0
/
   4      0.000
   5      0.000
/
```

6

Finite elements for integral operators

1.   Introduction

The equation for the static scalar potential at distance r from a point charge q in a uniform medium of permittivity ϵ is

$$V = q/(4\pi\epsilon r). \tag{1.1}$$

This relation gives a lead to an important approach, alternative to the partial differential equation method which has been applied so far to the solution of problems in electromagnetics. If it is supposed that a volume charge density ρ within a given volume Ω and a surface charge density σ over a given surface S are the sources of a static electric field, then from Eq. (1.1) the scalar potential due to such sources is

$$V(\mathbf{r}) = \int_{\Omega} \frac{\rho(\mathbf{r}')}{4\pi\epsilon|\mathbf{r} - \mathbf{r}'|} \, d\Omega' + \int_{S} \frac{\sigma(\mathbf{r}')}{4\pi\epsilon|\mathbf{r} - \mathbf{r}'|} \, dS'. \tag{1.2}$$

The notation in Eq. (1.2) refers to an element of charge at Q, position vector $\mathbf{r}'$, producing a field which is evaluated at the point P, position vector $\mathbf{r}$, as shown in Fig. 6.1. The position of the common origin O of $\mathbf{r}$ and $\mathbf{r}'$ is arbitrary and unimportant. The integrations are taken over the volume Ω and surface S supporting the charge distributions and with respect to the dummy variable $\mathbf{r}'$. Usually this is practical for finite volumes.

However, electrostatics problems are not ordinarily posed by specifying charge distributions. Rather, the potentials of conducting electrodes are given, the surface charge $\sigma(\mathbf{r}')$ on such electrodes being regarded as an unknown along with the potential distribution $V(\mathbf{r})$ in the interelectrode space. Thus in the three-dimensional case here, given the function $V_0(\mathbf{r})$ on S and supposing for simplicity that volume charge is absent, Eq. (1.2) can be written as an integral equation for the unknown σ,

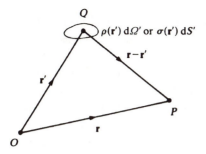

Fig 6.1 Source and field points for integral operations in electrostatics.

$$V_0(\mathbf{r}) = \int_S \frac{\sigma(\mathbf{r}')}{4\pi\epsilon|\mathbf{r} - \mathbf{r}'|}\, \mathrm{d}S, \tag{1.3}$$

the points $\mathbf{r}$ and $\mathbf{r}'$ both lying on S. Normally V_0 will be piecewise constant, representing a system of electrodes charged to known potentials. Then the unknown fields in the interelectrode space may be worked out using Eq. (1.2).

The function

$$G(\mathbf{r}, \mathbf{r}') = \frac{1}{4\pi\epsilon|\mathbf{r} - \mathbf{r}'|}, \tag{1.4}$$

which appears in Eqs. (1.2) and (1.3) and is physically the potential at one place due to unit point charge at another, is described as a Green's function, in this case corresponding to electrostatics in a three-dimensional, unbounded uniform medium. It should be noted that the Green's function (1.4) is singular at $\mathbf{r} = \mathbf{r}'$. Nevertheless, it is evident on physical grounds that the integrals (1.2) and (1.3) do exist and that σ is uniquely determined by Eq. (1.3) — the singularity due to $G(\mathbf{r}, \mathbf{r}')$ is *integrable*. Obviously, special care has to be exercised in numerical schemes which exploit integrals such as these.

The two-dimensional equivalent to Eq. (1.1) gives the potential at any point $P : (x_P, y_P)$ due to a line charge at point $Q : (x_Q, y_Q)$ as

$$V(P) = \frac{1}{2\pi\epsilon}\ln\left(\frac{r_0}{r_{PQ}}\right)q(Q), \tag{1.5}$$

where $q(Q)$ is the charge density per unit length of a line source extending to infinity in both z-directions, and r_{PQ} is the distance from P to Q. The apparently arbitrary distance r_0 is introduced to establish the reference zero of potential. When considering the three-dimensional, point-source case this reference was tacitly chosen to be at infinity. Here, however, it is noted that V in Eq. (1.5) becomes infinite as r_{PQ} increases without limit.

Such nonphysical behaviour corresponds to the impossibility of realising the truly two-dimensional mode. Note that the Green's function now takes the different, logarithmic form embodied in Eq. (1.5).

In general it may be said that any electromagnetics problem described by a partial differential equation, with or without source terms and subject to prescribed boundary conditions, can alternatively be posed in integral equation form. Such formulation requires the finding of an appropriate Green's function. These are unfortunately problem dependent but, nevertheless, a number of cases are discussed in this chapter where the Green's function is known. The advantages to be gained in numerical schemes exploiting integral operators are principally a reduction in the number of variables which need to be handled in matrix operations and the elimination of the necessity to extend models to the infinite boundaries of open systems. The reduction in number of variables simplifies programming, but rarely saves time or memory because each finite element is usually related to every other, so that the resulting matrices are dense.

2.   One-dimensional finite elements for electrostatics

In this section the problem of a simple strip transmission line is examined to introduce the use of integral-equation methods in finite element analysis. Subsequently, this problem will be generalized to cover more complicated shapes and more difficult problems.

2.1   *Stationary functionals for transmission lines*

The finite element approach to solving integral-equation problems follows the same path as already explored in some detail for differential equations. The essential point remains not to tackle the equations head-on, but to find a physical or mathematical quantity whose minimum or stationary point corresponds to the solution of the integral equation. The stationary point is then sought by approximating the problem variables over a set of finite elements. Not surprisingly, the stationary functional applicable to the electrostatics of a transmission line, such as that shown in Fig. 6.2, is closely related to stored energy.

The electrostatically stored energy in any static field is easily calculated as

$$W = \frac{1}{2} \int V(P)q(P) \, d\Omega_P, \tag{2.1}$$

the integral being taken over all portions of space that contain charges; $q(P)$ is the local charge density at the point P where the volume element

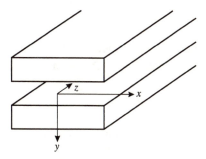

Fig 6.2 Two-conductor transmission line structure.

$d\Omega_P$ is located. In structures consisting of parallel conductors and good insulators, as in Fig. 6.2, all the charges must reside on the conductors as surface charges σ, so it suffices to integrate over all the conductor surfaces:

$$W = \frac{1}{2} \int V(P)\sigma(P)\, ds_P. \tag{2.2}$$

In Fig. 6.2, this integral is a contour integral, taken over the conductor perimeter in the x–y plane.

If the potential $V(P)$ is due solely to the surface charges in the transmission line itself, as is ordinarily the case, then it can be calculated by superposing all the potential contributions of all the charges, as given by Eq. (1.5):

$$V(P) = \int \frac{\ln(r_0/r_{PQ})}{2\pi\epsilon} \sigma(Q)\, ds_Q. \tag{2.3}$$

Combining Eq. (2.2) with Eq. (2.3), the stored energy is given by

$$W = \frac{1}{2} \int \int \frac{\ln(r_0/r_{PQ})}{2\pi\epsilon} \sigma(P)\sigma(Q)\, ds_Q\, ds_P. \tag{2.4}$$

As is well known, the stored energy must reach a constrained extremum at the correct solution. The constraint to be imposed here is that the potential V must assume certain prescribed values on the conductor surfaces. The quantity to be rendered stationary is related to stored energy,

$$\mathfrak{F}(\sigma) = \frac{1}{2} \int \int \frac{\ln(r_0/r_{PQ})}{2\pi\epsilon} \sigma(P)\sigma(Q)\, ds_Q\, ds_P - \int V_0\sigma(Q)\, ds_Q. \tag{2.5}$$

This functional is seen to resemble its counterpart for Poisson's equation: the difference between the stored energy and a source term that embodies the prescribed potential values V_0. To show that this quantity does indeed have a stationary point at the correct solution Σ, write

$$\sigma = \Sigma + \theta h, \tag{2.6}$$

where h is some error function and θ is a parameter. Substituting Eq. (2.6) into (2.5), there results a lengthy expression of the form

$$\mathfrak{F}(\sigma) = \mathfrak{F}(\Sigma) + \theta\delta\mathfrak{F}(\Sigma) + \frac{1}{2}\theta^2\delta^2\mathfrak{F}(\Sigma). \tag{2.7}$$

Taking the terms one at a time, the first variation $\delta\mathfrak{F}$ is given by

$$\delta\mathfrak{F}(\sigma) = \int\int \frac{\ln(r_0/r_{PQ})}{2\pi\epsilon}\sigma(P)h(Q)\,\mathrm{d}s_Q\,\mathrm{d}s_P - \int V_0 h(Q)\,\mathrm{d}s_Q$$
$$= \int h(Q)\left\{\int \frac{\ln(r_0/r_{PQ})}{2\pi\epsilon}\sigma(P)\,\mathrm{d}s_P - V_0\right\}\mathrm{d}s_Q. \tag{2.8}$$

The integrand in braces is easily recognized as echoing Eq. (2.3). It must therefore vanish at the correct solution, $\sigma = \Sigma$, with the consequence that

$$\mathfrak{F}(\sigma) = \mathfrak{F}(\Sigma) + \frac{1}{2}\theta^2\delta^2\mathfrak{F}(\Sigma). \tag{2.9}$$

The quantity $\mathfrak{F}(\sigma)$ is therefore *stationary* at the correct solution. Is it also external? To find out, examine the second variation,

$$\delta^2\mathfrak{F}(\sigma) = \int\int \frac{\ln(r_0/r_{PQ})}{2\pi\epsilon}h(P)h(Q)\,\mathrm{d}s_Q\,\mathrm{d}s_P. \tag{2.10}$$

This is immediately recognized as twice the energy associated with a charge distribution h, so it must always be positive. Consequently, $\mathfrak{F}(\sigma)$ is not merely stationary, but extremal. It therefore provides the right foundation for constructing finite element solutions.

2.2   Finite element modelling of the integral equation

Finding a stationary functional for the physical problem at hand is the first step in finite element methodology. The remainder of the procedure follows exactly the same formal steps as were carried out for problems initially posed in terms of differential equations. The problem region is divided into finite elements and an approximation is written on each element for the quantity to be determined. Finally, the elements are assembled and the energy minimized.

To keep the discussion simple, consider a transmission line of two thin flat strips, as in Fig. 6.3(*a*). Let the two conductors be divided into a set of M finite elements. The elements will of course be one-dimensional, since only surface charges need be dealt with. On each element, the surface charge density is modelled as a linear combination of conventional finite element interpolation functions $\alpha_i(\xi)$. The approximating functions for

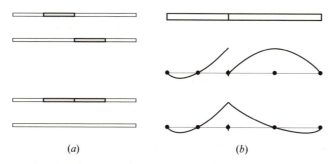

Fig 6.3   Transmission line of two flat metal strips. (*a*) No singularities occur when integrating over disjoint elements, only adjacent elements. (*b*) Adjacent elements: $\alpha_j \alpha_k$ vanishes unless both functions have unity value at element join.

all the elements are then combined into a single numbering scheme, so that the charge density anywhere on the conductors is given by

$$\sigma(P) \simeq \sum_k \sigma_k \alpha_k(\xi_P), \tag{2.11}$$

just exactly as was done in the one-dimensional transmission line example of Chapter 1. Substitution of (2.11) into (2.5) results in an approximation $F(\sigma)$ to the functional, $F(\sigma) \simeq \mathfrak{F}(\sigma)$:

$$F(\sigma) = \frac{1}{2} \sum_j \sum_k \sigma_k \sigma_j \int \int \frac{\ln(r_0/r_{PQ})}{2\pi\epsilon} \alpha_j(P)\alpha_k(Q) \, ds_Q \, ds_P$$
$$- \sum_k \sigma_k \int V_0 \alpha_k(Q) \, ds_Q. \tag{2.12}$$

Minimization with respect to each of the numeric coefficients σ_k,

$$\frac{\partial F}{\partial \sigma_k} = 0, \tag{2.13}$$

then yields a matrix equation in wholly standard form,

$$\mathbf{S}\boldsymbol{\sigma} = \mathbf{V}, \tag{2.14}$$

where

$$S_{jk} = \int \int \frac{\ln(r_0/r_{PQ})}{2\pi\epsilon} \alpha_j(P)\alpha_k(Q) \, ds_Q \, ds_P, \tag{2.15}$$

$$V_k = \int V_0 \alpha_k(Q) \, ds_Q. \tag{2.16}$$

It remains to carry out some manipulative work to put the matrix equations in a computable form, and to solve them.

It is important to note that the integrals here are taken only over the conductor surfaces, whereas in the differential-equation formulation the element integrals cover the dielectric volume. Not only are the finite elements of lower dimensionality — triangles instead of tetrahedra in three-dimensional problems, or line segments instead of triangles in two dimensions — but the number of elements is lowered correspondingly. This permits a large reduction in matrix size. Unfortunately, the appearance of a Green's function under the integral sign in Eq. (2.15) implies that the integral does *not* vanish if points P and Q are on different elements. Consequently, the matrix **S** will be full, with no zero elements.

To appreciate the extent of matrix size reduction when differential-equation boundary-value problems are transformed into integral-equation form, it may be helpful to examine a model problem. Consider, for example, the square region uniformly subdivided into $N \times N$ elements of Fig. 6.4 (*a*), or the cube subdivided into $N \times N \times N$ elements, as shown in Fig. 6.4(*b*). Solving a boundary-value problem in the interior of the square region involves $2N^2$ triangular elements, while its integral-equation formulation involves only $4N$ one-dimensional elements (line segments). The corresponding three-dimensional problem on a cubic region with N-fold division of each edge would involve $5N^3$ tetrahedral elements for the differential-equation formulation, or $12N^2$ triangular elements to model the surface if the problem is recast as an integral equation. The matrix size is thus reduced considerably. Unfortunately, this reduction is not always accompanied by a corresponding decrease in memory and computing time, because sparsity is lost. The exact memory and time requirements depend on the methods used to exploit sparsity in the differential-equation formulation, but a rough estimate is easily found. If the widely popular scheme is adopted of profile storage and solution by Gaussian elimination (as discussed in some detail in Chapter 10 of this book), the square-region problem will require $O(N^3)$ memory and $O(N^4)$

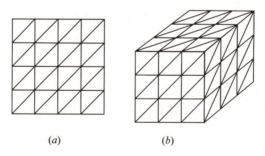

(*a*)   (*b*)

Fig 6.4   Integral and differential formulations of problems. (*a*) Square region divided into $2N^2$ triangles. (*b*) Cube modelled by $5N^3$ tetrahedra.

computing time, while its three-dimensional counterpart takes $O(N^5)$ memory and $O(N^7)$ time. Because the integral-equation problems yield full matrices, there is less choice about methods; any scheme for storing and solving will require $O(N^4)$ memory and $O(N^6)$ time in two dimensions, $O(N^6)$ memory and $O(N^9)$ time in three. The differential-equation formulations clearly have the advantage in both cases, so the integral formulations are rarely used for interior boundary-value problems. Their true value is realized in solving *exterior* problems such as the conductors of Fig. 6.2, where an extremely large region would need to be discretized in order to ensure accuracy in the differential-equation solution.

In general, the element matrices that arise from integral equations will be more difficult to compute than their counterparts in the differential equation case, because most Green's functions contain singularities. On the other hand, there are no differential operators under the integral sign, so the approximating functions need not be continuous. Elements of order *zero*, i.e., approximation functions simply constant on an element, are therefore acceptable — though whether they are particularly attractive is debatable given that such rough approximations may force an increase in matrix size.

2.3   *Finite element matrices*

The finite element matrices described by Eqs. (2.15)–(2.16) are directly analogous to the matrices encountered in the differential-equation treatment of similar physical problems, but there are some mathematical differences.

Consider a pair of elements, as shown in Fig. 6.3(*a*). On each element, let local normalized coordinates ξ be introduced, and let the approximating functions α_j on each element be the familiar Lagrangian interpolation polynomials. Three distinct situations arise: points P and Q of Eq. (2.3) may be on the same element, or on two elements distant from each other, or on abutting elements that just touch at their ends. The integration problems in the three cases are different, and need to be treated differently.

Integration over two disjoint elements (not touching) in Eq. (2.15) means that the distance r_{PQ} is never zero. The integrand is therefore continuous and bounded. It is not always easy to integrate, even if the Green's function is a simple transcendent as is the case here; but numerical integration is certain to do the job. Both the practice and theory of numerical integration are well established for continuous functions. In fact, numerical integration is usually preferable, even if the integral of Eq. (2.15) can be expressed in closed form; double integration of transcendental functions quickly leads to very lengthy expressions whose evaluation actually takes more computer time than numerical integration. Perhaps surprisingly, it can even produce greater error when many terms, often nearly equal but opposite in sign, are added and roundoff error accumulates.

At the other extreme, when P and Q are on the same element, the logarithmic function contains a singularity at $r_{PQ} = 0$, and numerical integration is less reliable. In the present case, these integrals are fortunately expressible in terms of elementary transcendental functions, so they are best evaluated analytically. For the logarithmic function, Table 6.1 gives the results for interpolation polynomials of orders $N = 0$ to 3. Although the task of computing these integrals is formidable, the answers are surprisingly compact.

When P and Q are on neighbouring elements, the singularity at $r_{PQ} = 0$ arises if, and only if, P and Q are located at the endpoints of their respective elements. This situation is shown in Fig. 6.3(*b*). Note, however, that the element endpoints are also interpolation nodes of the usual Lagrangian interpolation functions. On each element, all the inter-

Table 6.1 Logarithmic integrals on Lagrangian polynomials,

$$\int_0^a \int_0^a \alpha_m(x)\alpha_n(\xi) \ln|x - \xi| \, dx \, d\xi = K_1 a^2 \ln a + K_0 a^2$$

N m n	K_1	K_0	N m n	K_1	K_0
0 0 0	1	−3/2	3 0 0	1/64	−23/512
			3 0 1	3/64	−47/512
1 0 0	1/4	−7/16	3 0 2	3/64	−5/512
1 0 1	1/4	−5/16	3 0 3	1/64	−5/512
1 1 0	1/4	−5/16	3 1 0	3/64	−47/512
1 1 1	1/4	−7/16	3 1 1	9/64	−207/512
			3 1 2	9/64	−45/512
2 0 0	1/36	−5/48	3 1 3	3/64	−5/512
2 0 1	1/9	−5/36	3 2 0	3/64	−5/512
2 0 2	1/36	+1/48	3 2 1	9/64	−45/512
2 1 0	1/9	−5/36	3 2 2	9/64	−207/512
2 1 1	4/9	−7/9	3 2 3	3/64	−47/512
2 1 2	1/9	−5/36	3 3 0	1/64	−5/512
2 2 0	1/36	+1/48	3 3 1	3/64	−5/512
2 2 1	1/9	−5/36	3 3 2	3/64	−47/512
2 2 2	1/36	−5/48	3 3 3	1/64	−23/512

polation functions therefore vanish at the endpoint except for one. The logarithmic singularity is weaker than a polynomial zero, so that their product is regular. Thus, a singularity can only occur when two conditions are met: first, that $r_{PQ} = 0$, and second, that both j and k take on the values that correspond to the interpolation functions of value 1 at *both* element ends. For any other pair $\alpha_j(P)\alpha_k(Q)$, one function or the other vanishes, so that the integral is no longer singular. It follows that the neighbouring-element integrals may be computed numerically, except in the one case illustrated at the bottom of Fig. 6.3(*b*), where both interpolation polynomials have unity value at the element join. In the conventional double-subscript notation for the $\alpha_{ij}(\zeta_1, \zeta_2)$, these are α_{0N} on the left and α_{N2} on the right. For example, for $N = 2$, there are three interpolation polynomials on each element; a singularity is only encountered with the combination of α_{02} and α_{20} on the two elements.

Numerical integration by Gaussian quadratures will be dealt with in a subsequent chapter; for the moment, it suffices to mention that it is both stable and accurate for continuous and bounded integrands. The coincident-element integrals are given in Table 6.1. The corresponding expressions for the singular adjacent-element integrals appear in Table 6.2, to third order. It will be noted that the coincident-element integrals all have the same form, but differ in multiplying constants; the constants are given in the table. The adjacent-element integrals also have similar forms, but the element width cannot be factored out this time because the two widths are different. Table 6.2 therefore gives the coefficients of the various logarithmic terms as polynomials.

2.4   *Numerical integration*

Analytic evaluation of the adjacent-element integrals is all very well if the elements lie along the same straight line, as in Fig. 6.3(*b*). On the other hand, if they are inclined with respect to each other, analytic integration becomes extremely complicated. Fortunately, it turns out that numerical integration suffices. Singularities can only occur at the element join, but that is a point where Guassian quadrature never actually evaluate the integrand. Practical analysts therefore often simply ignore the singularity and integrate numerically, despite the fact that no guarantees can be given as to the error level.

For the completely general case of two elements arbitrarily placed and arbitrarily located with respect to each other, the analytic integrals are complicated in the extreme, and only a few special cases have ever been worked out. For the collinear case, however, the results of Table 6.2 can be used to compare with numerical integration. Figure 6.5 shows the results obtained for such a test, with relative element widths $2a = 3b$.

Table 6.2 Adjacent-element integrals of Lagrange polynomials,

$$I_N = \int_a^{a+b} \int_0^a \alpha_{N0}(x)\alpha_{0N}(\xi)\ln|x-\xi|\,\mathrm{d}x\,\mathrm{d}\xi$$

$$I_0 = \frac{(a+b)^2}{2}\ln(a+b) - \frac{a^2}{2}\ln a - \frac{b^2}{2}\ln b - \frac{3ab}{2}$$

$$I_1 = \frac{(a+b)^4}{24ab}\ln(a+b) - \frac{a^2(a+4b)}{24b}\ln a - \frac{b^2(b+4a)}{24a}\ln b$$
$$- \frac{2(a^2+b^2)+25ab}{48}$$

$$I_2 = -\frac{(4a^2-7ab+4b^2)(a+b)^4}{360a^2b^2}\ln(a+b) + \frac{a^3(4a+9b)}{360b^2}\ln a$$
$$+ \frac{b^3(9a+4b)}{360a^2}\ln b + \frac{8(a^4+b^4)+14(a^3b+ab^3)-63a^2b^2}{720ab}$$

$$I_3 = \frac{(99a^4-108a^3b+146a^2b^2-108ab^3+99b^4)(a+b)^4}{13440a^3b^3}\ln(a+b)$$
$$- \frac{a^2(99a^3+288a^2b+308ab^2+224b^3)}{13440b^3}\ln a$$
$$- \frac{b^2(224a^3+308a^2b+99b^3+288ab^2)}{13440a^3}\ln b$$
$$- \frac{396(a^6+b^6)+954(a^5b+ab^5)+788(a^4b^2+a^2b^4)+2875a^3{}^3}{53760a^2b^2}$$

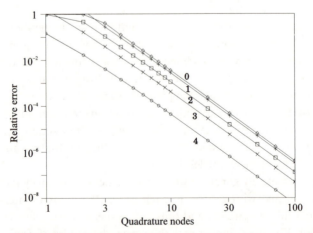

Fig 6.5 Error in numerical integration over adjacent elements of
width ratio 3:2 and polynomial orders 0 through 4, using Gauss–
Legendre quadratures and ignoring the singularity.

Elements of orders 0 and through 4 were integrated numerically, using Gauss–Legendre quadrature formulae from 100 nodes down to 1 node, and the results were compared with the known correct results.

It would appear from Fig. 6.5 that the results obtained from numerical integration are very good, despite the lack of theoretical foundation for this procedure. Note that the errors given there are the error levels in the element integrals, not the computed charge densities; the error sensitivity of charge turns out to be lower still. As a practical matter, therefore, numerical integration is often used even where it is not strictly applicable, such as the analysis of the two-wire transmission line of Fig. 6.2 where the conductors are thick and all four sides of each line must be fully modelled by finite elements.

3.   Green's function for a dielectric slab

One of the drawbacks encountered with integral operator methods is the fact that the Green's functions involved depend very much upon the particular problem being tackled. So far only the very simple situation of a uniform medium has been dealt with. Now a dielectric slab is considered and it is shown how a Green's function may be found for such a geometry by the method of partial images.

3.1   *Dielectric half-space*

The simple case of a half-space, $x < 0$, filled with a homogeneous dielectric of permittivity ϵ_1, is considered. Two-dimensional symmetry is assumed and a line charge q coulombs per metre is supposed lying at distance a from the surface $x = 0$. This charge results in electric flux radiating uniformly from its length. Consider a typical tube of flux ψ as shown in Fig. 6.6. At the dielectric surface, some fraction of the flux $K\psi$ will fail to penetrate, whilst the remainder $(1 - K)\psi$ must continue into the dielectric material. The fraction K and the direction taken by the flux lines which fail to penetrate the interface may be found from the conditions of continuity for electrostatic fields. These require that the normal component of flux density and the tangential component of electric field should be continuous across the interface. From Fig. 6.6 it is seen that the normal flux density requirement gives

$$(1 - K)\psi \sin \theta_1 = \psi \sin \theta_1 - K\psi \sin \theta_2. \tag{3.1}$$

It follows that $\theta_1 = \theta_2$ always; that is, the angles subtended by the incident and returned flux must be equal. On the other hand, continuity of the tangential electric field component is possible only if

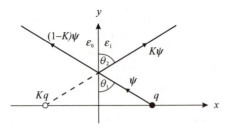

Fig 6.6 Typical flux lines associated with a line charge near a semi-infinite dielectric half-space.

$$\frac{1}{\epsilon_1}(1 - K)\psi \cos \theta_1 = \frac{1}{\epsilon_0}\psi \cos \theta_1 + \frac{1}{\epsilon_0}K\psi \cos \theta_2, \tag{3.2}$$

so that K must have the value

$$K = \frac{\epsilon_0 - \epsilon_1}{\epsilon_0 + \epsilon_1}. \tag{3.3}$$

The analysis above shows that the geometrical relationships governing the behaviour of flux lines near a dielectric interface are analogous to those of optics and that the image coefficient K plays a role corresponding to the optical reflection coefficient. The equality of angles found leads to an apparent existence of images as shown in Fig. 6.6. Flux lines on the right-hand side of the interface appear to be due to two different line sources, the original source q and an image source Kq located in the dielectric region at a distance a behind the interface. Now considering the region $x < 0$, any measurement performed in this dielectric region would indicate a single source of strength $(1 - K)q$ located in the right half-space. Thus, for the simple half-space dielectric slab, the potential of the line source of Fig. 6.6 is as follows:

$x > 0$:

$$V(x,y) = -\frac{q}{4\pi\epsilon_0}\left(\ln\{(x - a)^2 + y^2\} + K \ln\{(x + a)^2 + y^2\}\right). \tag{3.4}$$

$x < 0$:

$$V(x,y) = -\frac{q}{4\pi\epsilon_1}(1 - K)\ln\{(x - a)^2 + y^2\}. \tag{3.5}$$

3.2   *The dielectric slab*

The method used here to describe image formation in the single interface of the dielectric half-space is immediately applicable to a slab of finite thickness. Figure 6.7 shows how multiple images arise, with a dif-

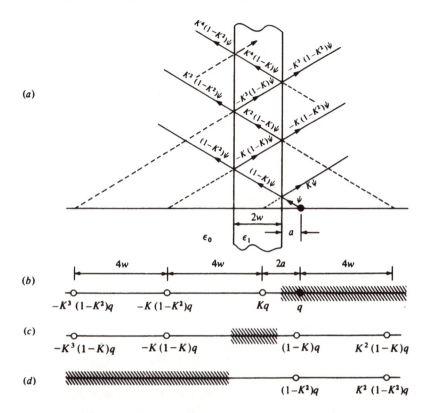

Fig 6.7 (a) Flux lines due to a line charge near a dielectric slab. (b) Image representation valid in the region containing the charge. (c) Image representation valid in the central region. (d) Image representation valid in the left region.

ferent imaging for each of the three regions corresponding to the slab. The strengths and positions of the image charges may readily be checked by the reader using the principles which were established for imaging in the infinite half-space case (Fig. 6.6). Having obtained the image pattern for a dielectric slab, Green's functions, valid in each of the three regions separated by the slab faces, are easily written down from the resulting infinite-series potential function

$$V(\mathbf{r}) = -\sum_{i=1}^{\infty} \frac{1}{2\pi\epsilon} q_i \ln |\mathbf{r} - \mathbf{r}_i|, \tag{3.6}$$

where q_i is the charge associated with an ith image (or the original line charge where applicable) located at position $\mathbf{r}_i$. Note that the appropriate value of ϵ, either ϵ_1 or ϵ_0, corresponds to the location of the observation point $\mathbf{r}$, and not to the image point $\mathbf{r}_i$. A particularly important case arises when both the source point and the observation point lie on one surface of the slab. If the source charge q is located at $Q:(w,0)$ and the observation point P is at (w,y), then

$$V(P) = \frac{q}{4\pi\epsilon} \ln|y^2|$$

$$+ \frac{q}{4\pi\epsilon} \sum_{i=1}^{\infty} K^{2i-1}(K^2 - 1) \ln\left(\frac{y^2}{y^2 + 4w^2 - 16w^2 i + 16w^2 i^2}\right).$$

(3.7)

This case is of practical importance, for it describes various printed circuit boards and semiconductor chips quite accurately.

3.3   *Microstrip transmission-line analysis*

The calculation of parameters associated with microstrip lines can be achieved by the methods outlined here. It can be seen immediately how the basic microstrip problem, that of calculating the capacitance between dielectric-separated ribbons as illustrated in Fig. 6.8, might be tackled. The conducting ribbons are discretized in the same fashion as the line (Fig. 6.3). The matrix relationship, Eq. (2.14), is set up as before, this time with a Green's function corresponding to the infinite series of line charges of Eq. (3.6). The actual line capacitance may be determined as the ratio of charge to voltage by summing the individual charges calculated from solution of Eq. (2.14) for one or the other electrode separately.

4.   A strip transmission-line program

The principles set out relating to the integral equations of two-dimensional electrostatics (also applicable to other two-dimensional

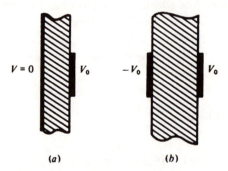

(a)   (b)

Fig 6.8   Microstrip transmission line. (*a*) Ribbon conductor and dielectric sheet with conducting backing. (*b*) Electrically equivalent two-ribbon line.

Poisson equation problems) are illustrated here. A simple finite element program suitable for solving the problem of a strip transmission line located a distance s above a ground plane (see Fig. 6.9) is given. Consider a transmission line formed by placing a flat, thin conducting ribbon of width $2w$ parallel to, and some separating distance s away from, a perfectly conductive ground plane. For purposes of analysis, the ground plane may be replaced by an image of the strip, similar but charged oppositely, a distance $2s$ away. Thus the original strip lies at $y = +s$, $-w \leq x \leq +w$, while its image is located at $y = -s$, $-w \leq x \leq +w$.

At the point $P:(x,s)$ the electric potential $V(x)$ is determined by all charges on the strip and its image. If the strip carries positive charge and its image carries the corresponding negative charge, then

$$V(x) = -\int_0^w \frac{\sigma(\xi)}{2\pi\epsilon} \ln \frac{|x-\xi|}{\sqrt{\{(x-\xi)^2 + (2s)^2\}}} \frac{|x+\xi|}{\sqrt{\{(x+\xi)^2 + (2s)^2\}}} d\xi$$

(4.1)

because the line charge $\sigma(\xi)d\xi$ is accompanied by one exactly like it at $(-\xi, s)$, and opposite but equal line charges at $(\xi, -s)$ and $(-\xi, -s)$.

Coding the problem solution discussed above is a straightforward matter in any appropriate programming language. Fortran, Pascal or C are all suitable for the task, permitting a complete program to be constructed in a few pages of code. An illustrative program STRIPLIN is given at the end of this chapter.

Like all finite element programs in this book, STRIPLIN is organized as a loosely coupled set of subprograms, united by a common pool of data. Many of the subprograms are either common, or very closely related. The general structure of STRIPLIN parallels that of LOSSYLIN, the program for solving leaky direct-current transmission

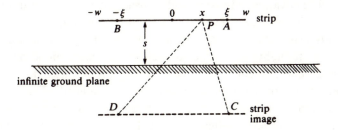

Fig 6.9 Strip transmission line above infinite ground plane. $PA = |x - \xi|$; $PB = |x + \xi|$; $PC = \sqrt{\{(x - \xi)^2 + (2s)^2\}}$; $PD = \sqrt{\{(x + \xi)^2 + (2s)^2\}}$.

line problems; indeed the input routine `inprob` is quite similar to subroutine `prblem` of the transmission line program LOSSYLIN.

Program STRIPLIN is organized very much like LOSSYLIN. It begins with the obvious steps: opening input and output files, initializing of matrices, and reading the input data file. The main mathematical work is done by subroutine `mkcoef`, which creates the potential coefficient matrix that models the integral operator. Its key portion is almost a direct copy of the integral equation (4.1) above:

```
C  Compute entries of coeffs from Green's
   functions
   do 70 j = 1, i
      xmidj = (x(j+1) + x(j))/2.
      dist1 = abs(xmidi - xmidj)
       if (i.eq. j) dist1 = exp(alog(widti) - 1.5)
      dist2 = xmidi + xmidj
      dist3 = sqrt((xmidi + xmidj)**2 +
      (2*space)**2)
      dist4 = sqrt((xmidi - xmidj)**2 +
      (2* space)**2)
      coeffs(i,j) = alog((dist1*dist2)/
      (dist3*dist4))
   *            / (-2. * PI)
      coeffs(j,i) = coeffs(i, j)
   70 continue
```

Here XMIDI and XMIDJ are the midpoint locations of elements *i* and *j*; WIDTI is the width of element *i*.

5.   Integral equations in magnetostatics

The advantages to be gained by using integral-equation formulations of field problems become important in magnetostatics, where numerous problems involve open boundaries. Integral equations have also been thought preferable to differential-equation formulations because a finite element mesh is only needed for surfaces, not volumes, so that mesh construction is considerably simplified. This property has traditionally been thought particularly valuable for magnetic field problems of electric machines, where extremely complicated geometric shapes are apt to be encountered. Its importance has decreased as the art of automatic mesh construction has progressed; but the ability of integral formulations to deal well with exterior regions remains valid.

5.1   *Magnetization integral equation*

The approach considered here is worked in terms of the magnetization vector

$$\mathbf{M} = \frac{\mathbf{B}}{\mu_0} - \mathbf{H}. \tag{5.1}$$

The vector $\mathbf{M}$ is a measure of the internal magnetism present in a material, either as permanent magnetism or induced through the presence of current sources. It may be noted that $\mu = \mu_0$, $\mathbf{B} = \mu_0\mathbf{H}$ corresponds to a nonmagnetic medium and Eq. (5.1) confirms that $\mathbf{M}$ vanishes under these circumstances. Equation (5.1) may be written

$$\mathbf{M} = \chi\mathbf{H}, \tag{5.2}$$

from which it follows that the susceptibility χ is given by

$$\chi = \mu/\mu_0 - 1. \tag{5.3}$$

This parameter may be regarded as a known material property. In many cases χ will be dependent upon $\mathbf{H}$. From Chapter 3, Section 7, the magnetic field intensity in Eq. (5.1) can be separated into two parts

$$\mathbf{H} = \mathbf{H}_c + \mathbf{H}_M, \tag{5.4}$$

the first part $\mathbf{H}_c$ being directly the result of the current sources present, whereas $\mathbf{H}_M$ is regarded as being due to magnetization induced in the material. The source-field $\mathbf{H}_c$ is entirely independent of the material properties and determined by the relationship applying to distributed currents in free space,

$$\mathbf{H}_c(\mathbf{r}) = \frac{1}{4\pi} \int_{\Omega_J} \frac{\mathbf{J}(\mathbf{r}') \times (\mathbf{r} - \mathbf{r}')}{|\mathbf{r} - \mathbf{r}'|^3} \, d\Omega'. \tag{5.5}$$

Equation (5.5) is effectively a statement of the well-known Biot-Savart law and in fact a solution of the equation

$$\nabla \times \mathbf{H}_c = \mathbf{J}. \tag{5.6}$$

The vectors $\mathbf{r}$ and $\mathbf{r}'$ are, respectively, field and source point vectors OP and OQ referred to a common origin O, as depicted in Fig. 6.1. The integration is over the space region Ω_J which contains all of the current sources and the operation is performed with respect to the variable $\mathbf{r}' = (x', y', z')$. Since the Maxwell equation applying to time-invariant situations,

$$\nabla \times \mathbf{H} = \mathbf{J}, \tag{5.7}$$

must hold, it follows that $\mathbf{H}_M$ is irrotational, $\nabla \times \mathbf{H}_M = 0$, so that $\mathbf{H}_M$ may be expressed as the gradient of reduced scalar potential,

$$\mathbf{H}_M = -\nabla \mathscr{P}_c, \tag{5.8}$$

see Chapter 3. From Eq. (5.1) it follows that

$$\mathbf{H}_c + \mathbf{H}_M = \frac{\mathbf{B}}{\mu_0} - \mathbf{M}. \tag{5.9}$$

The divergence of $\mathbf{H}_c$ vanishes since Eq. (5.5) represents a magnetic field in free space. Thus, taking the divergence of Eq. (5.9) reveals that

$$\nabla \cdot \mathbf{H}_M = -\nabla \cdot \mathbf{M}. \tag{5.10}$$

Then from Eqs. (5.8) and (5.10) it is seen that the governing partial differential equation for the reduced scalar potential is the Poisson equation

$$\nabla^2 \mathscr{P}_c = \nabla \cdot \mathbf{M}. \tag{5.11}$$

Equation (5.11) may be compared with Poisson's equation as it appears in electrostatics for a uniform material:

$$\nabla^2 V = -\rho/\epsilon. \tag{5.12}$$

The electrostatics solution of Eq. (5.12) has already been discussed and, for a volume Ω bounded by a surface S, is

$$V(\mathbf{r}) = \int_\Omega \frac{\rho(\mathbf{r}')}{4\pi\epsilon|\mathbf{r} - \mathbf{r}'|} \, d\Omega' + \oint_S \frac{\sigma(\mathbf{r}')}{4\pi\epsilon|\mathbf{r} - \mathbf{r}'|} \, dS'. \tag{5.13}$$

The substitutions $\rho = \nabla \cdot \mathbf{D}$ and $\sigma dS' = -\mathbf{D} \cdot d\mathbf{S}'$ may now be made in Eq. (5.13); the minus sign is attached to $\mathbf{D} \cdot d\mathbf{S}'$ because the surface charge σ must absorb the lines of electric flux emanating from ρ. Thus

$$V(\mathbf{r}) = \int_\Omega \frac{\nabla' \cdot \mathbf{D}(\mathbf{r}')}{4\pi\epsilon|\mathbf{r} - \mathbf{r}'|} \, d\Omega' - \oint_S \frac{\mathbf{D}(\mathbf{r}') \cdot d\mathbf{S}'}{4\pi\epsilon|\mathbf{r} - \mathbf{r}'|}. \tag{5.14}$$

The symbol ∇' denotes the vector operator $(\partial/\partial x', \partial/\partial y', \partial/\partial z')$ acting with respect to the primed coordinate system $\mathbf{r}'$. Thus, observing that $\nabla \cdot \mathbf{M}$ corresponds to $-\rho/\epsilon = -\nabla \cdot \mathbf{D}/\epsilon$ and $\mathbf{M}$ to $-\mathbf{D}/\epsilon$, it is clear that the solution of Eq. (5.11) for the reduced scalar magnetic potential is

$$\mathscr{P}_c(\mathbf{r}) = -\int_{\Omega_M} \frac{\nabla' \cdot \mathbf{M}(\mathbf{r}')}{4\pi|\mathbf{r} - \mathbf{r}'|} \, d\Omega' + \oint_{S_M} \frac{\mathbf{M}(\mathbf{r}') \cdot d\mathbf{S}'}{4\pi|\mathbf{r} - \mathbf{r}'|}, \tag{5.15}$$

where all of the magnetic material is contained within the volume Ω_M enclosed by the surface S_M.

The divergence theorem of vector calculus may be used to transform the second integral of Eq. (5.15) so that

$$\mathscr{P}_c(\mathbf{r}) = -\int_{\Omega_M} \frac{\nabla' \cdot \mathbf{M}(\mathbf{r}')}{4\pi|\mathbf{r} - \mathbf{r}'|} \, d\Omega' + \int_{\Omega_M} \frac{1}{4\pi} \nabla' \cdot \frac{\mathbf{M}(\mathbf{r}')}{|\mathbf{r} - \mathbf{r}'|} \, d\Omega'. \tag{5.16}$$

The second integral in Eq. (5.16) above may be expanded using the vector calculus rule

$$\nabla(g\mathbf{F}) = g\nabla \cdot \mathbf{F} + \nabla g \cdot \mathbf{F} \tag{5.17}$$

so that, observing a cancellation of terms arising after the expansion, there results

$$\mathscr{P}_c(\mathbf{r}) = \frac{1}{4\pi} \int_{\Omega_M} \mathbf{M}(\mathbf{r}') \cdot \nabla' \left(\frac{1}{|\mathbf{r} - \mathbf{r}'|}\right) d\Omega'. \tag{5.18}$$

Equation (5.18) is written more explicitly

$$\mathscr{P}_c(\mathbf{r}) = \frac{1}{4\pi} \int_{\Omega_M} \frac{\mathbf{M}(\mathbf{r}') \cdot (\mathbf{r} - \mathbf{r}')}{|\mathbf{r} - \mathbf{r}'|^3} d\Omega'. \tag{5.19}$$

Since $\mathbf{H}_M$ may be expressed by the gradient of $\mathscr{P}_c$, Eq. (5.19) yields

$$\mathbf{H}_M(\mathbf{r}) = -\frac{1}{4\pi}\nabla \int_{\Omega_M} \frac{\mathbf{M}(\mathbf{r}') \cdot (\mathbf{r} - \mathbf{r}')}{|\mathbf{r} - \mathbf{r}'|^3} d\Omega', \tag{5.20}$$

the gradient operator ∇ here of course being with respect to the unprimed coordinates $\mathbf{r}$. Thence, Eq. (5.2) can be written, with the aid of Eqs. (5.4), (5.5) and (5.20),

$$\mathbf{M}(\mathbf{r}) = \frac{\chi(\mathbf{r})}{4\pi} \left\{ \int_{\Omega_J} \frac{\mathbf{J}(\mathbf{r}') \times (\mathbf{r} - \mathbf{r}')}{|\mathbf{r} - \mathbf{r}'|^3} d\Omega' \right.$$
$$\left. - \nabla \int_{\Omega_M} \frac{\mathbf{M}(\mathbf{r}') \cdot (\mathbf{r} - \mathbf{r}')}{|\mathbf{r} - \mathbf{r}'|^3} d\Omega' \right\}. \tag{5.21}$$

This integral equation determines the unknown vector $\mathbf{M}$, given χ and $\mathbf{J}$ everywhere.

To take into account the nonlinear material properties which commonly occur in magnetics, Eq. (5.21) can be solved iteratively starting from a guessed solution which, given B–H curves for the materials, enables $\chi(\mathbf{r})$ to be estimated initially. This method is fully described by Newman et al. (1972). A number of commercial computer programs are now in existence, based on the formulation of (5.21). These exploit the reduction in dimensionality and the simplification with respect to open boundary conditions which may be obtained using integral methods in comparison with the differential-operator approach.

5.2   *Boundary integral method*

In cases where a magnetic circuit is made up of homogeneous regions, each containing only linear magnetic material (that is, regions of constant and uniform permeability), driven by current excitation regions in air, a further economy may be effected by employing the boundary element method. Only the surface of the magnetic material need be discretized. Consider a configuration of constant-permeability regions driven by source currents located separately within regions Ω_J (Fig. 6.10). The problem will be shown to reduce to that of finding the total scalar potential $\mathscr{P}$ on S_M, the boundary of Ω_M. Subsequently the magnetic field anywhere can be derived from the values of $\mathscr{P}$ on S_M. A finite element implementation of the integral problem so defined is readily set up.

In Ω_M the total **H**-field is expressed by

$$
\begin{aligned}
-\nabla\mathscr{P} &= \mathbf{H}_M + \mathbf{H}_c \\
&= -\nabla\mathscr{P}_c + \mathbf{H}_c
\end{aligned}
\tag{5.22}
$$

Thus

$$
\mathscr{P} = \mathscr{P}_c - \int_{r_0}^{r} \mathbf{H}_c \cdot d\mathbf{l}
\tag{5.23}
$$

where $\mathbf{r}_0$ is some convenient point chosen to define the zero of both potentials $\mathscr{P}$ and $\mathscr{P}_c$. Using the notation $R = |\mathbf{r} - \mathbf{r}'|$, $\mathscr{P}_c$ may be expressed in the standard fashion from Eq. (5.18)

$$
\mathscr{P}_c = \frac{1}{4\pi} \int_{\Omega_M} \mathbf{M}(\mathbf{r}') \cdot \nabla'\left(\frac{1}{R}\right) d\Omega'.
\tag{5.24}
$$

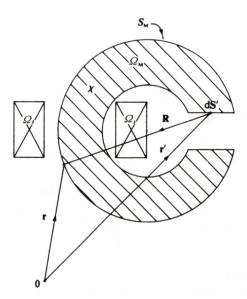

Fig 6.10 A two-dimensional magnetic problem for solution by the boundary integral method.

But

$$\mathbf{M}(\mathbf{r}) = \chi\mathbf{H}(\mathbf{r}) = -\chi\nabla\mathscr{P}(\mathbf{r}) \tag{5.25}$$

where here, since regions of constant permeability only are being considered, χ is a known constant. Thus from Eqs. (5.23)–(5.25),

$$\mathscr{P}(\mathbf{r}) = -\frac{1}{4\pi}\int_{\Omega_M} \chi\nabla\mathscr{P}(\mathbf{r})\cdot\nabla'\left(\frac{1}{R}\right) d\Omega' - \int_{\mathbf{r}_0}^{\mathbf{r}} \mathbf{H}_c\cdot d\mathbf{l}. \tag{5.26}$$

Equation (5.26) can be transformed as follows: consider the vector

$$\mathbf{F} = \chi\mathscr{P}\nabla\left(\frac{1}{R}\right) \tag{5.27}$$

and form

$$\nabla\cdot\mathbf{F} = \chi\nabla\mathscr{P}\cdot\nabla\left(\frac{1}{R}\right) + \chi\mathscr{P}\nabla^2\left(\frac{1}{R}\right) \tag{5.28}$$

whilst it may be noted that $\nabla^2(1/R)$ vanishes. (Note, however, that the expression above becomes more complicated if χ is not constant.) Writing the divergence theorem as

$$\int_{\Omega_M} \nabla'\cdot\mathbf{F}(\mathbf{r}')\,d\Omega' = \oint_{S_M} \mathbf{F}(\mathbf{r}')\cdot d\mathbf{S}' \tag{5.29}$$

it is seen that Eq. (5.26) may be rewritten

$$\mathscr{P}(\mathbf{r}) = -\frac{1}{4\pi}\oint_{S_M} \chi\mathscr{P}(\mathbf{r}')\nabla'\left(\frac{1}{R}\right)\cdot d\mathbf{S}' - \int_{\mathbf{r}_0}^{\mathbf{r}} \mathbf{H}_c\cdot d\mathbf{l}. \tag{5.30}$$

Equation (5.30) is valid for $\mathbf{r}$ anywhere. However, if $\mathbf{r}$ and $\mathbf{r}_0$ are restricted to points lying upon S_M, the integral equation is expressed entirely with respect to an unknown $\mathscr{P}$ on the boundary S_M. The input data requirements are:

■ $\mathbf{H}_c$ on S_M (easily evaluated from the source-currents $\mathbf{J}_c$ using the Biot–Savart law);
■ χ inside S_M;
■ the geometric size and shape of S_M.

Equation (5.30) can be solved by finite element methods similar to those described for two-dimensional electrostatic problems. A three-dimensional implementation is also possible. Once $\mathscr{P}$ is determined upon S_M, Eq. (5.30) may then be used in its more general role to specify $\mathscr{P}$ (and hence $\mathbf{H}$) anywhere.

6.   Integral operators in antenna theory

Two numerical techniques have traditionally dominated the field of antenna analysis: collocation (also known as point-matching) and projectively-formulated finite elements. Both have come to be known as 'the method of moments' ever since in 1968 Harrington attempted to demonstrate that projection and collocation are the same thing (Harrington, 1993). The attempt was unsuccessful, for the reasons alluded to in Chapter 3 of this book; but enough antenna engineers were misled to confuse the terminology of this field considerably, and perhaps permanently. Only the projective formulation will be dealt with here, for the collocation method cannot be considered a finite element method in any sense.

This section deals with the simplest practical problem in antenna theory, that of finding the current distribution on a single-wire antenna; the numerical treatment of more complex problems is similar, though the electromagnetic formulation of the problem may be more demanding. Readers seriously interested in the electromagnetics of antennas should consult one of the specialist works on the subject, for instance the text by Balanis (1982).

6.1   *Pocklington's equation*

This integral equation is one of the most useful in the numerical analysis of linear antennas and, remarkably, dates from a time when electromagnetics theory was still in its infancy. Figure 6.11 shows a straight cylindrical conducting wire whose radius a is much smaller than its length l, which in turn is comparable to the free-space wavelength λ. The very reasonable assumption is made that the current is entirely axially directed with azimuthal symmetry. Further, the wire conductivity is assumed high enough (or the frequency high enough) for all currents to flow in a thin tubular sheet on the cylindrical surface. Under this assumption, the current density **J** possesses only a longitudinal component J_z, so that the vector potential **A** due to current on the cylinder must be entirely z-directed also. Using the retarded potential solution of Maxwell's equations indicated in Chapter 3, and referring to Fig. 6.11, this sole component A_z is given at some observation point P by an integral over the volume occupied by all currents,

$$A_z(r_P, \theta_P, z_P) = \mu_0 \int \frac{\exp(jkR_{PQ})}{4\pi R_{PQ}} J_z(z_Q) \, d\Omega_Q. \qquad (6.1)$$

All variables may be taken as sinusoidal in time, hence represented in phasor form with angular frequency ω. In Eq. (6.1), $k = \omega\sqrt{\mu_0\epsilon_0}$ is the conventional wave number, while R_{PQ} represents the distance of the

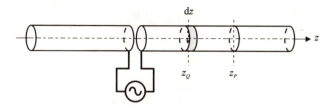

Fig 6.11   Cylindrical antenna.

observation point r_P, θ_P, z_P) from the source point (r_Q, θ_Q, z_Q) where the element of integration volume $d\Omega_Q$ is located,

$$R_{PQ} = \sqrt{r_P^2 - r_Q^2 - 2r_P r_Q \cos(\theta_P - \theta_Q) + (z_P - z_Q)^2}. \qquad (6.2)$$

Cylindrical coordinates are the natural choice here, for all fields in this problem clearly enjoy axial symmetry. The integral equation that describes currents on the antenna is derived starting from the electric field **E** which is, as always, given by

$$\mathbf{E} = -\nabla V - j\omega \mathbf{A}. \qquad (6.3)$$

To make the vector potential unique, the Lorentz gauge is commonly chosen,

$$\nabla \cdot \mathbf{A} = -j\omega\mu_0\epsilon_0 V. \qquad (6.4)$$

Since **A** only possesses one component, Eqs. (6.3) and (6.4) may be combined to yield

$$\mathbf{E} = \frac{1}{j\omega\mu_0\epsilon_0} \left\{ \nabla\left(\frac{\partial A_z}{\partial z_P}\right) + \omega^2\mu_0\epsilon_0 A_z \right\}. \qquad (6.5)$$

On the wire axis, the electric field must vanish everywhere, except where a gap has been cut to accommodate the driving source; there, it has some value $1_z E_a h(z_P)$. Thus, Eq. (6.5) implies

$$\frac{\partial^2 A_z}{\partial z_P^2} + k^2 A_z = j\omega\mu_0\epsilon_0 E_a h(z_P). \qquad (6.6)$$

Combining this with Eq. (6.1), there results

$$-\frac{j}{\omega\epsilon_0}\left(\frac{\partial^2}{\partial z_P^2} + k^2\right) \int \frac{\exp(-jkR_{PQ})}{4\pi R_{PQ}} J_z(z_Q)\, d\Omega_Q = E_a h(z_P). \qquad (6.7)$$

To simplify this expression, the volume integral of Eq. (6.1) may be treated as an iterated integral, $d\Omega_Q = r_Q\, d\theta_Q\, dr_Q\, dz_Q$. Along the wire axis, $r_P = 0$, Eq. (6.1) may be written as

$$A_z(z_P) = \mu_0 \int_0^{2\pi} \int_{a-\delta/2}^{a+\delta/2} \int_0^l \frac{\exp(-jkR_{PQ})}{4\pi R_{PQ}} J_z(z_Q) \, dz_Q r_Q \, dr_Q \, d\theta_Q,$$

$$(6.8)$$

where δ is the (very small) thickness of the layer of current flowing on the wire surface, and l is the length of the wire. Integration in the azimuthal direction is easy, since the integrand is not a function of θ_Q. To integrate over the very small distance δ in the radial direction, the Green's function under the integral sign is expanded in a Taylor's series about $r_Q = a$. Because r_Q is very nearly equal to a everywhere, only the dominant term is of interest. After expansion, integration, and taking the limit as δ grows vanishingly small, there finally results

$$A_z(z_P) = \mu_0 a \int_0^l \frac{\exp\{-jk\sqrt{a^2 + (z_Q - z_P)^2}\}}{2\sqrt{a^2 + (z_Q - z_P)^2}} \delta J_z(z_Q) \, dz_Q. \qquad (6.9)$$

The current density J_z is invariant with azimuthal position θ_Q, so it is often thought best to write this equation in terms of the total longitudinal current $I(z_Q)$,

$$I(z_Q) = 2\pi a \delta J_z(z_Q). \qquad (6.10)$$

Substituting this expression, and writing $1/\omega\epsilon_0$ in terms of the wavelength λ and the intrinsic impedance η_0 of free space as

$$\frac{1}{\omega\epsilon_0} = \frac{1}{\omega\sqrt{\mu_0\epsilon_0}}\sqrt{\frac{\mu_0}{\epsilon_0}} = \frac{\eta_0}{k} = \frac{\eta_o\lambda}{2\pi}, \qquad (6.11)$$

Eq. (6.7) takes the form

$$\int_0^l P(z_Q - z_P) I(z_Q) \, dz_Q = j\frac{E_a}{\eta_0} h(z_P), \qquad (6.12)$$

where

$$P(z_Q - z_P) = \frac{\lambda}{2\pi}\left(\frac{\partial^2}{\partial z_P^2} + k^2\right)\left(\frac{\exp\{-jk\sqrt{a^2 + (z_Q - z_P)^2}\}}{4\pi\sqrt{a^2 + (z_Q - z_P)^2}}\right).$$

$$(6.13)$$

Equation (6.12) is known as *Pocklington's* equation. It forms the key element in the formulation of almost any wire antenna problem, because the Pocklington kernel $P(z_P - z_Q)$ remains a very good approximation even for bent wires or arrays of multiple wires, where rotational symmetry no longer strictly applies.

For computation, it is most convenient to recast the Pocklington equation, this time entirely in dimensionless quantities. Introduce the normalized variables

$$
\left.
\begin{aligned}
\bar{z}_P &= z_P/\lambda, \\
\bar{z}_Q &= z_Q/\lambda,
\end{aligned}
\right\} \tag{6.14}
$$

$$
\bar{R} = \sqrt{(a/\lambda)^2 + (\bar{z}_P - \bar{z}_Q)^2}. \tag{6.15}
$$

In terms of these, the Pockington equation reads

$$
\frac{1}{\lambda}\int_0^{\bar{l}} P(\bar{z}_Q - \bar{z}_P)I(\bar{z}_Q)\,\mathrm{d}\bar{z}_Q = \mathrm{j}\frac{E_a}{\eta_0}h(z_P), \tag{6.16}
$$

where the kernel is now, in full detail with the differentiation explicitly carried out,

$$
\begin{aligned}
P(\bar{z}_Q - \bar{z}_P) = \frac{\exp(-\mathrm{j}2\pi\bar{R})}{8\pi^2\bar{R}^5} \Big((3 - 4\pi^2\bar{R}^2)(\bar{z}_P - \bar{z}_Q)^2 \\
- (1 - 4\pi^2\bar{R}^2)\bar{R}^2 + \mathrm{j}2\pi\bar{R}\big\{ 3(\bar{z}_P - \bar{z}_Q)^2 - \bar{R}^2 \big\} \Big),
\end{aligned}
\tag{6.17}
$$

and the upper limit of integration $\bar{l} = l\lambda$ is normalized in the same way as all other dimensional variables.

6.2 *Projective formulation*

The principle of projective solution here is exactly the same as in the electrostatics problem considered earlier. The wire that constitutes the antenna is subdivided into elements and the unknown $I(z_P)$ is approximated on each element by suitable functions, such as the one-dimensional Lagrangian interpolation polynomials,

$$
I(z_Q) = \sum_i I_i \alpha_i(z_Q). \tag{6.18}
$$

Under this supposition, the Pocklington equation reads

$$
\sum_i \left\{ \int_0^{\bar{l}} P(\bar{z}_Q - \bar{z}_P)\alpha_i(\bar{z}_Q)\,\mathrm{d}\bar{z}_Q \right\} I_i = \mathrm{j}\lambda\frac{E_a}{\eta_0}h(z_P). \tag{6.19}
$$

A projective approximation is now obtained in the usual fashion. Since the approximating functions $\alpha_j(\bar{z}_P)$ are square integrable, a valid inner product definition is

$$\langle \alpha_i, \alpha_j \rangle = \int_0^l \alpha_i(z_P)\alpha_j(z_P)\,\mathrm{d}z_P = \lambda \int_0^{\bar{l}} \alpha_i(\bar{z}_P)\alpha_j(\bar{z}_P)\,\mathrm{d}\bar{z}_P. \qquad (6.20)$$

Taking inner products with $\alpha_j(\bar{z}_P)$ and integrating, (6.19) becomes

$$\sum_i \left\{ \int_0^{\bar{l}} \int_0^{\bar{l}} P(\bar{z}_Q - \bar{z}_P)\alpha_i(\bar{z}_Q)\,\mathrm{d}\bar{z}_Q \alpha_j(\bar{z}_P)\,\mathrm{d}\bar{z}_P \right\} I_i =$$
$$\mathrm{j}\frac{E_a}{\eta_0} \int_0^l h(z_P)\alpha_j(z_P)\,\mathrm{d}z_P. \qquad (6.21)$$

For a wire antenna with a feed gap located at $z_P = z_{Pf}$, it is usual to take the gap to be very small and the electric field in the gap very intense, i.e., to set $h(z_P) = \delta(z_P - z_{Pf})$. The right-hand side of (6.21) is then easy to evaluate,

$$\mathrm{j}\frac{E_a}{\eta_0} \int_0^l h(z_P)\alpha_j(z_P)\,\mathrm{d}z_P = \mathrm{j}\frac{E_a}{\eta_0} \int_0^l \delta(z_P - z_{Pf})\alpha_j(z_P)\,\mathrm{d}z_P$$
$$= \mathrm{j}\frac{V_a}{\eta_0}\alpha_j(z_{Pf}). \qquad (6.22)$$

It should perhaps be noted that smallness of the feed gap is less an assumption than a close description of reality. After all, the gap cannot be large unless there are feed-wires attached, but feed-wires of any significant length surely must be modelled as a part of the antenna itself. With the right side thus evaluated, the final approximate form of the Pocklington equation reads

$$\sum_i \left\{ \int_0^{\bar{l}} \int_0^{\bar{l}} P(\bar{z}_Q - \bar{z}_P)\alpha_i(\bar{z}_Q)\,\mathrm{d}\bar{z}_Q \alpha_j(\bar{z}_P)\,\mathrm{d}\bar{z}_P \right\} I_i = \mathrm{j}\frac{V_a}{\eta_0}\alpha_j(\bar{z}_{Pf}). \qquad (6.23)$$

Although the integration limits are here shown as covering the entire antenna, each polynomial α_i is nonzero only over a single element. Consequently, each integration need be performed only over one finite element at a time. If w_i denotes the normalized width of the element on which α_i differs from zero, and w_j is the width of the element over which $\alpha_j \neq 0$, then (6.23) becomes

$$\sum_i \left\{ \int_0^{w_i} \int_0^{w_j} P(\bar{z}_Q - \bar{z}_P)\alpha_j(\bar{z}_P)\,\mathrm{d}\bar{z}_P \alpha_i(\bar{z}_Q)\,\mathrm{d}\bar{z}_Q \right\} I_i = \mathrm{j}\frac{V_a}{\eta_0}\alpha_j(\bar{z}_{Pf}). \qquad (6.24)$$

Numerical integration may therefore be used for one finite element pair at a time.

The Pocklington integral operator, like most integral operators of electromagnetics, produces distant responses from local causes. Consequently, the integrations cover the entire antenna and the resulting matrix is full. For this reason, a global numbering of the approximating functions $\alpha_i(\bar{z}_P)$ and their unknown coefficients I_i is often adopted from the outset. This contrasts with finite element methods based on differential operators, where effects are local, a great deal of the computational arithmetic is best expressed locally, and the usual procedure is to work in local numberings and local coordinates first, then to tie together the elements by constraint transformations. A second reason for immediate adoption of global numbering is the lack of continuity requirements: there are none, so no constraint transformation is required to enforce continuity. This permits use even of elements of order zero, as already demonstrated in the strip-line problem. To be fair, it should be pointed out that many workers in the field do impose continuity of current at element joints when higher-order elements are used. This rarely does harm to the solution, which should be nearly continuous anyway if it is nearly correct; but it can reduce the number of variables substantially, an important consideration when working with full matrices where the solution time is of order $O(N^3)$.

Various authors have discussed at length whether the Pocklington functional obtained by projection is extremal, or merely stationary, at the correct solution; after all, the quantities involved are complex but the inner product definition does not include complex conjugation. In the finite element formulation, this question may be dismissed because (6.21) results from taking an inner product with a purely real function $\alpha_j(z_P)$, which is anyway its own conjugate.

6.3   *Finite element matrices*

The finite element methodology here follows much the same principles as in the strip-line problem, with the difference that elements of arbitrary order are developed while only zero-order elements were considered for the strip line. The wire antenna is subdivided into elements, taking N_1 elements to the left of the feed gap (i.e., in the interval $0 \leq z \leq z_f$) and N_r elements to its right. Let ζ_P and ζ_Q represent local coordinates in two elements,

$$\left. \begin{aligned} \zeta_P &= \frac{1}{w_j}\left(\bar{z}_P - \bar{z}_P|_{Lj}\right) \\ \zeta_Q &= \frac{1}{w_i}\left(\bar{Q} - \bar{z}_Q|_{Li}\right), \end{aligned} \right\} \quad (6.25)$$

where Lj and Li denote the left-hand ends of the elements on which α_j and α_i are nonzero, respectively. The typical matrix entry C_{ij} may then be written

$$C_{ij} = w_i w_j \int_0^1 \int_0^1 P(\bar{z}_Q - \bar{z}_P)\alpha_j(\zeta_P)\,d\zeta_P \alpha_i(\zeta_Q)\,d\zeta_Q. \tag{6.26}$$

This double integral presents no difficulty, so long as the coordinate values ζ_Q and ζ_P lie in different elements. It is therefore readily handled by almost any numerical integration method, such as Gaussian quadrature[1]. A suitable order of quadrature may be chosen on the following basis. An N-point Gaussian formula is capable of integrating a polynomial of degree $2N - 1$ without error. Equation (6.26) requires integration over a rectangle in the $\bar{z}_Q$–$\bar{z}_P$ plane, where a Gaussian formula may be applied in each coordinate direction. If elements of order M are used, then N-point quadrature is sufficient wherever the Pocklington kernel P can reasonably be approximated by a polynomial of degree $2N - M - 1$ in each direction. This is not a stringent requirement for matrix entries that relate approximating functions on different elements. For example, 7-point quadrature applied to third-order elements is adequate if the Pocklington kernel can be well approximated by a polynomial of order 10 in each direction. This is very likely to prove an excellent approximation for well-separated elements, indeed even 5-point quadrature may prove sufficient.

6.4   *Quasi-singular terms*

When both integrations in Eq. (6.26) cover the same element, Gaussian quadrature is not entirely satisfactory. Unlike the strip-line problem, the Pocklington equation does not actually involve any singular integrals. However, from a numerical analyst's point of view all integrals that include $\bar{z}_Q = \bar{z}_P$ are almost as difficult as integrals containing true singularities. The imaginary part of the Pocklington kernel function is quite smooth and bounded, but its real part, though bounded, is not at all well behaved. Near $\bar{z}_Q = \bar{z}_P$ it rises very rapidly to high, though finite, values. Most numerical integration methods, including Gaussian quadrature, may fail to detect the narrow but high peak and therefore return incorrect answers.

The problem of quasi-singular integration can be dealt with by a simple coordinate transformation. If the approximating functions α_i

---

[1] Gaussian quadrature is described in Chapter 7, Section 4.3, where a brief table of quadrature coefficients is also given.

and α_j are nonzero in the same element, the task is evidently to evaluate the integral

$$C_{ij} = w^2 \int_0^1 \int_0^1 P(|\zeta_Q - \zeta_P|)\alpha_j(\zeta_P)\,d\zeta_P\alpha_i(\zeta_Q)\,d\zeta_Q. \qquad (6.27)$$

The region of integration is now the square $0 \le \bar{z}_Q \le w$, $\le \bar{z}_P \le w$ (subscripts would be redundant since the two variables range over the same element anyway). Let coordinates u, v be introduced by

$$\left.\begin{aligned} u &= \zeta_Q - \zeta_P, \\ v &= \zeta_Q - \zeta_P + 1. \end{aligned}\right\} \qquad (6.28)$$

The region of integration in the u–v plane is still a square, but a square rotated by 45°, as in Fig. 6.12. The integral C_{ij} now reads

$$C_{ij} = \int\int \alpha_i\left(\frac{v + u + \bar{l}}{2}\right)\alpha_j\left(\frac{v - u + \bar{l}}{2}\right)P(|u|)\,du\,dv. \qquad (6.29)$$

Observe that the Pocklington kernel function is an even function of u, because it is exactly symmetric in $\bar{z}_Q$ and $\bar{z}_P$. It is (trivially) an even function in v also. Consequently, it is sufficient to integrate only that part of the integrand in (6.29) which is even in both u and v. To do so, form the function even in both arguments

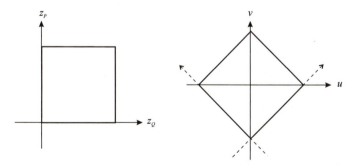

Fig 6.12 A rotation of the $z_Q - z_P$ coordinates maps all singularities and quasi-singularities onto the v-axis.

$$E_{ij}(u, v) = \alpha_i \left(\frac{v + u + \bar{l}}{2} \right) \alpha_j \left(\frac{v - u + \bar{l}}{2} \right)$$

$$+ \alpha_i \left(\frac{v - u + \bar{l}}{2} \right) \alpha_j \left(\frac{v + u + \bar{l}}{2} \right)$$

$$+ \alpha_i \left(\frac{-v + u + \bar{l}}{2} \right) \alpha_j \left(\frac{-v - u + \bar{l}}{2} \right)$$

$$+ \alpha_i \left(\frac{-v - u + \bar{l}}{2} \right) \alpha_j \left(\frac{-v + u + \bar{l}}{2} \right) \tag{6.30}$$

and integrate it over one quadrant only. In this way the integral becomes simply

$$C_{ij} = w^2 \int_0^1 \int_0^v E_{ij}(u, v) P(|u|) \, du \, dv. \tag{6.31}$$

The quasi-singularity in the integrand occurs for $u = 0$, i.e., it lies entirely on the v-axis. Any integration with respect to v is smooth and trouble-free, because $E_{ij}(u, v)$ is simply a polynomial (albeit somewhat complicated) while $P(|u|)$ does not vary with v at all. Gaussian quadrature is therefore a perfectly acceptable choice of procedure, and

$$C_{ij} \simeq w^2 \sum_k W_k \int_0^v E_{ij}(u, v_k) P(|u|) \, du, \tag{6.32}$$

where W_k and v_k are the Gaussian weights and quadrature nodes for approximate integration in $0 \leq v \leq 1$. The remaining troublesome integral is one-dimensional, so it can be evaluated using an adaptive quadrature algorithm. Such methods perform repeated integration and form an error estimate at each step, so they can reliably detect quasi-singularities. Romberg integration is one such method, and one adequate for this task. It is relatively slow, but fortunately the number of integrals that must be evaluated this way is small. If N_e elements of polynomial order N_p are used, only $O(N_e N_p^2)$ quasi-singular integrations are required, as against $O(N_e^2 N_p^2)$ integrations of all kinds.

7.   Surface-patch modelling for scattering and radiation

In this section, radiation from a general three-dimensional perfectly conducting body is considered. Particular cases of this problem are highly topical in the real world of high frequency engineering electromagnetics. The radiation may arise from scattering or from some internal driving source and can be analysed in terms of the surface currents and charges, say $\mathbf{J}_S$ and σ respectively, which become distributed over the body. Analytic treatment of such problems is seldom possible and recourse has to be made to numerical methods. However, it is possible to express the fields in free space associated with $\mathbf{J}_S$ and σ through one or other of a pair of relatively simple integral equations set up in terms of either $\mathbf{E}$ or $\mathbf{H}$.

7.1   *Representation of fields using scalar and vector potentials*

With the assumption that the radiation takes place into a uniform unbounded medium Ω, the corresponding electromagnetic field is very conveniently represented by the vector magnetic and scalar electric potentials, $\mathbf{A}$ and V, expressed as complex phasors at frequency ω with the common factor $\exp(\mathrm{j}\omega t)$ suppressed. The pair of equations

$$\mathbf{E} = -\mathrm{j}\omega\mathbf{A} - \nabla V \tag{7.1}$$

$$\mathbf{H} = \frac{1}{\mu}\nabla \times \mathbf{A} \tag{7.2}$$

may then be used to recover the fields themselves. As can be seen from Chapter 3, Section 3.6 the relevant Lorentz gauge phasor differential equations relating to free space pervaded by current density $\mathbf{J}$ and charge density ρ are

$$\nabla^2\mathbf{A} + k^2\mathbf{A} = -\mu_0\mathbf{J}, \tag{7.3}$$

$$\nabla^2 V + k^2 V = -\frac{\rho}{\epsilon_0}, \tag{7.4}$$

where $k^2 = \omega^2\mu_0\epsilon_0$.

7.2   *Integral expressions for the potentials*

In the case of radiation into uniform space from a perfect conductor, the current and charge distributions may be restricted to those on the surface S of the conductor, $\mathbf{J}_S$ and σ respectively. Then there are simple integral solutions to Eqs. (7.3) and (7.4), respectively

$$\mathbf{A}(\mathbf{r}) = \mu_0 \oint_S G(\mathbf{r},\mathbf{r}')\mathbf{J}_S(\mathbf{r}')\,\mathrm{d}S', \qquad \mathbf{r}\in\Omega,\ \mathbf{r}'\in S, \tag{7.5}$$

$$V(\mathbf{r}) = \frac{1}{\epsilon_0} \oint_S G(\mathbf{r},\mathbf{r}')\sigma(\mathbf{r}')\,\mathrm{d}S', \qquad \mathbf{r}\in\Omega,\ \mathbf{r}'\in S. \tag{7.6}$$

Here the scalar Green's function $G(\mathbf{r},\mathbf{r}')$ evidently satisfies

$$\nabla^2 G + k^2 G = -\delta(\mathbf{r} - \mathbf{r}') \tag{7.7}$$

and thus has the solution in three-dimensional space

$$G = \frac{\exp(-\mathrm{j}k|\mathbf{r}-\mathbf{r}'|)}{4\pi|\mathbf{r}-\mathbf{r}'|}. \tag{7.8}$$

The surface charge density σ is related to the surface current density by the continuity equation

$$\nabla_S \cdot \mathbf{J}_S + j\omega\sigma = 0, \tag{7.9}$$

where ∇_S represents the vector gradient operator confined to the surface S. The validity of Eqs. (7.3) and (7.4) depends upon the Lorentz gauge condition expressed in phasor form (see the equivalent equation in general time-varying form, Eq. (3.45), Chapter 3),

$$\nabla \cdot \mathbf{A} + j\omega\mu_0\epsilon_0 V = 0, \tag{7.10}$$

being satisfied. However it may be noted that enforcement of the continuity equation above amounts to the same thing; showing that this is so by means of some straightforward vector calculus manipulations is left as an exercise for the reader.

7.3   *The electric and magnetic field integral equations*

It is now possible to set down a pair of integral equations either of which can be used to relate the surface current $\mathbf{J}_S$ on the perfect conductor to its cause, an externally applied electromagnetic field expressed by the vectors $\mathbf{E}^{\text{ex}}$, $\mathbf{H}^{\text{ex}}$. Consider first the requirement that the surface components of $\mathbf{E}$ cancel with those of $\mathbf{E}^{\text{ex}}$, so as to satisfy the physical necessity that there shall be no surface component of the total $\mathbf{E}$-field on S. This requirement may be written as

$$\mathbf{1}_S \times (\mathbf{E} + \mathbf{E}^{\text{ex}}) = 0, \tag{7.11}$$

where $\mathbf{1}_S$ is the unit vector normal to the surface S.[2] Equation (7.1) may be used to express this constraint in terms of $\mathbf{A}$ and V. Then using Eqs. (7.5) and (7.6), at the same time eliminating σ through Eq. (7.9) and finally making the substitution $\nabla G(\mathbf{r}, \mathbf{r}') = -\nabla' G(\mathbf{r}, \mathbf{r}')$, gives the electric field integral equation (EFIE)

$$\mathbf{1}_S \times \mathbf{E}^{\text{ex}}(\mathbf{r}) = \mathbf{1}_S \times \oint_S \left(jk\eta \mathbf{J}_S(\mathbf{r}') G(\mathbf{r}, \mathbf{r}') \right.$$
$$\left. + \frac{\eta}{jk} \{\nabla'_S \cdot \mathbf{J}_S(\mathbf{r}')\} \nabla' G(\mathbf{r}, \mathbf{r}') \right) dS', \qquad \mathbf{r}, \mathbf{r}' \in S, \tag{7.12}$$

---

[2] Observe that in the notation adopted here, the plain vectors $\mathbf{E}$ and so forth represent the *scattered* field, not the total field; some writers carry through a labelling such as $\mathbf{E}^{\text{sca}}$.

where $\eta = (\mu_0/\epsilon_0)^{1/2}$. The integral equation here has been set down for a closed surface S. However it is equally applicable to an open surface representing a thin, perfectly conducting sheet of arbitrary shape. In that case $\mathbf{J}_S$ is to be interpreted as the sum of the surface currents on either side of the sheet. Currents cannot flow across any portion of the edge of the sheet from one side to the other; thus a constraint along the edge of the sheet forcing the current to be parallel to the edge there is required.

A dual integral equation expressed in terms of $\mathbf{H}^{ex}$ may be obtained by considering the magnetic field $\mathbf{H}$ due to a surface current $\mathbf{J}_S$,

$$\mathbf{H}(\mathbf{r}) = \oint_S \mathbf{J}_S(\mathbf{r}') \times \nabla' G(\mathbf{r}, \mathbf{r}') \, dS' \qquad \mathbf{r} \in \Omega, \mathbf{r}' \in S, \tag{7.13}$$

which follows directly from Eqs. (7.2) and (7.5). Just outside the surface S of the scatterer

$$\mathbf{1}_S \times (\mathbf{H}^{ex} + \mathbf{H}) = \mathbf{J}_S \tag{7.14}$$

so for $\mathbf{r}$ approaching S from the *outside*

$$\mathbf{J}_S(\mathbf{r}) = \mathbf{1}_S(\mathbf{r}) \times \mathbf{H}^{ex}(\mathbf{r}) + \mathbf{1}_S(\mathbf{r}) \times \oint_S \mathbf{J}_S(\mathbf{r}') \times \nabla' G(\mathbf{r}, \mathbf{r}') \, dS',$$
$$\mathbf{r}' \in S. \tag{7.15}$$

However the total field vanishes inside S,

$$(\mathbf{H} + \mathbf{H}^{ex}) = 0 \tag{7.16}$$

provided S is a closed surface. Thus for $\mathbf{r}$ approaching S from the *inside*

$$0 = \mathbf{1}_S(\mathbf{r}) \times \mathbf{H}^{ex}(\mathbf{r}) + \mathbf{1}_S(\mathbf{r}) \times \oint_S \mathbf{J}_S(\mathbf{r}') \times \nabla' G(\mathbf{r}, \mathbf{r}') \, dS',$$
$$\mathbf{r}' \in S. \tag{7.17}$$

In the limit of $\mathbf{r}$ lying on S it follows that

$$\frac{1}{2} \mathbf{J}_S(\mathbf{r}) = \mathbf{1}_S(\mathbf{r}) \times \mathbf{H}^{ex}(\mathbf{r}) + \mathbf{1}_S(\mathbf{r}) \times \oint_S \mathbf{J}_S(\mathbf{r}') \times \nabla' G(\mathbf{r}, \mathbf{r}') \, dS',$$
$$\mathbf{r}, \mathbf{r}' \in S. \tag{7.18}$$

This is the magnetic field integral equation (MFIE) relating to a perfectly conducting closed surface scatterer. The placement of both $\mathbf{r}$ and $\mathbf{r}'$ upon S raises some delicate issues and it turns out that the equation is not valid for open surfaces S. Moreover, when $\mathbf{r}'$ passes through $\mathbf{r}$ in the integration process, $G(\mathbf{r}, \mathbf{r}')$ and its derivatives become singular. For a thorough and lengthy discussion of such matters the reader is referred to Poggio and Miller (1973). In the usual way, once $\mathbf{J}_S$ has been determined the scattered field everywhere is easily recovered. An approximation to any general three-dimensional shape may be constructed from an interconnected set of planar patches, in which case it can be appreciated that there is then some hope of attacking the integral equations above to provide correspondingly approximate solutions to topical scattering and radiation problems. The preferred general shape for such patches is triangular, since these are conformable with any three-dimensional body, see Fig. 6.13. The unknown surface current $\mathbf{J}_S$, arising from incident radiation or a driving source, may be represented over the patches by unspecified parameters in some suitable fashion and either one of the integral equations, Eq. (7.12) or (7.18), solved by a weighted residual process to recover an approximation to $\mathbf{J}_S$.

7.4   *Solution of the EFIE and MFIE by triangular-patch modelling*

The electric field integral equation (EFIE) for radiation and scattering from a perfect conductor, Eq. (7.12), relates the current $\mathbf{J}_S$ induced upon the surface of the body to the external driving field represented by $\mathbf{E}^{ex}(\mathbf{r})$. However, the equation involves not only the surface current density $\mathbf{J}_S$ but its divergence as well, so that a piecewise-constant trial function vector representation is not adequate. Instead $\mathbf{J}_S$ is represented by linear basis functions associated with pairs of triangles sharing a single common edge (Rao, Wilton and Glisson, 1982). The unit basis functions are multiplied by amplitude coefficients I_n, eventually to be determined by the finite element process, each coefficient representing the normal current density flowing across the nth edge. To understand in detail this trial function representation of $\mathbf{J}_S$ consider the pair of adjacent triangular patches sharing the common edge of length l_n illustrated in Fig. 6.14, ordinarily there being three such interconnections associated with any given patch. Denote the two triangles T_n^+ and T_n^-; the designation $\pm$ is determined by the algebraic sign chosen for I_n so as to represent current flow proceeding across the nth edge from T_n^+ to T_n^-. Points in T_n^+ may be represented by the local vector $\boldsymbol{\rho}_n^+$ having as origin

Fig 6.13   Triangular-patch model of a circular cylinder section.

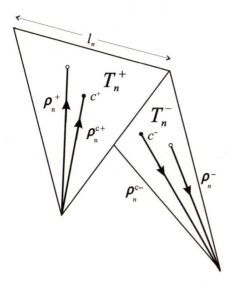

Fig 6.14   A pair of triangular patches with a common edge.

the free vertex of that triangle. The vector ρ_n^- is similarly chosen to represent points in T_n^-, except to note that it points *towards* the T_n^- free vertex, as in Fig. 6.14. Then a linear vector basis function $\mathbf{f}_n(\mathbf{r})$ associated with each and every edge l_n may be defined as

$$
\left.
\begin{aligned}
\mathbf{f}_n(\mathbf{r}) &= \frac{l_n}{2A_n^+}\rho_n^+, && \mathbf{r} \text{ in } T_n^+, \\[4pt]
\mathbf{f}_n(\mathbf{r}) &= \frac{l_n}{2A_n^-}\rho_n^-, && \mathbf{r} \text{ in } T_n^-, \\[4pt]
\mathbf{f}_n(\mathbf{r}) &= 0 && \text{otherwise.}
\end{aligned}
\right\}
\qquad (7.19)
$$

The function $\mathbf{f}_n(\mathbf{r})$ may be used to express a current density, $\mathbf{J}_n = I_n\mathbf{f}_n$, which clearly has a continuous normal component I_n across the edge l_n. The current has no component normal to any of the remaining edges of the pair of triangles and by definition flows from T_n^+ to T_n^-, so there is no current discontinuity to be associated anywhere with this model. Correspondingly, line charges are absent at the triangle boundaries. It is easily seen that $\nabla_S \cdot \rho_n^\pm = \pm 2$, so that using Eq. (7.19), a surface charge density

$$
\left.
\begin{aligned}
\sigma_n &= -\frac{1}{j\omega}I_n\nabla_S \cdot \mathbf{f}_n = \mp\frac{I_n l_n}{j\omega A_n^\pm}, && \mathbf{r} \text{ in } T_n^\pm, \\[4pt]
\sigma_n &= 0 && \text{otherwise,}
\end{aligned}
\right\}
\qquad (7.20)
$$

may be associated with each $\mathbf{J}_n$, consistent with the physical nature of high frequency radiating conductors which always support both surface current and charge. The current density vector lying in some particular triangle Δ, with edges of length l_1, l_2, l_3 and vertices $\mathbf{r}_1$, $\mathbf{r}_2$, $\mathbf{r}_3$ (numbered locally) may be written

$$\mathbf{J}_\Delta = \frac{1}{2A_\Delta}(l_1 I_1 \boldsymbol{\rho}_1 + l_2 I_2 \boldsymbol{\rho}_2 + l_3 I_3 \boldsymbol{\rho}_3), \tag{7.21}$$

where $\boldsymbol{\rho}_1 = r - r_1$ and so forth. Thus assuming temporarily that the origin of $\mathbf{r}$ lies in the plane Δ, it is clear that expression (7.21) has the necessary three degrees of freedom for it to represent an arbitrary linear vector variation of current in that plane

$$\mathbf{J}_\Delta = \mathbf{a}_0 + b\mathbf{r}. \tag{7.22}$$

In all, it can be appreciated that the surface current

$$\mathbf{J}_S = \sum_n I_n \mathbf{f}_n, \tag{7.23}$$

and surface charge

$$\sigma = -\frac{1}{j\omega} \sum_n I_n \nabla_S \cdot \mathbf{f}_n, \tag{7.24}$$

represent perfectly good trial functions for use in solving the EFIE, respectively being piecewise linear and piecewise constant. An inner product notation

$$\langle \boldsymbol{\alpha}, \boldsymbol{\beta} \rangle = \int_S \boldsymbol{\alpha} \cdot \boldsymbol{\beta} \, \mathrm{d}S, \tag{7.25}$$

where $\boldsymbol{\alpha}(\mathbf{r})$ and $\boldsymbol{\beta}(\mathbf{r})$ are any pair of vector functions defined upon S, is introduced. Assume that $\mathbf{A}$ and V are to be represented in terms of the basis functions $\mathbf{f}_n$ through Eqs. (7.5), (7.6), (7.23) and (7.24). Then a Galerkin weighted residual

$$R_m = j\omega\langle \mathbf{A}, \mathbf{f}_m \rangle + \langle \nabla V, \mathbf{f}_m \rangle - \langle \mathbf{E}^{\mathrm{ex}}, \mathbf{f}_m \rangle \tag{7.26}$$

can be formed to express the vanishing tangential total electric field on S, Eq. (7.11), with $\mathbf{E}$ given through Eq. (7.1); the weight functions $\mathbf{f}_m$ are chosen from the basis functions defined above. Note that since the $\mathbf{f}$'s are by definition vectors tangential to the patch model of S, the dot-product inherent in the operation here automatically picks out the tangential component of the total electric field required. Using the vector integration by parts theorem Eq. (2.3) of Appendix 2 in its two-dimensional form gives

$$\langle \nabla V, \mathbf{f}_m \rangle = \int_S \nabla V \cdot \mathbf{f}_m \, dS$$

$$= \oint_C V\mathbf{f}_m \cdot \mathbf{1}_n \, ds - \int_S V \nabla \cdot \mathbf{f}_m \, dS. \qquad (7.27)$$

Here C represents the closed curve corresponding to the edge of S if it is an open surface. Clearly the line integral around C in Eq. (7.27) vanishes, either because S is in fact closed or because in any case $\mathbf{f}_m$ is always parallel to free edges. The basis functions $\mathbf{f}_m$ have no component normal to the patch-modelled S, so $\nabla \cdot \mathbf{f}_m$ can be written as $\nabla_S \cdot \mathbf{f}_m$, giving finally

$$\langle \nabla V, \mathbf{f}_m \rangle = - \int_S V \nabla_S \cdot \mathbf{f}_m \, dS. \qquad (7.28)$$

Thus the difficulties which would have arisen from having to manipulate the vector gradient of the Green's function are avoided. Noting from Eq. (7.19) that

$$\nabla_S \cdot \mathbf{f}_m = \pm \frac{l_m}{A_m^{\pm}}, \qquad \mathbf{r} \in T_m^{\pm}. \qquad (7.29)$$

whence

$$\langle \nabla V, \mathbf{f}_m \rangle = -l_m \left(\frac{1}{A_m^+} \int_{T_m^+} V \, dS - \frac{1}{A_m^-} \int_{T_m^-} V \, dS \right)$$

$$\simeq -l_m \{ V(\mathbf{r}_m^{c+}) - V(\mathbf{r}_m^{c-}) \}. \qquad (7.30)$$

In Eq. (7.30) the integral of V over each triangle is approximated by assuming V to be constant at the value $V(r_m^{c\pm})$ found at its centroid. Making a similar approximation when evaluating the other inner products involved gives

$$\left\langle \begin{pmatrix} \mathbf{E}^{ex} \\ \mathbf{A} \end{pmatrix}, \mathbf{f}_m \right\rangle = \frac{l_m}{2} \left\{ \frac{1}{A_m^+} \int_{T_m^+} \begin{pmatrix} \mathbf{E}^{ex} \\ \mathbf{A} \end{pmatrix} \cdot \boldsymbol{\rho}_m^+ \, dS \right.$$

$$\left. + \frac{1}{A_m^-} \int_{T_m^-} \begin{pmatrix} \mathbf{E}^{ex} \\ \mathbf{A} \end{pmatrix} \cdot \boldsymbol{\rho}_m^- \, dS \right\}$$

$$\simeq \frac{l_m}{2} \left\{ \begin{pmatrix} \mathbf{E}^{ex}(\mathbf{r}_m^{c+}) \\ \mathbf{A}(\mathbf{r}_m^{c+}) \end{pmatrix} \cdot \boldsymbol{\rho}_m^{c+} + \begin{pmatrix} \mathbf{E}^{ex}(\mathbf{r}_m^{c-}) \\ \mathbf{A}(\mathbf{r}_m^{c-}) \end{pmatrix} \cdot \boldsymbol{\rho}_m^{c-} \right\}. \qquad (7.31)$$

Notice that this simplification of the integration process destroys the symmetry of the matrices ordinarily found when the Galerkin method is used. Finally, substituting Eqs. (7.30) and (7.31) into Eq. (7.26), having set each residual R_m to zero gives a matrix equation

$$\sum_{n=1}^{N} Z_{mn} I_n = V_m^{\text{ex}} \qquad = 1, 2, \ldots N, \tag{7.32}$$

where

$$Z_{mn} = l_m \left\{ \frac{j\omega}{2} \left(A_{mn}^+ \cdot \rho_m^{c+} + A_{mn}^- \cdot \rho_m^{c-} \right) + V_{mn}^- - V_{mn}^+ \right\} \tag{7.33}$$

$$V_m^{\text{ex}} = \frac{l_m}{2} \left(\mathbf{E}_m^+ \cdot \rho_m^{c+} + \mathbf{E}_m^- \cdot \rho_m^{c-} \right) \tag{7.34}$$

$$A_{mn}^{\pm} = \frac{\mu_0}{4\pi} \int_S \mathbf{f}_n(\mathbf{r}') \frac{\exp(-jkR_m^{\pm})}{R_m^{\pm}} \, dS' \tag{7.35}$$

$$V_{mn}^{\pm} = -\frac{1}{4\pi j\omega\epsilon} \int_S \nabla_S' \cdot \mathbf{f}_n(\mathbf{r}') \frac{\exp(-jkR_m^{\pm})}{R_m^{\pm}} \, dS' \tag{7.36}$$

$$R_m^{\pm} = |\mathbf{r}_m^{c\pm} - \mathbf{r}'| \tag{7.37}$$

and

$$\mathbf{E}_m^{\pm} = \mathbf{E}^{\text{ex}}(\mathbf{r}_m^{c\pm}). \tag{7.38}$$

Solution of the matrix equation Eq. (7.32) for I_n formally completes the exercise of modelling scattering from a perfectly conducting body by a numerical realization of the EFIE. As was remarked earlier, the EFIE may validly be applied to an open surface. The required edge constraint, that there shall be no surface current normal to the boundary of the open surface, is achieved naturally when using the triangular-patch functions described here. This is so since the boundary of S will be made up from triangle edges which are not joined to any others, forcing the trial surface current to be parallel to such edges.

For plane wave excitation the external source may be represented by

$$\mathbf{E}^{\text{ex}} = (E_\theta \mathbf{1}_\theta + E_\phi \mathbf{1}_\phi) \exp(j\mathbf{k} \cdot \mathbf{r}), \tag{7.39}$$

where $\mathbf{1}_\theta$ and $\mathbf{1}_\phi$ are the spherical coordinate angular unit vectors (aligned in the usual fashion relative to the Cartesian axes) and

$$\mathbf{k} = k(\sin\theta_0 \cos\phi_0 \mathbf{1}_x + \sin\theta_0 \sin\phi_0 \mathbf{1}_y + \cos\theta_0 \mathbf{1}_z) \tag{7.40}$$

is the wavenumber vector. This corresponds to radiation originating from a distant source positioned such that $\mathbf{k}$ points towards the origin in the $-\mathbf{1}_r$ direction defined by the spherical coordinate angles θ_0, ϕ_0.

The formulation here can also model the voltage source excitation which may exist as a voltage drop across a pair of antenna terminals. Consider the basic pair of triangular patches in a case where there is a slot along the common edge sustaining an excitation voltage. Then the field set up in such a slot (there may be more than one slot), and zero elsewhere on the conductor, will constitute the field $\mathbf{E}^{\text{ex}}$. Examine the last term in

Eq. (7.26) which determines the contribution of the driven slot to the right hand side of Eq. (7.31),

$$V_m^{\mathrm{ex}} = \langle \mathbf{E}^{\mathrm{ex}}, \mathbf{f}_m \rangle. \qquad (7.41)$$

Assume that the slot lies in the secondary triangle T_m^- and sustains a voltage V_m^{source} directed towards the secondary apex (see Fig. 6.15). The field observed as $-\nabla V$ within the slot is to be considered the *scattered* field, $\mathbf{E}_m^{\mathrm{source}}$, so the nominal excitation field is *minus* this. It is then seen that

$$\langle \mathbf{E}^{\mathrm{ex}}, \mathbf{f}_m \rangle = \int_{\mathrm{slot}} \left(\frac{V_m^{\mathrm{source}}}{d} \mathbf{1}_m^- \right) \cdot \left(\frac{l_m}{2A_m^-} \boldsymbol{\rho}_m^- \right) \mathrm{d}S \qquad (7.42)$$

where d is the width of the slot and $\mathbf{1}_m^-$ is the unit vector normal to the base of T_m^-. Some straightforward geometry applied to the quantities in Eq. (7.51) then reveals that

$$V_m^{\mathrm{ex}} = l_m V_m^{\mathrm{source}} \qquad (7.43)$$

for every edge m associated with a voltage-driven slot but otherwise zero.

The MFIE, Eq. (7.18), may be solved in a similar fashion. As it happens, no derivative terms of $\mathbf{J}$ are involved under the integral sign here, so a piecewise-constant approximation for this vector variable is allowable. This then allows a point collocation approach, outside the scope of this text; the original description of solution of the practical solution of this integral equation given by Wang (1978), and many subsequent embellishments, follow this route. The MFIE is only strictly applicable to closed surfaces S.

7.5   *Internal resonances*

It may be observed that both the electric and magnetic field integral equations, Eqs. (7.12) and (7.18), are expressed in terms of the tangential current components on the surface. This implies that when

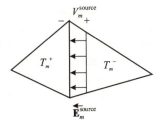

Fig 6.15   A voltage-driven slot constituting a source for an antenna modelled by triangular patches.

applied to the exterior of a closed, perfectly conducting body, each could equally well relate to the interior problem of the corresponding closed, perfectly conducting shell. Now although in general the external problem requires a nonzero driving function to yield a nontrivial solution, the same is not true of the shell interior; either homogeneous integral equation (that is with $\mathbf{E}^{ex} = 0$ or $\mathbf{H}^{ex} = 0$) applied to a closed surface has interior solutions which correspond to cavity resonance eigensolutions. Thus when the inhomogeneous exterior problem is posed at frequencies sufficiently high to allow physical resonances to occur within the closed shell, the exterior solutions are liable to be corrupted by internal resonance eigensolutions. As the frequency is taken higher and higher these unwanted resonances move closer together in the frequency domain so as to obscure the true external solution. Peterson (1990) surveys the remedies which may be used to eliminate internal resonances.

8.   Further reading

The book by Harrington (1993) is a reissue of a 1968 work and deals with both projective and collocative numerical methods. Regrettably, in that otherwise valuable early book the impression was created, and reinforced by an unsuccessful attempt at proof, that projective approximation and pointwise collocation are the same thing. This confusion has led many antenna engineers to call almost any numerical solution 'the method of moments'. Subject to this warning, the book can be recommended as an easy-to-read introduction to the integral equations of electrostatics as well as wire antennas. The solution of integral equations by finite element methods was given much more careful treatment subsequently by Johnson (1987). The treatment of electrostatics problems given here is based upon papers by Silvester (1968) and Benedek and Silvester (1972), which largely adopt the mathematical approach of Johnson. Finite element methods of high order are applied to wire-antenna problems by Silvester and Chan (1972) and to reflector antennas by Hassan and Silvester (1977). The research monograph edited by Chari and Silvester (1980) contains a chapter by C.W. Trowbridge, 'Applications of integral equation methods to magnetostatics and eddy-current problems', which is relevant to Section 5 here. In the same monograph McDonald and Wexler give some rather advanced theory concerning the treatment of mixed partial differential- and integral-equation formulations.

The electromagnetic theory of antennas is fully dealt with by Balanis (1982). The first ever solution of a radiation problem by means of an integral equation was derived in a paper by Pocklington (1897). It is interesting to note that in the same volume J.J. Thomson publishes a paper on cathode rays, from which it is clear that he was then just a little short of deducing the existence of the electron. Immediately following Pocklington's paper, C.T.R. Wilson writes on experiments in his cloud chamber, detecting rays emanating from uranium. The treatment of the MFIE by means of zero-order triangular patches was first given by Wang (1978), whilst the first-order triangular-patch function analysis applied to the EFIE and given here follows that given by Rao, Wilton and Glisson (1982). The book by Wang (1991) is the standard reference relating to the solution of antenna problems by means of integral equations.

9.   Bibliography

Balanis, C.A. (1982), *Antenna Theory, Analysis and Design*. New York: Harper & Row. xvii + 790 pp.

Benedek, P. and Silvester, P.P. (1972), 'Capacitance of parallel rectangular plates separated by a dielectric sheet', *IEEE Transactions on Microwave Theory and Techniques*, **MTT-20**, pp. 504–10.

Chari, M.V.K. and Silvester, P.P. (1980), *Finite Elements in Electric and Magnetic Field Problems*. Chichester: John Wiley. xii + 219 pp.

Harrington, R.F. (1993), *Field Computation by Moment Methods*. New York: IEEE Press. x + 229 pp. (Originally published: Macmillan, 1968.)

Hassan, M.A. and Silvester, P. (1977), 'Radiation and scattering by wire antenna structures near a rectangular plate reflector', *IEE Proceedings*, **124**, pp. 429–35.

Johnson, C. (1987), *Numerical Solution of Partial Differential Equations by the Finite Element Method*. Cambridge: Cambridge University Press. vi + 279 pp.

Newman, M.J., Trowbridge, C.W. and Turner, L.R. (1972), 'GFUN: an interactive program as an aid to magnet design', *Proceedings of the 4th International Conference on Magnet Technology*, pp. 617–26, Brookhaven, NY.

Peterson, A.F. (1990), 'The "interior resonance" problem associated with surface integral equations of electromagnetics: Numerical consequences and a survey of remedies', *Electromagnetics*, **10**, pp. 293–312.

Pocklington, H.C. (1897), 'Electrical oscillations in wires', *Cambridge Phil. Soc. Proc.*, **9**, pp. 324–32.

Poggio, A.J. and Miller, E.K. (1973), 'Integral equation solutions of three-dimensional scattering problems', in *Computer techniques for electromagnetics*, ed. R. Mittra. Oxford: Pergamon.

Rao, S.M., Wilton, D.R. and Glisson, A.W. (1982), 'Electromagnetic scattering by surfaces of arbitrary shape', *IEEE Trans. Antennas and Propagation*, **30**(3), pp. 409–18.

Silvester, P. (1968), 'TEM properties of microstrip transmission lines', *IEE Proceedings*, **115**, 43–8.

Silvester, P. and Chan, K.K. (1972), 'Bubnov-Galerkin solutions to wire-antenna problems', *IEE Proceedings*, **115**, 1095–9.

Stratton, J.A. (1941), *Electromagnetic Theory*. New York: McGraw-Hill, xv + 615 pp.

Wang, J.J.H. (1978), 'Numerical analysis of three-dimensional arbitrarily-shaped conducting scatterers by trilateral surface cell modelling', *Radio Science*, **13**(6), pp. 947–52.

Wang, J.J.H. (1991), *Generalized moment methods in electromagnetics*, Wiley: New York. xiii + 553 pp.

10.   Programs

Program STRIPLIN follows the organization of other programs in this book and uses some of the utility programs of Appendix 4. The listing given opposite is followed by three numbers, which constitute a trial data set for testing.

```
C*************************************************************
C********                                           ********
C******** Strip and ground plane:  integral elements ********
C********                                           ********
C*************************************************************
C      Copyright (c) 1995  P.P. Silvester and R.L. Ferrari
C*************************************************************
C
C      The subroutines  that make up this program  communicate
C      via named common blocks.  The variables in commons are:
C
C      Problem definition and solution
C      common /problm/
C        space  =  spacing, strip to ground plane
C        nelmt  =  number of elements in half-strip
C        x      =  nodal coordinates
C        charg  =  charge density on substrip
C
C      Global matrices and right-hand side
C      common /matrix/
C        coeffs  =  global coefficient matrix (connected)
C        rthdsd  =  right-hand side of system of equations
C
C      Predefined problem parameters (array dimensions):
C        MAXELM  =  maximum number of elements in problem
C
C===============================================================
C      Global declarations -- same in all program segments
C===============================================================
       parameter (MAXELM = 100, PI = 3.141596)
       common /problm/ space, nelmt, x(MAXELM+1), charg(MAXELM)
       common /matrix/ coeffs(MAXELM,MAXELM), rthdsd(MAXELM)
C===============================================================
C
C      Initialize matrices to zeros.
       call matini(coeffs, MAXELM, MAXELM)
       call vecini(charg, MAXELM)
C
C      Problem definition:
C              spacing between strip and ground plane
C              number of elements, x-value (1 or more times)
       call inprob
C
C      Create coefficient matrix
       call mkcoef
C
C      Construct right-hand side vector
       do 20 j = 1,nelmt
   20    rthdsd(j) = 1.
C
C      Solve the assembled finite element equations
       call eqsolv(coeffs, charg, rthdsd, nelmt, MAXELM)
C
C      Write the resulting charge densities to output file
```

```
      call output
C
      stop
      end
C
C**************************************************************
C
      Subroutine inprob
C
C**************************************************************
C
C     Reads in problem details from input file.
C
C==============================================================
C     Global declarations -- same in all program segments
C==============================================================
      parameter (MAXELM = 100, PI = 3.141596)
      common /problm/ space, nelmt, x(MAXELM+1), charg(MAXELM)
      common /matrix/ coeffs(MAXELM,MAXELM), rthdsd(MAXELM)
C==============================================================
C
C     Set up initial values
      nelmt = 1
      call vecini(x, MAXELM+1)
C
C     Read problem data
C        (1) spacing from ground plane to strip
      read (*, *) space
C        (2) number of elements, x at end of element group
C            (may be repeated any number of times)
   30 read (*, *, end=60) nelm, xelmt
      if (nelm .le. 0 .or. nelmt+nelm .gt. MAXELM) then
        call errexc('INPROB', 1000 + nelm)
      else
        xleft = x(nelmt)
        nleft = nelmt
        xelmt = xleft + xelmt
        nelmt = nelmt + nelm
        segms = nelmt - nleft
        do 40 nelm=nleft,nelmt
          x(nelm) = (nelmt-nelm)*xleft - (nleft-nelm)*xelmt
   40     x(nelm) = x(nelm) / segms
      endif
      go to 30
C
C     Data successfully read.  Set up problem values.
   60 nelmt = nelmt - 1
      if (nelmt .le. 1) call errexc('INPROB', 2)
C
      return
      end
C
C**************************************************************
C
```

```
      Subroutine mkcoef
C
C****************************************************************
C
C     Establishes coefficient matrix coeffs.
C
C================================================================
C     Global declarations -- same in all program segments
C================================================================
      parameter (MAXELM = 100, PI = 3.141596)
      common /problm/ space, nelmt, x(MAXELM+1), charg(MAXELM)
      common /matrix/ coeffs(MAXELM,MAXELM), rthdsd(MAXELM)
C================================================================
      do 80 i = 1,nelmt
C
C        Determine element i width and midpoint, quit if < 0
         widti = x(i+1) - x(i)
         xmidi = (x(i+1) + x(i))/2.
         if (widti .le. 0.) then
           call errexc('MKCOEF', i)
         else
C          Compute entries of coeffs from Green's functions
           do 70 j = 1,i
             xmidj = (x(j+1) + x(j))/2.
             dist1 = abs(xmidi - xmidj)
             if (i .eq. J) dist1 = exp(alog(widti) - 1.5)
             dist2 = xmidi + xmidj
             dist3 = sqrt((xmidi + xmidj)**2 + (2*space)**2)
             dist4 = sqrt((xmidi - xmidj)**2 + (2*space)**2)
             coeffs(i,j) = alog((dist1*dist2)/(dist3*dist4))
     *                     / (-2. * PI)
             coeffs(j,i) = coeffs(i,j)
  70       continue
         endif
  80   continue
C
      return
      end
C




C****************************************************************
C
      Subroutine output
C
C****************************************************************
C
C     Outputs problem and results to default output stream.
C
C================================================================
C     Global declarations -- same in all program segments
C================================================================
```

```
      parameter (MAXELM = 100, PI = 3.141596)
      common /problm/ space, nelmt, x(MAXELM+1), charg(MAXELM)
      common /matrix/ coeffs(MAXELM,MAXELM), rthdsd(MAXELM)
C================================================================
C
      totchg = 0.
      do 10 I = 1,nelmt
         width = x(i+1) - x(i)
         denst = charg(i) / width
         totchg = totchg + charg(i)
         write (*, 1000, err=890) i, x(i), denst
         write (*, 1000, err=890) i+1, x(i+1), denst
   10    continue
 1000 format (1x, i3, 2f12.6)
      write (*, 1000, err=890) nelmt, totchg
      return
  890 call errexc('OUTPUT', 1)
      end
```

7

Curvilinear, vectorial and unbounded elements

1. Introduction

Triangular and tetrahedral elements as described in the foregoing chapters are widely used because any polygonal object can be decomposed into a set of simplexes without approximation. In fact, such decomposition can be carried out by computer programs without human intervention, since there do exist mathematical techniques guaranteed to produce correct decomposition into simplexes. Unfortunately, simplex elements also have some serious shortcomings. They do not lend themselves well to the modelling of curved shapes, so that the intrinsically high accuracy of high-order elements may be lost in rather rough geometric approximation. They use polynomial approximation functions throughout, so that fields containing very rapid variations, or even singularities, cannot be well approximated. A third disadvantage of scalar simplex elements is precisely that they do model scalar quantities; they are not well suited to describing vector fields. Finally, they are unable to model large (or infinite) regions economically; yet many field problems of electromagnetics are 'open', in the sense that the region of principal interest is embedded in an infinitely extending homogeneous exterior space.

Alternative element shapes and alternative types of approximating functions can avoid most of the problems inherent in the simplex elements. The use of so-called *isoparametric* elements, which have curved sides, can often alleviate the problems encountered in geometric modelling, by shaping the elements to fit the real geometry. Vector fields require vector approximating functions, and while their theory is neither so simple, nor quite so completely developed as for scalars, useful vector-valued functions do exist. Singularities are at times well dealt with by including in the element formulation some approximating functions which have more or less the correct behaviour at the locations where singularities are expected. Field regions of infinite extent turn out to be surprisingly easy to deal with, provided they are homogeneous, linear and source-free. While these requirements may seem restrictive, they are in fact met by large classes of engineering problems.

Many different special-purpose elements have been developed and described in the literature. This chapter describes the principal types of element that address the four classes of restriction noted above: simplicial shape, scalar-valued approximation, bounded derivatives and finite geometric extent.

2.   Rectangular elements

A simple but aesthetically appealing element shape is the rectangle. Its three-dimensional counterpart, the rectangular parallelepiped, is often referred to as a *brick* in the finite element literature, whose name alludes to the easy-to-visualize regularity of a brick wall. Such elements have found wide use in civil engineering applications; in the literature of electrical engineering they are most frequently encountered in connection with nonlinear materials. Their properties and the common methods for deriving their element matrices are set out in this section.

2.1   *Cartesian product elements*

For rectangles and bricks, approximating functions can be derived by repeated application of the functions applicable to a line segment, already derived in Chapter 4, Section 2.3. In terms of the homogeneous coordinates (simplex coordinates) ζ_1, ζ_2 of a line segment, families of interpolation polynomials may thus be constructed to serve as approximation functions,

$$\alpha_{ij}(\zeta_1, \zeta_2) = R_i^{(N)}(\zeta_1) R_j^{(N)}(\zeta_2) \tag{2.1}$$

where the auxiliary polynomials $R_k^{(N)}(\zeta_n)$ are as defined in Chapter 4. Using the line segment functions twice, suitable interpolation polynomials of order N for the rectangular region of Fig. 7.1 are then

$$\alpha_{ijkl}(\xi_1, \xi_2, \eta_1, \eta_2) = R_i^{(N)}(\xi_1) R_j^{(N)}(\xi_2) R_k^{(N)}(\eta_1) R_l^{(N)}(\eta_2). \tag{2.2}$$

The subscripts are constrained by the requirements $i + j = N$ and $k + l = N$. Two distinct symbols have been used for the homogeneous coordinates, ξ to denote the two homogeneous coordinates parallel to the x-axis, η for those parallel to the y-axis. Analogous interpolation functions appropriate to the brick element are given by

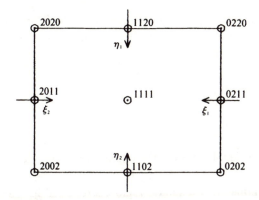

Fig 7.1 Nine-noded rectangular element, showing node numbering and orientation of homogeneous coordinates.

$$\alpha_{ijklmn}(\xi_1, \xi_2, \eta_1, \eta_2, \zeta_1, \zeta_2) =$$
$$R_i^{(N)}(\xi_1) R_j^{(N)}(\xi_2) R_k^{(N)}(\eta_1) R_l^{(N)}(\eta_2) R_m^{(N)}(\zeta_1) R_n^{(N)}(\zeta_2). \qquad (2.3)$$

Because $\xi_1, \xi_2, \eta_1, \eta_2, \zeta_1, \zeta_2$ are pairs of homogeneous coordinates, they are related by the rules

$$\xi_1 + \xi_2 = 1 \qquad (2.4)$$
$$\eta_1 + \eta_2 = 1 \qquad (2.5)$$
$$\zeta_1 + \zeta_2 = 1 \qquad (2.6)$$

and the subscripts to $R^{(N)}$ must satisfy $i + j = N$, $k + l = N$ and $m + n = N$. There can only be three independent coordinates in three-space, so the existence of six coordinate quantities necessarily requires three rules of interdependence.

Generation of the element matrices for a rectangular or brick element is easy because differentiation and integration are carried out in one dimension at a time. For example, a typical entry in the **T** matrix of a rectangular element X units by Y units in size will be given by

$$T_{(ijkl)(pqrs)} = XY \int_0^1 R_i(\xi_1) R_p(\xi_1) R_j(\xi_2) R_q(\xi_2) \, \mathrm{d}\left(\frac{x}{X}\right)$$
$$\times \int_0^1 R_k(\eta_1) R_r(\eta_1) R_l(\eta_2) R_s(\eta_2) \, \mathrm{d}\left(\frac{y}{Y}\right). \qquad (2.7)$$

The homogeneous coordinates ξ_1 and ξ_2 are parallel and antiparallel to x respectively, so the first integral is really one-dimensional; similarly, the second integral is effectively one-dimensional along the y-direction. Because both integrations are in one variable only, the algebraic task here is quite straightforward. In fact, there is not even any integration to be done if one-dimensional **T** matrices are available. The right-hand side of (2.7) can be rewritten as the product of integrals of one-dimensional simplex interpolation functions,

$$T_{(ijkl)(pqrs)} = XY \int_0^1 \alpha_{ij}\alpha_{pq} \, d\left(\frac{x}{X}\right) \int_0^1 \alpha_{kl}\alpha_{rs} \, d\left(\frac{y}{Y}\right)$$

$$= XY \, T^{(1)}_{(ij)(pq)} T^{(1)}_{(kl)(rs)} \tag{2.8}$$

where the superscripts denote that the one-dimensional matrices are to be understood.

The eight-way symmetry of a square region is reflected in index symmetry in (2.7). Hence **T** as given in (2.7) is invariant under index permutations corresponding to the following geometric transformations: (1) mirror reflection of the square about the x-axis, (2) mirror reflection about the y-axis, (3) reflection about the line $x = y$, (4) reflection about $x = -y$. For the nine-noded element of Fig. 7.1, the matrix **T** thus contains only three distinct diagonal entries and nine distinct off-diagonal entries, out of a total of 81. Symmetry rules for the matrix **S** are only a little more complicated.

2.2   *Completeness and continuity*

An important property of the polynomials associated with triangles and tetrahedra is that they form *complete* polynomial families. A family of polynomials is said to be *complete to degree N* if a linear combination of its members can exactly express any polynomial function of degree not exceeding N, but no polynomial of a higher degree. For example, the two-dimensional family of monomials

$$\mathbb{M}^{(N)} = \left\{1, x, y, x^2, xy, y^2, \dots, x^N, x^{N-1}y, x^{N-2}y^2, \dots, y^N\right\} \tag{2.9}$$

is complete to degree N, because any polynomial of this or lower degree can be written (trivially) as a sum of monomial terms with appropriate coefficients,

$$p(x, y) = a_{00} + a_{10}x + a_{01}y + a_{20}x^2 + \dots a_{0N}y^N. \tag{2.10}$$

To choose another familiar example, the family of Nth-order triangle interpolation polynomials $\alpha_{ijk}(\zeta_1, \zeta_2, \zeta_3)$ is complete to degree N because any two-dimensional polynomial of degree N or lower can be written in the form

$$p(x, y) = \sum_{ijk} p_{ijk}\alpha_{ijk}(\zeta_1, \zeta_2, \zeta_3) \tag{2.11}$$

but no polynomial of higher degree can. The important point about complete families of polynomials is that any coordinate translation

$$\begin{bmatrix} x' \\ y' \end{bmatrix} = \begin{bmatrix} a \\ b \end{bmatrix} + \begin{bmatrix} x \\ y \end{bmatrix} \tag{2.12}$$

or rotation

$$\begin{bmatrix} x' \\ y' \end{bmatrix} = \begin{bmatrix} \cos\theta & \sin\theta \\ -\sin\theta & \cos\theta \end{bmatrix} \begin{bmatrix} x \\ y \end{bmatrix} \tag{2.13}$$

will map any polynomial of degree N in x, y into another in x', y', also of degree N. Functions expressible in terms of coordinates (x, y) therefore remain exactly expressible in the coordinate quantities (x', y') also. Similarly, any function which can be expressed exactly in terms of triangle polynomials can still be so expressed after a linear coordinate transformation (translation or rotation). The orientation and placement of a triangular element, relative to the global x–y system of axes, is therefore irrelevant; coordinate transformations do not change the properties of the triangle functions as approximating functions. The simplex elements are said to be *geometrically isotropic*, a property which is the direct consequence of algebraic completeness.

Rectangular and brick elements differ from triangles and tetrahedra in one major respect: they do not enjoy the same property of geometric isotropy as the simplexes. The nine-noded rectangle of Fig. 7.1, for example, has an associated family of nine polynomials. If the element is centred on the origin of the x-y coordinate system, then the monomial terms making up this family all belong to the set

$$\mathbb{M}^{(R)} = \left\{ 1, x, y, x^2, xy, y^2, xy^2, x^2y, x^2y^2 \right\}. \tag{2.14}$$

This function set contains all six possible monomial terms to degree 2. Although it contains some members of degree 4 in x and y jointly, this set is clearly not complete to degree 4, nor even to degree 3. It is, for example, unable to express the simple cubic polynomial $(x^3 - 1)$. Along any line $x = $ constant, such an element is capable of modelling a quadratic function only, yet along the line $x = y$ a cubic or quartic can be expressed exactly. The set of functions (2.14) is therefore not complete. The orientation of such an element within the global coordinate system may affect its accuracy; the element is *geometrically anisotropic*. To make it isotropic, its function set must be made complete, either by reducing it to contain no functions of degree 3 or higher, or else enlarging it by $\{x^3, y^3\}$ to be complete to degree 3, and additionally by the set $\{x^3y, xy^3, x^4, y^4\}$ to be complete to degree 4.

Along the edges of the nine-noded element of Fig. 7.1, the function variation is exactly quadratic. Since there are three nodes along each edge, precisely the number required to determine a quadratic, such elements will yield continuous solutions in all cases. A similar argument applies to rectangles with more than three nodes along each edge, and to the corresponding brick elements. The error properties of such elements are similar to those of triangles of equal complete order, the element of Fig. 7.1 being similar to a second-order (six-noded) triangular element. This conclusion may be drawn readily from an analysis similar to that given for triangles: error will be dominated by the first missing Taylor's series term, which generally corresponds to the lowest-order term omitted from otherwise complete families of polynomials.

Because the function behaviour along rectangle edges is precisely polynomial of degree N, it is identical to function behaviour along the edges of a triangle with the same number of edge nodes. Consequently rectangular and triangular elements of like order (i.e., with the same number of edge nodes) can be intermixed in problem solving, without loss of function continuity at interelement edges. Exactly the same arguments can be made for tetrahedra and bricks in three dimensions. Indeed it is in three-dimensional problems that Cartesian product elements become particularly useful, for visualization of collections of tetrahedra is often very difficult.

2.3   *Supercompleteness and redundant nodes*

Cartesian product elements are sometimes said to be *supercomplete*; their error behaviour is substantially determined by the number of edge nodes, but the elements generally contain many more nodes, hence possess more degrees of freedom in their sets of approximating functions, than would be necessary and justified by the number of edge nodes. For example, a rectangular element with four nodes along each edge has a total of 16 nodes. Its convergence properties are substantially those of cubic triangles, which also have four edge nodes. A full family of cubic polynomials, however, contains 10 degrees of freedom rather than the 16 possessed by rectangular elements of degree 3.

Supercompleteness is an element property of no great value or practical utility. Various attempts have therefore been made to construct elements which still have the correct number of edge nodes to guarantee solution continuity, but have fewer nodes in the element interior. In general, the Cartesian product rectangles complete to degree N have $(N + 1)^2$ nodes. Interelement continuity can be guaranteed with only $4N$ nodes, $N + 1$ along each edge (including the corners), while polynomial completeness to degree N requires $(N + 2)(N + 1)/2$. Similarly, the brick elements obtained by Cartesian product rules have $(N + 1)^3$ nodes, many more than needed for either completeness or interelement continuity.

The number of nodes that a polynomial finite element must have to be at least complete to degree N is clearly dictated by two considerations: it must have exactly $N + 1$ nodes along each edge, and it must contain at least $(N + 2)(N + 1)/2$ nodes altogether, otherwise there are not enough degrees of freedom for the necessary number of independent polynomials. The Cartesian product elements have more nodes than necessary in all cases except $N = 1$. The minimum permissible number of nodes is determined in the low-order elements by the need for enough nodes along each edge, while in high-order elements the number needed for completeness predominates, as may be seen from Table 7.1.

Rectangular elements containing only boundary nodes can be built for element orders up to 4, since the number of nodes needed to guarantee completeness is automatically provided by satisfying the requirements of edge continuity. For fifth- and higher-order elements, the number of nodes to ensure polynomial completeness rises more rapidly than the number necessary for continuity at edges. However, the edge-continuity requirement is generally the more stringent one in practice, since the use of elements of fifth- or higher-order is comparatively uncommon.

2.4   *Boundary-node elements: construction*

There are two good ways of constructing rectangular finite elements with the minimally required number of nodes. One approach is direct construction: the necessary approximating functions are defined so as to be interpolative on the element nodes and provide continuity along element edges. The other method is sometimes referred to as *nodal condensation*. This procedure begins with a Cartesian product element and

Table 7.1 Nodes needed on rectangular elements

N	$(N + 1)^2$	$4N$	$\dfrac{(N + 2)(N + 1)}{2}$	Minimum complete	Interior nodes
1	4	4	3	4	–
2	9	8	6	8	–
3	16	12	10	12	–
4	25	16	15	16	–
5	36	20	21	21	1
6	49	24	28	28	4
7	64	28	36	36	8

eliminates the unwanted nodes by making their potential values dependent on the others in a prescribed way.

Taking the direct approach to element construction, the low-order boundary node elements may be built up very simply. To illustrate, consider the twelve-noded element of Fig. 7.2(*a*). In order for the polynomials in question to be interpolative on the point set chosen, all polynomials but one must vanish on any one node; and at that node, the selected polynomial must have unity value. For the node indicated in Fig. 7.2(*b*), such a polynomial can be constructed by multiplying four linear terms,

$$\alpha_2(u, v) = -\frac{27}{16}(u + 1)(u - 1)(v + 1)\left(u - v + \frac{4}{3}\right). \qquad (2.15)$$

The leftmost factor $(u + 1)$ ensures that the polynomial $\alpha_2(u, v)$ assumes zero value along the left element edge, $u = -1$; the second factor forces the polynomial to zero at the right edge, and $(v + 1)$ makes it vanish along the bottom edge. The fourth factor vanishes along a diagonal line passing through the remaining point, as in Fig. 7.2(*b*). The numeric factor $-27/16$ normalizes the resulting product so as to give it unity value at the relevant interpolation node. In a similar way, the interpolation function associated with the corner node in Fig. 7.2(*c*) is

$$\alpha_1(u, v) = -\frac{27}{256}(u + 1)(v + 1)\left(u - v + \frac{4}{3}\right)$$
$$\left(u - v - \frac{4}{3}\right)\left(-u + v + \frac{4}{3}\right). \qquad (2.16)$$

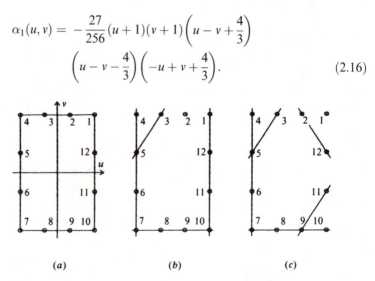

(*a*)   (*b*)   (*c*)

Fig 7.2   Twelve-noded boundary-node element. (*a*) Reference coordinate orientation and node numbering. (*b*) Construction of interpolation function for node 2. (*c*) Construction of interpolation function for node 1, a corner node.

Again the five factors supply four nodal lines of zero value, and the leading constant $-27/256$ ensures unity value at interpolation node 1. For the twelve-noded rectangle, no further polynomial functions are needed — coordinate transformations readily generate the remaining ten. As with the triangular elements and Cartesian product rectangles, a great many symmetries can be exploited in element matrix construction, so that only 16 entries in the 12×12 matrix $\mathbf{T}$ are independent.

Completeness to degree 3 of the polynomials obtained by the procedure of Fig. 7.2 can be proved by enumeration, showing that each of the ten monomial terms $u^i v^j$ making up the complete cubic family in u, v has an exact representation in terms of the twelve interpolation polynomials $\alpha_k(u, v)$ of the boundary-node element. To do so, twelve sets of coefficients c_k^{ij} are calculated, such that

$$u^i v^j = \sum_k c_k^{ij} \alpha_k(u, v). \tag{2.17}$$

These are found by straightforward (though tedious) algebraic substitution. If such a set of coefficients exists, every monomial is exactly expressible, so every polynomial must be also (since it is a linear combination of the monomials).

Similar systematic constructive procedures can be used to derive boundary-node rectangle matrices of higher orders. For order 5 and above, interior nodes are necessary, as indicated in Table 7.1. Their placement is not in all cases obvious, nor even unique, so there is some latitude for variation between similar but not altogether identical elements whose convergence properties are virtually the same. In fact, the boundary-node family of rectangular elements originally developed quite unsystematically and they were called *serendipity* elements for that reason.

2.5   *Boundary-node elements: condensation*

The second approach to constructing rectangular elements with minimal numbers of nodes begins with the full Cartesian product element and eliminates all unwanted nodes by a procedure usually called *condensation*. Suppose the element in question is embedded in a finite element mesh for some problem. Let the complete matrix equation of the entire problem be written in partitioned form as

$$\begin{bmatrix} \mathbf{M}_{rr} & \mathbf{M}_{rw} & 0 \\ \mathbf{M}_{wr} & \mathbf{M}_{ww} & \mathbf{M}_{wu} \\ 0 & \mathbf{M}_{uw} & \mathbf{M}_{uu} \end{bmatrix} \begin{bmatrix} \mathbf{U}_r \\ \mathbf{U}_w \\ \mathbf{U}_u \end{bmatrix} = \begin{bmatrix} \mathbf{R}_r \\ \mathbf{R}_w \\ \mathbf{R}_u \end{bmatrix}, \tag{2.18}$$

where the subscript w identifies the *wanted* nodes of the element, u identifies the *unwanted* nodes, and r identifies the *remaining* nodes, which do not belong to the element under investigation at all. The symbols $\mathbf{U}_r$, $\mathbf{U}_w$, and $\mathbf{U}_u$ denote the submatrices of potential values at the three classes of nodes, while the right-hand (source) vector $\mathbf{R}$ and the square coefficient matrix $\mathbf{M}$ are partitioned conformably. $\mathbf{M}$ of course depends on the physical nature of the problem; for example, $\mathbf{M} = \mathbf{S}$ in solving a problem of Laplace's equation, while $\mathbf{M} = \mathbf{S} + k^2\mathbf{T}$ for the Helmholtz equation. It should be noted that the unwanted nodes are all in the element interior and therefore not directly connected to nodes in the rest of the mesh; hence the submatrices $\mathbf{M}_{ru}$ and $\mathbf{M}_{vr}$ vanish.

Condensation proceeds by separating out that portion of Eq. (2.18) which deals with unwanted nodes, creating two distinct matrix equations:

$$\begin{bmatrix} \mathbf{M}_{rr} & \mathbf{M}_{rw} \\ \mathbf{M}_{wr} & \mathbf{M}_{ww} \end{bmatrix} \begin{bmatrix} \mathbf{U}_r \\ \mathbf{U}_w \end{bmatrix} + \begin{bmatrix} 0 \\ \mathbf{M}_{wu} \end{bmatrix} \mathbf{U}_u = \begin{bmatrix} \mathbf{R}_r \\ \mathbf{R}_w \end{bmatrix}, \tag{2.19}$$

$$\begin{bmatrix} 0 & \mathbf{M}_{uw} \end{bmatrix} \begin{bmatrix} \mathbf{U}_r \\ \mathbf{U}_w \end{bmatrix} + \mathbf{M}_{uu}\mathbf{U}_w = \mathbf{R}_u. \tag{2.20}$$

The latter equation is easily solved for the unwanted potentials $\mathbf{U}_u$ in terms of the wanted ones:

$$\mathbf{U}_u = \mathbf{M}_{uu}^{-1}\mathbf{R}_u - \mathbf{M}_{uu}^{-1}\mathbf{M}_{uw}\mathbf{U}_w. \tag{2.21}$$

Note that the potentials at other elements do not enter into this equation, only the wanted potentials in the element being condensed. Substituting Eq. (2.21) into (2.19), the unwanted potentials are eliminated:

$$\begin{bmatrix} \mathbf{M}_{rr} & \mathbf{M}_{rw} \\ \mathbf{M}_{wr} & \mathbf{M}_{ww} - \mathbf{M}_{wu}\mathbf{M}_{uu}^{-1}\mathbf{M}_{uw} \end{bmatrix} \begin{bmatrix} \mathbf{U}_r \\ \mathbf{U}_w \end{bmatrix} = \begin{bmatrix} \mathbf{R}_r \\ \mathbf{R}_w - \mathbf{M}_{wu}\mathbf{M}_{uu}^{-1}\mathbf{R}_u \end{bmatrix}. \tag{2.22}$$

It must be emphasized that the submatrices $\mathbf{M}_{rr}$, $\mathbf{M}_{rw}$ and $\mathbf{M}_{wr}$, as well as the source subvector $\mathbf{R}_r$, remain unmodified. In other words, no matrix entry is affected by the condensation process in any way so long as it is not directly connected to the element being modified. It is therefore not actually necessary to possess any knowledge of the remaining parts of the finite element mesh while the condensation proceeds; it can be carried out on one element in isolation of all the others. To state the matter briefly: the element matrix for the modified element is obtained by partitioning the node set of its unmodified parent, and using $\mathbf{M}_{ww} - \mathbf{M}_{wu}\mathbf{M}_{uu}^{-1}\mathbf{M}_{uw}$ as the describing matrix of its reduced offspring. The source term, should there be any, must of course also be replaced by $\mathbf{R}_w - \mathbf{M}_{wu}\mathbf{M}_{uu}^{-1}\mathbf{R}_u$.

The condensation process may be carried out either algebraically, thus arriving at explicit expressions for the new approximating functions, or numerically. Although both approaches are perfectly proper, most finite analysts tend to reserve their algebraic energies for the direct construction of functions and to perform condensation numerically.

3. Quadrilaterals and hexahedra

The rectangular elements described above are of limited use as they stand, for rectangles are a very restrictive geometric shape. Much more interest attaches to the possibility of constructing general, including even curvilinear quadrilaterals in two dimensions and convex hexahedra ('squashed bricks') in three. This section outlines the general method of deriving such elements from rectangles by coordinate transformations, and shows how rectilinear quadrilaterals can be obtained and used. The next section then generalizes this technique to apply to curvilinear elements as well.

3.1 *Transformation of rectangular elements*

General quadrilaterals and hexahedra can be created out of true rectangles and parallelepipeds ('bricks') by applying coordinate transformations that turn squares in the u–v coordinate system into general quadrilaterals in x–y, and cubes in u–v–w into general hexahedra in x–y–z. The coordinate transformations then carry the approximating functions used on each square or cubic element into a family of interpolation functions on the transformed elements, and it only remains to carry through the work of calculating the element $\mathbf{S}$ and $\mathbf{T}$ matrices.

When constructing finite element matrices for quadrilateral or hexahedral elements, it is necessary to differentiate and integrate potential functions with respect to the global x–y (or x–y–z) coordinates. However, derivatives with respect to u and v are much easier to calculate than derivatives with respect to the global coordinates x–y because the element interpolation functions are known in terms of the local coordinates u–v of each element. To form global derivatives of some function f, recourse may be had once more to the chain rule of differentiation. In two dimensions, the chain rule leads

$$
\begin{bmatrix} \dfrac{\partial}{\partial x} \\[2mm] \dfrac{\partial}{\partial y} \end{bmatrix} f = \begin{bmatrix} \dfrac{\partial u}{\partial x} & \dfrac{\partial v}{\partial x} \\[2mm] \dfrac{\partial u}{\partial y} & \dfrac{\partial v}{\partial y} \end{bmatrix} \begin{bmatrix} \dfrac{\partial}{\partial u} \\[2mm] \dfrac{\partial}{\partial v} \end{bmatrix} f. \tag{3.1}
$$

The square matrix on the right will be recognized immediately as the Jacobian matrix $\mathbf{J}$ of the coordinate transformation $\mathfrak{J}$ which maps points in the x–y plane into corresponding points in the u–v plane. It is often given the abbreviated notation

$$\mathbf{J} = \frac{\partial(u, v)}{\partial(x, y)} \tag{3.2}$$

in the literature of differential geometry and analysis. Because it describes the relationship of two coordinate systems, the order of $\mathbf{J}$ is always equal to the dimensionality of the geometric space in question, i.e., 2×3 or 3×3. Of couse, this matrix contains position functions rather than numeric values in all but exceptional cases.

The coordinate transformation $\mathfrak{J}$ that maps global coordinates x, y into local coordinates u, v is rarely available in a form convenient for setting up its Jacobian matrix. The usual procedure in finite element analysis is to transform standard rectangular elements into shapes prescribed by the boundary-value problem, so the transformations are set up in opposite sense, as prescriptions $\mathfrak{J}^{-1}$ which show how the standard values of u, v are to be mapped into the given values of x, y on the irregular rectangular (or brick) element. Inverting $\mathfrak{J}^{-1}$ analytically is possible in principle but rarely practical. Fortunately, the inverse transformation can be avoided. It is known from differential geometry that the Jacobian $\mathbf{J}(\mathfrak{J}^{-1})$ of the inverse transformation $\mathfrak{J}^{-1}$ is the inverse of the Jacobian of the forward transformation $\mathfrak{J}$:

$$\mathbf{J}^{-1}(\mathfrak{J}) = \left[\frac{\partial(x, y)}{\partial(u, v)}\right]^{-1} = \left[\frac{\partial(u, v)}{\partial(x, y)}\right] = \mathbf{J}(\mathfrak{J}^{-1}). \tag{3.3}$$

Hence the most common procedure for setting up finite element matrices is to compute the derivatives the easy way around, then to invert the Jacobian of the inverse transformation numerically. Derivatives are therefore calculated by the rule

$$\begin{bmatrix} \dfrac{\partial}{\partial x} \\[2mm] \dfrac{\partial}{\partial y} \end{bmatrix} f = \begin{bmatrix} \dfrac{\partial x}{\partial u} & \dfrac{\partial y}{\partial u} \\[2mm] \dfrac{\partial x}{\partial v} & \dfrac{\partial y}{\partial v} \end{bmatrix}^{-1} \begin{bmatrix} \dfrac{\partial}{\partial u} \\[2mm] \dfrac{\partial}{\partial v} \end{bmatrix} f. \tag{3.4}$$

Inverting the Jacobian is a simple matter, since only a 2×2 matrix is encountered in setting up rectangular elements; the matrix order is independent of the number of nodes on the element. Correspondingly, a 3×3 Jacobian matrix applies to brick elements of any kind.

Various element shapes can be generated by coordinate transformations. For example, the fundamental square element which spans $-1 \geq u \geq +1$, $-1 \geq v \geq +1$, as in Fig. 7.3(*a*), can be transformed easily into an arbitrary convex quadrilateral, as in Fig. 7.3(*b*), or into even more

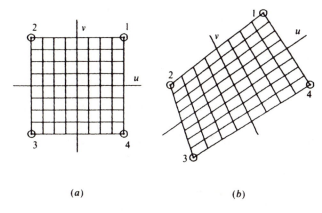

Fig 7.3   Coordinate transformation to produce general quadrilateral elements. (*a*) Square reference element in *u*–*v* plane. (*b*) Corresponding element in *x*–*y* plane; note that the *u*, *v* axes map into straight but not orthogonal lines.

complicated shapes. Corresponding transformations can be defined for bricks and (less commonly) for triangles and tetrahedra, indeed for any finite element. The finite element model of a given boundary-value problem will often be easiest to build using a variety of element shapes, so that each element in the final model will have been generated from a basic prototypal element (e.g., the square) by a different coordinate transformation. Of course, the transformations which are admissible for this purpose cannot be completely arbitrary. The following restrictions must be observed:

(1) The transformations must be unique: a given point (x, y) must map into exactly one point (u, v).
(2) The inverse transformations must also be unique: a given point (u, v) must be the image of exactly one point (x, y).
(3) Transformations for neighbouring elements must be compatible. Adjacent element sides must match, and the principal continuity conditions of the problem must be satisfied.

In addition, the transformation equations should be simple enough to require no great computing effort. While desirable, this clearly does not define a basic requirement, only a desirable property.

Two closely related types of transformation are popularly employed in finite element programs, the quadrilateral and isoparametric transformations. These will be examined in the following.

3.2   *Quadrilateral elements*

A useful element shape easily derivable from rectangles is the general quadrilateral. Quadrilateral elements provide flexibility in geometric modelling that is comparable to that of triangles, and are used by many analysts in preference to triangles, where triangular elements might be an alternative reasonable choice. Since both are rectilinear figures, there is no problem solvable with quadrilaterals that cannot be solved using triangular elements. However, there are many circumstances, especially in three dimensions, where visualization is greatly aided by topological regularity, leading to a preference for quadrilaterals and hexahedra over simplex elements.

Quadrilateral elements are readily created by transforming square elements in a *u–v* reference space into quadrilaterals in *x–y*, using the bilinear coordinate transformation

$$
\left.
\begin{aligned}
x &= \frac{1}{4}(x_1 - x_2 + x_3 - x_4)uv + \frac{1}{4}(x_1 - x_2 - x_3 + x_4)u \\
&\quad + \frac{1}{4}(x_1 + x_2 - x_3 - x_4)v + \frac{1}{4}(x_1 + x_2 + x_3 + x_4), \\
y &= \frac{1}{4}(y_1 - y_2 + y_3 - y_4)uv + \frac{1}{4}(y_1 - y_2 - y_3 + y_4)u \\
&\quad + \frac{1}{4}(y_1 + y_2 - y_3 - y_4)v + \frac{1}{4}(y_1 + y_2 + y_3 + y_4).
\end{aligned}
\right\} \tag{3.5}
$$

The subscripted quantities refer to the vertex numbering of Fig. 7.3(*a*); the four corners of the square are numbered in the usual counterclockwise fashion, beginning in the first quadrant. It is easy to verify that the four corners of the square reference element in the *u–v* coordinate plane map into the four corners $(x_i, y_i), i = 1, \ldots, 4$. The transformation is bilinear, i.e., linear in *u* and *v* taken individually; straight lines in the *u* plane therefore map into straight lines in the *x–y* plane. Its Jacobian is easily calculated as

$$
\mathbf{J} = \begin{bmatrix} (x_1 - x_2 + x_3 - x_4)v & (y_1 - y_2 + y_3 - y_4)v \\ +(x_1 - x_2 - x_3 + x_4) & (y_1 - y_2 - y_3 + y_4) \\[2mm] (x_1 - x_2 + x_3 - x_4)u & (y_1 - y_2 + y_3 - y_4)u \\ +(x_1 + x_2 - x_3 - x_4) & +(y_1 + y_2 - y_3 - y_4) \end{bmatrix}. \tag{3.6}
$$

This matrix reduces to constant values if the coefficients of *u* and *v* vanish, i.e., if

$$
\left.
\begin{aligned}
x_1 - x_2 &= x_4 - x_3, \\
y_1 - y_2 &= y_4 - y_3,
\end{aligned}
\right\} \tag{3.7}
$$

which says, in effect, that opposite sides should be of the same length. This simplification occurs for parallelograms, geometric figures far more flexible than squares or rectangles.

Approximating functions for quadrilateral elements are easy to construct, using the general procedure already set out. The key point is to define the elements initially in the u–v plane where the construction is easy, then to transform into the x–y plane. In fact, the reference element in the u–v coordinates is a square, so it is only necessary to define one approximating function, the other three are readily obtained by coordinate interchanges. The function associated with node 1 of Fig. 7.3(a) is

$$\alpha_1(u, v) = \frac{1}{4}(u + 1)(v + 1) \tag{3.8}$$

and the other three immediately follow:

$$\left.\begin{aligned}
\alpha_2(u, v) &= \alpha_1(-u, v) = -\frac{1}{4}(u - 1)(v + 1) \\[4pt]
\alpha_3(u, v) &= \alpha_2(u, -v) = +\frac{1}{4}(u - 1)(v - 1) \\[4pt]
\alpha_4(u, v) &= \alpha_3(-u, v) = -\frac{1}{4}(u + 1)(v - 1).
\end{aligned}\right\} \tag{3.9}$$

Explicit algebraic expressions for the approximating functions are not so easy to state in terms of the x–y coordinates, for it is necessary to invert the coordinate transformation of Eq. (3.6) to obtain them.

The Jacobian **J** of a coordinate transformation $\mathfrak{J}$ expresses the local geometric properties of $\mathfrak{J}$. Its magnitude det(**J**) denotes the local area magnification while its individual components show the relative twisting and stretching in the different coordinate directions. If the Jacobian reaches zero value somewhere in a transformed finite element, the area surrounding that point in u–v shrinks to nothingness in x–y, the transformation is not unique, and no useful results are likely to be obtained. The question of major interest is therefore: when does the Jacobian (3.6) vanish?

The Jacobian (3.6) of a bilinear transformation vanishes whenever an elementary area $du\,dv$ is transformed into zero area in the x–y plane. Since a bilinear transformation maps straight lines into straight lines, this can only happen if two coordinate lines coalesce into one, or cross. How this can happen is illustrated in Fig. 7.4, which shows three mappings of the reference element into the x–y plane. The mapping of Fig. 7.4(a) turns u–v coordinate lines into straight but nonparallel lines in the x–y plane; however, all coordinate lines are clearly distinct and all square area patches between coordinate lines map into distinct, though unequal, patches in x–y. Figure 7.4 (b) shows a mapping which is similar except

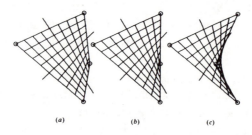

Fig 7.4 Quadrilateral elements obtained from square reference element by coordinate transformations. (*a*) Stable transformation. (*b*) Mutually perpendicular sides in *u–v* become collinear in *x*, potentially unstable. (*c*) Singular coordinate transformation.

that the rightmost vertex of the element in 7.4(*a*) has moved left until it is exactly collinear with two others. In this circumstance, the coordinate lines must cross at that vertex, and the Jacobian becomes singular there. A still more extreme case is illustrated in Fig. 7.4(*c*), where the quadrilateral in the *x–y* plane is no longer convex. There are clearly multiple crossings of coordinate lines, the Jacobian taking on positive, zero and negative values in the element. Clearly, the general quadrilateral element — and by a similar argument, the general hexahedral element — must be restricted to be convex. No such restriction was needed for triangles and tetrahedra, since those figures are convex anyway.

4.   Isoparametric elements

Of the many conceivable coordinate transformations which come close to fulfilling the requirements already enumerated, one family is especially close in spirit and form to the finite element method, and is therefore widely used. This family uses coordinate mappings derived from the element approximating functions themselves. It produces curvilinear elements very well suited to modelling curved shapes so that quite complicated geometric shapes can often be represented with a surprisingly small number of elements.

4.1   *Isoparametric transformations*

The central idea of isoparametric transformations is simple enough. Over any finite element of given order N in the *u–v* plane, the potential function f, or indeed any other scalar function, is approximated interpolatively in terms of the element functions,

$$f = \sum_i f_i \alpha_i(u, v), \qquad (4.1)$$

exactly as has been done in all the elements considered so far. The key idea of isoparametric elements lies in a simple observation: The global coordinate values x, y are functions of the local coordinates u, v on the element. A given pair of transformation functions $x(u, v)$, $y(u, v)$ are nothing more than scalar functions defined over the element, and may therefore be described in terms of the element approximating functions, exactly like any other scalar function:

$$\left.\begin{aligned} x &= \sum_i x_1 \alpha_i(u, v), \\ y &= \sum_i y_i \alpha_i(u, v). \end{aligned}\right\} \quad (4.2)$$

The functions $\alpha_i(u, v)$ are polynomials. The transformation (4.2) is therefore nonlinear and the resulting element shape is not rectilinear. Figure 7.5 shows the result: the square reference element of Fig. 7.5(a) is mapped into the curved shape of Fig. 7.5(b).

If x and y are exactly expressible as polynomials of suitably low degree in u, v then the transformation (4.2) is exact. Such transformations, in which the same family of approximation functions is used to approximate the potential and also to express the element shape transformation, are called *isoparametric*, and the resulting elements are known as *isoparametric elements*.

The name *isoparametric* refers to the fact that the same degree of approximation, indeed the same space of approximating functions, is used to model the potential function and the geometric shape (via the element shape transformation). It will be immediately evident that the functions used to model the unknown potential in Eq. (4.1) need not necessarily be the same as the functions used in (4.2), though as a matter of convenience they often are. It is perfectly acceptable in prin-

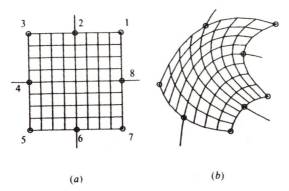

(a) (b)

Fig 7.5 Isoparametric mapping. (a) Square reference element. (b) The corresponding curvilinear quadrilateral.

ciple to model the geometric shape by some lower-order transformation. The resulting elements are sometimes called *subparametric*; conversely, the term *superparametric* describes elements whose shape transformation functions are of a higher order than the functions used to approximate the potential. For example, high-order elements of general quadrilateral shape are subparametric: the quadrilateral element transformation (3.5) is of lower degree than the potental approximating functions. In fact, subparametric elements are very common, for they include squares, straight-sided quadrilaterals and all the high-order simplex elements. Superparametric elements are very rarely encountered in practice.

The isoparametric elements most widely used are probably the eight-noded curvilinear quadrilaterals shown in Fig. 7.5. These elements are supercomplete beyond degree 2 in the reference coordinates u–v and therefore show some geometric anisotropy. Nevertheless, they yield good approximations to the potential function and also allow a great deal of geometric flexibility, while still being reasonably simple to program and fast to compute. Approximating functions appropriate to eight-noded isoparametric elements are constructed by a straightforward process of placing nodal lines in the u–v plane so as to ensure the right distribution of zeros, then normalizing to unity value at the remaining node:

$$
\left.
\begin{aligned}
\alpha_1(u, v) &= \frac{1}{4}(u + 1)(v + 1)(u + v - 1)\\[4pt]
\alpha_2(u, v) &= \frac{1}{2}(u + 1)(v + 1)(-u + 1)\\[4pt]
\alpha_3(u, v) &= \frac{1}{4}(-u + 1)(v + 1)(-u + v - 1)\\[4pt]
\alpha_4(u, v) &= \frac{1}{2}(-v + 1)(v + 1)(-u + 1)\\[4pt]
\alpha_5(u, v) &= \frac{1}{4}(u - 1)(-v + 1)(u + v + 1)\\[4pt]
\alpha_6(u, v) &= \frac{1}{2}(u + 1)(-v + 1)(-u + 1)\\[4pt]
\alpha_7(u, v) &= \frac{1}{4}(u + 1)(-v + 1)(u - v - 1)\\[4pt]
\alpha_8(u, v) &= \frac{1}{2}(u + 1)(v + 1)(-v + 1).
\end{aligned}
\right\}
\tag{4.3}
$$

The square eight-noded element is algebraically as well as geometrically symmetric about three axes, so a good deal of work can be saved. Only two of its approximating functions actually need to be constructed, one associated with a corner node and one with a mid-side node. The remaining six are obtained by coordinate rotations and reflections.

Eight-noded isoparametric elements are able to follow arbitrary curves surprisingly well, as a simple example will illustrate. Figure 7.6 shows two finite element models of a circular rod embedded in a rectangular region (e.g., a dielectric rod in a waveguide or a round conductor in a hollow cavity). The upper model comprises only three elements: a single eight-noded isoparametric element to represent one quarter of the circular rod, two more to fill the space outside it. The lower model still uses two elements in the exterior region, but subdivides the circular rod into three elements. The circular boundary is accurately represented in both cases.

Isoparametric curvilinear elements, unlike straight-sided quadrilaterals, need not be convex; indeed the element shown in Fig. 7.5(*b*) is not. From the smooth behaviour of its coordinate lines, however, it is

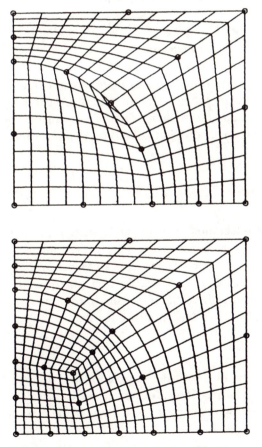

Fig 7.6 Two models for a circular object in a rectangular space. Collinear sides may lead to marginal numerical stability of upper model; lower one is stable.

clear that the Jacobian in this case does not vary much across the element, and certainly does not approach zero anywhere. On the other hand, the upper model of Fig. 7.6 is likely to lead to numerical instability and large error because the transformation that models the quarter-circle by a single element is singular at its upper right vertex — even though the element is clearly convex. The difficulty becomes clear on closer examination. At the node in question, two element sides meet; these are orthogonal to each other in the u–v reference plane, but have become collinear in the x–y plane. Such a transformation is only possible if the Jacobian is singular at the point where the coordinate lines meet, i.e., at the element node. The lower model in Fig. 7.6, on the other hand, maintains rough local orthogonality of coordinate lines everywhere. The transformation is thus more nearly conformal, the Jacobian is bounded both above and below, and any computation undertaken is likely to prove stable and accurate.

4.2   *Isoparametric element matrices*

Formation of the element matrices for isoparametric elements follows the usual principles but involves more computation (and less algebraic analysis) than is the case for simplexes. Consider again the transformation that maps the square reference element of Fig. 7.5(a) into the curvilinear element of Fig. 7.5(b). As usual, the matrix **T** is formed by evaluating

$$T_{ij} = \int_{\Omega_e} \alpha_i \alpha_j \, d\Omega, \tag{4.4}$$

where the region of integration Ω_e is the finite element in the real geometric space described by coordinates x–y or x–y–z (not the reference space u–v–w). Integration over this generally curvilinear region is often quite difficult; it is much easier to integrate over the square in the u–v plane. An area element $du \, dv$ of the u–v plane transforms into an area element in the x–y plane in accordance with the rule

$$dx \, dy = \det(\mathbf{J}) \, du \, dv, \tag{4.5}$$

where $\det(\mathbf{J})$ is the determinant of the Jacobian matrix **J** of the coordinate transformation. The area integral over the element surface Ω_e may therefore be evaluated by computing the corresponding integral in the u–v plane instead,

$$T_{ij} = \int_{-1}^{+1} \int_{-1}^{+1} \alpha_i \alpha_j \det(\mathbf{J}) \, du \, dv. \tag{4.6}$$

This integral is usually evaluated numerically, for although it is not in principle difficult, analytic integration is often tedious. The Dirichlet matrix $\mathbf{S}$ is constructed by a procedure based on similar principles but a little more complicated in its details. The typical matrix entry again takes the usual form,

$$S_{ij} = \int_{\Omega_e} \nabla \alpha_i \cdot \nabla \alpha_j \, d\Omega, \tag{4.7}$$

the differentiations being understood to refer to the local geometric co-ordinates x, y. The surface integration can be referred to the u–v plane in the same way as for the matrix $\mathbf{T}$,

$$S_{ij} = \int_{-1}^{+1} \int_{-1}^{+1} \left(\frac{\partial \alpha_i}{\partial x} \frac{\partial \alpha_j}{\partial x} + \frac{\partial \alpha_i}{\partial y} \frac{\partial \alpha_j}{\partial y} \right) \det(\mathbf{J}) \, du \, dv \tag{4.8}$$

but this does not suffice, since differentiation with respect to x and y raises problems in its turn. To move the differentiations onto u and v instead, the chain rule (3.1) may be invoked once again. For brevity in notation, define differentiation operators

$$\mathfrak{D}_{xy} = \begin{bmatrix} \dfrac{\partial}{\partial x} \\ \dfrac{\partial}{\partial y} \end{bmatrix} \quad \text{and} \quad \mathfrak{D}_{uv} = \begin{bmatrix} \dfrac{\partial}{\partial u} \\ \dfrac{\partial}{\partial v} \end{bmatrix}. \tag{4.9}$$

In this notation, Eq. (3.4) reads

$$\mathfrak{D}_{xy}f = \mathbf{J}^{-1}\mathfrak{D}_{uv}f. \tag{4.10}$$

Substituting in Eq. (4.8), the matrix $\mathbf{S}$ is thus expressible entirely in the reference coordinates u, v:

$$S_{ij} = \int_{-1}^{+1} \int_{-1}^{+1} (\mathfrak{D}_{uv}\alpha_i)^{\mathrm{T}} \mathbf{J}^{-\mathrm{T}} \mathbf{J}^{-1} (\mathfrak{D}_{uv}\alpha_j) \det(\mathbf{J}) \, du \, dv. \tag{4.11}$$

Here the superscript T denotes the transpose, $-\mathrm{T}$ the inverse transpose. Since only the reference coordinates u, v are involved, all the required calculations are easy to state and formulate. The integrand of (4.11) is unfortunately a little complicated, involving as it does the inverse Jacobian matrix $\mathbf{J}^{-1}$ whose four components are ratios of polynomials, not polynomials, and therefore difficult to integrate. There is not even any guarantee the Jacobian is invertible at all, for there can be no certainty that the isoparametric mappings are always one-to-one. In fact, there exist many nonunique isoparametric and even quadrilateral transformations, so the problem is potentially much more serious than the mere absence of formal proof.

The inverse Jacobian of a coordinate transformation can be found in analytic terms by computational symbolic algebra; but it is rarely practical to do so. When evaluated and printed, each entry in $\mathbf{J}^{-1}$ for the eight-noded isoparametric element occupies about two pages! Further, even when the inverse has been found, the analytic task has only begun, for the integration of rational fractions is much more difficult still. All practical general-purpose finite element programs therefore integrate numerically. Even when the inverse Jacobian is not required, as in forming the matrix $\mathbf{T}$, numerical methods are preferred on grounds of efficiency; the evaluation of complicated algebraic expressions can require prodigious amounts of computer time.

4.3   *Integration over isoparametric elements*

Even when the finite element integrals of Eqs. (4.6) and (4.11) can be evaluated algebraically, the resulting expressions are usually of such great length and complexity that their evaluation costs more computing time than the equivalent numerical integration. Numerical integration is fortunately eased by the fact that (4.11) contains two one-dimensional integrals (with respect to u and v), so one-dimensional numerical integration techniques are wholly adequate to the task. If the coordinate transformation is one-to-one, the integrands in the $\mathbf{S}$ and $\mathbf{T}$ matrix components, Eqs. (4.6) and (4.11), are smooth, continuous and bounded throughout the element. In such cases, it is appropriate and efficient to employ Gaussian integration formulae to perform the integration numerically. An M-point Gaussian quadrature formula[1] integrates some function $f(s)$ approximately in one space direction, by calculating a weighted sum of its values at a set of coordinate values $\{s_m, m = 1, \ldots, M\}$:

$$\int_{-1}^{+1} f(s)\,\mathrm{d}s \simeq \sum_{m=1}^{M} w_m f(s_m). \tag{4.12}$$

The coordinate values s_m are referred to as *quadrature points* or *quadrature nodes*, while the numerical coefficients w_m are called *quadrature weights* or *quadrature coefficients*. Components of the matrices $\mathbf{S}$ and $\mathbf{T}$ are evaluated by substituting Gaussian integration formulae for all one-dimensional integrations:

[1] One-dimensional numerical integration is often called *quadrature* in the literature of numerical analysis; one therefore speaks of *quadrature formulae*.

$$T_{ij} = \sum_{m=1}^{M} \sum_{n=1}^{M} w_m w_n \alpha_i(u_m, v_n) \alpha_j(u_m, v_n) \det\{\mathbf{J}(u_m, v_n)\}, \qquad (4.13)$$

$$S_{ij} = \sum_{m=1}^{M} \sum_{n=1}^{M} w_m w_n \{\mathfrak{D}_{uv}\alpha_i(u_m, v_n)\}^{\mathsf{T}} \mathbf{J}^{-\mathsf{T}}(u_m, v_n)$$

$$\mathbf{J}^{-1}(u_m, v_n)\{\mathfrak{D}_{uv}\alpha_j(u_m, v_n)\}\det\{\mathbf{J}(u_m, v_n)\}. \qquad (4.14)$$

The summations in (4.13) and (4.14) run over the M Gaussian quadrature nodes in the u and v directions. Note that the word *nodes* here refers to points relevant to the integration technique, it has nothing to do with the nodes of finite elements.

A fundamental and highly desirable property of Gaussian integration is that an M-point Gaussian formula integrates exactly any polynomial of degree $2M - 1$ or lower. A reasonable choice of M can therefore be made by examining the functions to be integrated and judging what degree of polynomial might be necessary to approximate them reasonably well. For example, consider an eight-noded boundary-node element. To form its $\mathbf{T}$ matrix as given by Eq. (4.13), integration in two directions is required. The element interpolation functions are quadratic in each direction, so the product $\alpha_i\alpha_j$ is at most quartic in either u or v. By a similar argument, the determinant of the Jacobian $\det(\mathbf{J})$ is at most quartic in each variable also. The integrand is therefore always a polynomial of degree at most eight in each variable, so a five-point Gaussian quadrature formula will yield exact integration. The corresponding argument for $\mathbf{S}$ involves judgment as well as analysis, for the inverse Jacobian is not a polynomial function and the precision of approximation depends on the element shape. For reasonably smoothly curved, not very contorted, elements it usually suffices to evaluate $\mathbf{S}$ with a quadrature formula that integrates $\mathbf{T}$ exactly.

Sets of Gaussian quadrature weights and nodes are tabulated in handbooks on numerical analysis, for values of M ranging from 1 to 1024. Table 7.2 gives a short selection of formulae adequate for most finite element analysis.

4.4 *Isoparametric elements in practice*

The construction of computer programs based on isoparametric finite elements differs little from programs based on triangles, other than in the element formation itself. Indeed the high-order triangular finite element program discussed in Chapter 4 lends itself to the use of isoparametric elements as well. In fact, only two changes are needed: a subroutine to compute the element matrices must be provided, and the input routines must be made to recognize the new element type. The

Table 7.2 Gaussian integration $\displaystyle\int_{-1}^{+1} f(x)\,dx \simeq \sum_{i=1}^{M} w_i f(x_i)$

Nodes	w_i	$\pm x_i$
$M = 2$	1.000000000	0.577350269
$M = 3$	0.888888889	0.000000000
	0.555555556	0.774596669
$M = 4$	0.652145155	0.339981044
	0.347854845	0.861136312
$M = 5$	0.568888889	0.000000000
	0.478628670	0.538469310
	0.236926885	0.906179846
$M = 6$	0.467913935	0.238619186
	0.360761573	0.661209386
	0.171324492	0.932469514
$M = 7$	0.417959184	0.000000000
	0.381830051	0.405845151
	0.279705391	0.741531186
	0.129484966	0.949107912
$M = 8$	0.362683783	0.183434642
	0.313706646	0.525532410
	0.222381034	0.796666477
	0.101228536	0.960289856

HITRIA2D program recognizes element types by their number of nodes, so the latter requires adding exactly one program line:

```
C    Construct element s and t matrices
     if (nve(ie) .eq. 3  .or. nve(ie) .eq. 6  .or.
*       nve(ie) .eq. 10 .or. nve(ie) .eq. 15)
*       call trielm(ie)
     if (nve(ie) .eq. 8) call iso8el(ie)
```

The element matrix evaluation of course takes a good deal more, some 150–200 lines of source code in Fortran or C. While the techniques for programming Eqs. (4.13) and (4.14) are straightforward in principle, much of the detail is best appreciated by studying an actual working program. To help in doing so, a subroutine `iso8el` appears at the end of this chapter.

The isoparametric element subroutine given at the end of this chapter is conservatively designed and not very efficiently implemented. Numerous programming short-cuts leading to reduced computing time or storage will no doubt suggest themselves to the reader. Most of these lead to somewhat reduced legibility of the program and have therefore been avoided by the authors. Again, five-point Gaussian quadrature is a conservative choice and the interested reader may wish to experiment with integration formulae of lower orders, which results in greater program efficiency.

Isoparametric elements obviously require a great deal more computation than would be involved in a corresponding number of triangular elements. Estimates of the total work required are therefore of some interest; they can fortunately be obtained without much trouble. The evaluation of one matrix (either $\mathbf{S}$ or $\mathbf{T}$) for an N-noded isoparametric element requires computation of $(N + 1)(N + 2)/2$, or approximately $N^2/2$, matrix entries. Numerical integration over the element area requires M^2 evaluations of the integrand, each involving a number p of arithmetic operations determined by the type of element. For the eight-noded isoparametric element, about 20 multiply-and-add operations are needed to evaluate the summand in $\mathbf{T}$, another 50 or so to evaluate the somewhat more complicated expressions in $\mathbf{S}$. For a finite element mesh of E elements, the number of total arithmetic operations A is thus approximately

$$A = \frac{EpM^2N^2}{2} \tag{4.15}$$

Computing both the $\mathbf{T}$ and $\mathbf{S}$ matrices of an eight-noded isoparametric element, using five-point Gaussian quadratures, thus involves some 50 000–60 000 floating-point arithmetic operations, plus indexing calculations. This number obviously far exceeds the few hundreds of operations needed for a second-order triangular element. However, the operation count A of Eq. (4.15) is directly proportional to the number of elements E, while the computing effort eventually required to solve the finite element equations is proportional to a higher power of E, at least $E^{3/2}$ and frequently E^2 or higher. Consequently, computing expense for large problems is always dominated by the cost of solving simultaneous equations. The extra arithmetic needed to compute element matrices is therefore not a significant barrier to the use of isoparametric elements.

The behaviour of solutions obtained with reasonably good isoparametric element models may be assessed by studying representative examples. Figure 7.7 shows a static potential solution obtained for the five-element subdivision of Fig. 7.6. The physical model here comprises a round cylindrical object in a tunnel cut into a block of homogeneous material; it might arise, for example, when analysing a rectangular enclosure around a current-carrying wire, or perhaps a rectangular waveguide containing a dielectric rod. Not only do the eight-noded elements provide for close following of the round wire boundary, they also yield flux lines of great smoothness. However, these elements do not provide derivative continuity, merely function continuity. At the diagonal join between two large elements (upper right portion of Fig. 7.7) the flux lines therefore exhibit clearly visible kinks, much as would be expected in working with high-order triangular elements. There is little to choose between triangular and isoparametric elements on grounds of potential modelling accuracy; however, the curvilinear elements have an obvious advantage in reproducing curved shapes.

5.   Vector-based simplexes

Up to this point the finite element interpolation functions considered have exclusively been scalars, $\alpha_m(\mathbf{r})$, used directly in nodal expansions. For three-dimensional time or frequency-domain problems it is generally necessary to employ a vector variable. With such a variable the only option available using nodal finite elements (i.e., elements using approximation functions interpolative on element nodes) would be an expansion of the vector trial function such as

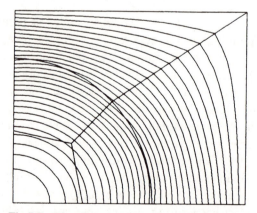

Fig 7.7 Flux lines around a circular conductor in a superconductive shielding cavity, computed using eight-noded isoparametric boundary-node elements.

$$\mathbf{U}(\mathbf{r}) = \sum_{m=1}^{M} \mathbf{U}_m \alpha_m(\mathbf{r}) \tag{5.1}$$

where $\mathbf{U}_m$ represents an array of M unknown nodal vectors, that is to say $3M$ scalars, to be solved for in the usual finite element fashion. The expansion (5.1) may be appropriate if the vector variable is one which can remain fully continuous everywhere throughout the problem domain, as is the case for vector potential $\mathbf{A}$ when employed in solving for eddy currents. On the other hand, if a field vector $\mathbf{E}$ or $\mathbf{H}$ is sought in regions with discontinuous material properties, there is a physical requirement for the normal vector component to be discontinuous at a material interface; alternatively, if a flux vector $\mathbf{D}$ or $\mathbf{B}$ is to be found, it is the tangential components which become discontinuous. In either case the expansion Eq. (5.1) is inappropriate. However, when implementing a variational or Galerkin finite element formulation, the interface conditions for normal $\mathbf{E}$ and $\mathbf{H}$ at a spatial step in either or both of ϵ and μ turn out to be natural, that is not requiring specifically to be enforced. Therefore rather than using the expansion (5.1) there is reason to seek a representation of the trial function $\mathbf{U}$, representing either $\mathbf{E}$ or $\mathbf{H}$, which conforms to these continuity rules. The working variable is expanded instead using vector interpolation functions τ,

$$\mathbf{U}(\mathbf{r}) = \sum_{m=1}^{M} U_m \tau_m(\mathbf{r}), \tag{5.2}$$

defined within finite elements, where each τ_m exhibits tangential but not normal continuity between elements. In a finite element analysis the approximations representing a field variable change abruptly from element to element. So even when modelling a uniform region there is in a sense a fictitious material discontinuity between every element and it is no bad thing to maintain this tangential continuity everywhere.

The mathematical properties of finite elements based upon the expansion Eq. (5.2) were fully explored by Nedelec (1980). Since 1980 the ideas contained in Nedelec's paper have gradually filtered into the practical finite elements community, particularly amongst those concerned with electromagnetics. The following sections set out methods of applying Nedelec's elements to the general vector electromagnetic problem.

5.1 *The natural interface condition*

The freedom allowable for the normal components of any field vector at a material discontinuity when the subject of a variational analysis may be explained as follows. Consider a functional, say in a trial electric field variable $\mathbf{E}'$,

$$\mathfrak{F}(\mathbf{E}') = \frac{1}{2} \int_\Omega (\nabla \times \mathbf{E}' \cdot \mu_r^{-1} \nabla \times \mathbf{E}' - k^2 \mathbf{E}' \cdot \epsilon_r \mathbf{E}') \, \mathrm{d}\Omega \qquad (5.3)$$

whereas $\mathbf{E}$ is the *true* solution of the vector Helmholtz equation

$$\nabla \times \mu_r^{-1} \nabla \times \mathbf{E} - k^2 \epsilon_r \mathbf{E} = 0, \qquad (5.4)$$

$k^2 = \mu_0 \epsilon_0$, within the region Ω for appropriate boundary conditions on $\partial\Omega$. From Chapter 3, Section 4.7, the first variation of $\mathfrak{F}$ for a perturbation $\mathbf{E}' = \mathbf{E} + \theta \delta\mathbf{E}$ is

$$\delta\mathfrak{F} = -\oint_{\partial\Omega} \mu_r^{-1} (\nabla \times \mathbf{E}) \times \delta\mathbf{E} \cdot \mathbf{1}_n \, \mathrm{d}S. \qquad (5.5)$$

Using the Maxwell electric curl equation gives

$$\delta\mathfrak{F} = \mathrm{j}\omega\mu_0 \oint_{\partial\Omega} \mathbf{H} \times \delta\mathbf{E} \cdot \mathbf{1}_n \, \mathrm{d}S. \qquad (5.6)$$

In the Chapter 3 discussion it was shown that $\delta\mathfrak{F}$ given by Eq. (5.6) vanishes and the functional Eq. (5.3) is made stationary about $\mathbf{E}$, providing that $\mathbf{E}'$ in Ω is sufficiently continuous and that the external problem boundaries are properly looked after. However as well, Eq. (5.6) gives valuable information about the relationship between adjacent finite elements Ω_e which are always defined in a piecewise, element-by-element fashion. The vectors $\mathbf{E}$ and $\mathbf{H}$ are true solutions of Maxwell's equations; thus by definition, their tangential components are continuous across any interelement boundary, even if ϵ and μ are discontinuous. Suppose that in all finite element operations, trial functions $\mathbf{E}'$ are chosen such that tangential components remain continuous between elements; then tangential $\delta\mathbf{E}$ is similarly continuous. Writing $\mathbf{H} = \mathbf{H}_t + \mathbf{H}_n$, $\delta\mathbf{E} = \delta\mathbf{E}_t + \delta\mathbf{E}_n$ and applying elementary rules of vector algebra, noting that the normal components $\mathbf{H}_n$, $\delta\mathbf{E}_n$ and the unit normal $\mathbf{1}_n$ are all collinear whereas in general the tangential components $\mathbf{H}_t$ and $\delta\mathbf{E}_t$ are not, gives

$$\delta\mathfrak{F}_e = \mathrm{j}\omega\mu_0 \oint_{\partial\Omega_e} \mathbf{H}_t \times \delta\mathbf{E}_t \cdot \mathbf{1}_n \, \mathrm{d}S. \qquad (5.7)$$

Now consider a pair of adjacent finite elements e_1, e_2 sharing a common surface Γ_{e12}, of course with equal and opposite normals $\pm\mathbf{1}_{ne}$. The contributions to the first variation from e_1 and e_2 are additive and may separately be determined from Eq. (5.7). Because of the continuity of $\mathbf{H}_t$ and $\delta\mathbf{E}_t$ and the opposition of the normals, these contributions cancel one another out. Hence regardless of any discontinuity of $\mathbf{E}_n$, $\mathbf{H}_n$ (as a result of spatial jumps in material properties) or of $\mathbf{E}_n'$ (from deliberate choice of vector interpolation function, regardless as to whether there is a material discontinuity or not at $\partial\Omega_e$) no net $\delta\mathfrak{F}$ arises from internal finite element interfaces. Thus all of the interelement requirements other than continuity of tangential components of the trial vector are natural to $\mathfrak{F}$. Providing that the proper external boundary constraints are maintained, namely that $\mathbf{E}_t'$ is specified on the driving planes, set to zero on electric walls and left unconstrained at magnetic walls, the choice of vector interpolation functions maintaining just tangential continuity between finite elements allows the stationary functional Eq. (5.3) to be used globally over a patchwork of finite elements. A dual argument shows that the magnetic variable version of this functional may similarly be employed, whilst the same conclusions are reached from a Galerkin weighted residual analysis.

As with nodal expansions, it is convenient first to think in terms of a single finite element; the presence of neighbouring elements is acknowledged simply by ensuring that the appropriate tangential continuity is maintained. In the first instance, $\mathbf{U}_m$ and U_m in the two types of trial function expansion are merely undetermined coefficients without any particular physical significance. In the case of an expansion about nodes $\mathbf{r}_n$, by specifying that $\alpha_m(\mathbf{r}_n) = \delta_{mn}$ shall hold, $\mathbf{U}_m$ is immediately identified with a nodal field vector. However the physical associations for U_m in the vector expansion Eq. (5.2) are not so obvious.

5.2   *Hierarchal tetrahedral elements*

Once again there is the choice of simplex or brick elements and of the trial function polynomial order; in this section the simplex will be considered. Since the topic here is primarily applicable to three-dimensional modelling, the appropriate simplex is a tetrahedron. The hierarchal scheme of vector interpolation functions introduced by Webb and Forghani (1993), allowing different orders of tetrahedra to be juxtaposed in the same mesh, is described here. Such a scheme allows selective improvement of finite element solution accuracy in particular regions of a problem space to be implemented very simply. At the same time as dealing with the vector elements it is appropriate briefly to describe the hierarchal nodal interpolation functions which logically accompany the vector ones. The two families of finite elements being fully compatible, the nodal functions can easily be introduced in cases where vector and scalar trial variables need to be used in the same problem.

Consider the general tetrahedral element defined by the nodes $(x_i, y_i, z_i), i = 1, \ldots, 4$. Simplex coordinates for such an element are defined explicitly by

$$\zeta_1 = \frac{1}{6V} \begin{vmatrix} 1 & x & y & z \\ 1 & x_2 & y_2 & z_2 \\ 1 & x_3 & y_3 & z_3 \\ 1 & x_4 & y_4 & z_4 \end{vmatrix}$$

$$\text{where } 6V = \begin{vmatrix} 1 & x_1 & y_1 & z_1 \\ 1 & x_2 & y_2 & z_2 \\ 1 & x_3 & y_3 & z_3 \\ 1 & x_4 & y_4 & z_4 \end{vmatrix}, \tag{5.8}$$

similarly $\zeta_2, \zeta_3, \zeta_4$, whilst it may be noted that $\zeta_1 + \zeta_2 + \zeta_3 + \zeta_4 = 1$. These coordinates have already been used to set up general high-order nodal interpolation functions (see Chapter 4). In order to construct vector interpolation functions examine the gradients $\nabla\zeta_i$, perhaps the simplest set of general vector functions relating to the tetrahedron simplex. An explicit expression for $i = 1$ is

$$\nabla\zeta_1 = \frac{1}{6V} \begin{vmatrix} 0 & \mathbf{1}_x & \mathbf{1}_y & \mathbf{1}_z \\ 1 & x_2 & y_2 & z_2 \\ 1 & x_3 & y_3 & z_3 \\ 1 & x_4 & y_4 & z_4 \end{vmatrix}, \tag{5.9}$$

and so forth for $i = 2, 3, 4$. Observing that the tetrahedron edge defined say by vertices 2 and 3 is the vector

$$\mathbf{e}_{23} = (x_3 - x_2)\mathbf{1}_x + (y_3 - y_2)\mathbf{1}_y + (z_3 - z_2)\mathbf{1}_z \tag{5.10}$$

it is seen by inspection that $\nabla\zeta_1 \cdot \mathbf{e}_{23} = 0$, a similar result being obtained for edges $\mathbf{e}_{24}$ and $\mathbf{e}_{34}$. Thus the vector $\nabla\zeta_1$ is perpendicular to the plane $\zeta_1 = 0$ defined by the vertices 2, 3, 4. If instead the scalar product is taken of $\nabla\zeta_i$ with the vector for an edge including the vertex i, the result is ± 1. The properties of $\nabla\zeta_i$ may be summarized as:

(i) $\nabla\zeta_i$ is a constant vector whose magnitude depends upon all four vertex coordinates.

(ii) $\nabla\zeta_i$ is perpendicular to the face $\zeta_i = 0$ (opposite to vertex i).

(iii) The component of $\nabla\zeta_i$ along an edge passing through vertex i depends only upon the end coordinates of that edge.

(iv) Any component of $\nabla\zeta_i$ parallel to a face including vertex i depends only upon the vertex coordinates of that face.

The simplex coordinates ζ themselves are C_0-continuous functions on passing from one element to another. Hence it follows that any scalar polynomial function of the ζ's multiplied by a vector $\nabla\zeta_i$ obeys the tangential continuity requirements established for vector interpolation functions to be used in variational finite element schemes. Each of the vector functions constructed in this way exhibits a specific interelement discontinuity in its normal component across edges and faces common to any of the tetrahedron vertices i used in its construction. However noting (ii) above it may be observed that there is also a normal component contribution to such edges and faces from the noncommon vertices; this gives sufficient freedom for the correct normal interelement discontinuity (or continuity, as the case may be) to be established as an approximation, exploiting the natural continuity conditions of the variational process in this respect. Of course, the accuracy of the normal component modelling achieved will be commensurate with that of the overall finite element procedure.

It remains to specify the combinations of the ζ's and $\nabla\zeta$'s suitable for use as tangential vector interpolation functions τ_m. An overriding consideration is that each τ_m must be linearly independent of every other one. The polynomial completeness of any set of τ's is also significant. For interpolation of orders 1, 2 and 3 the numbers of functions necessary for a complete scalar expansion in three dimensions (sometimes referred to as the *degrees of freedom*) are respectively 4, 10 and 20; for vector expansions, obviously three times as many are needed. The scheme given by Webb and Forghani (1993) for the tangential interpolation functions τ_m is set out in Table 7.3; also shown are the scalar interpolation functions α_m. For an element which is complete to an order n_p all of the m functions down to a heavy line are required. Some of the natural groupings of functions τ_m shown are *not* complete to an integer order. These have been assigned a notational fractional order of completeness n_p, defined from the number of functions actually involved compared with that required for full completeness. For example first group of vector edge elements τ_1 to τ_6 comprise the Nedelec–Bossavit *edge elements* (Nedelec, 1980: Bossavit, 1982) and has order $n_p = 0.5$. By inspection, the functions tabulated are indeed linearly independent of each other; if combinations of the ζ's and $\nabla\zeta$'s are to be employed there is no possibility of any other choice for the vector interpolation functions shown, except for those of order $n_p = 1.5$ involving the functions $(\nabla\zeta_i)\zeta_j\zeta_k$ and so forth. In that instance, the combinations $(\nabla\zeta_k)\zeta_i\zeta_j$ are arbitrarily ruled out because their inclusion would violate the linear independence requirement when the combinations $\nabla(\zeta_i\zeta_j\zeta_k)$ are used in making up the $n_p = 2$ vector set.

The simplex coordinate scalar functions ζ_i, designated by a single index clearly are associated with the tetrahedron vertices, functions employing combinations of just a pair of indices i, j with the edges whilst those using i, j, k represent tetrahedron faces. In order, at the assembly stage of a full

Table 7.3 Hierarchal scalar and vector interpolation functions: n_p =polynomial order; m = single index function label. V, E, F indicate association with tetrahedron vertices, edges and faces respectively.

	Scalar		Type		Vector	
n_p	m	α_m		n_p	m	τ_m
	1–4	ζ_i	V			
			E*	1–6		$\zeta_i \nabla \zeta_j - \zeta_j \nabla \zeta_i$
1		ζ_i		0.5		
	5–10	$\zeta_i \zeta_j$	E*		7–12	$\nabla(\zeta_i \zeta_j)$
2				1		
			F†		13–16	$(\nabla \zeta_i)\zeta_j \zeta_k$
			F††		17–20	$(\nabla \zeta_j)\zeta_k \zeta_i$
				1.5		
	11–16	$\zeta_i \zeta_j(\zeta_i - \zeta_j)$	E*		21–26	$\nabla\{\zeta_i \zeta_j(\zeta_i - \zeta_j)\}$
	17–20	$\zeta_i \zeta_j \zeta_k$	F**		27–30	$\nabla(\zeta_i \zeta_j \zeta_k)$
3				2		

Notes:

* Derived from edges $i,j = (1,2),\ (1,3),\ (1,4)\ (2,3),\ (2,4)\ (3,4)$

** Derived from faces $i,j,k = (1,2,3),\ (1,2,4),\ (1,3,4),\ (2,3,4)$

† Derived from faces $(i),j,k = ((1),2,3),\ ((2,3,4),\ ((3,4,1),\ ((4,1,2)$

†† Derived from faces $(j),k,i = ((2,3,1),\ ((3,4,2),\ ((4,1,3),\ ((1),2,4)$

The third possible set of combinations $(k),i,j$ is ruled out because the associated vector functions, added to those of the previous two listed combinations, create the $\Delta(\zeta_i \zeta_j \zeta_k)$ set.

programming procedure, to connect the hierarchal vector and scalar finite elements described here, a new approach must be taken. Instead of merely ensuring the uniqueness of interpolation function coefficients at nodes common to different elements, now the connection of edges and faces has to be considered as well. All pairs of tetrahedral elements are examined for their connectivity, which may consist of common vertices, edges or a face, and the coefficients of the corresponding interpolation functions, shown in Table 7.3 identified by the labels V, E and F respectively, made equal. Notice that for the hierarchal scalar interpolation functions, the property $\alpha_m(\mathbf{r}_n) = \delta_{mn}$ only applies to the first-order functions, so that edge and face connectivity must be imposed here too. The only nodal coordinates now involved are those of the tetrahedron vertices. Of course these do fundamentally define the geometry of any given problem, but now no intermediate nodes are assigned. Observe also that the vector degrees of freedom do not have assigned position other than being associated with edges or faces, neither do the scalar ones intermediate to the vertices. The hierarchal scheme for the scalar and vector functions described here is shown schematically in Fig. 7.8.

The interpolation functions described are indeed *hierarchal*, that is different orders of them can be used abutting each other within an overall mesh, because the lower order collections of functions are subsets of the higher ones. Suppose elements A and B abut one another as illustrated in Fig. 7.9. The global vertex numbering scheme shown is allowable as it achieves the edge and face connectivity required. Thus, using Table 7.3, suppose the first six τ_m complete to order 0.5 represent an element A, abutting an element B represented by the first twelve τ's complete to order 1. Then, in a self-explanatory notation, the trial functions in the elements will be respectively

$$\mathbf{E}_A = \sum_{m=1}^{6} E_{Am} \tau_{Am}^{(0.5)} \qquad \text{(in tetrahedron } A) \quad (5.11)$$

$$\mathbf{E}_B = \sum_{m=1}^{6} E_{Bm} \tau_{Bm}^{(0.5)} + \sum_{m=7}^{12} E_{Bm} \tau_{Am}^{(1)} \qquad \text{(in tetrahedron } B) \quad (5.12)$$

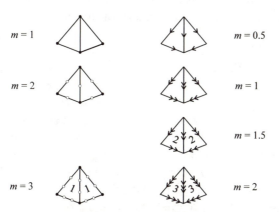

Fig 7.8 Scalar (left) and vector (right) hierarchal elements. Degrees of freedom for scalar elements are shown as solid circles (for vertices), open circles (edges) and numerals (faces); for vector elements they are arrows (for edges) and numerals (faces). [After Webb and Forghani (1993).] © 1993 IEEE.

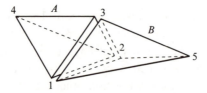

Fig 7.9 A pair of connected tetrahedra.

where the E_{Am} and E_{Bm} are unknown coefficients associated with the tetrahedron edges, to be determined by the finite element process. All of the τ's above take the form

$$\tau_{Xm} = \zeta_{Xi}\nabla\zeta_{Xj} \mp \zeta_{Xj}\nabla\zeta_{Xi} \qquad (5.13)$$

where X is either A or B and i, j represents some edge in one or other of the tetrahedra. (Note that there is no edge corresponding to $i = 4, j = 5$.) With reference to Fig. 7.9 and its global vertex numbering suppose $i = 1$ and $j = 5$ (and therefore $X = B$), so that

$$\tau_{Xm} = \zeta_{B1}\nabla\zeta_{B5} \mp \zeta_{B5}\nabla\zeta_{B1}. \qquad (5.14)$$

In the plane of triangle $(1, 2, 3)$ common to A and B, ζ_{B5} vanishes whereas $\nabla\zeta_{B5}$ is a constant vector normal to that plane. Thus in this instance, τ_{Xm} is a vector function of position which in the said plane is entirely normal to it. The same is obviously true for any τ_{Xm} corresponding to any i or j not belonging to the common triangle. Now consider the τ_{Xm} corresponding to edges (i, j) of the common plane $(1, 2, 3)$. As already remarked, the six interpolation functions, $\tau_{Xm}^{(0.5)}$ have continuous interelement tangential components, so in the interests of maintaining continuity of tangential trial field $\mathbf{E}$ between A and B, the strategy of identifying the unknown (but to be determined later) coefficients $E_{Xm}, m = 1, \ldots, 6$ with their corresponding edges irrespective of association with A or B should be followed. Notice that both edges i, j, and faces i, j, k, are associated with a *sense* as well as direction, so that in some instances depending upon the numbering scheme used, $E_{Am} = -E_{Bm}$ should be set (this possible reversal of sign also has to be heeded when connecting elements of the same order). It can further be seen that the three coefficients amongst E_{Bm}, $m = 7, \ldots, 12$, which correspond in B to the three edges common to A and B but are absent in the A-expansion, should be set to zero. Then full tangential continuity of the series representations Eqs. (5.11) and (5.12) across the A–B interface is achieved. At the same time it is noted that there is discontinuity of the corresponding normal component, but with sufficient degrees of freedom available to allow the variational process to set up a good approximation to the correct jump in this component. Notice that there is loss of completeness due to setting three out of the twelve E_{Bm}'s to zero; full $n_p = 1$ completeness is only achieved for elements in the B-region not connected directly to A. A similar argument can be given to justify the hierarchal nature of all of the groups of vector functions (and separately the scalar ones) shown in Table 7.3; appropriate coefficients in the higher-order expansion have to be set to zero and, as always in the case of vector expansions, due regard to edge and face senses taken.

5.3   *Nedelec–Bossavit edge elements*

The first six vector elements $\tau_m = \zeta_i \nabla \zeta_j - \zeta_j \nabla \zeta_i$, introduced by Bossavit (1982) and essentially based upon the fundamental work of Nedelec (1980), have a special physical significance, particularly in relation to electromagnetics. This is seen by taking the scalar product of τ_m with its corresponding tetrahedron edge vector $e_m, m = 1, \ldots, 6$. Using expression (5.9) for $\nabla \zeta_1$ and a similar expression for $\nabla \zeta_2$ gives

$$\mathbf{e}_1 \cdot \nabla \zeta_2 = \frac{1}{6V} \begin{vmatrix} 1 & x_1 & y_1 & z_1 \\ 0 & x_2 - x_1 & y_2 - y_1 & z_2 - z_1 \\ 1 & x_3 & y_3 & z_3 \\ 1 & x_4 & y_4 & z_4 \end{vmatrix} = 1 \qquad (5.15)$$

$$\mathbf{e}_1 \cdot \nabla \zeta_1 = \frac{1}{6V} \begin{vmatrix} 0 & x_2 - x_1 & y_2 - y_1 & z_2 - z_1 \\ 1 & x_3 & y_3 & z_3 \\ 1 & x_2 & y_2 & z_2 \\ 1 & x_4 & y_4 & z_4 \end{vmatrix} = -1 \qquad (5.16)$$

so that

$$\mathbf{e}_1 \cdot \tau_1 = \mathbf{e}_1 \cdot (\zeta_1 \nabla \zeta_2 - \zeta_2 \nabla \zeta_1) = \zeta_1 + \zeta_2. \qquad (5.17)$$

Noting that on the edge $(1, 2)$ itself ζ_3 and ζ_4 both vanish and therefore $\zeta_1 + \zeta_2 = 1$ it is deduced that, also on that edge,

$$\mathbf{e}_1 \cdot \tau_1 = 1. \qquad (5.18)$$

A similar relation is true for every edge of the tetrahedron. Moreover, as has already been established,

$$\mathbf{e}_m \cdot \tau_n = 0, \qquad m \neq n. \qquad (5.19)$$

Thus if, in any given tetrahedral element, a trial function $\mathbf{U}$ is represented by the edge element interpolation functions $m = 1, \ldots, 6$ alone

$$\mathbf{U}(\mathbf{r}) = \sum_{m=1}^{6} U_m \tau_m(\mathbf{r}), \qquad (5.20)$$

it is clear that the line integral of $\mathbf{U}$ along the tetrahedron edge $\mathbf{e}_m$ is

$$\int_{\mathbf{e}_m} \mathbf{U}(\mathbf{r}) \cdot \mathbf{dr} = U_m \tag{5.21}$$

whilst the line integral along any continuous path of tetrahedron edges is just the algebraic sum of the appropriate expansion coefficients U_m. The line integrals of **E** and **H** have a special significance (see Chapter 3, Section 2.2). Thus, as will be seen in Chapter 8, Section 5 describing eddy-current analysis, this form of expansion, although of low order, is particularly appropriate in some cases.

6.   Vector-based hexahedra

In the finite element analysis of problems requiring just a scalar variable, isoparametric hexahedral nodal 'brick' elements are a valuable addition to the three-dimensional toolbox (as outlined earlier in this chapter); they are particularly useful in modelling curved boundaries. Thus one is led to enquire whether there are curvilinear hexahedral vector finite elements equivalent to the scalar ones. An appropriate hexahedral vector element was first introduced, in a rudimentary low-order form with straight sides, by van Welij (1985). Subsequently, Crowley, Silvester and Hurwitz (1988) generalized the hexahedral form with their mixed-order covariant projection elements, which could be of high order and have curved sides. Examine, as an example, the 27-node curved brick of Fig. 7.10(*a*) defined by the nodal Cartesian coordinates $(x_i, y_i, z_i) i = 1, 27$ mapped from a reference cube occupying the (u, v, w)

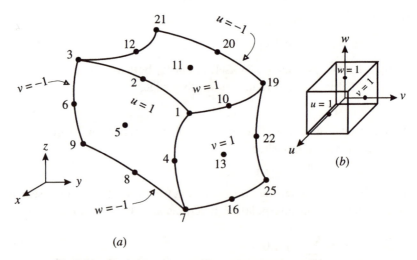

(*a*)

Fig 7.10   (*a*) A 27-node curvilinear hexahedron in (x, y, z)-space; (*b*) Reference cube in (u, v, w)-space.

space, $-1 \leq u \leq 1$, $-1 \leq v \leq 1$, $-1 \leq w \leq 1$, Fig. 7.10(b). Vectors referred to the rectangular (x, y, z) axes are not going to be particularly useful for such a curvilinear element, because the continuity rules for the vector variables, either **E** or **H**, at material interfaces will be awkwardly expressed. Material interfaces, and finite element boundaries will ordinarily be arranged to coincide; thus a means of referring vectors to axes related to the element boundary planes is sought. Then in accordance with variational rules, tangential vector continuity at material interfaces can easily be enforced and the discontinuous normal components allowed to adjust naturally.

6.1 *General mixed-order covariant projection elements*

The mapping from $\mathbf{r} = (x, y, z)$ to $\boldsymbol{\rho} = (u, v, w)$ for each element with M modes is very conveniently done using scalar nodal interpolation functions $\alpha_m(u, v, w)$ of polynomial form,

$$x = \sum_{m=1}^{M} x_m \alpha_m(u, v, w) \tag{6.1}$$

$$y = \sum_{m=1}^{M} y_m \alpha_m(u, v, w) \tag{6.2}$$

$$z = \sum_{m=1}^{M} z_m \alpha_m(u, v, w). \tag{6.3}$$

Here the usual Kronecker delta relation, $\alpha_m(u_n, v_n, w_n) = \delta_{mn}$, holds. The polynomial order of the $\alpha_m(u, v, w)$ chosen for this transformation between the real finite element in **r**-space and the reference cube in $\boldsymbol{\rho}$-space is a side issue, determined by the geometrical boundaries which it is wished to model. However to keep something specific in mind, consider again the 27-node brick, Fig. 7.10. Equation (6.4) shows a selection of the 27 nodes $\boldsymbol{\rho}_m$ in (u, v, w)–space alongside interpolation functions $\alpha_m(u, v, w)$ having the required behaviour $\alpha_m(\boldsymbol{\rho}_m) = \delta_{mn}$:

$$\boldsymbol{\rho}_1 = (1, 1, 1), \quad \alpha_1 = u(1 + u)v(1 + v)w(1 + w)/8,$$

$$\vdots$$

$$\boldsymbol{\rho}_4 = (1, 1, 0), \quad \alpha_4 = u(1 + u)v(1 + v)(1 + w)(1 - w)/4,$$

$$\boldsymbol{\rho}_5 = (1, 0, 0), \quad \alpha_5 = u(1 + u)(1 + v)(1 - v)(1 + w)(1 - w)/2, \tag{6.4}$$

$$\vdots$$

$$\boldsymbol{\rho}_{14} = (0, 0, 0), \quad \alpha_{14} = (1 + u)(1 - u)(1 + v)(1 - v)(1 + w)(1 - w),$$

$$\vdots$$

Appropriate sign changes made in the coefficients of u, v, w yield the rest of the 27 interpolation functions. Note that a single-valued inverse of the relations Eqs. (6.1)–(6.3), say

$$\boldsymbol{\rho} = \boldsymbol{\rho}(\mathbf{r}) \tag{6.5}$$

may be assumed to exist for $\mathbf{r}$ and $\boldsymbol{\rho}$ within their respective element domains, although it will not usually be easy to set this down explicitly. Consider the point P in $\mathbf{r}$-space but designated by its mapped point (u, v, w) in (or on the boundary of) the reference cube. Equations (6.1)–(6.3) can be written more compactly as

$$\mathbf{r} = \mathbf{r}(u, v, w) \tag{6.6}$$

whilst an infinitesimal perturbation about P represented by the reference cube perturbation (du, dv, dw) yields

$$d\mathbf{r} = \frac{\partial \mathbf{r}}{\partial u}\, du + \frac{\partial \mathbf{r}}{\partial v}\, dv + \frac{\partial \mathbf{r}}{\partial w}\, dw. \tag{6.7}$$

Keeping v and w constant gives $d\mathbf{r} = \partial \mathbf{r}/\partial u\, du$, defining a vector

$$\mathbf{a}_u = \frac{\partial \mathbf{r}}{\partial u} \tag{6.8}$$

which is readily evaluated through Eqs. (6.1)–(6.3) given the curved brick coordinates and some corresponding set of interpolation functions, say Eqs. (6.4); the vectors $\mathbf{a}_v$ and $\mathbf{a}_w$ may similarly be constructed. These so-called *unitary vectors*, described in some detail by Stratton (1941), have a simple geometrical interpretation; keeping v and w constant, the parameter u defines a curve in $\mathbf{r}$-space to which $\mathbf{a}_u$ is a tangent, see Fig. 7.11. Thus the vectors $\mathbf{a}_v$ and $\mathbf{a}_w$ are both tangential to any curved surface $u =$ constant; in particular they are tangential to the faces $u = \pm 1$ of a brick element. Tangents to the other curved faces arise similarly. It is easily appreciated that, since the $\mathbf{r}$-space finite element is defined by an arbitrary

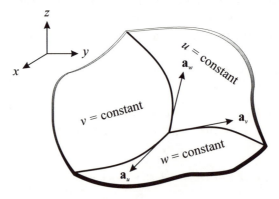

Fig 7.11   Unitary vectors for curvilinear hexahedra.

set of nodal coordinates, in general the unitary vectors **a** are neither of unit length nor are they mutually perpendicular.

Suppose some polynomial trial function representation of the electric field vector, say $\mathbf{E}'(u, v, w)$, nominally related back to **r**-space by means of Eq. (6.5), is constructed for the curvilinear brick element e. Then the components of $\mathbf{E}'$ tangential to the curved face $u = -1$ will there be represented by their so-called *covariant projections*

$$E_v^e = \mathbf{E}' \cdot \mathbf{a}_v, \ E_w^e = \mathbf{E}' \cdot \mathbf{a}_w. \tag{6.9}$$

Extending the polynomial representation of the vector $\mathbf{E}'$ in the usual piecewise fashion to a collection of elements e, using as parameters the E_u^e, E_v^e, E_w^e's corresponding to the various hexahedral faces, it is required to arrange that interelement tangential components of $\mathbf{E}'$ remain continuous whilst normal components are left unconstrained. This can be done by ensuring that the appropriate covariant projections E^e are the same for neighbouring elements. The subsequent reconstruction of $\mathbf{E}'$ from E_u^e, E_v^e and E_w^e, necessary in order to set up the finite element matrix equations, is not now as straightforward as it would have been had the **a**'s been a mutually orthogonal trio of unit vectors. However, consider the element vector trial function

$$
\begin{aligned}
\mathbf{E}' = {}& E_u^e(u, v, w) \frac{\mathbf{a}_v \times \mathbf{a}_w}{V} \\
& + E_u^e(u, v, w) \frac{\mathbf{a}_w \times \mathbf{a}_u}{V} + E_w^e(u, v, w) \frac{\mathbf{a}_w \times \mathbf{a}_u}{V}
\end{aligned} \tag{6.10}
$$

where, recalling the vector algebra rules concerning mixed scalar-vector products, the normalizing parameter V introduced is given by any of the alternatives

$$V = \mathbf{a}_u \cdot \mathbf{a}_v \times \mathbf{a}_w = \mathbf{a}_v \cdot \mathbf{a}_w \times \mathbf{a}_u = \mathbf{a}_w \cdot \mathbf{a}_u \times \mathbf{a}_v. \tag{6.11}$$

The E_u^e's and so forth may be considered in the first instance to be arbitrary polynomial expansions in u, v, w, so that   Eq. (6.10) indeed does represent some sort of vector expansion for $\mathbf{E}'$. Taking the scalar product of both sides of Eq. (6.10) and applying elementary rules of vector algebra, it is seen that

$$\mathbf{E}' \cdot \mathbf{a}_u = E_u^e(u, v, w), \tag{6.12}$$

similarly for the other two covariant projections. A notation commonly used for the *reciprocal unitary vectors*, in Eq. (6.10) is

$$\mathbf{a}^u = \frac{\mathbf{a}_v \times \mathbf{a}_w}{V}, \mathbf{a}^v = \frac{\mathbf{a}_w \times \mathbf{a}_u}{V}, \mathbf{a}^w = \frac{\mathbf{a}_u \times \mathbf{a}_v}{V}. \tag{6.13}$$

Now suppose that

$$E_u^e = \sum_{i=0}^{m} \sum_{j=0}^{n} \sum_{k=0}^{n} E_{uijk}^e h_i(u) h_j(v) h_k(w), \tag{6.14}$$

$$E_v^e = \sum_{i=0}^{n} \sum_{j=0}^{m} \sum_{k=0}^{n} E_{vijk}^e h_i(u) h_j(v) h_k(w), \tag{6.15}$$

$$E_w^e = \sum_{i=0}^{n} \sum_{j=0}^{n} \sum_{k=0}^{m} E_{wijk}^e h_i(u) h_j(v) h_k(w). \tag{6.16}$$

Here E_{uijk}^e, E_{vijk}^e, E_{wijk}^e are unknown coefficients, $m > 0$, $n > 0$ and

$$h_0(\zeta) = 1 - \zeta, \quad h_1(\zeta) = 1 + \zeta, \quad h_i(\zeta) = (1 - \zeta^2)\zeta^{i-2},$$
$$i \geq 2, \tag{6.17}$$

that is for $i > 0$, h_i is a polynomial of order i. Note that the possibility of $m \neq n$ is allowed for, that is the order of the polynomial approximations in u assigned to the u-covariant projection E_u^e may be different from that assigned in v and w; similarly the polynomial order of v in E_v^e and w in E_w^e may differ from that of the remaining two reference cube variables. The significance of choosing such a *mixed polynomial order* will be discussed presently; again it must be stressed that the vector polynomial orders m and n are unconnected with the order of the scalar (x, y, z) to (u, v, w) mapping. The former polynomial orders have a fundamental bearing upon the accuracy of the final solution approximation whereas the latter only affect the accuracy of modelling the problem boundaries. If the boundaries were say entirely polygonal, it would suffice to use say an eight-node trilinear transformation between ρ and $\mathbf{r}$.

Consider the trial expansion for $\mathbf{E}'$, Eq. (6.10) resulting from using the covariant projection polynomial expansions. Eqs. (6.14)–(6.17) for some permissible pair m, n (neither $m = 0$ nor $n = 0$ is allowed). In particular examine how tangential continuity between vectors in the common faces of adjacent hexahedra might be ensured, whilst leaving free the other components. Two of the curved faces of the tetrahedral element are represented by $u = \pm 1$, $(-1 \leq v \leq 1, -1 \leq w \leq 1)$, similarly the other four faces. For the moment, confine attention to $\mathbf{E}'$ derived from the covariant projections, Eqs. (6.14)–(6.17), putting say $u = -1$. Now all of the $h(u)'s$ vanish at the curvilinear face $u = -1$ apart from $h_0 = 2$, giving the projections there as

$$E_u^e(-1, v, w) = \sum_{j=0}^{n} \sum_{k=0}^{n} E_{u0jk}^e 2h_j(v)h_k(w) \tag{6.18}$$

$$E_v^e(-1, v, w) = \sum_{j=0}^{m} \sum_{k=0}^{n} E_{v0jk}^e 2h_j(v)h_k(w) \tag{6.19}$$

$$E_w^e(-1, v, w) = \sum_{j=0}^{n} \sum_{k=0}^{m} E_{w0jk}^e 2h_j(v)h_k(w). \tag{6.20}$$

Clearly the coefficients E_{v0jk}^e, E_{w0jk}^e define the tangential field at $u = -1$. At the face $u = +1$ it is $h_1 = 2$ which alone survives, so that in the same way the coefficients E_{v1jk}^e, E_{w1jk}^e give the tangential field there. Similar arguments apply to the coefficients associated with the surfaces $v = \pm 1$ and $w = \pm 1$. In a vector finite element finite assembly procedure therefore it becomes necessary to identify such tangential field coefficients relating to any given element face and match them up with the corresponding ones on the neighbouring element face connecting to it. Having already satisfied the requirement for tangential continuity of $\mathbf{E}'$ at the face $u = -1$, the remaining $i = 0$ coefficient there, E_{u0jk}^e, may therefore be left free and unmatched. Similarly E_{u1jk}^e, belonging to the face $u = +1$ and the corresponding coefficients associated with v and w are left to their own devices. The components of $\mathbf{E}'$ on $u = \pm 1$ relating to all E_{uijk}^e, $i > 1$, similarly for $v = \pm 1$ and $w + \pm 1$, obviously vanish, so the corresponding coefficients are also left unconstrained. It is clear that the whole finite element assembly procedure consists of examining elements face by face, identifying the coefficients E^e in neighbouring elements which have to be matched and leaving the others unpaired. In the general mixed-order case, the lower-order interpolation functions $h(\zeta)$ are subsets of the higher-order ones. Thus if it is desired, different order vector hexahedra can be employed in the same mesh in a similar fashion to the hierarchal scheme described for tetrahedral elements.

6.2 *Covariant projection elements of mixed order $m = 1, n = 2$*

A description of the mixed order $m = 1$, $n = 2$ elements, first introduced by Crowley, Silvester and Hurwitz (1988), representing a practical embodiment of the general hexahedral case now follows. For the sake of argument, suppose the 27-node $\boldsymbol{\rho}$ to $\mathbf{r}$ transformation is employed. Then the unitary vectors are

$$\mathbf{a}_u = \sum_{i=1}^{27} \mathbf{r}_i \frac{\partial \alpha_i}{\partial u}, \quad \mathbf{a}_v = \sum_{i=1}^{27} \mathbf{r}_i \frac{\partial \alpha_i}{\partial v}, \quad \mathbf{a}_w = \sum_{i=1}^{27} \mathbf{r}_i \frac{\partial \alpha_i}{\partial w} \tag{6.21}$$

where the $\alpha_i(u, v, w)$ are given by Eq. (6.4); these in turn give rise to the reciprocal unitary vectors $\mathbf{a}^u$, $\mathbf{a}^v$, $\mathbf{a}^w$. There is no difficulty in writing down these **r**-space vectors as functions of their mapping coordinates (u, v, w). Expressing them in terms of (x, y, z) would be more awkward but is never required explicitly; when finally evaluating the finite element matrices, the **r**-space integrations are, as usual, performed in **ρ**-space with the aid of a Jacobian transformation. However, note that a 20-node isoparametric curvilinear brick or even a simple 8-node brick with straight sides could have been employed.

Now consider the scalar functions multiplying the reciprocal unitary vectors, to yield the vector interpolation functions required here. These scalars are the polynomial terms of the covariant projections. Eqs. (6.18)–(6.20), which show that for $m = 1$, $n = 2$ there are $2 \times 3 \times 3$ of them, with a similar number of as yet undetermined coefficients E^e, associated with each reciprocal unitary vector. Thus there are a total of 54 degrees of freedom overall. Table 7.4 sets out the details of the (u, v, w)-functions which have to multiply the vector $\mathbf{a}^u$, itself already some known function of local coordinates, in order to recover the associated full vector interpolation function. Reconstruction of the corresponding multipliers for the $\mathbf{a}^u$ and $\mathbf{a}^w$ vectors is left to the reader. Figure 7.12 shows schematically how these vectors relate to the faces and edges of the curvilinear hexahedron; so as to facilitate the assembly of a collection of connected hexahedra, each vector is considered to point inwards. In the diagram, every vector is shown as if acting at the point on the edge or face where it becomes a maximum. However it should be appreciated that these are continuous vector functions which span the entire finite element and that the schematic points of action do not have the same significance as do the nodes in a scalar finite element interpolation. That they coincide with some (not all) of the 27 **ρ** to **r** transformation nodes used here is fortuitous.

With reference to Table 7.4, it is easy to see how a complete assembly procedure now could be specified. The most efficient way would be to search first looking for common edges to match up the appropriate undetermined type E coefficients E^e_{uijk}, E^e_{vijk} and E^e_{wijk} belonging to the associated elements e; up to four hexahedra could share any given edge. Then the connectivity of faces would be examined, each face of course being common to not more than two elements; the corresponding coefficients E^e of type F would be matched up whereas those of type U are left alone. The modification of the procedure here when the element boundary face is homogeneous Neumann, homogeneous Dirichlet or inhomogeneous Dirichlet is obvious.

Table 7.4 Functions $h_i(u)h_j(v)h_k(w)$ multiplying $E^e_{uijk}\mathbf{a}^u$ for an $m = 1, n = 2$ covariant projection element; E, F and U denote association with tangential edge, tangential face and unconstrained transverse face vectors respectively. The functions representing the other two projections follow similarly

i, j, k	$h_i(u)h_j(v)h_k(w)$	maximum at (u, v, w)	node	
0 0 0	$(1 - u)(1 - v)(1 - w)$	$(1, 1, 1)$	1	E
0 0 1	$(1 - u)(1 - v)(1 + w)$	$(1, 1, -1)$	7	E
0 0 2	$(1 - u)(1 - v)(1 - w^2)$	$(1, 1, 0)$	4	F
0 1 0	$(1 - u)(1 + v)(1 - w)$	$(1, -1, 1)$	3	E
0 1 1	$(1 - u)(1 + v)(1 + w)$	$(1, -1, -1)$	9	E
0 1 2	$(1 - u)(1 + v)(1 - w^2)$	$(1, -1, 0)$	6	F
0 2 0	$(1 - u)(1 - v^2)(1 - w)$	$(1, 0, 1)$	2	F
0 2 1	$(1 - u)(1 - v^2)(1 + w)$	$(1, 0, -1)$	8	F
0 2 2	$(1 - u)(1 - v^2)(1 - w^2)$	$(-1, 0, 0)$	5	U
1 0 0	$(1 + u)(1 - v)(1 - w)$	$(-1, 1, 1)$	19	E
1 0 1	$(1 + u)(1 - v)(1 + w)$	$(-1, 1, -1)$	25	E
1 0 2	$(1 + u)(1 - v)(1 - w^2)$	$(-1, 1, 0)$	22	F
1 1 0	$(1 + u)(1 + v)(1 - w)$	$(-1, -1, 1)$	21	E
1 1 1	$(1 + u)(1 + v)(1 + w)$	$(-1, -1, -1)$	27	E
1 1 2	$(1 + u)(1 + v)(1 - w^2)$	$(-1, -1, 0)$	24	F
1 2 0	$(1 + u)(1 - v^2)(1 - w)$	$(-1, 0, 1)$	20	F
1 2 1	$(1 + u)(1 - v^2)(1 + w)$	$(-1, 0, -1)$	26	F
1 2 2	$(1 + u)(1 - v^2)(1 - w^2)$	$(-1, 0, 0)$	23	U

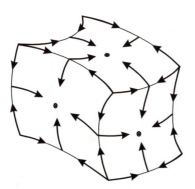

Fig 7.12 Covarient projection element of mixed order $m = 1$, $n = 2$.

7.   Spurious modes

The point has already been made that the tangentially continuous vector interpolation functions described above conform to the continuity requirements of the general electromagnetic problem relating to electromagnetic disturbances in a medium whose properties change discontinuously with position. Thus they would seem to be the natural choice for representing either **E** or **H** in time-varying (or equivalently complex phasor) situations where a scalar variable alone is inadequate. However one can be more positive than this; they *should* be used, otherwise spurious, nonphysical results are likely to turn up even in cases where the problem region is entirely homogeneous.

7.1   *Spurious modes in eigenproblems*

Spurious modes were first noticed when modelling uniform waveguides with inhomogeneous cross-sections using a vector working variable approximated by nodal finite elements; they also appeared when solving for cavity modal solutions. The easily recognizable nonphysical modes would occur interspersed with the real modes being sought. It was observed that they were characterized by a substantially nonzero divergence, whereas the wanted modes were good approximations to the real thing which, at least within uniform regions, is solenoidal. The unwanted modes could be pushed up amongst the high-order modes by adding a *penalty function* to the stationary functional (Rahman, Fernandez and Davies, 1991), for instance the magnetic version of Eq. (5.3) is replaced by

$$\mathfrak{F}(\mathbf{H}') = \frac{1}{2}\int_{\Omega} \left\{ \nabla \times \mathbf{H}' \cdot \epsilon_r^{-1} \nabla \times \mathbf{H}' - k^2 \mathbf{H}' \cdot \mu_r \mathbf{H}' \right.$$
$$\left. + q(\nabla \cdot \mathbf{H}^2) \right\} d\Omega, \tag{7.1}$$

where q is a constant chosen to be sufficiently large that the unwanted (but recognizable) modes become out of range.

7.2   *Spurious modes in deterministic problems*

It subsequently became apparent that solutions to deterministic problems were also liable to be contaminated by spurious modes, though it was not so easy to recognize the contamination or to choose q appropriately. A typical deterministic problem in electromagnetics to be solved might correspond to the vector Helmholtz equation

$$\nabla \times \frac{1}{\mu_r} \nabla \times \mathbf{E} - \epsilon_r k^2 \mathbf{E} = 0, \tag{7.2}$$

either boundary driven or subject to a nonzero right-hand side as the driving function. Whatever, the countably infinite set of eigensolutions $\mathbf{E}_i$

and k_i associated with Eq. (7.2) as it stands, subject to natural boundary conditions as if the problem-space was a closed cavity, must be regarded as complementary solutions, linear combinations of which will be expected to appear in the final solution. Any numerical solution procedure must be capable of reproducing such eigensolutions accurately and consistently. In particular there are infinitely many eigensolutions of Eq. (7.2) corresponding to $k_i = 0$. These represent a special set in that they alone are *irrotational*, that is they can be represented by the gradient of a scalar function,

$$\mathbf{E} = -\nabla\phi. \tag{7.3}$$

It has been recognized that inadequate modelling of these irrotational solutions, so that they appear with $k_i \neq 0$, lies at the root of the spurious-mode problem. Although it is apparent that they are not wanted, in order for them to be absent the finite element modelling of say Eq. (7.2) should be capable of reproducing the irrotational modes accurately. Then when tackling the waveguide or cavity-modal problem, a finite number of truly zero k_i turn up whilst the associated $\mathbf{E}_i$ are readily recognized and discarded. On the other hand in the deterministic case $k_i \neq 0$ will have been set, so the numerical process, now being capable of reproducing an irrotational mode spectrum, instead ensures that such a modal content is absent! Nedelec (1980) shows that in order to be able to do just this, you need to employ the tangentially continuous vector interpolation functions described above.

7.3   *A numerical experiment*

Rather than attempting to reproduce Nedelec's sophisticated mathematical argument, the results of a simple numerical experiment which very clearly confirms his theory are discussed. Dillon, Liu and Webb (1994) examined the stationary functional

$$F = \int_\Omega \{(\nabla \times \mathbf{E})^2 - k_0^2 \mathbf{E}^2\} \, d\Omega \tag{7.4}$$

corresponding to the oscillation modes of a rectangular, perfectly conducting empty cavity, Fig. 7.13. The eigensolutions for which the field does not vary with z can be dealt with very efficiently in terms of E_z alone (as if it were a scalar variable), the other components subsequently being derived from E_z. On the other hand you can solve rather less efficiently for the two-dimensional vector $\{E_x(x,y), E_y(x,y)\}$ and such a procedure then typifies the working required in the general three-dimensional vector finite element case. The calculation also corresponds to finding the propagation modes of a hollow waveguide whose cross-section is in the

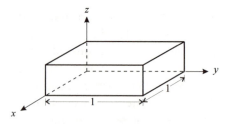

Fig 7.13　Rectangular cavity for numerical testing of vector finite elements.

(x, y) plane here, the values of k_0 being the cut-off wavenumbers for the transverse electric case. For a square cross-section of unit side, these wavenumbers are well known to be represented by

$$k_0^2 = \pi^2(M^2 + N^2) \tag{7.5}$$

where M and N correspond, in order of increasing k_0, to the integer pairs (0,1), (1,0), (1,1), (0,2), (2,0), and so forth. Being two-dimensional, the problem can be worked in terms of triangles or quadrilaterals instead of the tetrahedral and hexahedral finite elements so far discussed.

The actual implementation of the finite element procedure is not fundamentally different from that employed with scalar interpolation functions. That is to say the appropriate vector trial function expanded in terms of the vector interpolation functions, as in Eq. (5.2), is substituted into the functional Eq. (7.4). The integrations there are carried out numerically, element by element, with the aid of a Jacobian transformation between the (u, v) and (x, y) coordinates. This yields a quadratic form equivalent of the functional but expressed in terms of the unknown coefficients U_m of Eq. (5.2), here representing the electric field. Finally rendering the quadratic form stationary with respect to each and every unconstrained member of the set U_m yields the sparse matrix equation representing the system.

7.4　Triangular vector elements

The results for the tangentially continuous triangular vector elements were clear cut; tests were made with the equivalents of the elements listed in Table 7.3 for orders up to $n_p = 2$. For the square cross-section divided into up to 32 triangular elements, the modes were reproduced quite precisely with no spurious modes at all, provided only that tangential and not normal continuity was imposed. On the other hand, imposing normal continuity as well is equivalent to the employment of nodal finite elements; computation of the waveguide modes in such cases is well known to give results riddled with spurious solutions.

7.5   *Quadrilateral vector elements*

Equivalents of the hexahedral covariant projection elements are obtained by mapping quadrilateral elements in (x, y)-space onto the normalized (u, v)-square, $-1 \leq u \leq +1$, $-1 \leq v \leq +1$. As in the three-dimensional case, the electric field is represented in terms of its components $E_u = \mathbf{E} \cdot \mathbf{a}_u$, $E_v = \mathbf{E} \cdot \mathbf{a}_v$. Now the vector $\mathbf{E}$ is reconstructed as

$$\mathbf{E} = E_u \left(\frac{\mathbf{a}_v \times \mathbf{a}_z}{|\mathbf{a}_u \times \mathbf{a}_v|} \right) + E_v \left(\frac{\mathbf{a}_z \times \mathbf{a}_u}{|\mathbf{a}_u \times \mathbf{a}_v|} \right), \tag{7.6}$$

where $\mathbf{a}_u$ and $\mathbf{a}_v$ are the unitary vectors as defined in Eq. (6.21) whilst $\mathbf{a}_z$ is the unit vector in the z-direction. Then it follows that

$$E_u = \sum_{i=0}^{m} \sum_{j=0}^{n} E_{uij} h_i(u) h_j(v), \tag{7.7}$$

$$E_v = \sum_{i=0}^{m} \sum_{j=0}^{n} E_{vij} h_i(u) h_j(v), \tag{7.8}$$

where the h-functions are defined in Eq. (6.17) and as before, there is the possibility that the polynomial orders m and n are not the same. Figure 7.14 shows schematically the first three quadrilateral vector elements Q_{mn} here. Dillon, Liu and Webb (1994) tried out a number of different possibilities using up to 32 of the quadrilateral elements to model the 1 metre side square waveguide in comparison with the known eigensolutions Eq. (7.5). Testing up as far as $m = 5$, $n = 5$ it was found that unless $m < n$, and furthermore that only tangential and not normal continuity was maintained between the quadrilaterals, spurious modes would always appear in the vector finite element solutions. (It was however found necessary to distort the finite element mesh slightly from being perfectly regular within the quadrilateral boundary in order to induce the spurious modes.) The results of a test using a single element whilst changing the

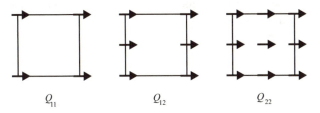

$$Q_{11} \qquad\qquad Q_{12} \qquad\qquad Q_{22}$$

Fig 7.14   Three quadrilateral edge elements. Each arrow represents one of the u vector interpolation functions making up the summation in Eq. (7.7). The v-functions of Eq. (7.8) are not shown.

values of m and n are summarized in Table 7.5, showing conclusively that $m < n$ is a necessary requirement in order to avoid spurious modes.

7.6   Solenoidality

The lowest-order vector interpolation functions, both tetrahedral and hexahedral, have built-in zero divergence. Spurious modes are always found to have relatively high divergence whilst the essence of the penalty function method lies in the elimination of such highly divergent solutions. Thus at one time it was considered that zero divergence was an essential property required of the vector interpolation functions. However none of the higher-order interpolation functions here is solenoidal; it is found that nevertheless these are capable of generating highly accurate, spurious-node free solutions, so the issue of solenoidality in the choice of vector interpolation functions appears to be a 'red herring'.

8.   Corner singularity elements

The construction of finite elements as set out so far has relied on the assumption that the potential (or other) function to be analysed is continuous and bounded, and has derivatives which are also continuous and bounded. Why the latter requirement should be necessary emerges from the error bounds given in Chapter 4 Section 4.2. These rely on Taylor's series expansions to characterize the convergence of solutions. Such error analyses are very useful in determining relative behaviour of different types of elements, and in assessing which type of element should be employed for any particular class of problem. However, they strictly apply only to boundary-value problems in convex regions, because a Taylor's series does not exist for the potential near a sharp corner. For such regions, special element functions may be called into service, functions chosen to incorporate the appropriate form of field singularity.

Table 7.5 Number of spurious modes of a square waveguide: electric field analysis using a single Q_{mn} element

m	$n = 2$	$n = 3$	$n = 4$	$n = 5$
1	0	0	0	0
2	2	0	0	0
3	4	2	0	0
4	6	4	2	0
5	8	6	4	2

The electric field **E** becomes singular at corners with included angles in excess of 180°, such as that in Fig. 7.15, which shows one-quarter of a rectangular coaxial transmission line. This behaviour is not restricted to right-angled corners but occurs at any sharp reentrant corner where the included angle θ exceeds 180°. Although the rectangular transmission line does not easily yield to purely analytic methods, the simpler problem of an isolated obtuse-angled corner (e.g., finding the field in three quadrants of the plane) can be solved by a conformal transformation. It is then reasonable to assume that the local field behaviour at the reentrant corner is not much affected by the geometric shape of boundaries at distant points; in other words, to suppose that the local behaviour of the potential at and near the corner is dictated principally by the shape of the corner itself. Near the corner, the conformal mapping solution indicates that the potential varies radially in accordance with

$$u = U\, r^{\pi/\theta}. \tag{8.1}$$

While this potential function is continuous and bounded, the corresponding electric field is not. Taking the gradient of the potential, the field near the corner is found to be

$$E = -\frac{\partial u}{\partial r} = -\frac{\pi}{\theta} U\, r^{(\pi/\theta-1)}, \tag{8.2}$$

so it follows that the field is locally unbounded, has no Taylor's series, and cannot be well approximated by polynomials. Solutions based on triangular or isoparametric elements, as indicated in Fig. 7.15, will suffer

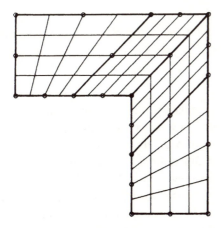

Fig 7.15   One-quarter of a rectangular coaxial transmission line modelled by boundary-node elements. A field singularity is expected at the reentrant corner.

inaccuracy principally from their inability to model the local field behaviour near the corner.

The finite element method can be made strictly valid for problems of this type by using elements which contain interpolation functions with the right type of derivative singularities. This new set of interpolation functions must be used instead of the standard ones. There are several procedures by which a set of polynomial interpolation functions can be made to incorporate singularities of the right sort. The following approach, due to Akin, is simple and has the strong advantage of fitting the special elements easily into a mesh of ordinary ones.

On an ordinary finite element, for example one of the parallelogram elements shown in Fig. 7.15, let a new set of functions $\beta_i^{(k)}(u, v)$ be defined in terms of the standard approximating functions $\alpha_i(u, v)$:

$$\left. \begin{aligned} \beta_i^{(k)} &= \frac{\alpha_i}{(1 - \alpha_k)^p}, \qquad i \neq k \\ \beta_k^{(k)} &= 1 - (1 - \alpha_k)^{1-p}. \end{aligned} \right\} \quad (8.3)$$

The superscript k indicates at which node in the finite element the field singularity occurs. In other words, the function $\beta_k^{(k)}$ provides for gradient singularity, while all the other functions in the set $\{\beta_i^{(k)}, i = 1, \ldots, N\}$ remain regular. The value of p is yet to be chosen; it follows from considerations of continuity and asymptotic behaviour.

Consider first any element edge $\mathscr{E}$ which does not intersect node k, i.e., any edge along which no singularity occurs. Along such an edge α_k vanishes, so that

$$(1 - \alpha_k)_{\mathscr{E}} = 1. \quad (8.4)$$

Along such an edge, the modified interpolation functions therefore reduce to the standard ones,

$$\beta_i^{(k)}\big|_{\mathscr{E}} = \alpha_i\big|_{\mathscr{E}}. \quad (8.5)$$

At edges not containing the singularity, interelement continuity of potential is thus guaranteed between any pair of abutting special elements as well as between a special and an ordinary element. Along edges that do contain node k, the modified functions $\beta_i^{(k)}$ have a shape different from the unmodified functions α_i. However, the $\beta_i^{(k)}$ only depend on the values of α_i and the parameter p. The standard functions α_i are continuous across element interfaces, so the condition for the modified functions $\beta_i^{(k)}$ to be continuous is simple: p must be the same in adjoining elements. In other words, the value of p must be associated with a particular corner, and must have that value in all modified elements that abut on the corner.

Near the singular node k, the unmodified approximating function α_k is a regular polynomial; at the node itself, it has the value 1. Let a system of polar coordinates ρ, ϕ be centered on node k. In the immediate neighbourhood of node k, the function α_k may be expanded in a polynomial series in the coordinates; retaining only the leading terms,

$$\beta_k^{(k)} = 1 - (1 - \alpha_k)^{1-p} \tag{8.6}$$

becomes

$$\begin{aligned}
\beta_k^{(k)} &= 1 - (1 - 1 - a_1 p - \ldots)^{1-p}, \\
&= 1 - (-a_1)^{1-p}\rho^{1-p} + \ldots
\end{aligned} \tag{8.7}$$

so

$$\frac{\partial}{\partial\rho}\beta_k^{(k)} = -(-a)^{1-p}\rho^{-p} + \ldots \tag{8.8}$$

Clearly, the derivative singularity is $O(-p)$. Comparing with Eq. (8.2) above, the field behaviour will be asymptotically correct provided one chooses

$$p = 1 - \frac{\pi}{\theta}. \tag{8.9}$$

This result hinges on the Taylor series expansion introduced in Eq. (8.7) and its subsequent differentiation. Both operations become much more complicated, and (8.9) will not lead to the same result, if the Jacobian of the coordinate transformation underlying the finite element in question varies in the neighbourhood of node k. Appropriate basic elements for such singularity treatment are therefore rectangles, parallelograms and triangles; curvilinear isoparametric elements cannot lead to the correct result.

Singularity elements find application in specialized programs more frequently than in general-purpose packages. The main reason readily emerges from the argument given here: the exponent p is not local to an element, but characterizes a group of elements surrounding a singular point. It is quite difficult to devise general-purpose programs to spot reentrant corners unerringly and to determine which elements touch upon the singular point. Thus data handling and interpretation, not finite element mathematics, places a limit on the use of elements with field singularities.

9.   Recursive methods for infinite finite elements

Practical electromagnetics problems often involve finding fields near objects placed in an infinitely extending surrounding space. Such problems can be tackled successfully by finite element methods, provided that the elements used are capable of modelling the infinitely extending region. Though they still contain finite energy or power, such elements have infinite geometric extent and may therefore be called *infinite* finite elements. The principles of constructing such elements will be briefly stated in this section, followed by details of some particular methods.

9.1   *A characteristic problem*

The general approach to infinite finite elements is best introduced by means of a particular class of two-dimensional problem. This class typically contains the field problem illustrated in Fig. 7.16, which shows a transmission-line structure composed of two circular wires supported by a dielectric covering that also serves to fix the spacing between wires. The electromagnetic energy in such a line is clearly bound to the guiding structure and propagates in the dielectric materials that support the conductors, as well as in the immediately surrounding free space. The analyst will therefore be interested in a region of space whose diameter may perhaps be a few times larger than the structural parts of the line, but there is no clear limit to the region of interest. Such problems, often referred to as *open-boundary problems*, are sometimes solved by introducing a totally artificial boundary at some distance, effectively converting the open-boundary problem into a closed one by main force. This approach may yield acceptable results, but there is no clear way to tell whether it has, nor what error may be involved.

Open-boundary problems may be solved without approximation by several techniques, all having in common the idea that the open (infinitely extending) region is subdivided into *interior* and *exterior* portions so that

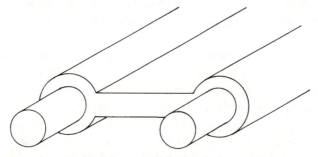

Fig 7.16   A popular form of ribbon transmission line. The wires are held by a dielectric support; energy transport takes place both in and around the dielectric.

the interior part contains the structures and fields of principal interest. An artificial, more or less arbitrary, boundary Γ is placed around the *interior* region Ω_i, separating it from the rest of space, the *outer* region Ω_o; clearly, $\Gamma = \Omega_i \cap \Omega_o$. Such separation is shown in Fig. 7.17 for two-space, as would be appropriate to the transmission line problem of Fig. 7.16. As in finite regions, the essence of the finite element method is still to render stationary a functional, typically

$$\mathfrak{F}(U) = \int_{\Omega} (\nabla U \cdot \nabla U - k^2 U^2 - gU)\, d\Omega \qquad (9.1)$$

the integral being taken over all space. The integral divides naturally into integrals over the two regions Ω_i and Ω_o,

$$\mathfrak{F}(U) = \int_{\Omega_i} (\nabla U \cdot \nabla U - k^2 U^2 - gU)\, d\Omega$$
$$+ \int_{\Omega_o} (\nabla U \cdot \nabla U - k^2 U^2 - gU)\, d\Omega. \qquad (9.2)$$

The inner region is amenable to treatment by standard finite element techniques and therefore will not be further considered here. The exterior element must adhere to the same general principles as the interior ones, but applied to an infinite geometric extent. Thus, it must possess suitable approximating functions, able to model the function behaviour in the exterior adequately. At the exterior-interior interface Γ, the principal continuity conditions must be satisfied. In scalar problems, this means the potential function U must be continuous across Γ, while in vector problems the *tangential* component of the vector unknown **U** must be continuous.

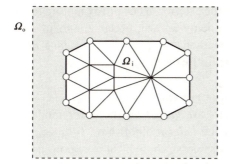

Fig 7.17   All space is subdivided into an *inner* region Ω_i where conventional finite elements may be used, and an *outer* infinite finite element Ω_o.

Three distinct techniques are available for constructing approximating functions for the exterior region: recursive growth, geometric transformation and implicit construction. These will be considered in some detail in this chapter and the chapters to follow.

The *recursive growth* approach constructs an infinitely extending mesh of standard finite elements but discards all exterior nodes by a process of condensation. It owes its name to the sequence of computation employed. In each step of the recursion, the exterior region is enlarged, the element matrices **S** and **T** modified to take account of the enlarged region, and the newly added nodes in the exterior are removed. In its usual implementation, this process does not really build an infinite region, merely a very large one; but since it is easy to build exterior regions with a diameter 10^{10} or 10^{20} times that of the inner region, the difference can be made invisible even in multiple-precision computation. Compatibility between the elements used in the inner and outer regions is automatically satisfied in this process.

The *transformation* methods are exactly what their name implies: geometric transformations are sought that turn the exterior region into a finite region. The functional $\mathfrak{F}(U)$ is transformed in this process, so the construction of element matrices for the infinite region is replaced by the construction of element matrices for the transformed functional, but over a finite region. Provided the geometric transformation is well chosen, the compatibility conditions can be satisfied and the transformed region can be treated by a standard or nearly standard finite element method. The approximating functions used in the exterior are not explicitly written down, but they are known; they are simply the transforms of the standard finite element approximating functions.

The *implicit* modelling methods are sometimes called *hybrid* methods. The differential equation boundary-value problem in the outer region is reformulated as an integral equation problem, so that integral-operator finite elements are used in the exterior. The approximating functions used in the exterior are not explicitly known, except for their values on the boundary Γ itself. These methods are popular in antenna problems and will be considered in detail in further chapters.

9.2   *The recursive growth process*

The family of *recursive growth* methods for modelling infinite regions, often also called the *balloon* algorithms, forgoes reaching true infinity and settles for a very large though finite region instead. It uses standard finite elements as its fundamental building blocks and it effectively constructs a huge mesh without writing down the corresponding huge matrices.

The principle of balloon algorithms is straightforward. Suppose that a satisfactory finite element model has been built for the region of principal interest $\Omega^{(\mathrm{p})}$. For purposes of illustration, it will be assumed that in this model the boundary Γ_0 comprises three straight element edges with quadratic potential variation, say second-order triangles or eight-noded isoparametric elements. Such an interior region is sketched in Fig. 7.8 (*a*), in outline only; its interior details are not important to this discussion. The problem now is to construct a valid representation of the outside region — an *infinite element*, as it were — whose interior edge will match both the shape and the type of approximating functions of the interior model, making the complete geometric model seamless and the potential continuous across Γ_0. A modest first step will be to make the solution cover a larger, though still finite, region by attaching to $\Omega^{(\mathrm{p})}$ a border of additional elements, as shown in Fig. 7.18(*a*). This border will be made up so as to satisfy three requirements:

(1) The interior edge Γ_0 of the bordering region matches the boundary of $\Omega^{(\mathrm{p})}$ both as to node number and potential variation.

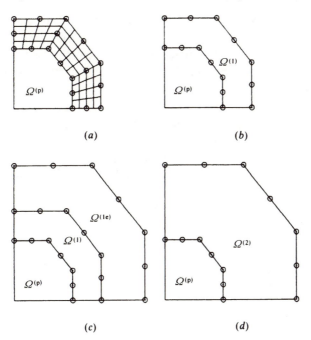

(*a*)   (*b*)

(*c*)   (*d*)

Fig 7.18   The ballooning algorithm. (*a*) The region of principal interest $\Omega^{(\mathrm{p})}$ is augmented by a border of elements. (*b*) Bordering elements are condensed to form a single superelement $\Omega^{(1)}$. (*c*) The superelement is scaled and attached to itself. (*d*) Condensation produces an enlarged superelement $\Omega^{(2)}$.

(2) The exterior edge Γ_1 of the bordering region is an enlarged (scaled) but otherwise geometrically similar replica of the interior edge Γ_0.

(3) The element subdivision within the bordering region is such that the approximation functions along the exterior boundary Γ_1 are the same as along its interior edge Γ_0.

There is no need to use the same type of element in the bordering region as in the exterior; all that is required is compatibility of approximating functions along the joining edges.

The finite element nodes in the border can be classified into three groups: nodes on the interior boundary Γ_0, nodes on the exterior boundary Γ_1, and nodes which lie on neither. The border may be regarded as a finite element in its own right (it is, after all, a defined geometric region with well-defined approximating functions associated with nodes), but this *superelement* contains unwanted interior nodes. These can be eliminated using a condensation process, exactly as used to eliminate excess interior nodes in rectangular elements (see Section 2.5 *Boundary-node elements: condensation*). Condensation produces a describing matrix for the border region, which may be regarded as a large finite element $\Omega^{(1)}$ that enlarges the region of solution beyond the interior region $\Omega^{(p)}$, as illustrated in Fig. 7.18(*b*). This new special-purpose element $\Omega^{(1)}$ fills the space between the boundary Γ_0 of the interior region $\Omega^{(p)}$ and the new exterior boundary Γ_1.

The next step in the procedure is to enlarge the bordering region. This purpose is achieved by making an enlarged copy $\Omega^{(1e)}$ of the border region, scaled so that its inner edge just coincides with Γ_1, and attaching it to the existing model, as in Fig. 7.18(*c*). The element matrices applicable to $\Omega^{(1e)}$ are easily derived from those of $\Omega^{(1)}$, so no extensive arithmetic is involved in this step. The precise scaling law by which the matrices are obtained depends on the type of problem, but is easily derived. For the Helmholtz equation in the plane, two matrices are involved,

$$T_{ij} = \int_{\Omega^{(1e)}} \alpha_i \alpha_j \, d\Omega \tag{9.3}$$

and

$$S_{ij} = \int_{\Omega^{(1e)}} \nabla \alpha_i \cdot \nabla \alpha_j \, d\Omega, \tag{9.4}$$

where the region of integration $\Omega^{(1e)}$ is the enlarged superelement. It is a proportional enlargement of $\Omega^{(1)}$, all linear dimensions being multiplied by some scale factor ρ,

$$\int_{\Omega^{(1e)}} \alpha_i \alpha_j \, d\Omega = \frac{|\Omega^{(1e)}|}{|\Omega^{(1)}|} \int_{\Omega^{(1)}} \alpha_i \alpha_j \, d\Omega. \tag{9.5}$$

In two-dimensional problems, **T** thus scales by ρ^2,

$$T_{ij}\big|_{\Omega^{(1e)}} = \rho^2 T_{ij}\big|_{\Omega^{(1)}}, \tag{9.6}$$

while in three dimensions, the scaling factor is ρ^3,

$$T_{ij}\big|_{\Omega^{(1e)}} = \rho^3 T_{ij}\big|_{\Omega^{(1)}}. \tag{9.7}$$

Correspondingly, the gradient operators in Eq. (9.4) are multiplied by ρ^{-1}. As a result, the Dirichlet matrix **S** in two dimensions is invariant,

$$S_{ij}\big|_{\Omega^{(1e)}} = S_{ij}\big|_{\Omega^{(1)}}, \tag{9.8}$$

but in three-dimensional problems it scales with length,

$$S_{ij}\big|_{\Omega^{(1e)}} = \rho S_{ij}\big|_{\Omega^{(1)}}. \tag{9.9}$$

In other words, the matrices **T** and **S** for the enlarged superelement are obtained by a simple scalar multiplication; and for the particularly important case of **S** in two dimensions, the matrix of $\Omega^{(1)}$ is simply copied! Enlargement of the problem region to the extent of $\Omega^{(1e)}$ is therefore easy. The resulting larger region includes unwanted interior nodes, as illustrated in Fig. 7.18(*c*) where 7 such nodes appear. They are eliminated by condensation, as in Fig. 7.18(*d*), yielding a larger superelement $\Omega^{(2)}$. It should be noted that $\Omega^{(2)}$ includes the geometric space occupied by $\Omega^{(1)}$; in other words, the new large superelement *replaces* its predecessor, it is not an additional element. This process of replacement continues recursively. The superelement $\Omega^{(2)}$ is next scaled to yield an enlarged copy $\Omega^{(2e)}$ which is attached to $\Omega^{(2)}$; its unwanted interior nodes are eliminated by condensation to produce a still larger superelement $\Omega^{(3)}$, and so on. The process may be summarized as follows.

(1) Construct a finite element mesh for the interior region $\Omega^{(p)}$, so that any radial line from the origin intersects the outer boundary Γ_0 only once.

(2) Create the boundary Γ_1 by scaling Γ_0, taking each point of Γ_0 radially outward by a scaling factor $\rho > 1$.

(3) Create a finite element mesh in the region bounded by Γ_0 and $\Gamma - 1$, using elements compatible with those in $\Omega^{(p)}$.

(4) Eliminate all interior nodes between Γ_0 and Γ_1 by condensation, thereby creating a border element $\Omega^{(1)}$.

(5) For $i = 1, 2, \ldots$ do the following:

(5.1) Scale the region $\Omega^{(i)}$ to produce region $\Omega^{(ie)}$ bounded by Γ_{i+1}, taking each point of $\Omega^{(i)}$ radially outward so the inner boundary of $\Omega^{(ie)}$ matches the outer boundary Γ_i of $\Omega^{(i)}$.

(5.2) Form the union of $\Omega^{(i)}$ and $\Omega^{(ie)}$. Eliminate interior nodes by condensation to produce $\Omega^{(i+1)}$.

(6) At a very great distance (large i), terminate the process by setting the potentials on the exterior boundary, Γ_i to zero.

It should be noted that the number of finite element nodes is the same at each step of the procedure. As the outer region grows, its effect is immediately incorporated in the matrix description of the problem; there is no growth in matrix size. Further, the new matrices at each step are generated by geometric scaling transformations from the matrices of the preceding step, so there is no need to recompute them.

9.3   *Balloon recursion: the algorithm*

The recursive growth or 'balloon' process has now been described in considerable geometric detail; it remains to turn the geometric argument into algebra fit for programming. It will now be developed in detail, assuming that the exterior region is governed by Laplace's equation so that only the Dirichlet matrix **S** needs to be computed and no source distributions need be considered. If required, the procedure is easily generalized to include sources and to obtain the metric **T**.

Consider the first step of superelement scaling, where the original superelement bounded by Γ_0 and Γ_1 is scaled to form the next, bounded by Γ_1 and Γ_2. Let the nodes on the inner boundary Γ_0 be numbered in some sequence; let the nodes on Γ_1 and Γ_2 be numbered in the corresponding sequence, so that the kth node on Γ_0 is mapped into the kth node on Γ_1, which is in turn mapped into the kth node on Γ_2. Let the **S** matrix of the two superelements $\Omega^{(1)}$ and $\Omega^{(2)}$, taken together, be formed. It may be written as the sum of two terms, corresponding to the inner and outer superelements:

$$\left(\begin{bmatrix} \mathbf{S}_{00} & \mathbf{S}_{01} \\ \mathbf{S}_{10} & \mathbf{S}_{11} \\ & & 0 \end{bmatrix} + \sigma \begin{bmatrix} 0 \\ & \mathbf{S}_{00} & \mathbf{S}_{01} \\ & \mathbf{S}_{10} & \mathbf{S}_{11} \end{bmatrix} \right) \begin{bmatrix} \mathbf{U}_0 \\ \mathbf{U}_1 \\ \mathbf{U}_2 \end{bmatrix} = \begin{bmatrix} \mathbf{G}_0 \\ 0 \\ 0 \end{bmatrix}.$$

$$(9.10)$$

Here the matrices are partitioned, with the indices attached to the submatrices indicating the contour numbers. In other words, $\mathbf{U}_i$ is the subvector containing the nodal potentials on Γ_i, $\mathbf{S}_{ij}$ is the submatrix that relates the potentials on Γ_i to those on Γ_j. The factor σ is a scalar introduced to allow the same development to apply to any spatial dimensionality; if the geometric scaling multiplies linear dimensions by ρ, then in N space dimensions

$$\sigma = \rho^{N-2}. \tag{9.11}$$

Combining the two terms,

$$\begin{bmatrix} \mathbf{S}_{00} & \mathbf{S}_{01} & \\ \mathbf{S}_{10} & \mathbf{S}_{11} + \sigma\mathbf{S}_{00} & \sigma\mathbf{S}_{01} \\ & \sigma\mathbf{S}_{10} & \sigma\mathbf{S}_{11} \end{bmatrix} \begin{bmatrix} \mathbf{U}_0 \\ \mathbf{U}_1 \\ \mathbf{U}_2 \end{bmatrix} = \begin{bmatrix} \mathbf{G}_0 \\ 0 \\ 0 \end{bmatrix}. \tag{9.12}$$

The unwanted variables are all associated with nodes on the boundary Γ_1, i.e., those partitioned as the middle row of Eq. (9.12). Extracting this row from the matrix equation,

$$[\mathbf{S}_{10} \quad \sigma\mathbf{S}_{01}] \begin{bmatrix} \mathbf{U}_0 \\ \mathbf{U}_2 \end{bmatrix} + (\mathbf{S}_{11} + \sigma\mathbf{S}_{00})\mathbf{U}_1 = 0. \tag{9.13}$$

Continuing the classical process of condensation, this equation is easily solved for $\mathbf{U}_1$ since the square matrix that multiplies it is known to be nonsingular:

$$\mathbf{U}_1 = -(\mathbf{S}_{11} + \sigma\mathbf{S}_{00})^{-1}[\mathbf{S}_{10} \quad \sigma\mathbf{S}_{01}] \begin{bmatrix} \mathbf{U}_0 \\ \mathbf{U}_2 \end{bmatrix}. \tag{9.14}$$

Substitution of this result into the remaining two rows of Eq. (9.12),

$$\begin{bmatrix} \mathbf{S}_{00} & \mathbf{S}_{01} & \\ & \sigma\mathbf{S}_{10} & \sigma\mathbf{S}_{11} \end{bmatrix} \begin{bmatrix} \mathbf{U}_0 \\ \mathbf{U}_1 \\ \mathbf{U}_2 \end{bmatrix} = \begin{bmatrix} \mathbf{G}_0 \\ 0 \end{bmatrix}, \tag{9.15}$$

then produces

$$\left(\begin{bmatrix} \mathbf{S}_{00} & \\ & \sigma\mathbf{S}_{11} \end{bmatrix} - \begin{bmatrix} \mathbf{S}_{01} \\ \sigma\mathbf{S}_{10} \end{bmatrix} (\mathbf{S}_{11} + \sigma\mathbf{S}_{00})^{-1} [\mathbf{S}_{10} \quad \sigma\mathbf{S}_{01}] \right) \begin{bmatrix} \mathbf{U}_0 \\ \mathbf{U}_2 \end{bmatrix} = \begin{bmatrix} \mathbf{G}_0 \\ 0 \end{bmatrix}. \tag{9.16}$$

The second term in parentheses may be multiplied and expressed in detail,

$$\begin{bmatrix} \mathbf{S}_{00} - \mathbf{S}_{01}(\mathbf{S}_{11} + \sigma\mathbf{S}_{00})^{-1}\mathbf{S}_{10} & -\sigma\mathbf{S}_{01}(\mathbf{S}_{11} + \sigma\mathbf{S}_{00})^{-1}\mathbf{S}_{01} \\ -\sigma\mathbf{S}_{10}(\mathbf{S}_{11} + \sigma\mathbf{S}_{00})^{-1}\mathbf{S}_{10} & \sigma\mathbf{S}_{11} - \sigma^2\mathbf{S}_{10}(\mathbf{S}_{11} + \sigma\mathbf{S}_{00})^{-1}\mathbf{S}_{01} \end{bmatrix} \begin{bmatrix} \mathbf{U}_0 \\ \mathbf{U}_2 \end{bmatrix}$$
$$= \begin{bmatrix} \mathbf{G}_0 \\ 0 \end{bmatrix}. \tag{9.17}$$

The right-hand vector remains unchanged in this process. Note that the components of the matrix $\mathbf{S}$ are expressed solely in terms of the sub-matrices of the original superelement matrix, and the scalar multiplier σ.

This initial step is easily generalized to apply to the entire recursive growth process. Every recursion step in building the infinite region is similar. Thus, at the kth step, the argument begins with

$$
\left(
\begin{bmatrix}
\begin{bmatrix} \mathbf{S}_{00}^{(k)} & \mathbf{S}_{0k}^{(k)} \\ \mathbf{S}_{k0}^{(k)} & \mathbf{S}_{kk}^{(k)} \end{bmatrix} & \\ & 0 \end{bmatrix}
+ \sigma
\begin{bmatrix}
0 & \\ & \begin{bmatrix} \mathbf{S}_{00}^{(k)} & \mathbf{S}_{0k}^{(k)} \\ \mathbf{S}_{k0}^{(k)} & \mathbf{S}_{kk}^{(k)} \end{bmatrix}
\end{bmatrix}
\right)
\begin{bmatrix} \mathbf{U}_0 \\ \mathbf{U}_k \\ \mathbf{U}_{k+1} \end{bmatrix}
=
\begin{bmatrix} \mathbf{G}_0 \\ 0 \\ 0 \end{bmatrix}
$$

$$(9.18)$$

and leads directly to the result

$$
\mathbf{S}^{(k+1)} =
\begin{bmatrix}
\mathbf{S}_{00}^{(k)} - \mathbf{S}_{01}^{(k)}\left[\mathbf{B}^{(k)}\right]^{-1}\mathbf{S}_{10}^{(k)} & -\sigma\mathbf{S}_{01}^{(k)}\left[\mathbf{B}^{(k)}\right]^{-1}\mathbf{S}_{01}^{(k)} \\
\left[-\sigma\mathbf{S}_{10}^{(k)}\left[\mathbf{B}^{(k)}\right]^{-1}\mathbf{S}_{10}^{(k)}\right] & \sigma\mathbf{S}_{11}^{(k)} - \sigma^2\mathbf{S}_{10}^{(k)}\left[\mathbf{B}^{(k)}\right]^{-1}\mathbf{S}_{01}^{(k)}
\end{bmatrix},
$$

$$(9.19)$$

where

$$
\mathbf{B}^{(k)} = \mathbf{S}_{11}^{(k)} + \sigma\mathbf{S}_{00}^{(k)}.
$$

$$(9.20)$$

For practical computation, it should be noted that this matrix occurs four times in Eq. (9.19), and that each occurrence involves a matrix-valued quadratic form. The matrix $\mathbf{B}^{(k)}$ is definite, so that a Cholesky decomposition is readily possible,

$$
\mathbf{B}^{(k)} = \mathbf{D}\mathbf{D}^{\mathrm{T}}.
$$

$$(9.21)$$

This permits rewriting the quadratic forms in a computationally efficient manner. For example, the submatrix $\mathbf{S}_{01}^{(k+1)}$ may be written

$$
\mathbf{S}_{01}^{(k+1)} = -\sigma\mathbf{S}_{01}^{(k)}\left[\mathbf{D}^{(k)}\mathbf{D}^{(k)\mathrm{T}}\right]^{-1}\mathbf{S}_{01}^{(k)}
$$

$$(9.22)$$

which is easily restated as

$$
\mathbf{S}_{01}^{(k+1)} = -\sigma\left[\mathbf{D}^{(k)-1}\mathbf{S}_{10}^{(k)}\right]^{\mathrm{T}}\left[\mathbf{D}^{(k)-1}\mathbf{S}_{01}^{(k)}\right].
$$

$$(9.23)$$

A product of the form $\mathbf{D}^{-1}\mathbf{S}_{01}$ is most efficiently formed by computing the triangular factor $\mathbf{D}$, then performing one forward elimination per column of $\mathbf{S}_{01}$. This strategy is invariably cheaper than explicitly computing the inverse of $\mathbf{B}^{(k)}$.

It might be worth noting that superelements as used in the balloon algorithm do not always have the same number of nodes; every superelement is constructed anew for a specific problem and possesses a distinctive number of nodes. This fact suggests that it is not generally convenient to store its node list, or its describing matrices, in fixed arrays; dynamically allocated arrays or linked lists are likely to be better. Despite its great popularity for finite element work, Fortran 77 is probably not the best language for this purpose; C, C++ or Fortran 90 are likely to prove superior in this regard.

9.4   *Recursive growth: convergence rate*

A crucial question in setting up recursive growth procedures is the rate at which the matrices converge to their final values that represent the infinite region — or, what is equivalent, how fast the outer boundary diverges toward infinity. The latter rate is solely dependent on the geometric construction and is easily derived.

At each step of the ballooning procedure, the newly generated region $\Omega^{(i)}$ is scaled to yield its enlargement $\Omega^{(ie)}$. If the exterior boundary Γ_i of $\Omega^{(i)}$ is q times more distant from the origin than Γ_0, then Γ_{i+1} must be q^2 times more distant than Γ_0, because Γ_{i+1} is obtained by scaling Γ_0 to coincide with Γ_i. To begin the process, Γ_0 is scaled by a factor ρ to yield Γ_1, so the distance r_1 from the origin to a point on $\Gamma - 1$ is ρ times greater than the distance r_0 to the corresponding point on Γ_0. Thus if a given point on Γ_0 is at a distance r from the origin, its counterpart on Γ_1 is at a distance $\rho^2 r$, that on Γ_2 at a distance $\rho^4 r$, and the corresponding point on Γ_i will be $\rho^{2^{i-1}} r$. This rate of growth borders on the unimaginable as may be seen from Table 7.6.

For magnification ratios ρ of 1.4 or 1.2, the first few recursive steps produce modest though worthwhile increases, but subsequent growth is

Table 7.6 Linear dimension ratio growth in ballooning

Step	$\rho = 1.1$	$\rho = 1.2$	$\rho = 1.4$	$\rho = 1.7$	$\rho = 2.0$
1	1.100E+00	1.200E+00	1.400E+00	1.700E+00	2.000E+00
2	1.210E+00	1.440E+00	1.960E+00	2.890E+00	4.000E+00
3	1.464E+00	2.074E+00	3.842E+00	8.352E+00	1.600E+01
4	2.144E+00	4.300E+00	1.476E+01	6.976E+01	2.560E+02
5	4.595E+00	1.849E+01	2.178E+02	4.866E+03	6.544E+04
6	2.111E+01	3.418E+02	4.743E+04	2.368E+07	4.295E+09
7	4.458E+02	1.168E+05	2.250E+09	5.607E+14	1.845E+19
8	1.987E+05	1.365E+10	5.063E+18	3.144E+29	3.403E+38
9	3.949E+10	1.864E+20	2.563E+37	9.884E+58	1.158E+77
10	1.560E+21	3.474E+40	6.570E+74	9.77E+17	1.34E+154

explosive. Even with small values of ρ, which might be appropriate if the bordering region is a thin strip of many elements, more than seven or eight steps are rarely needed. Keeping in mind that a light-year is about 10^{16} metres, magnification by thirty or forty orders of magnitude really does turn submicroscopic distances into intergalactic!

The balloon algorithm works extremely well for static problems in precisely the form stated here. It has the interesting strength that the distant field (the potential values placed at the infinitely distant boundary) need not be zero. If they are set to describe a dipolar potential distribution, for example, the interior region is subjected to a uniform field, permitting the calculation of such classical quantities as depolarization coefficients. Generalizations of the balloon algorithm to diffusion and wave problems, with both matrices **T** and **S** appearing, are straightforward. Diffusion problems are just as stable as static ones, but wave-propagation problems can stumble in computation if one of the partial regions added in the recursion chances to resonate at the operating frequency. This limits the permissible magnification ratios ρ. This problem has been addressed in detail, and a full mathematical theory is available.

10.   Exterior mappings

Mapping methods are widely used in two-dimensional problems, as yet there is little evidence in the literature on their use in three dimensions. The theoretical development, on the other hand, is similar for both.

10.1   *Coordinate and functional transformation*

For the sake of simplicity, the method will again be developed in terms of a scalar field problem describable by a single potential function U. A coordinate transformation $\mathcal{T}$ is found that maps the exterior region Ω_o into a finite transformed region Ω_t, subject to the requirement that the separating boundary Γ must map into itself: $\mathcal{T}\Gamma = \Gamma$. One such transformation is graphically depicted in Fig. 7.19.

The finite element functional $\mathfrak{F}(U)$ is once again written as the sum of two integrals, over the inner and outer regions. Under the assumption that standard finite element techniques will be used in the inner region, attention will be concentrated on the outer region only. Thus

$$\mathfrak{F}(U) = \mathfrak{F}_i(U) + \mathfrak{F}_o(U), \tag{10.1}$$

$$\mathfrak{F}_o(U) = \int_{\Omega_o} (\nabla V \cdot \nabla U + k^2 U^2 - gU)\, d\Omega. \tag{10.2}$$

This integral is transformed to an integral over Ω_t under the coordinate transformation $\mathcal{T}$ which carries the original coordinates x, y into new

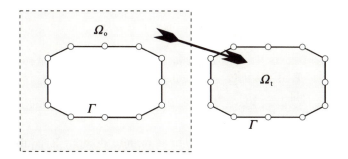

Fig 7.19   Coordinate transformation $\mathscr{T}$ maps the outer region Ω_o into a finite region Ω while preserving the separating boundary Γ.

coordinates x', y'. As always, the derivatives in ∇U must be written in the transformed coordinates:

$$
\begin{bmatrix} \dfrac{\partial U}{\partial x} \\[2ex] \dfrac{\partial U}{\partial y} \end{bmatrix} = \begin{bmatrix} \dfrac{\partial x'}{\partial x} & \dfrac{\partial y'}{\partial x} \\[2ex] \dfrac{\partial x'}{\partial y} & \dfrac{\partial y'}{\partial y} \end{bmatrix} \begin{bmatrix} \dfrac{\partial U}{\partial x'} \\[2ex] \dfrac{\partial U}{\partial y'} \end{bmatrix} = \mathbf{J} \begin{bmatrix} \dfrac{\partial U}{\partial x'} \\[2ex] \dfrac{\partial U}{\partial y'} \end{bmatrix}, \tag{10.3}
$$

where $\mathbf{J}$ denotes the Jacobian of the transformation. The surface (or volume) integral transforms by the usual rule,

$$
\mathrm{d}\Omega = \mathrm{d}x\,\mathrm{d}y = |\mathbf{J}|\mathrm{d}x'\,\mathrm{d}y' = |\mathbf{J}|\,\mathrm{d}\Omega'. \tag{10.4}
$$

The squared gradient term in the integrand of Eq. (10.2) may deserve a little extra attention. Combining the derivatives given by Eq. (10.3), it reads in full

$$
\nabla U \cdot \nabla U = \left(\mathbf{J} \begin{bmatrix} \dfrac{\partial U}{\partial x'} \\[2ex] \dfrac{\partial U}{\partial y'} \end{bmatrix} \right)^{\mathrm{T}} \left(\mathbf{J} \begin{bmatrix} \dfrac{\partial U}{\partial x'} \\[2ex] \dfrac{\partial U}{\partial y'} \end{bmatrix} \right)
$$

$$
= \begin{bmatrix} \dfrac{\partial U}{\partial x'} & \dfrac{\partial U}{\partial y'} \end{bmatrix} \mathbf{J}^{\mathrm{T}}\mathbf{J} \begin{bmatrix} \dfrac{\partial U}{\partial x'} \\[2ex] \dfrac{\partial U}{\partial y'} \end{bmatrix}. \tag{10.5}
$$

This quadratic form in the components of the gradient embeds the square and symmetric matrix $\mathbf{J}^{\mathrm{T}}\mathbf{J}$. In form, this is indistinguishable from an anisotropic (tensor) medium. For example, in electrostatics the energy density W is given by

$$W = \frac{1}{2} \mathbf{E} \cdot \mathbf{D} = \frac{1}{2} \mathbf{E} \cdot (\bar{\bar{\varepsilon}} \mathbf{E}) \tag{10.6}$$

where $\bar{\bar{\varepsilon}}$ represents the tensor permittivity of the medium. Rewriting the electric field $\mathbf{E}$ as the negative gradient of potential u,

$$W = \frac{1}{2} \nabla \phi \cdot (\bar{\bar{\varepsilon}} \nabla \phi). \tag{10.7}$$

Clearly, integration over the original infinite domain is equivalent to integration over the transformed domain, provided the material property is also transformed. The effect is that of introducing an anisotropic medium. Thus the gradient term in the integrand of Eq. (10.2) becomes

$$\nabla U \cdot \nabla U = \nabla' U \cdot (\bar{\bar{\mathbf{p}}} \nabla' U) |\mathbf{J}|, \tag{10.8}$$

where $\bar{\bar{\mathbf{p}}}$ represents the apparent tensor material property in x', y' coordinates,

$$\bar{\bar{\mathbf{p}}} = \frac{\mathbf{J}^{\mathrm{T}} \mathbf{J}}{|\mathbf{J}|}. \tag{10.9}$$

The transformed exterior functional contribution $\mathfrak{F}_o(U)$ can now be stated explicitly and compactly as

$$\mathfrak{F}_o(U) = \int_{\Omega_o} \left(\nabla' U \cdot \bar{\bar{\mathbf{p}}} \nabla' U + \frac{k^2}{|\mathbf{J}|} U^2 - \frac{g}{|\mathbf{J}|} U \right) |\mathbf{J}| \, \mathrm{d}\Omega, \tag{10.10}$$

where the primes indicate that differentiation is to be carried out with respect to the primed coordinates. In view of Eq. (10.4), it is now an easy matter to express $\mathfrak{F}_o$ as an integral over the transformed region,

$$\mathfrak{F}_o(U) = \int_{\Omega_t} \left(\nabla' U \cdot \bar{\bar{\mathbf{p}}} \nabla' U + \frac{k^2}{|\mathbf{J}|} U^2 - \frac{g}{|\mathbf{J}|} U \right) \mathrm{d}\Omega. \tag{10.11}$$

This integral is easy to treat numerically, for it covers the finite (transformed) region of integration Ω_t.

In summary, the transformation approach to exterior elements requires finding a coordinate transformation $\mathscr{T}$ which maps the exterior into an interior region one-to-one in such a way that the boundary Γ is preserved. Once this transformation is known, the exterior element matrices can be computed by any standard finite element program capable of dealing with anisotropic materials, for the region Ω_t is finite and may be treated in the same way as any finite region.

10.2   *Kelvin transformation*

The simplest coordinate transformation $\mathscr{T}$ that satisfies the needs of finite element analysts is one that predates the finite element method itself by nearly a century, the *inversion* or *Kelvin transformation*. This transforms the exterior of a circle into its interior, as sketched in Fig. 7.20. Using the Kelvin transformation, the entire x–y plane can be transformed onto the two faces of a circular disk: the circular interior region onto one face, the exterior of this circle onto the other. The transformation is easiest stated in terms of polar coordinates, where it carries coordinates r, θ into r', θ':

$$r' = \frac{R^2}{r}, \qquad \theta' = -\theta. \tag{10.12}$$

To develop the details of the Kelvin transformation, it is easiest to continue in polar coordinates, where the gradient is given by

$$\nabla U = 1_r \frac{\partial U}{\partial r} + \frac{1_\theta}{r} \frac{\partial U}{\partial \theta}. \tag{10.13}$$

The Jacobian of the Kelvin transformation is easily obtained. In keeping with Eqs. (10.3) and (10.4), the Jacobian relates the incremental distances in the two coordinate systems,

$$\begin{bmatrix} dr' \\ r'\,d\theta' \end{bmatrix} = \mathbf{J} \begin{bmatrix} dr \\ r\,d\theta \end{bmatrix} = \begin{bmatrix} \dfrac{\partial r'}{\partial r} & \dfrac{1}{r}\dfrac{\partial r'}{\partial \theta} \\[2mm] r'\dfrac{\partial \theta'}{\partial r} & \dfrac{r'}{r}\dfrac{\partial \theta'}{\partial \theta} \end{bmatrix} \begin{bmatrix} dr \\ r\,d\theta \end{bmatrix}, \tag{10.14}$$

so its value is obtained by straightforward differentiation as

$$\mathbf{J} = \begin{bmatrix} -\dfrac{R^2}{r^2} & 0 \\[2mm] 0 & -\dfrac{R^2}{r^2} \end{bmatrix} = -\frac{R^2}{r^2} \begin{bmatrix} 1 & 0 \\ 0 & 1 \end{bmatrix}. \tag{10.15}$$

To within a multiplier, this matrix equals the unit matrix! This property simplifies both mathematical development and computation. Indeed the

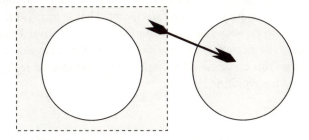

Fig 7.20 The Kelvin transformation maps the exterior of a circle onto the interior of another circle of the same radius.

noteworthy point about the Kelvin transformation is that it makes the fictitious anisotropic material property $\bar{\bar{p}}$ turn into

$$\bar{\bar{p}} = \frac{\mathbf{J}^T\mathbf{J}}{|\mathbf{J}|} = \begin{bmatrix} 1 & 0 \\ 0 & 1 \end{bmatrix}, \tag{10.16}$$

the identity matrix. The transformed exterior functional contribution $\mathfrak{F}_0(U)$ therefore becomes

$$\mathfrak{F}_0(U) = \int_{\Omega_i} \left\{ \nabla'U \cdot \nabla'U + k^2\left(\frac{r'}{R}\right)^2 U^2 - g\left(\frac{r'}{R}\right)^2 U \right\} d\Omega. \tag{10.17}$$

Clearly, Laplace's equation is invariant under this transformation. Consequently, the same finite element method can be used for both interior and exterior regions. The Poisson and Helmholtz equations are not invariant, but they can still be treated quite simply. It is particularly worth noting that because the Jacobian is scalar, the terms involving U and U^2 in Eq. (10.17) vary with position but remain isotropic throughout.

Although very straightforward and easy to use, the Kelvin transformation has shortcomings. The circular shape is not easy to model with triangular or quadrilateral elements, it is always necessary to make approximations. Further, the circle is not always a convenient shape for embedding the region of interest in a problem; for example, the transmission line of Fig. 7.16 would be better served by a rather flat ellipse or a rectangle. Other transformations are therefore desirable.

10.3   *General inversions*

A family of coordinate transformations that may be viewed as a generalization of the Kelvin transformation is obtained by taking the interior region perimeter to be some shape other than a circle, but *star-shaped* with respect to some origin or *star-point*. That is to say, the shape will be restricted to be such that any radial line drawn outward from the origin intersects the perimeter exactly once. The closed curve shown in Fig. 7.21 satisfies this criterion. The shape of any star-shaped region may be described by the equation

$$R = R(\theta), \tag{10.18}$$

where $R(\theta)$ is a single-valued function. The exterior of such a figure is mapped onto a finite region by the family of transformations described by

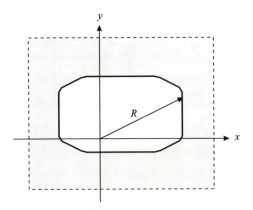

Fig 7.21 A general inversion maps the exterior of a star-shaped figure onto the interior of another figure bounded by the same closed curve.

$$\left.\begin{aligned} r' &= \frac{R^2}{r}(1-a) + Ra, \\ \theta' &= -\theta, \end{aligned}\right\} \quad (10.19)$$

where a is a scalar parameter. These resemble the Kelvin transformation, indeed the Kelvin transformation is included as the special case where $R(\theta) = $ constant, $a = 0$.

Like the Kelvin transformation, the family of general inversions has the fundamental property of transforming the boundary Γ of the inner region Ω_i into itself. To prove this assertion, it suffices to evaluate r' on Γ. On this boundary, $R = r$, so

$$r'|_\Gamma = \frac{r^2}{r}(1-a) + ra = r. \quad (10.20)$$

Interestingly, the point at infinity does not always transform into a point:

$$\begin{aligned} r' &= \lim_{r\to 0}\left\{ \frac{R^2(\theta)}{r}(1-a) + aR(\theta) \right\} \\ &= aR(\theta), \end{aligned} \quad (10.21)$$

which describes a curve similar to Γ, but scaled radially by the factor a. Clearly, the point at infinity transforms into the origin for any transformation with $a = 0$, including of course the Kelvin transformation. For $0 < a < 1$, the point at infinity transforms into a boundary that always lies nearer the origin than Γ, as illustrated in Fig. 7.22(*a*); for $a > 1$, it lies outside, as in Fig. 7.22(*b*).

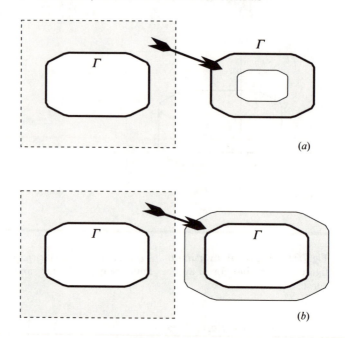

Fig 7.22   General inversions in two dimensions. (*a*) For $a < 1$, the point at infinity maps into a curve similar to Γ, but contained within it. (*b*) For $a > 1$, the point at infinity is mapped into a curve outside Γ.

The Jacobian of such a transformation is generally a little more complicated than that of a Kelvin transformation, but still reasonably easy to compute. Its algebraic derivation is tedious; the result, however, is simple enough:

$$\mathbf{J} = \begin{bmatrix} -\dfrac{(1-a)T^2(\theta)}{r^2} & \left\{ a + 2(1-a)\dfrac{R(\theta)}{r} \right\} \dfrac{\mathrm{d}R}{\mathrm{d}\theta} \\[4mm] 0 & -a\dfrac{R(\theta)}{r} - (1-a)\left\{ \dfrac{R(\theta)}{r} \right\}^2 \end{bmatrix}. \qquad (10.22)$$

For practical computing, it may be more useful to express it in terms of the transformed radial coordinates:

$$\mathbf{J} = \begin{bmatrix} -\dfrac{(1-a)(r')^2}{R^2(\theta')} & \dfrac{2(1-a)r' + aR(\theta')}{R(\theta')}\dfrac{\mathrm{d}R}{\mathrm{d}\theta'} \\[4mm] 0 & -\dfrac{(1-a)r' + aR(\theta')}{R^2(\theta')}r' \end{bmatrix}. \qquad (10.23)$$

A very special case is that of a complete inversion, i.e., one where the exterior is mapped into a region exactly similar to the interior. This corresponds to $a = 0$. The Jacobian then simplifies to

$$
\mathbf{J} = \begin{bmatrix} -\left\{\dfrac{r'}{R(\theta')}\right\}^2 & \dfrac{2r'}{R(\theta')}\dfrac{dR}{d\theta'} \\[2ex] 0 & -\left\{\dfrac{r'}{R(\theta')}\right\}^2 \end{bmatrix}
\tag{10.24}
$$

and the equivalent anisotropic material tensor becomes

$$
\bar{\bar{\mathbf{p}}} = \mathbf{J}^{\mathrm{T}}\dfrac{\mathbf{J}}{|\mathbf{J}|} = \begin{bmatrix} 1 & \dfrac{2r'}{R(\theta')}\dfrac{dR}{d\theta'} \\[2ex] \dfrac{2r'}{R(\theta')}\dfrac{dR}{d\theta'} & 1 + \left\{\dfrac{2r'}{R(\theta')}\dfrac{dR}{d\theta'}\right\}^2 \end{bmatrix}.
\tag{10.25}
$$

This tensor reduces to the identity if, and only if, $dR/d\theta' = 0$, i.e., if the interior region is a circle. This is of course the Kelvin transformation, which now appears as a special case. Are there any other values of the parameter a for which these transformations reproduce the very valuable property of the Kelvin transformation of leaving Laplace's equation invariant? Invariance requires first of all the matrix $\mathbf{J}^{\mathrm{T}}\mathbf{J}$ to be diagonal. Taking $\mathbf{J}$ in the general case, multiplying and setting the off-diagonal component to zero,

$$
\{2(1-a)r' + aR(\theta')\}\dfrac{dR}{d\theta'} = 0.
\tag{10.26}
$$

This differential equation has only one solution valid for any θ', the constant solution $R(\theta') = R_0$. In other words, true invariance is possible only for Kelvin transformations.

It is of interest to enquire under what circumstances these transformations may fail to be invertible, i.e., are there any circumstances in which the determinant of $\mathbf{J}$ vanishes? Evaluating the determinant and setting it to zero, only one solution emerges,

$$
a = 1.
\tag{10.27}
$$

Inspection of Fig. 7.22 reveals this to be a clearly degenerate case, one that maps every point of the exterior region into the bounding curve Γ.

Inversion techniques, described here for the two-dimensional case, work equally well in three (or even more!) dimensions.

11.   Further reading

A brief but very readable account of the approximating functions appropriate to various element shapes, including singularities, is given by Gallagher (1980). The methods used for programming elements of various shapes and types are dealt with in detail, and illustrated by programs, by Akin (1982), as well as the earlier book by Bathe and Wilson (1976). Isoparametric elements, particularly the eight-noded variety, are widely used in structural mechanics and are extensively illustrated by both, as well as Mori (1986). The excellent volume by Pepper and Heinrich (1992) not only treats quadrilateral and isoparametric elements at length in print, but includes their programs on diskette. Singularity elements lead to numerous mathematically interesting questions, some of which are dealt with by Mitchell and Wait (1977). Infinite elements and the ballooning technique have not yet been extensively treated in established textbooks but a detailed treatment will be found in Silvester *et al.* (1977).

12.   Bibliography

Akin, J.E. (1982), *Application and Implementation of Finite Element Methods.* London: Academic Press. xi + 325 pp.

Bathe, K.-J. and Wilson, E. L. (1976), *Numerical Methods in Finite Element Analysis.* Englewood Cliffs, NJ: Prentice-Hall. xv + 529 pp.

Bossavit, A. (1982), 'Finite elements for the electricity equation', in *The Mathematics of Finite Elements and Applications (MAFELAP IV)*, pp. 85–91, ed. J.R. Whiteman. London: Academic Press.

Bossavit, A. and Vérité, J-C (1982), 'A mixed FEM-BIEM method to solve 3-D eddy-current problems', *IEEE Transactions on Magnetics*, **MAG-18**, pp. 431–5.

Crowley, C.W., Silvester, P.P. and Hurwitz, H. (1988), 'Covariant projection elements for 3D vector field problems', *IEEE Transactions on Magnetics*, **24**, pp. 397–400.

Dillon, B.M., Liu, P.T.S. and Webb, J.P. (1994), 'Spurious modes in quadrilateral and triangular edge elements', *COMPEL*, **13**, Supplement A, pp. 311–16.

Gallagher, R.H. (1980), 'Shape functions', in *Finite Elements in Electrical and Magnetic Field Problems*, pp. 49–67 eds. M.V.K. Chari and P.P. Silvester. Chichester: John Wiley.

Mitchell, A.R. and Wait, R. (1977), *The Finite Element Method in Partial Differential Equations.* Chichester: John Wiley. x + 198 pp.

Mori, M. (1986), *The Finite Element Method and its Applications.* New York: Macmillan. xi + 188 pp.

Nedelec, J.C. (1980), 'Mixed finite elements in $\mathbb{R}^3$', *Numerische Mathematik*, **35**, pp. 315–41.

Pepper, D.W. and Heinrich, J.C. (1992), *The Finite Element Method: Basic Concepts and Applications.* Washington: Hemisphere. xi + 240 pp. + disk.

Rahman, B.M.A., Fernadez, F.A. and Davies, J.B. (1991), 'Review of finite element methods for microwave and optical waveguides', *Proceedings IEEE*, **79**, pp. 1442–8.

Silvester, P.P., Lowther, D.A., Carpenter, C.J. and Wyatt, E.A. (1977), 'Exterior finite elements for 2-dimensional field problems with open boundaries', *Institution of Electrical Engineers Proceedings*, **124**, 1267–70.

Stratton, J.A. (1941), *Electromagnetic Theory* pp. 38ff. New York: MCGraw-Hill.

van Welij, J.S. (1985), 'Calculation of eddy currents in terms of H on hexahedra', *IEEE Transactions on Magnetics*, **21**, pp. 2239–41.

Webb, J.P. and Forghani, B. (1993), 'Hierarchal scalar and vector tetrahedra', *IEEE Transactions on Magnetics*, **29**, pp. 1495–8.

13.   Programs

The following subroutine package computes the **S** and **T** matrices of an eight-noded isoparametric element. No main program is furnished with it here; the main program of HITRIA2D is entirely suitable and will easily incorporate this new element, provided one additional line is inserted, as discussed in Section 4.4. Intermixing of isoparametric and triangular elements is then possible.

One example data set follows the subroutine package. It was used to produce Fig. 7.7 and it may serve as a model, to clarify data formats and layout.

```
C***************************************************************
C
      Subroutine iso8el(ie)
C
C***************************************************************
C      Copyright (c) 1995  P.P. Silvester and R.L. Ferrari
C***************************************************************
C
C     Constructs  the element matrices s, t  for an isoparame-
C     tric element with 8 nodes.  ie = element number.
C
C===============================================================
C     Global declarations -- same in all program segments
C===============================================================
      parameter (MAXNOD = 50, MAXELM = 75, MAXNVE = 15)
      logical constr
      common /problm/ nodes, nelmts, x(MAXNOD), y(MAXNOD),
     *          constr(MAXNOD), potent(MAXNOD),
     *          nvtx(MAXNVE,MAXELM), source(MAXELM), nve(MAXELM)
      common /matrix/ s(MAXNOD,MAXNOD), t(MAXNOD,MAXNOD),
     *          rthdsd(MAXNOD)
      common /workng/ sel(MAXNVE,MAXNVE), tel(MAXNVE,MAXNVE),
     *          intg(MAXNVE), intg0(MAXNVE)
C===============================================================
C
      dimension alf(8), duv(2,8), tj(2,2), tji(2,2), tjti(2,2)
      dimension wq(5), sq(5)
C      common /gauss5/ wq, sq
      common /geom8/ alf, duv, tj, tji, tjti, detj
C     Gaussian quadrature formulae for m = 5
      data wq/0.2369269, 0.4786287, 0.5688889,
     &                              0.4786287, 0.2369269/
      data sq/-0.9061799, -0.5384693, 0.0000000,
     &                              0.5384693, 0.9061799/
C
C     Make sure there are 8 nodes
      if (nve(ie) .ne. 8)  call errexc('ISO8EL', nve(ie))
C
C     Clear arrays, then start the work
      do 30 i = 1,8
        do 20 j = 1,8
          sel(i,j) = 0.
          tel(i,j) = 0.
   20     continue
   30   continue
C
C     Compute element matrix entries sel(i,j), tel(i,j)
C
C     Indexing:  i counts matrix rows
C                j counts matrix columns
C                m counts quadrature nodes in u
C                n counts quadrature nodes in v
C
      do 280 m = 1,5
```

```
          do 270 n = 1,5
            u = sq(m)
            v = sq(n)
            call i8geom(ie, u, v)
C           Compute sel, tel contributions
            do 230 i = 1,nve(ie)
              do 220 j = 1,i
                tel(i,j) = tel(i,j) +
     &                          wq(m)*wq(n)*alf(i)*alf(j)*detj
                sum = 0.
                do 130 is = 1,2
                  do 120 js = 1,2
                    sum = sum + duv(is,i) *
     &                          tjti(is,js) * duv(js,j)
  120             continue
  130           continue
                sel(i,j) = sel(i,j) + wq(m)*wq(n)*sum*detj
  220         continue
  230       continue
  270     continue
  280   continue
C
C     Make sel, tel symmetric!
        do 310 i = 1,8
          do 300 j = 1,i
            tel(j,i) = tel(i,j)
            sel(j,i) = sel(i,j)
  300     continue
  310   continue
C
  900 return
      end
C
C*************************************************************
C
      Subroutine i8geom(ie, u, v)
C
C*************************************************************
C     Copyright (c) 1995  P.P. Silvester and R.L. Ferrari
C*************************************************************
C
C     Determines geometric  parameters and  derivatives of an
C     isoparametric element with 8 nodes. ie = element number.
C
C==============================================================
C     Global declarations -- same in all program segments
C==============================================================
      parameter (MAXNOD = 50, MAXELM = 75, MAXNVE = 15)
      logical constr
      common /problm/ nodes, nelmts, x(MAXNOD), y(MAXNOD),
     *        constr(MAXNOD), potent(MAXNOD),
     *        nvtx(MAXNVE,MAXELM), source(MAXELM), nve(MAXELM)
      common /matrix/ s(MAXNOD,MAXNOD), t(MAXNOD,MAXNOD),
     *        rthdsd(MAXNOD)
```

```
      common /workng/ sel(MAXNVE,MAXNVE), tel(MAXNVE,MAXNVE),
     *          intg(MAXNVE), intg0(MAXNVE)
C===============================================================
C
      dimension alf(8), duv(2,8), tj(2,2), tji(2,2), tjti(2,2)
      common /geom8/ alf, duv, tj, tji, tjti, detj
C
      Do 50 k = 1,8
C
C               Interpolation functions for 8-noded element
      alf(1) = (-1.+u+v)*(1.+u)*(1.+v)/4.
      alf(2) = (1.+u)*(1.+v)*(1.-u)/2
      alf(3) = (1.+v)*(1.-u)*(v-(1.+u))/4.
      alf(4) = (1.+v)*(1.-u)*(1.-v)/2
      alf(5) = -(1.+u+v)*(1.-u)*(1.-v)/4.
      alf(6) = (1.+u)*(1.-u)*(1.-v)/2
      alf(7) = (1.+u)*(1.-v)*(u-(1.+v))/4.
      alf(8) = (1.+u)*(1.+v)*(1.-v)/2.
C
C               Function derivatives with respect to u
      duv(1,1) = (2.*u+v*(1.+2.*u+v))/4.
      duv(1,2) = -u*(1.+v)
      duv(1,3) = (1.+v)*(2.*u-v)/4.
      duv(1,4) = (-1.+v)*(1.+v)/2
      duv(1,5) = (2.*u+v*(1.-(2.*u+v)))/4.
      duv(1,6) = u*(-1.+v)
      duv(1,7) = (2.*u+v*(v-(1.+2.*u)))/4.
      duv(1,8) = (1.+v)*(1.-v)/2.
C
C               Function derivatives with respect to v
      duv(2,1) = (1.+u)*(u+2.*v)/4.
      duv(2,2) = (1.+u)*(1.-u)/2
      duv(2,3) = (2.*v+u*(u-(1.+2.*v)))/4.
      duv(2,4) = v*(-1.+u)
      duv(2,5) = (1.-u)*(u+2.*v)/4.
      duv(2,6) = (-1.+u)*(1.+u)/2
      duv(2,7) = (1.+u)*(-u+2.*v)/4.
      duv(2,8) = -v*(1.+u)
   50 continue
C
C               Compute the Jacobian
      tj(1,1) = 0.
      tj(1,2) = 0.
      tj(2,1) = 0.
      tj(2,2) = 0.
      do 70 k = 1,8
      tj(1,1) = tj(1,1) + x(nvtx(k,ie))*duv(1,k)
      tj(1,2) = tj(1,2) + y(nvtx(k,ie))*duv(1,k)
      tj(2,1) = tj(2,1) + x(nvtx(k,ie))*duv(2,k)
      tj(2,2) = tj(2,2) + y(nvtx(k,ie))*duv(2,k)
   70 continue
C
C               Jacobian determinant, inverse, transpose-inverse
      detj = tj(1,1)*tj(2,2) - tj(1,2)*tj(2,1)
```

```
      tji(1,1) =  tj(2,2) / detj
      tji(1,2) = -tj(1,2) / detj
      tji(2,1) = -tj(2,1) / detj
      tji(2,2) =  tj(1,1) / detj
      do 100 is = 1,2
        do 90 js = 1,2
          tjti(is,js) = 0.
          do 80 ks = 1,2
            tjti(is,js) = tjti(is,js) +
     &                            tji(ks,is) * tji(ks,js)
  80          continue
  90        continue
 100      continue
C
      return
      end
```

```
 -3.000000   1.500000
 -3.000000   1.000000
 -1.851953   0.771640
 -0.878682   0.121322
  0.560659   1.060661
 -0.228362  -0.851948
  0.000000  -2.000000
  1.000000  -2.000000
 -3.000000  -0.500000
 -2.375000  -0.625000
 -1.750000  -0.750000
 -1.314341  -0.314339
 -3.000000   0.250000
 -1.500000  -2.000000
 -0.750000  -2.000000
 -2.250000  -2.000000
 -1.625000  -1.375000
 -3.000000  -1.250000
 -3.000000  -2.000000
  2.000000   2.000000
 -0.500000   2.000000
 -3.000000   2.000000
  2.000000  -2.000000
  2.000000   0.000000
/
   8   0.000      23 24 20   5   4   6   7   8
   8   0.000      20 21 22   1   2   3   4   5
   8   1.000       9 10 11  12   4   3   2  13
   8   1.000      14 15  7   6   4  12  11  17
   8   1.000      14 17 11  10   9  18  19  16
/
  20       0.000
  21       0.000
  22       0.000
  23       0.000
  24       0.000
/
```

8

Time and frequency domain problems in bounded systems

1.   Introduction

It is possible to take time variation into account in electro-magnetics finite element analysis at a number of different levels of complexity. If the rates of time variation are sufficiently slow, the pre-Maxwell versions of the fundamental electromagnetic laws neglecting the displacement-current term $\partial \mathbf{D}/\partial t$ in the Maxwell magnetic curl equation, may be assumed. This is allowable at power frequencies and may also apply at higher frequencies when considering miniature components; however microwave and optical configurations will ordinarily require displacement current to be included in any analysis. Another simplification arises if all of the materials concerned are linear, in which case analysis in the frequency domain is possible. Subsequently a transformation back to the time domain using standard inversion methods may be made, whilst in many cases the frequency domain solution is meaningful in itself. In this chapter attention is focussed upon two areas, eddy current modelling and the analysis of bounded high-frequency systems such as waveguides and cavities, for both of which very useful differential operator finite element computations can be accomplished.

2.   Eddy currents

When a time-varying magnetic field is set up within a conducting body, the resulting Faraday emf acting around closed paths enclosing the field causes circulating *eddy currents* to flow within the body. Many electrical machines and power devices depend heavily upon the effects of time-changing magnetic fields, so eddy currents are generated within them and give rise to unwanted power dissipation. On the other hand, there are some practical industrial heating methods which rely upon low-frequency currents induced by the eddy phenomenon. Thus it is very useful to be able to calculate the extent of and to optimize eddy currents.

In recent years, a considerable literature dealing with the numerical solution of problems relating to eddy currents has accumulated. Practical configurations are invariably irreducibly three-dimensional. No clear consensus appears to have emerged as to the best method of attack, although in many cases some finite element approach or other is used. The finite element methods mainly employ differential operators, using vector and scalar potentials together or one of the field vectors directly, whilst some approaches also employ an integral operator outside the eddy-current region. The differential finite element techniques are selected for attention here, applied to relatively simple configurations. For a comprehensive treatment covering more complicated practical geometrical arrangements, the reader should consult the literature; Kriezis *et al.* (1992) give review of this literature. Most practical configurations are essentially three-dimensional, whence it is necessary to employ a vector variable in the eddy-currents region. This in turn introduces difficulties associated with the fact that the variational (or equivalent Galerkin) weak forms of the electromagnetic equations leave some unwanted degrees of freedom. These difficulties have been overcome by the application of gauge relations (Chapter 3, Sections 3.4–3.6) in cases where vector potentials and nodal finite elements are employed, or by the use of low-order tetrahedral tangentially continuous vector elements (Chapter 7, Section 5.3) for the field vector variables. In nonconducting regions, the quasistatic form of Maxwell's equations cast in terms of a single scalar variable may be utilized to represent the fields more economically.

2.1 *Quasistatic electromagnetic equations*

In the main, eddy-current time-fluctuation rates are low enough for the quasistatic simplification of the Maxwell–Ampere equation neglecting displacement current,

$$\nabla \times \mathbf{H} = \mathbf{J}, \tag{2.1}$$

to apply. The other Maxwell curl-equation, representing the mechanism inducing the emf which in fact drives the eddy currents, remains as

$$\nabla \times \mathbf{E} = -\partial \mathbf{B}/\partial t. \tag{2.2}$$

The coupling of Eqs. (2.1) and (2.2) occurs via the constitutive relation

$$\mathbf{B} = \mu \mathbf{H} \tag{2.3}$$

and, for conducting regions,

$$\mathbf{J} = \sigma \mathbf{E}. \tag{2.4}$$

In Eq. (2.4) σ will ordinarily represent linear isotropic conductivity, this parameter being a time-independent scalar constant. On the other hand the **B–H** relationship, Eq. (2.3), is often required to represent saturable iron operated substantially into its nonlinear region, so that the permeability μ is by no means so simple. However here it will be assumed that the magnetic materials concerned are isotropic and that hysteresis may be neglected. Thus at worst, μ is a single-valued scalar whose value depends upon **B**. When there is nonlinearity it is not possible to work in the frequency domain and the partial time derivative in Eq. (2.2) must be retained.

Suppose the problem-space consists of an eddy-current region Ω_1, of permeability μ_1 and conductivity σ_1 but containing no sources, surrounded by a region Ω_2, of zero conductivity but which may have current sources $\mathbf{J} = \mathbf{J}_S$ (Fig. 8.1). An example here would be a machine connected to a current generator, the filamentary machine windings themselves having geometry and conductivity such that they may be regarded as current sources in the otherwise nonconducting region Ω_2.

Other portions of the machine, represented by Ω_1, are composed of conductors of one sort or another, not connected to the generator coils; eddy currents $\mathbf{J} = \mathbf{J}_E$ are induced in such conductors. In another typical situation, a fluctuating magnetic field is generated externally and applied by means of highly permeable pole pieces. Here Ω_2 need not contain internal current sources, the system being regarded as boundary driven, whilst Ω_1 supports eddy currents as before. Thus Eqs. (2.1)–(2.4), subject to appropriate boundary constraints and with $\mathbf{J}_S$ (if any) regarded as given whereas $\mathbf{J}_E$ is unknown, represent the fundamental set of equations which determine eddy-current behaviour. To keep matters here from becoming too involved, it will be stipulated that both regions are simply

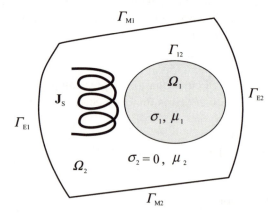

Fig 8.1   Simple eddy-current configuration.

connected, ruling out for instance Ω_1 representing a toroidal region.

The usual boundary and interface relations apply (see Chapter 3, Section 2.4). The quasistatic approximation whereby displacement current is neglected, leading to Eq. (2.1), implies that

$$\nabla \cdot \mathbf{J} = 0, \tag{2.5}$$

since the divergence of any curl vector must vanish. It should be observed that at the interface between a conductor and an insulator there can be no normal component of current density. Following from Eq. (2.2), the magnetic flux density has to be solenoidal,

$$\nabla \cdot \mathbf{B} = 0. \tag{2.6}$$

3. Eddy-current analysis using vector potential

In the manner already described in Chapter 3, Section 3.2, define a magnetic vector potential $\mathbf{A}$ by means of the relation

$$\mathbf{B} = \nabla \times \mathbf{A} \tag{3.1}$$

satisfying identically the solenoidal property of $\mathbf{B}$, Eq. (2.6). One out of the two Maxwell curl equations, namely Eq. (2.2), is automatically satisfied by setting

$$\mathbf{E} = -\partial \mathbf{A}/\partial t - \nabla V, \tag{3.2}$$

where V is a scalar potential function. The basic differential equation for $\mathbf{A}$ and V in the eddy-current region Ω_1 is obtained via the curl equation, Eq. (2.1), eliminating $\mathbf{J}$ through Eqs. (2.4) and (3.2):

$$\nabla \times \frac{1}{\mu} \nabla \times \mathbf{A} + \sigma \frac{\partial \mathbf{A}}{\partial t} + \sigma \nabla V = 0 \quad \text{in } \Omega_1. \tag{3.3}$$

In alternative schemes, not taken any further here, it is permissible to use instead electric vector and magnetic scalar potentials, $\mathbf{T}$ and P respectively, defined by $\mathbf{J} = \nabla \times \mathbf{T}$ and $\mathbf{H} = \mathbf{T} - \nabla P$, to give the dual of Eq. (3.3),

$$\nabla \times \mu \nabla \mathbf{T} - \frac{1}{\sigma} \frac{\partial \mathbf{T}}{\partial t} - \frac{1}{\sigma} \nabla P = 0 \quad \text{in } \Omega_1. \tag{3.4}$$

Notice that because of the nondivergence of all curl vectors, Eq. (3.3) incorporates the requirements, Eq. (2.5), that $\mathbf{J} = -\sigma(\partial \mathbf{A}/\partial t + \nabla V)$ shall be solenoidal. However, because of the Helmholtz indeterminacy arising when a vector is fixed purely by means of its curl, the gauge of $\mathbf{A}$ must be set by arbitrarily specifying its divergence. In eddy-current analysis the Coulomb gauge (Chapter 3, Section 3.4),

$$\nabla \cdot \mathbf{A} = 0, \tag{3.5}$$

is commonly chosen and is selected for the illustration here. In that case, it can be argued that Eq. (3.3) may be replaced by

$$\nabla \times \frac{1}{\mu}\nabla \times \mathbf{A} - \nabla \frac{1}{\mu}\nabla \cdot \mathbf{A} + \sigma \frac{\partial \mathbf{A}}{\partial t} + \sigma \nabla V = 0 \quad \text{in } \Omega_1. \tag{3.6}$$

The first two terms of Eq. (3.6) then reduce to the vector Laplacian operator if μ is piecewise constant; however now the automatic satisfaction of the solenoidal requirement of $\mathbf{J}$ is forfeited and must specifically be imposed through the equation

$$\nabla \cdot \sigma \left(\frac{\partial \mathbf{A}}{\partial t} + \nabla V \right) = 0 \quad \text{in } \Omega_1. \tag{3.7}$$

The nonconducting region Ω_2, can also, if desired, be represented in a directly compatible fashion through Ampere's law by

$$\nabla \times \frac{1}{\mu}\nabla \times \mathbf{A} = \mathbf{J}_S \quad \text{in } \Omega_2. \tag{3.8}$$

However, this is a vector equation used in circumstances where a scalar one can also be found. The scalar form for the governing equation, to be preferred on the grounds of computational economy, can be set up in the quasistatic approximation using a reduced scalar potential representation (see Chapter 3, Section 7.2), writing

$$\mathbf{H} = \mathbf{H}_S - \nabla \Phi \quad \text{in } \Omega_2. \tag{3.9}$$

Here the source-field $\mathbf{H}_S$ is calculable directly from the given $\mathbf{J}_S$ via the Biot–Savart law, Chapter 3, Eq. (7.12), so that enforcing the solenoidal property of $\mathbf{B}$, Eq. (2.6), the governing scalar equation for the nonconducting region Ω_2 becomes

$$\nabla \cdot \mu(\mathbf{H}_S - \nabla \Phi = 0 \quad \text{in } \Omega_2, \tag{3.10}$$

in which case the formulation may be described[1] as '$\mathbf{A}$, $V - \Phi$' Some workers have experienced numerical difficulties in using reduced scalar potentials to represent regions in Ω_2 of high permeability, where $\mathbf{H}$ becomes relatively small so that there tends to be near-cancellation of the terms on the right-hand side of Eq. (3.9). In general such regions include no source currents, thus the total scalar potential P, as in Chapter 3, Section 7.1, can be employed instead of Φ, where

---

[1] Note that there is very little consensus in the literature concerning the symbols assigned to the various scalar potentials applicable to eddy-currents problems.

$$\mathbf{H} = -\nabla P. \tag{3.11}$$

The governing equation for such a subregion of Ω_2 becomes

$$\nabla \cdot (\mu \nabla P) = 0, \tag{3.12}$$

whilst no particular difficulty arises in interfacing the total and reduced scalar potentials regions; the formulation may now be described as '$\mathbf{A}, V- \phi -P$'. However it may be argued that if adequate precision, say 64-bit, is available in the computational arithmetic operations, reduced scalar potentials can safely be used throughout a general nonconducting region to model practical materials with high permeability in the presence of a source field.

3.1 *Boundary and interface constraints*

The finite element procedures for dealing with Eqs. (3.10) and (3.12) are very similar, so only the latter equation is considered here, corresponding say to a system without internal sources but driven by an external fluctuating magnetic field. In this simplification of the general configuration the outside boundary Γ (Fig. 8.1) can be supposed to be made up from a magnetic wall Γ_{M}, on which tangential $\mathbf{H}$ has to vanish, and from an electric wall Γ_{E}, on which normal $\mathbf{B}$ vanishes. The surface Γ_{M} might be considered as representing a pair of infinitely permeable $(\mu \to \infty)$ pole pieces sustaining a given, time-varying scalar magnetic potential difference, say $P_0 = f(t)$ on one pole face, Γ_{M1}, and $P_0 = 0$ on the other Γ_{M2}, whilst Γ_{E}, here also represented by two distinct portions, constitutes a perfectly conducting boundary. Planes of symmetry can also be represented by appropriately located planar electric or magnetic walls. The situation is precisely analogous to magnetostatic cases and is the dual of electrostatic ones. Thus for variational or Galerkin-based procedures the requirement is

$$P = P_0 \quad \text{on } \Gamma_{\mathrm{M}}, \tag{3.13}$$

corresponding to an inhomogeneous essential boundary which must be enforced, whereas

$$\mathbf{1}_{\mathrm{n}} \cdot \mu \nabla P = - \quad \text{on } \Gamma_{\mathrm{E}}, \tag{3.14}$$

is required, representing a natural boundary which becomes satisfied in a weak sense if nothing is done about it.

Turning to the interface Γ_{12} between the nonconducting and eddy-current regions, the situation here is somewhat more complicated. In order that normal $\mathbf{B}$ and tangential $\mathbf{H}$ shall be continuous on the interface, it is required that

$$\mathbf{1}_{n1} \cdot (\nabla \times \mathbf{A}) - \mathbf{1}_{n2} \cdot \mu_2 \nabla P = 0 \quad \text{on } \Gamma_{12}, \tag{3.15}$$

$$\frac{1}{\mu_1} (\nabla \times \mathbf{A}) \times \mathbf{1}_{n1} - \nabla P \times \mathbf{1}_{n2} = 0 \quad \text{on } \Gamma_{12}, \tag{3.16}$$

where $\mathbf{1}_{n1}$ and $\mathbf{1}_{n2}$ are the (equal and opposite) outward normals of the regions concerned. In order that no electric current shall flow into the nonconducting region Ω_2, Eqs. (2.4) and (3.2) require as well that

$$\mathbf{1}_{n1} \cdot \left(\frac{\partial \mathbf{A}}{\partial t} + \nabla V \right) = 0 \quad \text{on } \Gamma_{12} \tag{3.17}$$

must be satisfied. Since V is defined only in region Ω_1 Eq. (3.17) in fact constitutes the boundary constraint on V for this region; being an entirely natural boundary condition, a numerical value for V must be fixed somewhere within Ω_1.

3.2   Condition for the uniqueness of **A** and V

In a numerical analysis it is particularly important to ensure that any proposed formulation is capable of producing unambiguous results. A careful examination of the Helmholtz theorem concerning the uniqueness of a vector defined by its curl and divergence in some *finite* region of space,[2] reveals that the normal component also needs to be specified on the region boundary if such uniqueness is to be assured. Thus the interface constraint

$$\mathbf{A} \cdot \mathbf{1}_{n1} = 0 \quad \text{on } \Gamma_{12} \tag{3.18}$$

is added to the set of boundary/interface relations Eqs. (3.15)–(3.16) and is shown below to have the desired effect of eliminating any ambiguity in the results obtained by applying the formulation here.

To establish the uniqueness of **A** and V, first it needs to be remarked that as far as **B**, **E** and P are concerned, it is well known that the equations given above would yield unique results for those variables (the scalar potential to within an arbitrary constant). Thus it is only necessary further to find that **A** is unique, when in view of Eq. (3.2), V becomes fixed to within an arbitary constant. Suppose two different values of vector potential $\mathbf{A}_1$ and $\mathbf{A}_2$ arise from the same physical situation, and form the vector

$$\mathbf{A}_0 = \mathbf{A}_1 - \mathbf{A}_2. \tag{3.19}$$

---

[2] The vector **A** is rendered unique by specifying its gauge alone only if its domain is infinite and it is known that $|\mathbf{A}| \sim 1/r$ as $r \to \infty$. Otherwise some extra condition is required.

From Eqs. (3.1) and (3.5) and the uniqueness of **B**, **E**, and P, there results

$$\nabla \times \mathbf{A}_0 = 0 \quad \text{in } \Omega_1, \tag{3.20}$$
$$\nabla \cdot \mathbf{A}_0 = 0 \quad \text{in } \Omega_1. \tag{3.21}$$

Equations (3.20) and (3.21) show that $\mathbf{A}_0$ can be represented by some scalar u, where

$$\mathbf{A}_0 = \nabla u \quad \text{in } \Omega_1, \tag{3.22}$$
$$\nabla^2 u = 0 \quad \text{in } \Omega_1 \tag{3.23}$$

whence using Eq. (3.18), the normal derivative of u satisfies

$$\frac{\partial u}{\partial n} = 0 \quad \text{on } \Gamma_{12}. \tag{3.24}$$

Equations (3.22) and (3.24) ensure that u is constant within Ω_1, hence $\mathbf{A}_0 = 0$ and the uniqueness of **A** is established. It may be noted that the imposition of the extra condition $\mathbf{A} \cdot \mathbf{1}_{n1} = 0$, Eq. (3.18), implicitly applies for all time and thus reduces the boundary constraint upon V, Eq. (3.17), to

$$\frac{\partial V}{\partial n} = 0 \quad \text{on } \Gamma_{12}. \tag{3.25}$$

A considerable simplification is seen to arise if σ is piecewise constant. By virtue of the Coulomb gauge also applying for all time, Eq. (3.7) now reduces to

$$\nabla^2 V = 0 \quad \text{in } \Omega_1, \tag{3.26}$$

the boundary constraint Eq. (3.25) still applying. Thus the solution of the Laplace problem represented by Eqs. (3.25) and (3.26) is V a constant or zero. In this case **A** is the sole variable inside Ω_1 and Eqs. (3.7) and (3.17) are no longer required.

3.3 *Conducting slab in a spatially uniform magnetic field*

An examination of what is probably the simplest of all eddy-current configurations, that of a conducting slab introduced into a sinusoidally varying spatially uniform magnetic field, illustrates some of the features of the general formulation given above. A magnetic field $H_0 \exp(\mathrm{j}\omega t)\mathbf{1}_z$ is assumed to be generated by a pair of infinitely permeable planar pole pieces, with a slab Ω_1, of uniform conductivity σ and thickness $2a$, interposed between them in an effectively two-dimensional arrangement, Fig. 8.2. The fundamental complex phasor equation for eddy currents expressed in terms of the **H**-field is readily obtained from Eqs. (2.1)–(2.4) as

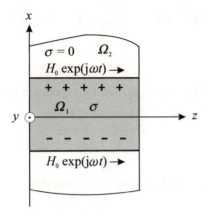

Fig 8.2 Conducting slab inserted into a sinusoidally varying spatially uniform magnetic field.

$$\nabla \times \nabla \times \mathbf{H} = -j\omega\mu\sigma\mathbf{H} \tag{3.27}$$

inside the slab. With the symmetry applying here and after some algebra, this reduces to the diffusion equation

$$\frac{\partial^2 H_z}{\partial x^2} = j\omega\mu\sigma H_z. \tag{3.28}$$

By inspection, the solution inside the slab is

$$H_z = H_0 \frac{\cosh \gamma x}{\cosh \gamma a}, \tag{3.29}$$

where $\gamma = (1+j)\sqrt{\omega\mu\sigma/2}$, satisfying the requirement that tangential $\mathbf{H}$ shall be continuous at $x = a$. Substituting back into Eq. (2.1) and then using Eq. (2.4) gives a wholly y-directed $\mathbf{E}$-field

$$E_y = -\frac{\gamma H_0 \sinh \gamma x}{\sigma \cosh \gamma a}. \tag{3.30}$$

Having arrived at the analytical field solution, it is now possible to reconstruct the $\mathbf{A}$, V solution. Using Eq. (3.1) in full component form it is required that

$$\frac{\partial A_z}{\partial y} - \frac{\partial A_y}{\partial z} = 0, \tag{3.31}$$

$$\frac{\partial A_x}{\partial z} - \frac{\partial A_z}{\partial x} = 0, \tag{3.32}$$

$$\frac{\partial A_y}{\partial x} - \frac{\partial A_x}{\partial y} = \mu H_0 \frac{\cosh \gamma x}{\cosh \gamma a}. \tag{3.33}$$

By inspection, Eqs. (3.31)–(3.33) have the solution

$$A_x = Dz + Ey, \tag{3.34}$$

$$A_y = \frac{\mu H_0 \sinh \gamma x}{\gamma \cosh \gamma a} + Ex + Cz, \tag{3.35}$$

$$A_z = Cy + Dx, \tag{3.36}$$

where C, D and E are arbitrary constants which mutually cancel out upon substitution back into Eqs. (3.31)–(3.33). The solution, with its awkward indeterminacy, does nevertheless satisfy the Coulomb gauge requirement, Eq. (3.5), that $\nabla \cdot \mathbf{A} = 0$. However, now examine the requirement for uniqueness, Eq. (3.18), that $\mathbf{A} \cdot \mathbf{1}_n = 0$ on the boundary of the eddy-current region. In the two-dimensional configuration of the slab problem this boundary comprises the lines $x = \pm a$, upon which $\mathbf{A} \cdot \mathbf{1}_n = A_x$ for all y and z, and the lines $z = 0$, d, where $\mathbf{A} \cdot \mathbf{1}_n = A_z$ for all x and y. Evidently $C = D = E = 0$ if those components are to vanish. Thus application of the constraint Eq. (3.18) has cleared the **A**-solution of any arbitrariness, as predicted. Having done this, substitution of Eqs. (3.30) and (3.35) into Eq. (3.2) relating **E** and **A** and V, produces the result that $\nabla V = 0$, confirming the prediction that if σ is piecewise constant, there is no need to introduce the scalar electric potential into a numerical eddy-current analysis. In the example here, the representation of the magnetic field outside the eddy-current region by a magnetic scalar potential and its reconciliation with the vector potential at the conductor boundary is trivial.

4.   Weak formulation of the **A**, *V–P* problem

The weak versions of the **A**, *V–P* equations suitable for finite element implementation are very conveniently reached by a Galerkin weighted residual procedure. The appropriate differential relations, here Eqs. (3.6), (3.7) and (3.12), are selected. Each differential expression, which has to vanish in its own particular volume domain, is multiplied by an arbitrary vector or scalar weighting function, **W** or W as appropriate, and integrand over that domain; the three integrals are then summed to give a weighted residual. So as to be able to employ the typical finite element expansion for **A**, V and P, that is a patchwork of low-order polynomial functions spanning nodal unknowns and connecting together with C^0-continuity, the second-order differentials are transformed to first-order ones by means of a generalized integration-by-parts procedure. This is done at the expense of bringing in the first-order derivatives of the weight functions and of introducing surface integrals.

The weight functions are then selected in turn to be the same as the interpolation functions employed in the finite element expansion but restricted to those associated with free nodes of the finite element mesh. The weighted residual for every set of W and W weight functions chosen is then equated to zero. This procedure sets up a matrix system which can be solved for the nodal unknowns, if necessary with Newton iteration (see Chapter 5) and time-stepping to reproduce transient phenomena in nonlinear systems, in terms of the system driving parameters.

4.1   *Galerkin procedure for the* **A**, *V–P formulation*

As sketched out above, the Galerkin treatment hinges upon the consideration of weighted residuals formed from the governing differential equations of the system. For region Ω_1, these equations are Eq. (3.6), derived from Ampere's law incorporating the Coulomb gauge and Eq. (3.7), representing the nondivergence of eddy currents, giving residuals

$$R_{1a} = \int_{\Omega_1} \left\{ \mathbf{W} \cdot \nabla \times \frac{1}{\mu} \nabla \times \mathbf{A} - \mathbf{W} \cdot \nabla \frac{1}{\mu} \nabla \cdot \mathbf{A} \right.$$
$$\left. + \mathbf{W} \cdot \sigma \left(\frac{\partial \mathbf{A}}{\partial t} + \nabla V \right) \right\} \mathrm{d}\Omega, \tag{4.1}$$

$$R_{1b} = - \int_{\Omega_1} W \left\{ \nabla \cdot \sigma \left(\frac{\partial \mathbf{A}}{\partial t} + \nabla V \right) \right\} \mathrm{d}\Omega, \tag{4.2}$$

whilst in region Ω_2, for the simplified boundary-driven problem with no sources, there is Eq. (3.12), a Laplace-type equation expressing the nondivergence of **B**, giving a weighted residual

$$R_2 = \int_{\Omega_2} W \nabla \cdot (\mu \nabla P) \, \mathrm{d}\Omega. \tag{4.3}$$

Here, prior to invoking the Galerkin choice for the weight functions, **W** and W may be considered to be entirely arbitrary. Nevertheless, provided appropriate boundary and interface constraints are satisfied, the residuals (4.1)–(4.3) must always vanish for any exact solution to the problem; making the residuals zero for finite element approximations of the variables **A**, V, P and a finite set of weight functions should yield a reasonable solution. Integrating by parts to transform the second order-differential terms of Eqs. (4.1) and (4.2), using the formulae given in Appendix 2, gives

$$R_{1a} = \int_{\Omega_1} \frac{1}{\mu} (\nabla \times \mathbf{W}) \cdot (\nabla \times \mathbf{A}) \, d\Omega - \oint_{\Gamma_{12}} \frac{1}{\mu} \mathbf{W} \times (\nabla \times \mathbf{A}) \cdot \mathbf{1}_{n1} \, dS$$

$$+ \int_{\Omega_1} \frac{1}{\mu} (\nabla \cdot \mathbf{W})(\nabla \cdot \mathbf{A}) \, d\Omega - \oint_{\Gamma_{12}} \frac{1}{\mu} (\nabla \cdot \mathbf{A}) \mathbf{W} \cdot \mathbf{1}_{n1} \, dS$$

$$+ \int_{\Omega_1} \mathbf{W} \cdot \sigma \left(\frac{\partial \mathbf{A}}{\partial t} + \nabla V \right) d\Omega, \tag{4.4}$$

$$R_{1b} = \int_{\Omega_1} \nabla W \cdot \sigma \left(\frac{\partial \mathbf{A}}{\partial t} + \nabla V \right) d\Omega$$

$$- \oint_{\Gamma_{12}} W \sigma \left(\frac{\partial \mathbf{A}}{\partial t} + \nabla V \right) \cdot \mathbf{1}_{n1} \, dS, \tag{4.5}$$

$$R_2 = \int_{\Omega_2} \mu \nabla W \cdot \nabla P \, d\Omega - \oint_{(\Gamma_H + \Gamma_B)} \mu W (\nabla P) \cdot \mathbf{1}_{n2} \, dS$$

$$- \oint_{\Gamma_{12}} \mu W \nabla P \cdot \mathbf{1}_{n2} \, dS, \tag{4.6}$$

where $\mathbf{1}_{n1}$ and $\mathbf{1}_{n2}$ are the outward normals to regions Ω_1 and Ω_2. Putting these R's to zero then gives weak versions of the original governing equations. The less restrictive smoothness required of the dependent variables because of the absence of second-order derivatives here makes the weak form very suitable for numerical applicatons.

Now consider **A**, V and P to be represented by nodal finite element expansions expressed in terms of known interpolation functions. It is convenient here to use a single subscript j to represent the nodal count for the whole problem space, but to distinguish between nodes in Ω_1 and Ω_2 by means of a superscript 1 or 2. Then the nodal expansion may be written

$$A = \sum_j (A_{xj} \alpha_j^{(1)} \mathbf{1}_x + A_{yj} \alpha_j^{(1)} \mathbf{1}_y + A_{zj} \alpha_j^{(1)} \mathbf{1}_z), \tag{4.7}$$

$$V = \sum_j V_j \alpha_j^{(1)}, \tag{4.8}$$

$$P = \sum_j P_j \alpha_j^{(2)}. \tag{4.9}$$

Note that the Cartesian vector expression (4.7) can be represented alternatively by vector interpolation function expansion

$$\mathbf{A} = \sum_{j'} A_{j'} \boldsymbol{\alpha}_{j'}^{(1)}, \tag{4.10}$$

where, in region Ω_1, the index j' assumes three times as many values as those of j, with

$$A_{j'} = A_{xj}, \text{ or } A_{yj}, \text{ or } A_{zj},$$

and

$$\boldsymbol{\alpha}_{j'}^{(1)} = \alpha_j^{(1)}\mathbf{1}_x, \text{ or } \alpha_j^{(1)}\mathbf{1}_y, \text{ or } \alpha_j^{(1)}\mathbf{1}_z. \tag{4.11}$$

However for brevity in what follows, the distinction between j and j' and between $\alpha^{(1)}$ and $\alpha^{(2)}$ is omitted.

In the Galerkin finite element form of the weighted residual procedure the weight functions are constructed from the nodal interpolation functions, that is to say the α and $\boldsymbol{\alpha}$'s here. The residuals (4.4)–(4.6) are made to vanish for as many different interpolation/weight functions as is appropriate for the degrees of freedom involved in the nodal expansions (4.8)–(4.10), giving enough equations to solve for the unknown nodal coefficients.

At this point some of the detail is given which finally allows the finite element matrix equations to be set down. The procedure is an extension of that relating to two-dimensional, scalar triangular elements as already described in connection with variational methods in Chapters 3 and 4. Notwithstanding that most nodes will be common to several elements, each nodal coefficient relating to the expansions (4.8)–(4.10) is single-valued, thus the trial variables $\mathbf{A}$, V and P all may be considered to be C_0-continuous in their domains. Also, the basic scalar α's are defined in the usual way within their own elements by polynomials such that

$$\alpha_i(\mathbf{r}_j) = 1 \; (i = j), \quad \alpha_i(\mathbf{r}_j) = 0 \; (i \neq j), \tag{4.12}$$

but are taken to be zero everywhere else. Furthermore, as in the two-dimensional scalar case, appropriate unions of functions defined by Eq. (4.12) are assumed when a nodal point is common to more than one finite element. These measures, with obvious extensions applying to the vector interpolation/weight functions (4.11), guarantee that the Galerkin weights are also C_0-continuous. Overall these assumptions ensure that the weak-form residuals Eqs. (4.4)–(4.6) nicely exist in spite of the rather angular functional form which a patchwork of finite elements ordinarily constitutes. Note that except when selecting weight functions corresponding to nodes actually on the boundaries/interfaces Γ, the surface integrals in the weak-form residual expressions Eqs. (4.4)–(4.6) contribute nothing to assembled finite element matrices.

Now examine the constraints Eqs. (3.13)–(3.18) relating to boundaries Γ_M, Γ_E and the interface Γ_{12} in relation to the surface integral terms in the residuals Eqs. (4.4)–(4.6) when weight functions corresponding to nodes upon these surfaces are selected. The scalar variable boundaries Γ_M and Γ_E are of the inhomogeneous essential and natural type respectively, see Eqs. (3.13) and (3.14), so the same considerations apply as in the simple two-dimensional scalar case. That is to say prescribed values of scalar magnetic potential P_0, are assigned to nodes on Γ_M and the corresponding interpolation functions are ignored for use as weight functions, whilst the boundary Γ_E is natural, nodes on it being treated in exactly the same way as any interior node. Accordingly, the first surface integral term in the residual expression Eq. (4.6) may be ignored when using the expression to set up the finite element matrices.

On the interface Γ_{12}, the $\mathbf{A} \cdot \mathbf{1}_n = 0$ condition, Eq. (3.18), is a homogeneous essential constraint; the Galerkin rules for this are that A_n should be prescribed as zero on Γ_{12} and the corresponding weight functions chosen such that $\mathbf{W} \cdot \mathbf{1}_n = 0$. This means that the second surface integral in the expression Eq. (4.4) should be ignored; furthermore Eq. (3.25) represents a natural boundary, so that V requires no constraint on Γ_{12}. Thus, with the restriction that V must be assigned an arbitrary value somewhere within Ω_1, the surface integral in the residual expression Eq. (4.5) should also be ignored. The remaining surface integrals in the expressions for R_{1a} and R_2 cannot be left out; they in effect represent the coupling of the potentials **A** and V between the regions expressed by the continuity of tangential **H** and normal **B**. First of all note that in the surviving surface integral of Eq. (4.4), vector algebra rules show that

$$\mathbf{W} \times (\nabla \times \mathbf{A}) \cdot \mathbf{1}_{n1}/\mu = \mathbf{W} \cdot \{(\nabla \times \mathbf{A}) \times \mathbf{1}_{n1}\}/\mu. \tag{4.13}$$

Now the tangential **H** coupling is clearly satisfied if the latter term is replaced by $\mathbf{W} \cdot \nabla P \times \mathbf{1}_{n2}$. Similarly the normal **B** coupling is accounted for by exchanging $\mu W \nabla P \cdot \mathbf{1}_{n2}$ for $W(\nabla \times \mathbf{A}) \cdot \mathbf{1}_{n1}$ in the second surface integral of R_2, Eq. (4.6). (Note in both cases here the minus sign in Eq. (3.11) and the fact that $\mathbf{1}_{n1} = -\mathbf{1}_{n2}$.) It is now seen that a finite element recipe can now be set down, defined by the residual equations

$$R_{1ai} = \int_{\Omega_1} \frac{1}{\mu}(\nabla \times \boldsymbol{\alpha}_i) \cdot (\nabla \times \mathbf{A}) \, d\Omega + \int_{\Omega_1} \frac{1}{\mu}(\nabla \cdot \boldsymbol{\alpha}_i)(\nabla \cdot \mathbf{A}) \, d\Omega$$
$$+ \int_{\Omega_1} \sigma \boldsymbol{\alpha}_i \cdot \left(\frac{\partial \mathbf{A}}{\partial t} + \nabla V\right) d\Omega - \oint_{\Gamma_{12}} \boldsymbol{\alpha}_i \cdot \nabla P \times \mathbf{1}_{n2} \, dS = 0, \tag{4.14}$$

$$R_{1bi} = \int_{\Omega_1} \sigma \nabla \alpha_i \cdot \left(\frac{\partial \mathbf{A}}{\partial t} + \nabla V\right) d\Omega = 0, \tag{4.15}$$

$$R_{2i} = \int_{\Omega_2} \mu \nabla \alpha_i \cdot \nabla P \, d\Omega - \oint_{\Gamma_{12}} \alpha_i (\nabla \times \mathbf{A}) \cdot \mathbf{1}_{n1} \, dS, \tag{4.16}$$

with **A** and P given by Eqs. (4.7)–(4.9), subject to the boundary/interface constraints

$$P = P_0 \quad \text{on } \Gamma_M, \tag{4.17}$$

$$\mathbf{A} \cdot \mathbf{1}_n = 0 \quad \text{on } \Gamma_{12}. \tag{4.18}$$

A distinct advantage is to be had in carrying out the surface integrations if the surface Γ_{12} happens to coincide with the Cartesian coordinate planes, otherwise rather awkward vector geometry transformations have to be made. At the expense of modelling rather more space than would otherwise be necessary, this can always be done by extending the region Ω_1 into a rectangular box surrounding the irregular-shaped conductors (then the latter may be allowed to be multiply connected).

It can now be seen specifically how to form the Galerkin finite element matrices; to keep the argument simple, the case where σ is constant within the eddy-current region is considered. In that case, as has been already seen, V may be chosen to be zero whilst the residual Eq. (4.16) is no longer required. The vanishing of the residuals may be written explicitly as

$$\sum_j A_j \int_{\Omega_1} \left\{ \frac{1}{\mu} (\nabla \times \boldsymbol{\alpha}_i) \cdot (\nabla \times \boldsymbol{\alpha}_j) + \frac{1}{\mu} (\nabla \cdot \boldsymbol{\alpha}_i)(\nabla \cdot \boldsymbol{\alpha}_j) \right\} \mathrm{d}\Omega$$

$$+ \sum_j \frac{\partial A_j}{\partial t} \int_{\Omega_1} \sigma \boldsymbol{\alpha}_i \cdot \boldsymbol{\alpha}_j \, \mathrm{d}\Omega$$

$$- \sum_j P_j \oint_{\Gamma_{12}} \boldsymbol{\alpha}_i \cdot \nabla \alpha_j \times \mathbf{1}_{n2} \, \mathrm{d}S = 0, \tag{4.19}$$

$$\sum_j P_j \int_{\Omega_2} \mu \nabla \alpha_i \cdot \nabla \alpha_j \, \mathrm{d}\Omega - \sum_j A_j \oint_{\Gamma_{12}} \alpha_i (\nabla \times \boldsymbol{\alpha}_j) \cdot \mathbf{1}_{n1} \, \mathrm{d}S = 0. \tag{4.20}$$

Subject to proper programmer's book-keeping in selecting the scalar and vector interpolation functions α and $\boldsymbol{\alpha}$, together with the building in of the essential boundary condition values P_0 and the interface constraint $\mathbf{A} \cdot \mathbf{1}_n = 0$, the integrals in Eqs. (4.19) and (4.20) represent the finite element matrix coefficients being sought. The coding to set up the coefficients proceeds element-by-element and node-by-node, systematized in a fashion similar to that described for simple scalar two-dimensional systems employing triangular or isoparametric quadrilateral elements (see the example programs of Chapters 4 and 7). Its clear that the result will be a sparse symmetric matrix equation to be solved for the vector arrays

$\mathcal{A}$, $\partial\mathcal{A}/\partial t$ and $\mathcal{P}$ representing the nodal values of $\mathbf{A}$, $\partial\mathbf{A}/\partial t$ in Ω_1 and P in Ω_2. The equation has the general form

$$
\begin{bmatrix} \mathbf{M}_{AA} & \mathbf{M}_{AP} \\ \mathbf{M}_{PA} & \mathbf{M}_{PP} \end{bmatrix} \begin{bmatrix} \mathcal{A} \\ \mathcal{P} \end{bmatrix} + \begin{bmatrix} \mathbf{N}_{AA} & 0 \\ 0 & 0 \end{bmatrix} \begin{bmatrix} \dfrac{\partial\mathcal{A}}{\partial t} \\ 0 \end{bmatrix} = \begin{bmatrix} 0 \\ \mathcal{R}_P \end{bmatrix}. \qquad (4.21)
$$

The matrices $\mathbf{M}_{AA}$, $\mathbf{N}_{AA}$ derive from the volume integrals in Ω_1, see Eq. (4.19) and $\mathbf{M}_{PP}$ from the volume integral in Ω_2, Eq. (4.20); they have the usual sparsity associated with the fact that any given node is influenced algebraically only by nodes in its own finite element (or in immediately bordering ones). Notice that the time derivative coefficient matrix $\mathbf{N}_{AA}$ involves σ but not μ, so that when saturable iron is present in an otherwise linear system, this matrix represents a linear term. The matrices $\mathbf{M}_{AP}$ and $\mathbf{M}_{PA}$ are particularly sparse, having only terms relating to the coupling of the nodal values of $\mathbf{A}$ and P on the interface Γ_{12}. The nonzero part of the right-hand side of Eq. (4.21), $\mathcal{R}_P$, is generated from the driving influence associated with the eddy-current configuration, here for simplicity limited to a prescribed magnetic potential $P = P_0$ on the outer problem boundary. The other essential constraint, $\mathbf{A} \cdot \mathbf{1}_n = 0$ on Γ_{12} shows itself only in the matrices $\mathbf{M}$ and $\mathbf{N}$. If all of the materials concerned in the eddy-current configuration are linear and the excitation is sinusoidal, then $\partial\mathbf{A}/\partial t$ can be replaced by $j\omega\mathbf{A}$ and Eq. (4.20) solved directly for complex phasor $\mathcal{A}$ and $\mathcal{P}$, otherwise some alternative, more lengthy means of solving the resulting nonlinear first-order differential equation in the time domain must be found.

5.   Eddy-current analysis using the field variables directly

As an alternative to the use of vector and scalar potentials, eddy-current problems may be worked directly in terms of either one of the field vectors, $\mathbf{H}$ or $\mathbf{E}$. This approach has been favoured in some quarters and has the advantage that only three individual scalar unknowns are required for the working variable in the eddy-current region, compared with the four corresponding to $\mathbf{A}$ and V together. On the other hand, normal components of $\mathbf{H}$ and $\mathbf{E}$ are discontinuous across material discontinuities, whereas $\mathbf{A}$ and V may be regarded as possessing C_0-continuity to match that commonly achieved in finite element modelling by using nodal interpolation functions. Indeed, if nonlinear magnetic materials are to be modelled by iteratively adjusting piecewise-constant values of μ assumed within individual finite elements according to a previously computed $\mathbf{B}$, then there has to be a discontinuity of normal $\mathbf{H}$ across *every* element interface. The difficulty in using the field vectors in the face of such a wholesale discontinuity can be overcome by employing tangentially continuous vector interpolation functions, leaving the components normal to the element faces free (Chapter 7, Sections 9 and 10). For the illustration here $\mathbf{H}$ is chosen as the working variable, continuity of tangential $\mathbf{H}$ being forced whereas continuity of the normal flux vector $\mathbf{B} = \mu\mathbf{H}$ becomes satisfied only in the weak sense, that is numerically approximated from the finite element procedure. As in the vector potential formulation, substantial economy is effected if the field outside the eddy-current region is represented by a scalar potential; appropriate measures then have to be taken to match the different representations at the interface Γ_{12}. To vary the illustration but still keep the details relatively simple, a description will be given for the case where the source field derives from a known current distribution in the region Ω_2 (Fig. 8.1). The outer boundary Γ may be assumed sufficiently remote to allow the induced magnetic field there to be taken as zero.

5.1   *Galerkin procedure for an* $\mathbf{H}$, Φ *formulation*

A source-field $\mathbf{H_S}$, derived from the given current distribution in Ω_2, is assumed to be known everywhere in the problem space, computed perhaps through the Biot–Savart law; to this field is added an induced eddy-current field, denoted by $\mathbf{H_E}$, so that the total magnetic field is now $\mathbf{H} = \mathbf{H_E} + \mathbf{H_S}$. In region Ω_1 the relevant governing equations are obtained from Eqs. (2.1)–(2.4) eliminating $\mathbf{E}$, bearing in mind that $\nabla \times \mathbf{H_S} = 0$ everywhere, and now $\mathbf{B} = \mu(\mathbf{H_E} + \mathbf{H_S})$:

$$\nabla \times \frac{1}{\sigma} \nabla \times \mathbf{H_E} + \frac{\partial}{\partial t} \mu(\mathbf{H_E} + \mathbf{H_S}) = 0 \quad \text{in } \Omega_1. \tag{5.1}$$

Equation (5.1) is multiplied by an arbitrary vector weight function $\mathbf{W}$ and integrated over Ω_1. After an integration by parts, following the procedure adopted in the $\mathbf{A}$–V, P case above, a weighted residual

$$R_1 = \int_{\Omega_1} \frac{1}{\sigma} (\nabla \times \mathbf{W}) \cdot (\nabla \times \mathbf{H_E})\, d\Omega - \oint_{\Gamma_{12}} \frac{1}{\sigma} \mathbf{W} \times (\nabla \times \mathbf{H_E}) \cdot \mathbf{1}_{n1}\, dS$$

$$+ \int_{\Omega_1} \mathbf{W} \cdot \frac{\partial}{\partial t} \mu(\mathbf{H_E} + \mathbf{H_S})\, d\Omega \tag{5.2}$$

for the conducting region Ω_1 ensues. This residual cannot be applied directly to the nonconducting region Ω_2, since $\sigma = 0$ there. However Eq. (2.2) still applies in Ω_2 and may be written

$$\nabla \times \mathbf{E} = -\frac{\partial}{\partial t} \mu(\mathbf{H_E} + \mathbf{H_S}) \quad \text{in } \Omega_2, \tag{5.3}$$

whereas Eq. (2.1) now becomes

$$\nabla \times \mathbf{H}_E = 0 \quad \text{in } \Omega_2. \tag{5.4}$$

A compatible weighted residual for Ω_2 can be set up from Eq. (5.3), which it may be noted is a first-order differential relation, giving

$$R_2 = \int_{\Omega_2} \left\{ \mathbf{W} \cdot \nabla \times \mathbf{E} + \mathbf{W} \cdot \frac{\partial}{\partial t} \mu(\mathbf{H}_E + \mathbf{H}_S) \right\} d\Omega. \tag{5.5}$$

Performing an integration by parts on the first term within the integrand above gives

$$R_2 = \int_{\Omega_2} \left\{ \mathbf{E} \cdot \nabla \times \mathbf{W} + \mathbf{W} \cdot \frac{\partial}{\partial t} \mu(\mathbf{H}_E + \mathbf{H}_S) \right\} d\Omega$$
$$- \oint_{\Gamma_{12}+\Gamma} \mathbf{W} \times \mathbf{E} \cdot \mathbf{1}_{n2} \, dS. \tag{5.6}$$

Now examine R_2 in the light of the fact that in a Galerkin procedure, $\mathbf{W}$ is going to be chosen from the same functional space as representing the vector unknown $\mathbf{H}_E$. In Ω_2 this will be the space for which $\nabla \times \mathbf{W} = 0$, see Eq. (5.4). The first term in the volume integral of the expression Eq. (5.6) thus vanishes. Moreover the final objective is formally to add the residuals R_1 and R_2; of course such an addition is only strictly necessary when the weight function corresponds to finite elements bordering on Γ_{12}, because otherwise $\mathbf{W}$ vanishes in one or other of Ω_1 and Ω_2. Now $\mathbf{E} = \nabla \times \mathbf{H}_E / \sigma$ in Ω_1, so in Eqs. (5.2) and (5.6) for R_1 and R_2 respectively, terms $\mathbf{W} \times \mathbf{E} \cdot \mathbf{1}_{n1,2}$ integrated over Γ_{12} arise and may be rewritten $\mathbf{E} \times \mathbf{1}_{n1,2} \cdot \mathbf{W}$. Since the tangential component of $\mathbf{E}$ is continuous whilst $\mathbf{1}_{n1} = -\mathbf{1}_{n2}$, it is seen that cancellation of the surface integrals involving these terms occurs, providing it can be ensured that the tangential components of $\mathbf{W}$ also are continuous on Γ_{12}. The latter is very simply ensured when $\mathbf{H}_E$ and $\mathbf{W}$ are represented by the gradient of scalar interpolation functions in Ω_2. Finally the surface integral upon the remote outer boundary may be safely neglected. Alternatively the boundary Γ, if not remote, is easily dealt with if it has the properties of either Γ_M or Γ_E as in Section 4. The convention, already described, is adopted where formally the weight functions are appropriately continuous functions over the whole of the problem Euclidean space but, chosen from the interpolation functions employed, ordinarily vanish outside their assigned finite elements. Thus in summary it is seen that a composite weighted residual may be written as

$$R = \int_{\Omega_1} \frac{1}{\sigma} (\nabla \times \mathbf{W}) \cdot \nabla \times \mathbf{H}_E) \, d\Omega$$

$$+ \int_{\Omega_1 + \Omega_2} \mathbf{W} \cdot \frac{\partial}{\partial t} u(\mathbf{H}_E + \mathbf{H}_S) \, d\Omega. \tag{5.7}$$

In the region Ω_2, in view of Eq. (5.4), it is permissible to employ the reduced scalar potential Φ as the working variable, substituting

$$\mathbf{H}_E = -\nabla \Phi \tag{5.8}$$

and correspondingly use a scalar weight function given by

$$\mathbf{W} = -\nabla W, \tag{5.9}$$

so that the weighted residual (5.7) may be rewritten as

$$R = \int_{\Omega_1} \left\{ \frac{1}{\sigma} (\nabla \times \mathbf{W}) \cdot (\nabla \times \mathbf{H}_E) + \mathbf{W} \cdot \frac{\partial}{\partial t} \mu (\mathbf{H}_E + \mathbf{H}_S) \right\} \, d\Omega$$

$$- \int_{\Omega_2} \nabla W \cdot \frac{\partial}{\partial t} \mu (-\nabla \Phi + \mathbf{H}_S) \, d\Omega. \tag{5.10}$$

Provided it is ensured that on Γ_{12} tangential components of $\mathbf{H}_E$ match the corresponding components of $-\nabla \Phi$, similarly $\mathbf{W}$ and $-\nabla W$, it is seen that nothing further need be done about the interface between the universally applicable but computationally expensive $\mathbf{H}_E$-variable in the conducting region Ω_1 and the more economical Φ only allowable in nonconductors Ω_2.

Under some circumstances, particularly to avoid the region modelled by Φ becoming multiply connected (and consequently Φ multivalued), a third region Ω_3 may be introduced representing some of the nonconducting space with the **H**-variable. It is clear, from the arguments used in setting up the residual R, that an integral term over Ω_3 may be added to Eq. (5.10) similar to the integral over Ω_1 but omitting the term containing $1/\sigma$. (Intuitively this might have been expected on the grounds that whilst $\sigma = 0$ in Ω_3, both $\mathbf{H}_E$ and $\mathbf{W}$ must have zero curl there.) Of course, appropriate steps must be taken to ensure that matching of appropriate tangential components occurs everywhere at Ω_3 interfaces. Some workers prefer to employ a boundary integral technique over the surface Γ constituting the outer extremity of the working-space. Provided the rest of the problem space is empty and unbounded, Laplace's equation holds there and the boundary element procedure required is relatively straightforward (see Chapter 6). Such an approach can be very efficient computationally, since the finite element region Ω_2 may be made quite small or even dispensed with entirely.

5.2   *Finite element implementation using vector edge elements*

Here will be described the approach appropriate for an $\mathbf{H} - \Phi$ formulation when the vector variables in tetrahedral finite elements are represented by a vector edge expansion (Chapter 7, Section 5.3). In any given eddy-current (Ω_1) tetrahedron, the trial $\mathbf{H_E}$-field is expanded as

$$\mathbf{H_E} = \sum_{m=1}^{6} h_{Em} \boldsymbol{\tau}_m, \tag{5.11}$$

where $\boldsymbol{\tau}_i$ constitute the six order-0.5 edge-associated vector interpolation functions for the tetrahedron given in Table 7.1, Chapter 7, whilst the coefficients h_{Ei} are to be determined. The edge functions are expressed in terms of the four simplex coordinates ζ_i by

$$\boldsymbol{\tau}_m = \zeta_i \nabla \zeta_j - \zeta_j \nabla \zeta_i, \tag{5.12}$$

it being noted that the ζ's are the functions both of position and the tetrahedron nodal coordinates given by Eq. (5.8), Chapter 7. Referring to Fig. 8.3, the convention usually adopted for associating m with i and j is

$$\left.\begin{aligned}
m &= 1, \quad (i,j) = (1,2), \\
m &= 2, \quad (i,j) = (1,3), \\
m &= 3, \quad (i,j) = (1,4), \\
m &= 4, \quad (i,j) = (2,3), \\
m &= 5, \quad (i,j) = (2,4), \\
m &= 6, \quad (i,j) = (3,4).
\end{aligned}\right\} \tag{5.13}$$

In the nonconducting region Ω_2, where $\mathbf{H_E} = -\nabla\Phi$, the unknown Φ is represented in the usual fashion in terms of nodal scalar interpolation functions by

$$\Phi = \sum_{m=1}^{4} \Phi_m \alpha_m. \tag{5.14}$$

It was established in Chapter 7, Section 5.3 that the significance of the

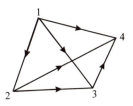

Fig 8.3   Tetrahedral element for vector edge interpolation functions.

h-coefficients in Eqs. (5.11) is that of a line integral of the trial function along a corresponding edge, that is to say

$$\int_1^2 \mathbf{H}_E \cdot d\mathbf{1} = h_{E1},$$ (5.15)

and so forth. Thus continuity of tangential **H** between the faces of adjoining tetrahedral finite elements is very simply ensured by associating the edge coefficients h_m with a unique value on every edge, apart from a possible discrepancy in sign, irrespective of which of the abutting tetrahedra to which it may belong. It is important to appreciate that whilst on the edge m represented by a vector $l_m \mathbf{1}_m$

$$\mathbf{H}_E \cdot \mathbf{1}_m = h_{Em},$$ (5.16)

the trial vector $\mathbf{H}_E$ in Eq. (5.16) and described by Eq. (5.11) itself does *not* lie along the edge. Hence associated with the Galerkin/variational procedure, there is scope for weak satisfaction of the requirement for continuous normal **B** at faces between elements whenever there is a discontinuity in μ.

Where an edge lies on Γ_{12}, the boundary between the eddy-current region Ω_1 and the nonconducting Ω_2, represented by $\mathbf{H} = \mathbf{H}_E + \mathbf{H}_S$ and $\mathbf{H} = -\nabla \Phi + \mathbf{H}_S$ respectively, evidently the requirement that tangential **H** shall be continuous is satisfied if

$$h_{Em} = \Phi_i - \Phi_j.$$ (5.17)

It can now be seen that using explicit vector and scalar interpolation functions τ_m and α_i, as appropriate, for the trial field expansions and their associated weight functions, also knowing $\mathbf{H}_S$ everywhere, the residual expression Eq. (5.10) forms the basis for setting up finite element matrix equations allowing solution of the unknown coefficients h and Φ. This is done in the usual Galerkin fashion, already described, by setting the residual R to zero for each and every allowable weight function chosen from the expansion functions $\mathbf{W} = \tau_m$ or $W = \alpha_i$.

5.3   *Use of a quasi source-field*

There is a final but important feature of the edge-element realization of the finite element recipe embodied in the residual expression Eq. (5.10). This is that whilst an edge-element expansion of the source-field $\mathbf{H}_S$,

$$\mathbf{H}_S = \sum_{m=1}^6 h_{Sm} \tau_m,$$ (5.18)

has to be found, it is not necessary that it should be the *true* free-space field as obtained from the Biot–Savart law (Webb and Forghani, 1990). There is no particular interest in that free-space field, since everywhere $\mathbf{H_S}$ is augmented by $\mathbf{H_E}$, whilst the Biot–Savart computation itself may turn out to be an expensive numerical operation. All that is required of the source term here is that it should satisfy Ampere's law,

$$\oint_C \mathbf{H_S} \cdot \mathrm{d}\boldsymbol{l} = \int_S \mathbf{J_S} \cdot \mathrm{d}\mathbf{S}, \tag{5.19}$$

around each and every closed path C of edge defining a surface S, or equivalently,

$$\nabla \times \mathbf{H_S} = \mathbf{J_S}. \tag{5.20}$$

There are many different choices of h_{Si} meeting this requirement, the distinguishing property of the unique Biot–Savart field being that in addition, $\nabla \cdot \mathbf{H_S} = 0$. It is sufficient to establish the h-coefficients of one of the infinitely many nonsolenoidal quasi source-fields. This may be done assigning values to h_{Sm} as follows.

Set up a tree of edges spanning the nodes in the finite element mesh of tetrahedra, that is to say form a trunk and nonoverlapping branches linking the nodes, eventually taking in every single node but never forming any closed loop. It is sufficient to remark here that algorithms do exist for the systematic and efficient setting-up of such trees, see Kettunen and Turner (1992). Assign an arbitrary value, say zero, to the h_{Si} for each of these edges. Now work through the co-tree of edges which remain. Each edge will be part of one or more triangular tetrahedron faces. If that edge closes a triangle whose other two sides are already assigned, work out h_{Si} so as to satisfy Eq. (5.19) and the given source current, otherwise leave it alone. After a few passes through the co-tree every edge will have been accounted for and a consistent set of h-coefficients, satisfying Ampere's law with respect to the source current and every possible closed path through the nodes, will have been created.

The quasi source-field edge-coefficients so created are then used in the setting-up of the finite element matrices as indicated. The fact that the h_{Sm} are to some extent arbitrary is of no consequence. The field $\mathbf{H_E}$ computed, like $\mathbf{H_S}$, is of no significance on its own but adjusts itself so that the sum $\mathbf{H} = \mathbf{H_S} + \mathbf{H_E}$ is approximated correctly and as accurately as the finite element discretization allows. Figures 8.4 and 8.5 illustrate a computation carried out by means of algorithms set up from the $\mathbf{H} - \Phi$ edge-element formulation given here. Arrow plots show the progression with time of the $\mathbf{H}$-field within a copper block placed between the pole-pieces of a C-magnet subject to a 50 Hz sinusoidal current excitation. It is

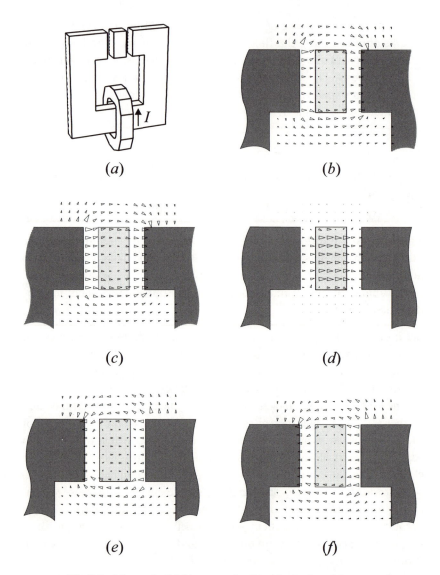

(a)   (b)

(c)   (d)

(e)   (f)

Fig 8.4 Magnetic field in a conducting block placed between the pole-pieces of a C-magnet subject to a sinusoidal excitation current I. (a) Magnet configuration, shown to scale in an isometric view; the block dimensions are $8 \times 4 \times 4$ skin-depths. (b)–(f) The field **H** in the vertical bisecting plane, shown at successive time intervals $\omega t = 0$, $\pi/4$, $\pi/2$, $3\pi/4$, π during a half-cycle; the arrowhead sizes are scaled to the maximum field occurring in the cross-section at any particular time instant. The actual fields occurring can be gauged from Fig. 8.5.

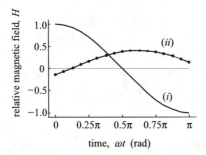

Fig 8.5   Relative magnetic field at the centre-point between the pole-pieces in the C-magnet of Fig. 8.4, (*i*) without the conducting block, (*ii*) with the conducting block in place (calculated at 17 discrete time instants). Linear materials with infinitely permeable pole pieces were assumed.

seen that the field inside the copper block diffuses in from that set up externally and then diffuses out again as the external field progresses through each half-cycle, but with phase lagging considerably behind the excitation.

6.   Solution of transient eddy-current problems by time-stepping

In the analysis given above, the time derivative $\partial/\partial t$ has been retained so as to admit the possibility of nonlinear materials, particularly magnetic ones, being present. If the materials being considered are all linear, then $\partial/\partial t$ may be replaced by $j\omega$ everywhere and the finite element matrix problems posed can be solved straight away using linear algebra. However, if transient and/or nonlinear problems are being considered, then a time-domain analysis becomes obligatory, the finite element matrix equations derived taking the general form

$$\mathbf{X}\mathcal{A} + \mathbf{Y}\partial\mathcal{A}/\partial t + \mathbf{Z} = 0. \qquad (6.1)$$

Here the symbol $\mathcal{A}$ will be used to represent the time varying array of unknown nodal potentials, both vector and scalar, arising in the analysis leading to Eq. (4.21) or perhaps the corresponding array of unknown edge coefficients in the **H**-variable analysis. The matrices **X** and **Y** arise from the problem geometry and materials whilst the vector array **Z** comes from the transient excitation and may be regarded as a prescribed function of time. In general, the time-domain problem represented by such a matrix problem can be solved by time-stepping (Wood, 1990). For illustration here, the **A**, *P* problem of Section 4 is selected for detailed examination. In such a case, the array **X** depends upon μ and therefore is time dependent, whereas **Y** has no such dependence and may be regarded as a known constant array.

A direct process whereby $\partial \mathcal{A}/\partial t$ is solved for from Eq. (6.1) and expressed in terms of $\mathcal{A}$ is not possible, because the matrix $\mathbf{Y}$ may correspond to elements where the conductivity σ is zero as well as to the eddy-current region. Instead an extension of the Galerkin process is called into play. First order approximations for $\mathcal{A}(t)$ and $\mathbf{Z}(t)$ spanning the time modes t_n and t_{n+1},

$$\mathcal{A}(t) = (1 - \tau)\mathcal{A}_n + \tau \mathcal{A}_{n+1}, \tag{6.2}$$

$$\mathbf{Z}(t) = (1 - \tau)\mathbf{Z}_n + \tau \mathbf{Z}_{n+1}, \tag{6.3}$$

where

$$\tau = (t = t_n)/(t_{n+1} - t_n)$$
$$= (t - t_n)/\Delta t, \tag{6.4}$$

are set up representing trial functions suitable for a weighted residual procedure. The space over which this particular weighted residual process is to be integrated is the one-dimensional region $0 \le \tau \le 1$. Selecting τ also as the weighting function to correspond with the Galerkin option and noting that

$$\frac{\partial \mathcal{A}}{\partial t} = \frac{\partial \mathcal{A}}{\partial \tau}\frac{\partial \tau}{\partial t} = (\mathcal{A}_{n+1} - \mathcal{A}_n)/\Delta t \tag{6.5}$$

gives a residual, to be set to zero:

$$\int_0^1 [\mathbf{X}\{(1 - \tau)\mathcal{A}_n + \tau \mathcal{A}_{n+1}\} + \mathbf{Y}(\mathcal{A}_{n+1} - \mathcal{A}_n)/\Delta t$$
$$+ (1 - \tau)\mathbf{Z}_n - \tau \mathbf{Z}_{n+1}]\tau \, d\tau = 0. \tag{6.6}$$

Performing the simple integrals indicated in Eq. (6.6) and then rearranging gives

$$(\mathbf{X}/3 + \mathbf{Y}/2\Delta t)\mathcal{A}_{n+1} + (\mathbf{X}/6 - \mathbf{Y}/2\Delta t)\mathcal{A}_n$$
$$- \mathbf{Z}_{n+1}/3 + \mathbf{Z}_n/6 = 0. \tag{6.7}$$

Equation (6.7) constitutes a recurrence relationship expressing $\mathcal{A}_{n+1}$ in terms of $\mathcal{A}_n$ and the given excitation vector $\mathbf{Z}(t)$, which latter can be discretized to provide $\mathbf{Z}_n$ for all values of the time-step index n. Given $\mathcal{A}_n$, Eq. (6.7) can be solved for $\mathcal{A}_{n+1}$ from

$$\mathcal{A}_{n+1} = (3\mathbf{Y} + 2\mathbf{X}\Delta t)^{-1}\{(3\mathbf{Y} - \mathbf{X}\Delta t)\mathcal{A}_n$$
$$+ (2\mathbf{Z}_{n+1} + \mathbf{Z}_n)\Delta t\}, \tag{6.8}$$

using essentially the same solver as employed for the linear harmonic problem. For accurate results it is important that the time-step interval Δt should be sufficiently fine. The common practice is to guess a reasonable first estimate for Δt. Subsequently at each time-step the new value of $\mathcal{A}$ is worked out both for Δt and for two successive steps of $\Delta t/2$. If the two estimates of the time-stepped $\mathcal{A}$ agree to within some prescribed tolerance the processor moves to the next step, otherwise a finer step-interval is chosen. Proceeding in this fashion, thus having solutions for t, $t + \Delta t/2$ and $t + \Delta t$, it is possible at the same time to approximate $\partial \mathcal{A}/\partial t$ and $\partial^2 \mathcal{A}/\partial t^2$. These differential coefficients may be used to estimate the value of $\mathcal{A}$ for any next following step. Such estimates are usefully employed by the matrix solver, if of the iterative variety, to speed up convergence to the proper new time-stepped vector $\mathcal{A}$. The solution of Eq. (6.8) can proceed following one of the standard nonlinear matrix solving routines such as described in Chapter 6, Sections 3 and 5. Naturally the repeated operation of the solver associated with time-stepping and nonlinearity makes such problems expensive to run.

7.   Guided waves

Electromagnetic fluctuations, representing either signal information or power, are very commonly channelled for appreciable distances along paths which maintain constant transverse properties. The channelling is usually brought about by means of metallic or dielectric members with a constant cross-section. The frequency range over which such *uniform waveguides* are of interest is exceedingly great, from those of power transmission up to optical frequencies. Thus it is of some importance to be able to perform design calculations and to make predictions relating to two-dimensional electromagnetic problems of this type. Quite a few configurations can be treated analytically, such treatments generally speaking being sufficient to characterize particular waveguide types. However when it comes to practical devices, operational and constructional convenience often leads to geometries which cannot be analysed in detail other than by numerical methods. Because of the ease in which arbitrary waveguide cross-sections can be modelled by it, the finite element method represents a very important means of such analysis.

7.1   *Waveguide equations*

In general the properties of materials making up waveguides may accurately be represented as linear, whereas the frequencies and dimensions employed together require that displacement current shall be taken into account. Because of the linearity, the analysis for waveguides is invariably carried out with respect to sinusoidal fluctuations at a single frequency ω, any more complicated time waveform being reconstructed by Fourier or Laplace transform methods. A complex phasor treatment becomes appropriate, whilst the uniformity in the axial (say z) direction allows the electromagnetic field at the frequency ω to be represented as

$$\mathbf{E}(x, y, z, t) = \mathbf{E}_0(x, y) \exp \mathrm{j}(\omega t \mp \beta z) \qquad (7.1)$$

(or similarly in terms of **H**). The determination of $\mathbf{E}_0$ (or $\mathbf{H}_0$) and β, in terms of ω and the waveguide parameters, is the goal which will be discussed here. The $\mp$ sign attached to β relates to propagation in the alternative $\pm z$-directions. At this stage it is sufficient to consider just the negative sign attached to β; the subsequent combination of waves proceeding in the positive and negative directions is merely a matter of elementary transmission line theory. It is important to appreciate that at any given frequency there can exist an infinite number of different but discrete values of β, corresponding to the modal field patterns allowable for waves in the particular guide. Mathematically this means that $\mathbf{E}_0$ and $\mathbf{H}_0$ are *eigensolutions* and, at any given frequency ω, the wavenumbers β are *eigenvalues*. A purely real β represents an unattenuated propagating mode with wavelength $\lambda = 2\pi/\beta$, whilst if purely imaginary it corresponds to an *evanescent* mode which decays very rapidly in the direction of propagation. In realistic situations the propagation constant will be $\gamma = \alpha + \mathrm{j}\beta$, a complex parameter representing practical devices with loss, whereas a lossless treatment is often sufficient to meet the waveguide analyst's need.

A simple piece of algebra takes the Maxwell curl equations, referred to Cartesian coordinates for a uniform region, and shows that the general wave, Eq. (7.1), can be expressed in terms of its longitudinal components alone:

$$H_x = \mathrm{j}\left(\omega\epsilon\frac{\partial E_z}{\partial y} - \beta\frac{\partial H_z}{\partial x}\right)/(k^2 - \beta^2), \tag{7.2}$$

$$H_y = \mathrm{j}\left(\omega\epsilon\frac{\partial E_z}{\partial x} + \beta\frac{\partial H_z}{\partial y}\right)/(k^2 - \beta^2), \tag{7.3}$$

$$E_x = \mathrm{j}\left(\omega\mu\frac{\partial E_z}{\partial y} + \beta\frac{\partial E_z}{\partial x}\right)/(k^2 - \beta^2), \tag{7.4}$$

$$E_y = \mathrm{j}\left(\omega\mu\frac{\partial H_z}{\partial x} - \beta\frac{\partial H_z}{\partial y}\right)/(k^2 - \beta^2), \quad k^2 = \omega^2\mu\epsilon. \tag{7.5}$$

The general vector Helmholtz wave equations, obtainable by eliminating **H** (or **E**) from the sourceless Maxwell curl equations, may be written as

$$\nabla \times \frac{1}{\mu}\nabla \times (\mathbf{E}, \mathbf{H}) - \omega^2\epsilon(\mathbf{E}, \mathbf{H}) = 0. \tag{7.6}$$

In the case where μ and ϵ are piecewise constant over the waveguide cross-section, these wave equations become

$$\nabla_t^2(\mathbf{E}, \mathbf{H} + (k^2 - \beta^2)(\mathbf{E}, \mathbf{H}) = 0, \tag{7.7}$$

where the transverse Laplacian operator

$$\nabla_t^2 = \frac{\partial^2}{\partial x^2} + \frac{\partial^2}{\partial y^2} \tag{7.8}$$

is introduced. In particular

$$\nabla_t^2(E_z, H_z) + (k^2 - \beta^2)(E_z, H_z) = 0, \tag{7.9}$$

representing the system differential equation when the waveguide fields are expressed entirely in terms of (E_z, H_z). Equations (7.2)–(7.5), Eq. (7.7) and Eq. (7.9) characterize many of the important properties of waveguide propagation. Subject to appropriate boundary conditions (and interface conditions between adjacent piecewise-constant regions) Eq. (7.9) shows that the problem here is indeed a modal one, β and (E_z, H_z) representing respectively eigenvalues and eigensolutions of the system equation.

7.2   *Classification of uniform waveguide modes*

Waveguides can broadly be classified according to the field mode pattern which propagates. The most basic of all the modes is the transverse electric and magnetic (TEM) mode which, as its name implies, has no field component whatsoever directed along the axis of propagation. Equations (7.2)–(7.5) imply that

$$\beta = \pm k, \tag{7.10}$$

in which case, from Eq. (7.8), the purely transverse field must be derivable from scalar potentials satisfying Laplace's equation:

$$\nabla_t^2(V, P) = 0, \tag{7.11}$$

$$\mathbf{E} = -\nabla V, \quad \mathbf{H} = -\nabla P. \tag{7.12}$$

It should be clear that such TEM propagation can arise only in two-conductor transmission lines, otherwise there is nothing but a trivial solution of Eq. (7.11), $(V, P) = 0$. (However any inhomogeneity of the transverse material profile will in general rule out true TEM propagation.) Unlike other waveguide modes, two-conductor propagation can occur at low frequencies such that the lateral waveguide dimensions are small compared with a free-space wavelength.

At microwave frequencies, the hollow conducting waveguide finds many applications. Here it is possible to propagate modes represented by E_z and H_z independently, thus having no component of one or other of these longitudinal fields. The so-called transverse electric (TE) and transverse magnetic (TM) modes are described in many elementary text books, for example Ramo, Whinnery and Van Duzer (1994). However there are a number of exotic cross-section shapes in practical use which do not have analytic solution, so numerical methods can be usefully applied to the hollow waveguide.

Lastly several important classes of waveguide propagate only modes which fall into none of the categories described above. These are the strip lines, inhomogeneously filled hollow pipes and optical waveguides. The modes which do propagate can be regarded as combinations of the three types above, inextricably coupled together by virtue of transverse inhomogeneities in the material supporting the waves. The finite element method has been applied extensively to such waveguides.

8.   TEM-mode waveguide analysis by finite elements

A complete field solution for any two-conductor cross-section may be worked out from Laplace's equation using the finite element techniques described in Chapters 2 and 4 for static problems. The solution parameters of interest, apart from the potentials and fields themselves, are the propagation constant (here trivially $\beta = k$) and the characteristic impedance Z_0. The latter may be expressed in terms of the capacitance and inductance per unit length, C and L respectively, again worked out as if for the static problem, as

$$Z_0 = \sqrt{(L/C)} \tag{8.1}$$

(see Ramo, Whinnery and Van Duzer, 1994). The parameters C and L themselves are easily computed by a finite element process. As it happens, because of the well-known fact that for all cases of parallel conductors embedded in a uniform medium

$$LC = \mu\epsilon \tag{8.2}$$

it is only necessary to calculate one of them, say C. The most convenient way of deriving this parameter is to exploit a facility which is built into most Laplace finite element packages, namely the ability to estimate the stored field energy

$$W = \frac{1}{2} \int_{\Omega} \epsilon \mathbf{E}^2 \, d\Omega. \tag{8.3}$$

The integral (8.3) may be evaluated from the vector of nodal potentials $\mathbf{U}$ (including those specified explicitly on the boundary as well as those worked out by the solver) and the assembled finite element matrix $\mathbf{S}$, as the quadratic form

$$W \approx \frac{1}{2} \mathbf{U}^T \mathbf{S} \mathbf{U}. \tag{8.4}$$

For solvers which have been set up using the variational procedure described in Chapter 4, Section 3 it is readily seen that this is equivalent to back-substitution into the functional for the Laplace solution and therefore the stored energy is indeed approximated by Eq. (8.4). Equation (8.4) is of course also valid if the finite element matrix $\mathbf{S}$ happens to have been set up by the Galerkin procedure. The capacitance being sought now may be estimated from the fact that the same stored energy is also expressed by

$$W = \frac{1}{2}CV_L^2,$$ (8.5)

where V_L is the voltage applied to the line.

8.1   *A coaxial line*

It is of interest to show how the full problem of determining both the electric and the magnetic field of a coaxial transmission line with intractable analytic solution, such as illustrated in Fig. 8.6, might be carried out. Setting up the electric field problem is the more straightforward of the two tasks. It is clear that the whole of the two-dimensional problem space could be modelled, with $V = V_L$ (= 1 say) on the inner conductor and $V = 0$ on the outer imposed as essential boundary conditions. With the symmetries existing for this particular geometry however, it is allowable and more economic to analyse just one-eighth of the problem space, say the region $ABCD$ indicated in Fig. 8.6. It is clear that the electric field lines must lie along AB and DC, so that these lines constitute boundaries upon which the natural condition, $\partial V/\partial n = 0$, applies. Figure 8.7 shows a mesh which might typically be employed for the finite element solution. The data of Fig. 8.7 together with the boundary constraints described is suitable for feeding into any typical two-dimensional finite element Laplace solver. The latter would normally return estimates of the nodal potentials V and allow post-processing of such potentials, for instance to provide corresponding estimates of the electric field vector everywhere within the problem region.

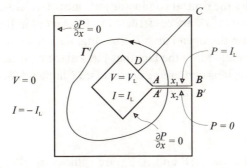

Fig 8.6   A coaxial transmission line.

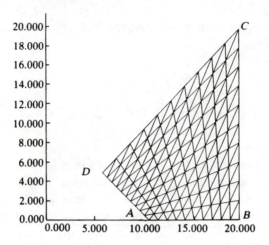

Fig 8.7   Finite element mesh for the coaxial line of Fig. 8.6, exploiting the problem symmetry.

The problem of determining the magnetic field requires out and return-currents I_L ($= 1$ say) to be specified in the coaxial conductors and the resulting scalar magnetic potential distribution P to be computed. Subsequent processing of P can yield

$$\mathbf{H} = -\nabla P \qquad (8.6)$$

if required.

   In order to be able to prescribe the boundary constraints for the Laplace problem here, observe that Ampere's rule requires that

$$\oint \mathbf{H} \cdot d\mathbf{l} = I_L \qquad (8.7)$$

for any closed curve integration path Γ which is within the problem space and encircles the central conductor just once. Two closely-spaced parallel lines, AB and $A'B'$, are constructed to form a cut in the problem domain and to create the open curve Γ' (see Fig. 8.6). Such a cut overcomes the difficulty presented by the fact that the scalar magnetic potential P is not defined in regions containing current (see Chapter 3, Section 7). It is clear that

$$\int_{\Gamma'} (-\nabla P) \cdot d\mathbf{l} = I_L, \qquad (8.8)$$

provided the lines AB and $A'B'$, are sufficiently close; carrying out the integration indicated in Eq. (8.8) then gives

$$P_{X1} - P_{X2} = I_L. \qquad (8.9)$$

Hence the essential boundaries for the Laplace problem in P for determining the magnetic field may be set as $P = I_L$ on AB and $P = 0$ on $A'B'$. The conducting walls of the transmission line coincide with magnetic field lines and hence must constitute natural boundaries $\partial P / \partial n = 0$, completing the closure of the problem domain required in order to effect the Laplace solution. Obviously because of the symmetries possessed by the configuration here, once again only the one-eighth problem region $ABCD$ need be set up for finite element solution, say $P = I_L/8$ being set on AB and $P = 0$ on CD. Notice that the only difference between the data required for the two alternative problems is that the natural and essential boundary constraints become interchanged, appropriate values for the essential boundary variable being set in each case. Figure 8.8 shows how the equipotential lines for the complementary problems form orthogonal families.

8.2 *Upper and lower bounds*

Having discussed how the problem might be worked to evaluate both the coaxial-line capacitance and inductance, an interesting and fundamental point now ensues. To calculate both C and L would in some sense amount to working the same problem twice, since the product LC is known in advance from Eq. (8.2). However, in evaluating, say, C by equating the capacitive stored energy to the electric field stored energy estimated by the finite element method, it would be correct to write

$$\frac{1}{2}CV_L \leq \frac{1}{2}\mathbf{U}_E^T\mathbf{S}_E\mathbf{U}_E, \tag{8.10}$$

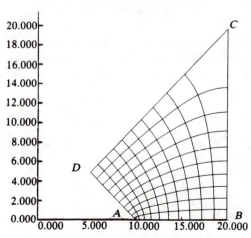

Fig 8.8 Complementary equipotential lines for the coaxial line problem of Fig. 8.6.

where the subscript E signifies finite element arrays appearing in the electric analysis. The inequality in Eq. (8.10) arises because the stationary point of the functional

$$\mathfrak{F}_E = \frac{1}{2} \int_\Omega \epsilon (\nabla U)^2 \, d\Omega \tag{8.11}$$

about the true solution $U = u$ is an absolute minimum. This fact may be seen from Eqs. (4.7)–(4.9) of Chapter 3 by altering appropriate terms in the inhomogeneous Helmholtz equation there, Eq. (4.2), to reduce it to a Laplace equation, giving

$$\mathfrak{F}_E(u + \theta h) = \mathfrak{F}_E(u) + \theta \int_\Omega \epsilon (\nabla u) \cdot (\nabla h) \, d\Omega$$
$$+ \frac{1}{2} \theta^2 \int_\Omega \epsilon (\nabla h)^2 \, d\Omega, \tag{8.12}$$

whilst noting that the finite element procedure is essentially a process of finding trial functions $U = u + \theta h$ such that the second term on the right-hand side of Eq. (8.12) vanishes. Combining Eqs. (8.1) and (8.2) gives Z_0 in terms of C alone:

$$Z_0 = \sqrt{\mu \epsilon} / C. \tag{8.13}$$

Thence, using Eq. (8.10), a corresponding inequality for the characteristic impedance is obtained:

$$Z_0 \geq \frac{\sqrt{\mu \epsilon} V_L^2}{\mathbf{U}_E^T \mathbf{S}_E \mathbf{U}_E}. \tag{8.14}$$

The quadratic form $\mathbf{U}_E^T \mathbf{S}_E \mathbf{U}_E$ is ordinarily computed using relative permittivities whilst Z_0 may be normalized with respect to the intrinsic wave impedance $\eta = \sqrt{\mu / \epsilon}$, so Eq. (8.14) is better written

$$Z_0 / \eta \geq \frac{\epsilon V_L^2}{\mathbf{U}_E^T \mathbf{S}_E \mathbf{U}_E}. \tag{8.15}$$

Now repeating the finite element procedure in terms of the magnetic variables **H** and P, a similar inequality,

$$\frac{1}{2} L I_L^2 \leq \frac{1}{2} \mathbf{U}_H^T \mathbf{S}_H \mathbf{U}_H, \tag{8.16}$$

is arrived at. But now instead of using Eq. (8.13), Z_0 must be expressed as

$$Z_0 = L / \sqrt{\mu \epsilon}. \tag{8.17}$$

This makes the corresponding inequality stand at

$$Z_0/\eta \leq \frac{\mathbf{U}_H^T \mathbf{S}_H \mathbf{U}_H}{\mu I_L^2}. \tag{8.18}$$

Thus absolute upper and lower bounds have been put around the characteristic impedance:

$$\frac{\epsilon V_L^2}{\mathbf{U}_E^T \mathbf{S}_E \mathbf{U}_E} \leq Z_0/\eta \leq \frac{\mathbf{U}_H^T \mathbf{S}_H \mathbf{U}_H}{\mu I_L^2}. \tag{8.19}$$

The placement of such bounds around numerical estimates is an objective always sought by the analyst, but all too seldom achieved.

9.   Analysis of the general waveguide by finite element methods

Uniformly filled perfectly conducting hollow waveguides can be analysed in terms of the TE and TM mode fields derived from a single longitudinal component, H_z or E_z, as is appropriate. These components satisfy a two-dimensional homogeneous Helmholtz equation, Eq. (7.9), subject to the necessary boundary constraints at the perfectly conducting walls. This equation, say for TM modes, is conveniently rewritten as

$$\nabla_t^2 E_z + k_c^2 E_z = 0, \tag{9.1}$$

where

$$k_c^2 = k^2 - \beta^2 \tag{9.2}$$

is the *cut-off wavenumber*. The variational procedure set out in Chapter 4 for constructing the corresponding finite element matrices $\mathbf{S}$ and $\mathbf{T}$ is evidently valid here. The resulting sparse matrix equation,

$$\mathbf{S}\mathscr{E} + k_c^2 \mathbf{T}\mathscr{E} = 0, \tag{9.3}$$

where $\mathscr{E}$ is a vector of unknown nodal values of E_z, is in standard eigen-equation form. It may routinely be solved for the eigenvalues k_{ci} and eigensolutions, $\mathscr{E}_i$, where $i = 1, 2, \ldots$ corresponds to a single index numbering scheme for the waveguide mode identification. Standard eigen-solvers generally return either a specified number of the lowest modes i in sequence or those lying between given values of k_c. A similar equation recovers $\mathscr{H}_i$ and k_{ci} for TE modes. Notice that although these two cases derive from one unique spatial configuration and the same partial differential equation, they are distinct because the boundary constraints imposed upon E_z and H_z and built into $\mathbf{S}$ and $\mathbf{T}$ are different. Specifically these constraints at the perfectly conducting waveguide boundary are $E_z = 0$ (an essential condition) and $\partial H_z/\partial n = 0$ (a natural condition). Once either $\mathscr{E}$ or $\mathscr{H}$ is determined it is a matter of routine postprocessing to recover the full modal field through Eqs. (7.2)–(7.5).

9.1   *The general waveguide with inhomogeneous cross-section*

Many important practical waveguides have an inhomogeneous cross-section, so that the analysis appropriate to hollow, perfectly conducting tubes considered up to this point does not apply. The commonest example of the more general form is the dielectric guide used for optical wave transmission, where there is no conducting tube at all, the guidance stemming from abrupt (or nearly so) changes in permittivity ϵ over the waveguide cross-section. The transverse inhomogeneity causes coupling between the TE and TM modes such that neither can exist separately. The resultant *hybrid* modes need at least two scalar variables for their representation. At first sight the solution would seem to be to employ E_z and H_z together, setting up an appropriate variational procedure. The intermodal coupling caused by the inhomogeneous cross-section results in quite a complicated scheme, but it can be worked out, see Csendes and Silvester (1970). However there are two objections to the approach. First, in optical waveguides, the longitudinal field components are usually very small compared with the transverse ones whilst k^2 and β^2 are very nearly the same. Equations (7.2)–(7.5) show that inaccurate numerical estimates of the transverse fields are likely to arise. Second, spurious modes have been found in finite element solutions using the (E_z, H_z) pair as working variables. These nonphysical solutions occur essentially because the zero divergence of $\epsilon\mathbf{E}$ and $\mu\mathbf{H}$, a necessary condition for any proper solution to the source-free Maxwell's equations, has not been assured. Although easily recognized, the presence of spurious modes is a nuisance which requires awkward diagnostics in any fully robust solution code. A number of alternative approaches exist and are compared by Dillon and Webb (1994). Here just two methods are singled out, on the grounds that between them, most of the objections and limitations which arise in tackling the inhomogeneous waveguide problem are covered in a reasonably efficient way.

9.2   *Transverse-field formulation*

The objective here is to employ the minimum number (two) of scalar variables but ensuring, in contrast to the (E_z, H_z) method, that the nondivergence requirement noted above is implicitly satisfied. The method has found the widest application in the analysis of optical waveguides, where the permittivity is inhomogeneous over the cross-section and may be anisotropic, whereas the permeability corresponds to a uniform nonmagnetic medium, $\mu = \mu_0$. To keep the analysis relatively simple assume isotropic dielectric materials, so that the relative permittivity ϵ_r is a simple scalar function of x and y, however for the moment retaining the possibility that there is also a scalar $\mu_r(x,y)$. Then the governing Helmholtz equation in the **H**-variable is

$$\nabla \times \epsilon_r^{-1}\nabla \times \mathbf{H} - k_0^2\mu_r\mathbf{H} = 0 \quad k_0^2 = \mu_0\epsilon_0. \tag{9.4}$$

Taking the waveguide axis to lie along the z-direction and propagation to be characterized by a complex wavenumber $\gamma = \alpha + j\beta$, corresponding to the general lossy case, the vector operator ∇ appearing in Eq. (9.4) may be divided into its transverse and longitudinal parts:

$$\nabla = \nabla_t - \mathbf{1}_z\gamma, \quad \nabla_t = \mathbf{1}_x\frac{\partial}{\partial x} + \mathbf{1}_y\frac{\partial}{\partial y}. \tag{9.5}$$

Omitting the common factor $\exp(j\omega t - \gamma z)$ throughout, the phasor vector magnetic field may be split up similarly:

$$\mathbf{H} = \mathbf{H}_t(x,y) + \mathbf{1}_z H_z(x,y). \tag{9.6}$$

This allows, after some algebra, Eq. (9.4) to be separated into its transverse and longitudinal components

$$\nabla_t \times \epsilon_r^{-1}\nabla_t \times \mathbf{H}_t - \gamma\epsilon_r^{-1}\nabla_t H_z - (k_0^2\mu_r + \gamma^2\epsilon_r^{-1})\mathbf{H}_t = 0 \tag{9.7}$$

$$\nabla_t \cdot \epsilon_r^{-1}\nabla_t H_z + \gamma\nabla_t \cdot \epsilon_r^{-1}\mathbf{H}_t + k_0^2\mu_r H_z = 0 \tag{9.8}$$

respectively. Now the condition for solenoidal magnetic field becomes

$$\nabla_t \cdot \mu_r\mathbf{H}_t - \gamma H_z = 0 \tag{9.9}$$

so that H_z can be eliminated between Eqs. (9.7) and (9.9) to give

$$\nabla_t \times \epsilon_r^{-1}\nabla_t \times \mathbf{H}_t - \epsilon_r^{-1}\nabla_t(\nabla_t \cdot \mu_r\mathbf{H}_t)$$
$$- (k_0^2\mu_r + \gamma^2\epsilon_r^{-1})\mathbf{H}_t = 0. \tag{9.10}$$

Equation (9.10) may be used as the system equation written in terms of a two-scalar working variable. In contrast with the (E_z, H_z) formulation it takes into account the requirement that $\nabla \cdot \mu\mathbf{H}$ shall vanish. However if μ_r varies abruptly in the transverse plane across a longitudinal interface, the component of $\mathbf{H}_t$ normal to that interface will be discontinuous and there will be difficulties in employing $\mathbf{H}_t$ as a working variable. Thus the transverse field formulation here is only suitable for modelling nonmagnetic waveguides, fortunately an important class which includes the optical variety. Putting $\mu_r = 1$, Eq. (9.10) is cast into a variational form suitable for finite element implementation; here the Galerkin weighted residual route is chosen. Thus, dot-multiply the equation by an arbitrary C_0-continuous function $\mathbf{W}_t(x,y)$ defined on the waveguide cross-section

Ω and integrate over that cross-section to obtain a weighted residual to be set to zero:

$$R = \int_\Omega \nabla_t \times (\epsilon_r^{-1} \nabla_t \times \mathbf{H}_t) \cdot \mathbf{W}_t \, d\Omega - \int_\Omega \epsilon_r^{-1} \nabla_t(\nabla_t \cdot \mathbf{H}_t) \cdot \mathbf{W}_t \, d\Omega$$

$$- \int_\Omega (k_0^2 + \gamma^2 \epsilon_r^{-1}) \mathbf{H}_t \cdot \mathbf{W}_t \, d\Omega = 0. \tag{9.11}$$

Using the two-dimensional equivalent of the vector integration by parts rules, Eqs. (2.5) and (2.3) in Appendix 2, applied respectively to the first and second integrals in Eq. (9.11), it is a simple matter to transform that weighted residual into its weak form

$$R = \int_\Omega (\nabla_t \times \mathbf{W}_t) \cdot (\epsilon_r^{-1} \nabla_t \times \mathbf{H}_t) \, d\Omega + \int_\Omega (\nabla_t \cdot \epsilon_r^{-1} \mathbf{W}_t)(\nabla_t \cdot \mathbf{H}_t) \, d\Omega$$

$$- \int_\Omega (k_0^2 + \gamma^2 \epsilon_r^{-1}) \mathbf{W}_t \cdot \mathbf{H}_t \, d\Omega + I_{C1} + I_{C2} = 0, \tag{9.12}$$

where

$$I_{C1} = - \int_C (\mathbf{1}_n \times \mathbf{W}_t) \cdot (\epsilon_r^{-1} \nabla_t \times \mathbf{H}_t) \, dC, \tag{9.13}$$

$$I_{C2} = \int_C (\mathbf{1}_n \cdot \mathbf{W}_t) \epsilon_r^{-1} (\nabla_t \cdot \mathbf{H}_t) \, dC. \tag{9.14}$$

Here in the first instance Ω may be regarded as a single two-dimensional finite element and C its boundary. The vectors $\mathbf{W}_t$ and $\mathbf{H}_t$ are continuous throughout the problem domain. From Maxwell's magnetic curl equation, Eq. (9.13) may be written

$$I_{C1} = - \int_C (\mathbf{1}_n \times \mathbf{W}_t) \cdot \mathbf{1}_z j\omega\epsilon_0 E_z \, dC. \tag{9.15}$$

However, E_z is to be continuous across all internal boundaries between elements, even when ϵ_r is discontinuous. Since the unit vectors $\mathbf{1}_n$ on the common boundary of any adjacent pair of elements are equal and opposite whilst $\mathbf{W}_t$ is continuous across the boundary, omitting I_{C1} in the finite element construction is equivalent to allowing such integrals to cancel at interelement boundaries, implying the correct continuity for E_z; thus the required continuity in E_z occurs as a natural condition. On the other hand, from Eq. (9.9), I_{C2}, Eq. (9.14), can be expressed as

$$I_{C2} = \int_C (\mathbf{1}_n \cdot \mathbf{W}_t) \epsilon_r^{-1} \gamma H_z \, dC \tag{9.16}$$

whereas H_z always has to be continuous. Hence in regions where ϵ is the same on either side of an internal boundary the omission of I_{C2} will also constitute a natural condition whereas the integral must be retained along any boundary between differing permittivities.

Having settled the issues which arise between finite elements, now suppose that Ω models the full waveguide cross-section. The closed curve C represents the waveguide boundary, which must consist of some combination of electric and/or magnetic walls (Chapter 3, Section 4.8) and perhaps a boundary at infinity. An electric wall requires E_z to vanish whereas H_z remains, so neglecting just I_{C1} sets up a natural constraint for such a boundary. On the other hand, for a magnetic wall the situation is reversed, so the natural condition omits I_{C2} alone. Paying attention to one or other of these conditions, at the price of performing a boundary integration, will satisfy the external boundary constraints in a natural fashion.

9.3   *The finite element matrices*

Consider now how the weighted residual, Eq. (9.12) can be used to set up a finite element matrix equation. First note that the Galerkin finite element procedure requires the selection of the weight functions $\mathbf{W}_t$ from the expansion functions used in approximating $\mathbf{H}_t$. Because vanishing divergence of $\mathbf{H}$ has been built into the formulation, a nodal expansion is adequate. In the first instance this may be written in terms of the unknown nodal vectors $\mathbf{H}_{tj}$, as

$$\mathbf{H}_t = \sum_{j=1}^{N} \mathbf{H}_{tj}\alpha_j(x, y). \tag{9.17}$$

However in order to identify the vector weights $\mathbf{W}_t$, this expansion is better written as

$$\mathbf{H}_t = \sum_{j=1}^{N} \sum_{m=1}^{2} H_{tjm}\alpha_j(x, y)\mathbf{1}_m, \tag{9.18}$$

where $\mathbf{1}_m$, $m = 1, 2$ represents two mutually perpendicular unit vectors in the waveguide cross-sectional plane and H_{tjm} are the two corresponding components of the nodal transverse field vector $\mathbf{H}_{tj}$. Evidently, up to $2N$ vector weight functions may be chosen from[3]

---

[3] The numbers i and j are alternative dummy indices identifying nodes whereas l and m similarly identify the orthogonal vector axes; indices i and l will be reserved to label a weight function associated with a particular node and coordinate axis, whereas j and m will be similarly used in the nodal trial vector expansion.

$$\mathbf{W}_t = \alpha_i(x,y)\mathbf{1}_l, \quad i = 1, N, l = 1, 2,$$

$$(x,y) \in \text{ any element common to node } i,$$

$$\mathbf{W}_t = 0, \qquad \text{otherwise.} \tag{9.19}$$

In the formulation here, employing one or other of the boundary integrals I_{C1} and I_{C2} to simulate electric and magnetic walls, none of the vectors $\mathbf{H}_{tj}$ is explicitly constrained. Thus there is no reason to use any orthogonal pair of axes other than some fixed Cartesian set,

$$\mathbf{1}_1 = \mathbf{1}_x, \mathbf{1}_2 = \mathbf{1}_y, \tag{9.20}$$

whilst all $2N$ weight vectors are in fact required. The trial expansion for $\mathbf{H}_t$, Eq. (9.18) and the weight functions (9.19) are now substituted into the residual expression Eq. (9.12) set to zero, to give a global, sparse matrix equation

$$(A^G_{ijlm} + \gamma^2 B^G_{ijlm})H_{jm} = 0, \quad i \text{ and } j = 1, N; \quad l \text{ and } m = 1, 2. \tag{9.21}$$

From the residual, Eq. (9.12), the local finite element matrices are

$$A^E_{ijlm} = \int_{\Omega_E} (\nabla_t \alpha_i \times \mathbf{1}_l) \cdot (\epsilon_r^{-1} \nabla_t \alpha_j \times \mathbf{1}_m) \, d\Omega$$

$$+ \int_{\Omega_E} (\nabla_t \epsilon_r^{-1} \alpha_i \cdot \mathbf{1}_l)(\nabla_t \alpha_j \cdot \mathbf{1}_m) \, d\Omega$$

$$- k_0^2 \delta_{lm} \int_{\Omega_E} \alpha_i \alpha_j \, d\Omega$$

$$+ \int_{C_E} \alpha_i (\mathbf{1}_n \cdot \mathbf{1}_l) \epsilon_r^{-1} (\nabla_t \alpha_j \cdot \mathbf{1}_m) \, dC, \tag{9.22}$$

$$B^E_{ijlm} = \delta_{lm} \int_{\Omega_E} \epsilon_r^{-1} \alpha_i \alpha_j \, d\Omega, \tag{9.23}$$

where δ_{lm} is the usual Kronecker delta. In Eq. (9.22), which refers to a single finite element, the line integral derives from I_{C2} whilst its boundary C_E is any portion of dielectric interface or electric wall which happens to belong to the element; otherwise it is neglected. If a portion of magnetic wall is included in the element boundary, the line integral is replaced by an expression appropriate to I_{C1} instead.

The global matrices may be assembled from the local matrix elements described by Eqs. (9.22) and (9.23) in exactly the same fashion as was done in the simpler case set out in Chapter 2. Even though the approach here has been via the Galerkin weighted residual method rather than through the minimization of a functional, because the two methods are exactly equivalent, the connection pattern typified by Eq. (2.27) in

Chapter 2 remains valid. The line integral term involved in Eq. (9.22) perhaps needs closer examination. Clearly the term here is nonzero only when the element concerned forms part of the dielectric interface or an electric wall. Furthermore the integral vanishes for any node i (but not j) not actually on C_E, since α_i correspondingly vanishes. The contribution to it from a neighbouring element on the other side of the interface, added in during the global matrix assembly process, has an equal but oppositely directed value for $\mathbf{1}_n$, and therefore tends to cancel with the first contribution. When an electric wall is being modelled by means of this integral, there is no contribution from the wall itself; this may be regarded as due to the wall being represented by $1/\epsilon_r \rightarrow 0$. Notice that if this integral were performed along a junction between a pair of finite elements *not* at a dielectric interface, the two neighbouring results would not cancel exactly, since whilst the α's are continuous functions from one element to the next, $\nabla\alpha$ is not. Such behaviour is entirely in accord with the notion that finite element approximations are C_0-continuous functions and that natural boundary or interface conditions are only satisfied in the limit. If instead or in addition there is a magnetic wall, then a similar line integration procedure is carried out using I_{C1} but ignoring I_{C2} in Eq. (9.12).

In practice, it is usual to employ first- or second-order triangular elements and to consider just piecewise-constant variations of ϵ. In that case the integrations involved in evaluating Eqs. (9.22) and (9.23) are straightforward using the analytic methods given in Chapter 4 and do not require numerical approximation.

9.4   *Alternative boundary constraint*

There is an alternative approach to dealing with the outer boundary of the waveguide, employing the constraints on $\mathbf{H}_t$ required at electric or magnetic walls, as the case may be, and no boundary integration. These constraints are that the normal and tangential components of $\mathbf{H}$ shall respectively vanish at electric and magnetic walls, applying also to $\mathbf{H}_t$. In the case of an electric wall, it has been seen that the omission of I_{C1}, Eq. (9.13), in Eq. (9.12) satisfies the electric wall condition as a natural consequence, whereas I_{C2} needs further attention. However if $\mathbf{H}_t$ is constrained such that its normal component vanishes, the Galerkin procedure requires that $\mathbf{W}_t$ shall also be so constrained. As a result of the term $\mathbf{1}_n \cdot \mathbf{W}_t$ in the expression for I_{C2}, this line integral will automatically vanish and therefore may be ignored as well as I_{C1}. On the other hand if there is a magnetic wall and both $\mathbf{H}_t$ and $\mathbf{W}_t$ are constrained such that their tangential components vanish, a similar argument establishes that both line integrals in Eq. (9.12) may again be ignored.

Evidently here is a situation where a natural boundary condition applies to H_t provided the line integrals I_{C1} and I_{C2} are properly incorporated into the finite element matrices, or the appropriate essential boundary constraint may be applied to H_t at electric or magnetic walls, in which case I_{C1} and I_{C2} may be ignored.

Applying the essential constraints is simple if the boundary coincides with one of the global coordinate axes in use, say it corresponds to the $l = 1$ direction, this in turn lying parallel to the x-axis. Then for an electric wall, requiring the normal component of H_t to vanish, the coefficient H_{tjy} is set to zero for each relevant node n, whilst the possible weight vectors $W_t = 1_y \alpha_i$ are ignored. For a magnetic wall, where instead the tangential component of H_t has to vanish, the procedure would obviously be switched around, setting H_{tjx} to zero and ignoring $W_t = 1_x \alpha_i$. For a boundary which does not coincide with one of the global coordinate axes (Fig. 8.9), the procedure is more complicated but still practical. Tangential and normal unit vectors at node j may be defined as

$$1_{\text{tang}} = 1_x \cos \theta_j + 1_y \sin \theta_j \tag{9.24}$$
$$1_{\text{norm}} = -1_x \sin \theta_j + 1_y \cos \theta_j \tag{9.25}$$

where θ_j is the inclination of the boundary with respect to the x-direction. It may be supposed that the general finite element computation is carried out in terms of the set of unknowns H_{tjx}, H_{tjy}, the Galerkin weights, Eq. (9.19), in general being chosen as the Cartesian unit vectors, Eq. (9.20). However at nodal points on the waveguide walls it is expedient instead to choose the weights from

$$W_t = 1_{\text{tang}} \alpha_i \quad \text{and} \quad W_t = 1_{\text{norm}} \alpha_i. \tag{9.26}$$

In the case of an electric wall, where the normal magnetic field is to vanish, this would require

$$H_{\text{norm}} = -H_{tjx} \sin \theta_j + H_{tjy} \cos \theta_j = 0 \tag{9.27}$$

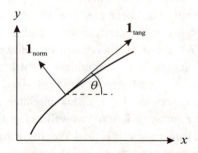

Fig 8.9 A general curvilinear boundary.

being set at every appropriate boundary point. Of the two boundary-orientated vector weight functions which might have been chosen, clearly the first in Eq. (9.26) should be used and the second ignored. The set of boundary equations (9.27) is then used to eliminate one or other of every boundary H_{tjx}, H_{tjy}, so that when the Galerkin procedure is invoked, there are not too many unknowns for the $\mathbf{H}_{tj}$ solution to be determinate. Conversely, a magnetic wall would be dealt with by setting

$$H_{\text{tang}} = H_{tjx} \cos \theta_j + H_{tjy} \sin \theta_j = 0 \tag{9.28}$$

and making the alternative choice from Eq. (9.26).

9.5   *Open boundaries*

A difficulty concerning dielectric waveguides is that they quite often have open boundaries, optical waveguides invariably so, whilst the finite element method clearly cannot cope with an infinite problem space. However the problem is not so acute as in the case of the finite element analysis of radiating structures, since here in general there *is* no transverse radiation, the waveguide fields in free space dying away exponentially with distance from the axis. Thus in general a full hybrid finite element/boundary integral approach (see Chapter 9) is not required, although it undoubtedly would result in excellent precision if applied. A simple solution to the problem is to place a notional closure around the waveguide at a safe distance from the axis where the waveguide fields may be supposed to have decayed sufficiently. Either an electric or magnetic wall condition may be imposed; if both can be tried, the change in solution parameters on going from one to the other is some measure of whether the artificial boundary has been placed far enough out or not. In any case, a procedure where the boundary is progressively enlarged until convergence is observed may be used. However such an approach can be expensive in terms of computing resources.

A method which has been applied with some success uses infinite elements. The open regions of the problem space are bordered with elements extending to infinity, such as depicted in Fig. 8.10, the border being a rectangle chosen with sides parallel to the Cartesian coordinate axes. Consider an infinite element forming part of the border parallel to the x-axis, so that it will be concerned with a finite range of x, say from x_1 to x_2 and an infinite range of y, say from y_1 to $+\infty$. A one-dimensional trial function $U(x)$ is constructed for each of the relevant $\mathbf{H}$-components so that it spans the nodes $x = x_1$ and $x = x_2$ (and any intermediate nodes, if high-order trial functions are being employed). Now within the infinite element an overall trial function $U(x) \exp\{-(y - y_1)/L\}$ is used, where L is an additional undetermined parameter. The integrations necessary to

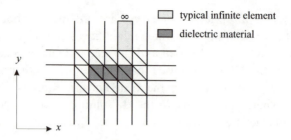

Fig 8.10 Dielectric waveguide with open boundaries modelled using infinite elements.

establish the matrices associated with an infinite element, Eqs. (9.22) and (9.23) but without the complication of the line integrals shown there, will extend to $y = +\infty$. However, with the presence of the exponential factor, such infinite integrals present no particular difficulty. Each x-border infinite element associated with $y \to \infty$ is given a similar treatment, the same or a different L-parameter being assigned. Cases where the infinite problem region extends instead to $y = -\infty$ are dealt with merely by changing the sign in the exponential part of the trial function. Infinite elements corresponding to a y-border are also similarly dealt with employing

$$U(x,y) = U(y)\exp\{\pm(x - x_1)/M\} \tag{9.29}$$

as the trial function whilst corner infinite elements are constructed with an appropriate $\exp\{\pm(x - x_1)/M\}\exp\{\pm(y - y_1)/L\}$ variation. The practice is to choose not too many different parameters M and L, but to leave those that are chosen undetermined so as to be fixed by the variational minimization process. The exponential form of the variation chosen to be built in may be justified on the grounds that in simple cases where an analytic solution can be found for open dielectric waveguides, the fields do indeed die away in this fashion. The procedure described is open to criticism as being somewhat crude, the counter-argument being that any plausible assumption is better than none if an artificial closure of an infinite system is to be made. Indeed the magnitudes of L and M themselves seem to be noncritical. Some workers use predetermined but plausible values of these parameters rather than go to the lengths required to obtain their variational estimates.

9.6   *Features of the transverse-field method*

The transverse-field formulation represents a very satisfactory approach for application to optical waveguides. This is because such waveguides are invariably composed of only dielectric materials so that a transverse magnetic field vector working variable modelling a fully continuous vector may be employed. Spurious modes are eliminated entirely whilst the method employs the minimum number of variables possible, namely two per node, and is therefore computationally very efficient. There is no particular difficulty in extending the approach to allow for anisotropic dielectric materials, provided the anisotropy is confined in a symmetric fashion to transverse-field effects so that the relative permittivity takes the tensor form[4]

$$
\varepsilon_r = \begin{bmatrix} \epsilon_{xx} & \epsilon_{xy} & 0 \\ \epsilon_{xy} & \epsilon_{yy} & 0 \\ 0 & 0 & \epsilon_{zz} \end{bmatrix} \tag{9.30}
$$

(see Lu and Fernandez, 1993). The final matrix eigenequation, Eq. (9.21) for the isotropic case worked through here, assumes a standard form, that is to say given a frequency ω, the unknown eigenvalue appears as γ^2 only — there are no terms in γ as well. It has to be observed that the symmetry often guaranteed for the matrices associated with finite element procedures is absent here, owing to the effect of the line integral term in Eq. (9.22). Less importantly, the second area integral in Eq. (9.22) also contributes to the lack of symmetry in cases where ϵ_r varies continuously instead of being piecewise constant, as usually may be assumed. However, efficient eigensolvers capable of dealing with the sparse matrix problem here have been developed; they are also applicable to cases where, because lossy dielectric materials are to be modelled, say $\epsilon_r = \epsilon_r' + j\epsilon_r''$, the matrices in Eq. (9.21) are complex requiring the propagation constant also to be complex, $\gamma = \alpha + j\beta$.

As well, in certain lossless waveguide configurations γ becomes complex for real ω. Such complex modes correspond to the evanescent modes which appear in simpler circumstances at frequencies below cut-off and have a purely imaginary propagation constant. Like the latter they actually represent no power transfer but do play a significant role in the performance of the waveguide at discontinuities. Thus it is of importance to be able to predict the characteristics of the complex modes, a task suited to methods determining γ^2 directly from frequency data. Figure 8.11 illustrates a case in which the transverse field method has been applied very effectively to the analysis of a laser optical waveguide; the effect of the active layer which provides the gain necessary for laser action is taken into account.

---

[4] This implies that the relation between electric flux density and electric field in Maxwell's equations takes the form $\mathbf{D} = \epsilon \cdot \mathbf{E}$, where the permittivity ϵ is now a 3×3 tensor array.

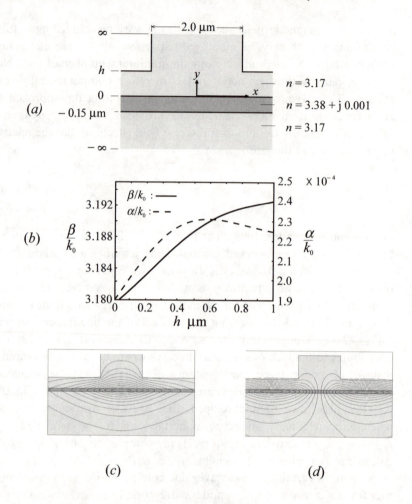

Fig 8.11 An integrated optical laser waveguide, refractive index n, with a thin active region. [Modified from Lu and Fernandez (1994).] (*a*) Basic geometry; (*b*) normalized propagation constants of the H_{11}^y dominant mode vs. the layer depth h, $\lambda = 1.5\,\mathrm{m}$; (*c*) contour plot of the H_y- component of the H_{11}^y mode, $h = 0.6\,\mathrm{m}$; (*d*) contour plot similarly of the H_{12}^y mode.

It is more difficult to employ the transverse field method to model situations where both magnetic and dielectric materials with abruptly varying properties are present. Each of the transverse field vectors $\mathbf{H_t}$ or $\mathbf{E_t}$ becomes discontinuous somewhere at an interface, so neither would represent a satisfactory finite element variable. Another instance where the method fails to produce good answers is the case where there is a highly conducting waveguide wall with reentrant sharp corners, the transverse field components there becoming singular.

9.7   *Using a full vector as the working variable*

The full vector Helmholtz equation governing wave propagation, expressed say in terms of the **H**-variable, is

$$\nabla \times \epsilon_r^{-1} \nabla \times \mathbf{H} - k_0^2 \mu_r \mathbf{H} = 0, \quad k_0^2 = \omega^2 \mu_0 \epsilon_0. \tag{9.31}$$

Its corresponding functional

$$\mathfrak{F}(\mathbf{H}) = \int_\Omega (\nabla \times \mathbf{H} \cdot \epsilon_r^{-1} \nabla \times \mathbf{H} - k_0^2 \mathbf{H} \cdot \mu_r \mathbf{H}) \, d\Omega \tag{9.32}$$

is stationary about the true solution provided the essential boundary constraint $\mathbf{H} \times \mathbf{1}_n = 0$ at magnetic walls is imposed, see Chapter 3, Sections 4.7 and 4.8. At an electric wall no constraint is required, the required boundary conditions there, namely that **H** is entirely tangential to the surface, being fulfilled to an accuracy appropriate to the finite element approximation as a natural result of making $\mathfrak{F}(\mathbf{H})$ stationary. The approach using the full vector as a working vector is frequently described as the 'standard formulation'. Assume, appropriate to a waveguide analysis, that

$$\mathbf{H} = \mathbf{H}_0(x, y) \exp(-\gamma z), \tag{9.33}$$

expanding the three-component vector $\mathbf{H}_0(x, y)$ as

$$\mathbf{H}_0 = \sum_{j=1}^{N} \mathbf{H}_{0j} \alpha_j \tag{9.34}$$

by means of scalar interpolation functions $\alpha_j(x, y)$ and vector unknowns $\mathbf{H}_{0j}$, defined at N nodes. Then it might be expected that a nodal finite element procedure based upon minimizing the functional Eq. (9.32) would be effective, at least in cases where $\mu_r = 1$ and **H** is therefore continuous everywhere. The finite element matrix equation would take the form

$$\mathbf{Ah}_0 + k_0^2 \mathbf{Bh}_0 = 0 \tag{9.35}$$

where $\mathbf{h}_0$ is a vector array made up from all of the components of the nodal field vectors $\mathbf{H}_{0j}, j = 1, 2, \ldots N$. Because the curl operator must be written as

$$\nabla \times = \left(\frac{\partial}{\partial x}, \frac{\partial}{\partial y}, -\gamma \right) \times \tag{9.36}$$

the finite element matrix **A** contains γ in a fairly involved fashion. Thus Eq. (9.35) represents an equation which nominally allows γ to be extracted as an eigensolution given k_0. However, this complicated dependence upon γ frustrates the efficient solution of the eigenproblem for γ, so that recourse has to be made to solving the problem the other way round, that is given γ, to find k_0, when Eq. (9.35) assumes a standard form. This rules out the possibility of working with lossy systems, which the functional Eq. (9.32) would permit, owing to the difficulty of guessing appropriate values of complex γ to correspond with the normalized frequency k_0, which naturally has got to be real. However another drawback to the possible method here is that the solutions are found to be heavily contaminated with spurious modes, nonphysical solutions for k_0 and $\mathbf{h}_0$. They arise through inadequate finite element modelling of $\nabla \times \mathbf{H}$ and characteristically have substantial values attached to $\nabla \cdot \mathbf{H}$ where zero is expected. An *ad hoc* solution to the spurious mode problem, briefly mentioned in Chapter 7, Section 7.1, has been put into practice by adding a penalty term to Eq. (9.32), using instead the modified functional

$$\mathfrak{F}_1(\mathbf{H}) = \mathfrak{F}(\mathbf{H}) + q \int_\Omega (\nabla \cdot \mathbf{H})^2 \, d\Omega, \tag{9.37}$$

where q is a purely numerical penalty weight factor with no particular value but chosen at will by the analyst. Since there is no question of treating other than lossless cases by this method, in fact it has been the practice to use the alternative functional,

$$\mathfrak{F}_2(\mathbf{H}) = \int_\Omega (\nabla \times \mathbf{H}^* \cdot \epsilon_{\mathrm{r}}^{-1} \nabla \times \mathbf{H} - k_0^2 \mathbf{H}^* \cdot \mu_r \mathbf{H}) \, d\Omega$$

$$+ q \int_\Omega (\nabla \cdot \mathbf{H})^* (\nabla \cdot \mathbf{H}) \, d\Omega, \tag{9.38}$$

the asterisk denoting a complete conjugate, which becomes valid for purely real μ_r, ϵ_r. Then putting $\gamma = j\beta$ and writing the magnetic variable as (H_x, H_y, jH_z), the analysis becomes formulated entirely in terms of real quantities, see Eqs. (9.7)–(9.10). The addition of the penalty term neither increases the size of the matrices involved nor affects their sparsity. The penalty function method has enjoyed a certain amount of success but suffers from the defect that the spurious modes are not completely eliminated merely pushed up in frequency, instead hopefully out of range of the wanted modes. There is little guidance available as to precisely what value of penalty factor q to choose, and it is still always necessary to check each mode to determine whether it is spurious or not.

9.8   Vector finite elements for the standard formulation

The full field waveguide formulation, when properly dealt with, nevertheless remains of some importance. The key to success here is the employment of tangentially continuous vector finite elements, which eliminates the spurious modes. The method is able to cope with situations where μ and ϵ both vary within the waveguide profile and also with conducting waveguides having sharp reentrant corners. Neither of these cases can be analysed effectively using the transverse field method. In waveguide analysis, the tangential continuity of H_z is already ensured, thus it is necessary only to model the transverse field $\mathbf{H}_t$ by means of the vector elements, H_z itself being expanded using nodal interpolation functions. In order to reproduce the detail of the standard formulation it is convenient once more to follow the Galerkin weighted residual route, equivalent to employing the stationary functional Eq. (9.32). The transverse and axial components of the vector Helmholtz equations, Eqs. (9.7) and (9.8) with both ϵ_r and μ_r retained, are multiplied by transverse and axial weight functions $\mathbf{W}_t$ and W_z respectively and integrated over the waveguide cross-section, giving the weighted residuals, to be set to zero, as

$$R_t = \int_\Omega \mathbf{W}_t \cdot (\nabla_t \times \epsilon_r^{-1} \nabla \times \mathbf{H}_t - \gamma \epsilon_r^{-1} \nabla_t H_z)\, d\Omega$$
$$- \int_\Omega (k_0^2 \mu_r + \gamma^2 \epsilon_r^{-1}) \mathbf{W}_t \cdot \mathbf{H}_t\, d\Omega, \tag{9.39}$$

$$R_z = \int_\Omega W_z (\nabla_t \cdot \epsilon_r^{-1} \nabla_t H_z + \gamma \nabla_t \cdot \epsilon_r^{-1} \mathbf{H}_t + k_0^2 \mu_r H_z)\, d\Omega. \tag{9.40}$$

In the usual way, using the vector integration by parts formulae derived in Appendix 2, Eqs. (9.39) and (9.40) may be put into the weak form

$$R_t = \int_\Omega \{ (\nabla_t \times \mathbf{W}_t) \cdot \epsilon_r^{-1} (\nabla_t \times \mathbf{H}_t) - (k_0^2 \mu_r + \gamma^2 \epsilon_r^{-1}) \mathbf{W}_t \cdot \mathbf{H}_t \}\, d\Omega$$
$$- \int_\Omega \gamma \epsilon_r^{-1} \mathbf{W}_t \cdot \nabla_t H_z\, d\Omega + I_{C1} \tag{9.41}$$

$$R_z = \int_\Omega (\nabla_t W_z \cdot \epsilon_r^{-1} \nabla_t H_z + \gamma \nabla_t W_z \cdot \epsilon_r^{-1} \mathbf{H}_t)\, d\Omega$$
$$- \int_\Omega k_0^2 \mu_r W_z H_z\, d\Omega + I_{C3} \tag{9.42}$$

where I_{C1} has already been defined by Eq. (9.13). It was argued there that the line integral I_{C1} corresponded to the natural continuity of E_z, so that it should be ignored when setting up the finite element matrices; provided $\mathbf{H}_t$ and $\mathbf{W}_t$ are tangentially continuous vectors and that both are constrained to have zero tangential components at magnetic walls, the same clearly applies here. On the other hand a new line integral

$$I_{C3} = - \int_C (W_z \epsilon_r^{-1} \nabla_t H_z + \gamma W_z \epsilon_r^{-1} \mathbf{H}_t) \cdot \mathbf{1}_n\, dC \tag{9.43}$$

now appears. In this case, with a little vector algebra and the help of Maxwell's magnetic curl equation, it is found that

$$
\begin{aligned}
I_{C3} &= -\int_C W_z (\mathbf{E}_t \times \mathbf{1}_z) \cdot \mathbf{1}_n \, dC \\
&= -j\omega\epsilon_0 \int_C W_z (\mathbf{E}_t \cdot \mathbf{1}_t) \, dC
\end{aligned}
\tag{9.44}
$$

where $\mathbf{1}_t$ is the unit tangential vector on C. It is readily seen that I_{C3} also should be ignored, so as to simulate continuous tangential $\mathbf{E}$ across inter-element boundaries and vanishing tangential electric field at an electric wall, as natural interface and boundary conditions respectively. On the other hand, if there is a magnetic wall, W_z should be set to zero to correspond with the essential constraint required there, namely that all tangential components of $\mathbf{H}$ should be made to vanish; this indicates that once again I_{C3} should be ignored.

9.9   *Matrix equations for the standard formulation*

The obvious way in which to embody the zero weighted residuals Eqs. (9.41) and (9.42) into a finite element procedure is to use tangentially continuous vector and scalar nodal interpolation functions respectively to represent $\mathbf{H}_t$ and H_z. The use of the former would be expected to, and indeed does, eliminate the possibility of spurious modes appearing. Furthermore, tangentially continuous interpolation functions conform to the physical requirement of tangential continuity in $\mathbf{H}$ (or $\mathbf{E}$) across material discontinuities and to the variational feature that the discontinuous normal components of these vectors may be left free to look after themselves.

Specifically for any two-dimensionally finite element in the waveguide cross-section, one may create an expansion for $\mathbf{H} = \mathbf{H}_t + H_z \mathbf{1}_z$ given by

$$
\mathbf{H}_t = \sum_{j=1}^{N_v^e} h_{tj} \mathbf{\tau}_j, \qquad H_z = \sum_{j=1}^{N_n^e} h_{zj} \alpha_j
\tag{9.45}
$$

where N_v^e two-dimensional vector interpolation functions $\mathbf{\tau}$ and N_n^e nodal interpolation functions α are to be associated with an element. The essential boundary constraint at magnetic walls is entered by setting to zero the appropriate coefficients h_{tj} and h_{zj} whilst the Galerkin weight functions are chosen as

$$
\mathbf{W}_t = \mathbf{\tau}_i, \qquad i = 1, \ldots, N_v^e
\tag{9.46}
$$
$$
W_z = \alpha_j, \qquad i = 1, \ldots, N_n^e.
\tag{9.47}
$$

There is no difficulty in setting up the two-dimensional equivalents of the tetrahedral vector interpolation functions τ given in Section 5, Chapter 7, whilst the corresponding quadrilateral covariant projection elements are given in Section 7.5 there. The nodal interpolation functions α are to be selected from the triangular simplex functions of Chapter 4 or the isoparametric quadrilateral ones of Chapter 7.

The expansions and Galerkin weights above may be substituted into the residuals, Eqs. (9.41) and (9.42) to provide finite element matrices, relating say to each individual element Ω_e,

$$A_{ij} = \int_{\Omega_e} (\nabla_t \times \tau_i) \cdot \epsilon_r^{-1} (\nabla_t \times \tau_j) \, d\Omega, \, i = 1, N_v^e, \quad j = 1, N_v^e, \quad (9.48)$$

$$B_{ij} = \int_{\Omega_e} \mu_r \tau_i \cdot \tau_j \, d\Omega, \qquad\qquad i = 1, N_v^e, \quad j = 1, N_v^e, \quad (9.49)$$

$$C_{ij} = \int_{\Omega_e} \epsilon_r^{-1} \tau_i \cdot \nabla \alpha_j \, d\Omega, \qquad\qquad i = 1, N_v^e, \quad j = 1, N_n^e, \quad (9.50)$$

$$D_{ij} = \int_{\Omega_e} \epsilon_r^{-1} \tau_i \cdot \tau_j \, d\Omega, \qquad\qquad i = 1, N_v^e, \quad j = 1, N_v^e, \quad (9.51)$$

$$E_{ij} = \int_{\Omega_e} \nabla_t \alpha_i \cdot \epsilon_r^{-1} \nabla_t \alpha_j \, d\Omega, \qquad i = 1, N_n^e, \quad j = 1, N_n^e, \quad (9.52)$$

$$F_{ij} = \int_{\Omega_e} \mu_r \alpha_i \alpha_j \, d\Omega, \qquad\qquad i = 1, N_n^e, \quad j = 1, N_n^e. \quad (9.53)$$

The element matrix coefficients above are evaluated in the usual fashion. That is to say, where the formulae involve low-order triangular vector elements whilst ϵ_r and μ_r are constant within each element, an algebraic treatment similar to that applying to the simplex elements described in Chapter 4 is possible. Where the elements are curvilinear quadrilaterals then a Gaussian quadrature numerical integration procedure, as used in Chapter 7 for isoparametric quadrilaterals, must be employed. The global matrices then can be assembled, element by element, in the usual way to form the system matrix equations

$$\mathbf{A}\mathbf{h}_t - k_0^2 \mathbf{B}\mathbf{h}_t - \gamma \mathbf{C}\mathbf{h}_z - \gamma^2 \mathbf{D}\mathbf{h}_t = 0, \qquad\qquad (9.54)$$

$$\mathbf{E}\mathbf{h}_z - k_0^2 \mathbf{F}\mathbf{h}_z + \gamma \mathbf{C}^T \mathbf{h}_t = 0, \qquad\qquad (9.55)$$

where $\mathbf{h}_t$ and $\mathbf{h}_z$ are column vectors representing the unknown expansion coefficients. Notice that in the first instance $\mathbf{h}_t$ and $\mathbf{h}_z$ include coefficients h_{ti} and h_{zi} which, by virtue of the essential boundary constraints, are known to be zero. However these may be considered as already eliminated from Eqs. (9.54) and (9.55). For example, the first term in Eq. (9.54) could, in this first instance, have been written in partitioned form as

$$\begin{bmatrix} \mathbf{A}_{ff} & \vdots & \mathbf{A}_{f0} \\ \cdots & \cdots & \cdots \\ \mathbf{A}_{0f} & \vdots & \mathbf{A}_{00} \end{bmatrix} \begin{bmatrix} \mathbf{h}_{tf} \\ \mathbf{0} \end{bmatrix} = \begin{bmatrix} \mathbf{A}_{ff}\mathbf{h}_{tf}\mathbf{A}_{f0}\mathbf{0} \\ \cdots \cdots \cdots \cdots \\ \mathbf{A}_{0f}\mathbf{h}_{tf} + \mathbf{A}_{00}\mathbf{0} \end{bmatrix} \tag{9.56}$$

where $\mathbf{h}_{tf}$ represents the column vector of h_{ti}'s which are free to vary and $\mathbf{0}$ the null vector of h_{ti}'s known to vanish. In the first partitioned entry on the right-hand side of Eq. (9.56), $\mathbf{A}_{ff}\mathbf{h}_{tf}$ corresponds to what has been written as $\mathbf{A}\mathbf{h}_t$ in Eq. (9.54) whereas $\mathbf{A}_{f0}\mathbf{0}$ is a null vector. The second horizontally partitioned entries of Eq. (9.56) are to be ignored completely; in the Galerkin analysis followed here they correspond to terms obtained using forbidden weights $\mathbf{W}_t$, chosen from interpolation functions belonging to degrees of freedom h_{ti} which have been already specified through essential boundary conditions. In the alternative variational approach which could have been used, the entries would have arisen through forbidden minimization with respect to degrees of freedom in the functional already so specified. With really efficient programming the lower partitioned $\mathbf{A}$'s are never calculated, whereas usually no great waste of computing resources is incurred in deriving and then subsequently discarding such terms.

Equations (9.54) and (9.55) allow γ to be extracted as an eigenvalue given k_0, or *vice versa*. From Eq. (9.55)

$$\mathbf{h}_z = -\gamma \mathbf{E} - k_0^2 \mathbf{F}^{-1}\mathbf{C}^{\mathrm{T}}\mathbf{h}_t \tag{9.57}$$

whereupon, substituting for $\mathbf{h}_z$ in Eq. (9.54) gives the standard eigen-equation form

$$\mathbf{A} - k_0^2 \mathbf{B}\mathbf{h}_t + \gamma^2 (\mathbf{C}\mathbf{E} - k_0^2 \mathbf{F}^{-1}\mathbf{C}^{\mathrm{T}} - \mathbf{D})\mathbf{h}_t = 0. \tag{9.58}$$

If $\gamma = 0$ the equations here are singular. In practice, solution of Eq. (9.58) usually does produce some precisely zero eigenvalues; however unlike the $\gamma \neq 0$ spurious modes produced elsewhere, they are easily recognizable and discarded.

9.10   *Features of the standard formulation*

The standard formulation given here worked in terms of tangentially continuous vector elements yields results which are free from spurious modes, as indeed does the transverse-field method. However unlike the latter, in addition it allows the analysis of cases where both μ and ϵ vary over the waveguide cross-section. This is so because such elements leave the normal component of the working vector variable, say $\mathbf{H}_n$, unconstrained at the element boundary, thus the jump in this component expected if the boundary is an interface between differing μ's is modelled in a nonspecific fashion. However in the variational/Galerkin process, such interfaces have been shown to constitute natural boundaries provided only tangential continuity is enforced (Chapter 7, Section 5.1). Hence the correct jump in normal component ensues, at least to an accuracy commensurate with the overall finite element approximation. If instead *continuity* of the normal component is expected, as at a plain interface or here between differing ϵ's, then it is this which is modelled to a good approximation by the vector finite element procedure.

When a waveguide with sharp reentrant conducting walls is being modelled, difficulties can arise when using the electric field as the working variable because at the sharp corners **E** is mildly singular, whilst its direction changes infinitely rapidly (Fig. 8.12). If instead **H** is being used, the latter remains finite but nevertheless does suffer an infinitely rapid change in direction. However in the standard formulation here only the tangential component of the field at the element boundaries adjacent to the wall is explicit in the formulation, to be set to zero or left free and finite, as the case may be. The infinitely rapid change in tangential direction is perfectly adequately modelled in the finite element geometry whilst a jump in the normal component (if any and albeit between side-by-side rather than face-adjacent elements) is natural; moreover the average value of the normal component over any element edge terminated by a sharp corner is expected to be finite (Webb, 1993). Thus with no more sophisticated justification than this, practical finite element analysts have tested the vector element standard formulation in such circumstances and found indeed that it works well where the nodally-based transverse-field method (Sections 9.2–9.5) does not.

10. Spatially periodic waveguides

Waveguides which have material properties and/or wall geometry repeating periodically in the axial coordinate, Fig. 8.13, find a number

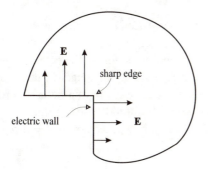

Fig 8.12 The singular electric field at sharp corner in a waveguide conducting wall. [After Webb (1993).] ©1993 IEEE.

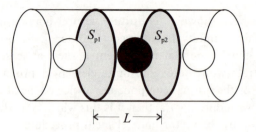

Fig 8.13 A spatially periodic waveguide unit cell; the arbitrary surfaces S_{p1} and S_{p2} are identical but spaced apart by the periodic length L.

of practical applications, both as filters and as delay lines. Thus consider a waveguide, with its axis aligned in the z-direction, defined by material characterized by $\epsilon(x, y, z)$ and $\mu(x, y, z)$ varying with spatial periodicity L such that

$$\epsilon(x, y, z + L) = \epsilon(x, y, z), \tag{10.1}$$
$$\mu(x, y, z + L) = \mu(x, y, z), \tag{10.2}$$

whilst the electric and/or magnetic boundary walls, if any, exhibit a similar geometric periodicity. Electromagnetic wave propagation in such an arrangement is governed by Floquet's theorem, which states that the field vectors in complex phasor form at frequency ω, typically $\mathbf{H}$ say, must vary as[5]

$$\mathbf{H}(x, y, = \mathbf{H}_p(x, y, z) \exp(-j\beta z) \tag{10.3}$$

where the complex wave amplitude has the same spatial periodicity as that of the waveguide itself,

$$\mathbf{H}_p(x, y, z + L) = \mathbf{H}_p(x, y, z). \tag{10.4}$$

10.1   *Finite element formulation for periodic waveguides*

It is possible to set up a finite element procedure which models this class of waveguide in rather a similar fashion to the nonperiodic case. That is to say a family of eigenvectors $\mathbf{H}_{pn}$ and eigenvalues β_n, given (normalized) frequency k_0, are sought in order to characterize propagation through the periodic waveguide. It should be clear that the spatial periodicity is such that just a single unit cell, say from $z = 0$ to $z = L$,

---

[5] In order to keep the analysis as simple as possible, the lossless case, $\gamma = j\beta$, is considered here.

needs to be employed as the computational working space. Furthermore, from Eq. (10.3) it can be seen that it is necessary only to investigate βL over a range of 2π, usually taken as

$$-\pi \leq \beta L \leq \pi, \tag{10.5}$$

since the Floquet periodicity is such that all results outside that range are simply related to those within.

Once more, the vector Helmholtz complex phasor equation, say in the H-variable,

$$\nabla \times \epsilon_r^{-1} \times \mathbf{H} - k_0^2 \mu_r \mathbf{H} = 0. \quad k_0^2 = \omega^2 \mu_0 \epsilon_0, \tag{10.6}$$

is to be solved subject to the wall constraints whatever these may be. Now additionally, the constraint corresponding to the periodicity from one end of the unit cell to the other needs to be worked in. In order to set up the finite element matrices, a variant of the Galerkin weighted residual procedure is adopted (Ferrari, 1991). Equation (10.6) is multiplied by a weight function $\mathbf{W}$ and integration over the problem space Ω, here a unit cell and the usual weak form of the weighted residual established using the vector integration-by-parts formula Eq. (2.5) of Appendix 2. The result is

$$R = \int_\Omega \{ (\nabla \times \mathbf{W}) \cdot (\epsilon_r^{-1} \nabla \times \mathbf{H}) - k_0^2 \mu_r \mathbf{W} \cdot \mathbf{H} \} \, d\Omega$$
$$- \oint_S \mathbf{W} \times (\epsilon_r^{-1} \nabla \times \mathbf{H}) \cdot \mathbf{1}_n \, dS. \tag{10.7}$$

Consider first the surface integral in Eq. (10.7). It vanishes over the waveguide walls in the usual fashion, either naturally at any electric wall or as a result of an essential boundary constraint,

$$\mathbf{H} \times \mathbf{1}_n = \mathbf{W} \times \mathbf{1}_n = 0 \tag{10.8}$$

at magnetic walls. If any part of the waveguide wall is at infinity and there are no transverse radiative modes, the associated surface integral may similarly be neglected. However the surface integral over the ends of the unit cell S_{p1} and S_{p2} (Fig. 8.13) defined by the Floquet periodicity is as yet unaccounted for. Now substitute $\mathbf{H}$ given by Eq. (10.3) and $\mathbf{W}$ given by

$$\mathbf{W} = \mathbf{W}_p(x, y, z) \exp(+j\beta z), \tag{10.9}$$

where the factor $\mathbf{W}_p$ has the same spatial periodicity as $\mathbf{H}_p$, into Eq. (10.7). The opposing signs assigned to $j\beta z$ in Eqs. (10.3) and (10.9) is crucial, for it can be seen that then the explicit factor $\exp(\pm j\beta z)$ cancels out everywhere in the residual. Adopt a notation

$$\nabla^{(\pm)} = \left(\frac{\partial}{\partial x}, \frac{\partial}{\partial y}, \frac{\partial}{\partial z} \pm j\beta \right) \qquad (10.10)$$

to account for the differing results of the curl operation upon $\mathbf{W}$ and $\mathbf{H}$ with their respective $\exp(\pm\beta z)$ factors. The surface integral remaining becomes

$$I_p = -\oint_{S_{p1}+S_{p2}} \left\{ \mathbf{W}_p \times (\epsilon_r^{-1} \nabla^{(-)} \times \mathbf{H}_p) \cdot \mathbf{1}_n \right\} dS. \qquad (10.11)$$

It is easy to see that this integral also disappears, since the integrand is periodic and exactly the same at one end of the unit cell as at the other, *except* for the unit normals $\mathbf{1}_n$, which are equal and opposite. Thus one finally ends up with a weighted residual

$$R = \int_{\Omega} \left\{ (\nabla^{(+)} \times \mathbf{W}_p) \cdot \epsilon_r^{-1} \nabla^{(-)} \times \mathbf{H}_p) - k_0^2 \mu_r \mathbf{W}_p \cdot \mathbf{H}_p \right\} d\Omega, \qquad (10.12)$$

to be put to zero with $\mathbf{H}_p$ expanded in terms of some suitable set of interpolation functions and $\mathbf{W}_p$ chosen from the same set, in the usual fashion.

10.2   Implementation using vector finite elements

Cases of practical interest will generally be irreducibly three-dimensional; as in the case of plain waveguides, steps must be taken in the finite element analysis to avoid the occurrence of spurious modes. Here, because the assumption of an $\exp(-j\beta z)$ dependence does not now fully define the field z-variation, a three-component vector working variable is obligatory, whilst the use of tangentially continuous finite elements is evidently required if nonphysical solutions are to be avoided. Since the geometries to be considered no longer have invariant properties with the z-coordinate, the employment of nodal finite elements to represent the z-field component is ruled out and a full three-dimensional tangentially continuous representation is required.

However adopting tangentially continuous elements the procedure for setting up the finite element matrix eigenequations is straightforward. Having decided upon the type of vector finite element τ to use, tetrahedral or covariant projection hexahedral (see Chapter 7, Sections 5 and 6 respectively), an expansion

$$\mathbf{H}_p = \sum_j h_{pj} \tau_j \qquad (10.13)$$

is made. The Galerkin weights are chosen in the usual fashion from the τ_j employed in Eq. (10.13)

$$\mathbf{W}_p = \tau_i. \qquad (10.14)$$

The interpolation and weight functions associated with any h_{pj} already set by an essential boundary constraint are ignored whilst the rest are substituted into the residual Eq. (10.12) set to zero to give the required matrix eigenequation. Thus the eigenequation which results is of the form

$$\mathbf{P}(j\beta)\mathbf{h_p} + k_0^2 \mathbf{Q}\mathbf{h_p} = 0, \qquad (10.15)$$

resulting in a countably infinite set of eigenvectors $\mathbf{h}_{pn}$ and eigenvalues $k_{0n}, n = 1, 2, \ldots$ given β, which may be obtained using relatively standard linear algebra routines. Because the use of the full three-component vector form of $\boldsymbol{\tau}$ is obligatory, the methods applying to plain waveguides which were used to cast the eigenequation into a standard form, returning β given k_0, are not available. Figure 8.14 shows a typical set of results relating to a periodic waveguide, here a benchmark case amenable to

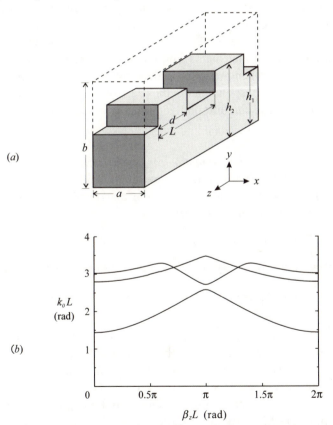

Fig 8.14   Finite element analysis of a three-dimensional periodic waveguide. (a) The geometry. (b) Plot of $\beta_z L$ vs. $k_0 L$, over a range of 0–2π in $\beta_z L$ for the three lowest propagating modes. Here $b = 2a$, $L = 1.26a$, $h_1 = a$ and $h_2 = 1.4a$; the relative permittivity of the dielectric material (shown shaded) is $\epsilon_r = 2.28$. [After Mias (1995).]

analytic treatment which agreed very well with the finite element results. Mixed-order hexahedral covariant projection elements $m = 1, n = 2$ (see Chapter 7, Section 6) were used in the computation.

11.   Cavity resonators

A closed volume with perfectly reflective walls enclosing free space and/or lossless material can sustain free electromagnetic oscillations at a countably infinite number of different discrete frequencies, each mode possessing its own characteristic field pattern. The inclusion of loss in such a system damps the oscillations and results in resonant-frequency peaks rather than a discrete frequency response; however if the damping is not too large its effects can be treated as a perturbation of the lossless case. Such resonators find many applications in electromagnetic engineering, ranging from frequency standards to the microwave oven. Thus it is of interest to be able to compute the resonant frequencies and field patterns of arrangements whose analytical treatment is intractable; the numerical problem presented is always that of finding eigenvalues and eigensolutions corresponding to the electromagnetic state of a closed region of space, a computation for which the finite element method is well suited.

11.1   *Finite element formulation for cavity analysis*

The formulation in the frequency domain for linear materials is relatively straightforward. Starting from the Helmholtz electromagnetic wave equation, say in the **H**-variable,

$$\nabla \times \epsilon_r^{-1} \nabla \times \mathbf{H} - k_0^2 \mu_r \mathbf{H} = 0, \quad k_0^2 = \omega^2 \mu_0 \epsilon_0, \tag{11.1}$$

and proceeding by the variational route requires the functional

$$\mathfrak{F}(\mathbf{H}) = \int_\Omega (\nabla \times \mathbf{H} \cdot \epsilon_r^{-1} \nabla \times \mathbf{H} - k_0^2 \mathbf{H} \cdot \mu_r \mathbf{H}) \, d\Omega \tag{11.2}$$

to be made stationary subject to any essential constraint $\mathbf{H} \times \mathbf{1_n} = 0$ at magnetic walls being imposed, the electric-wall constraint $\mathbf{E} \times \mathbf{1_n} = 0$ being natural (Chapter 3, Sections 4.7 and 4.8). In order to do this in a finite element context, the working variable **H** has to be expressed in terms of known interpolation functions and unknown coefficients in the familiar fashion. As in the earlier cases treated in this chapter, any use of scalar interpolation functions in a nodal expansion other than in trivial two-dimensional problems will result in troublesome, finite frequency spurious modes; in the general three-dimensional case the expansion has to be

$$\mathbf{H} = \sum_j h_j \tau_j, \tag{11.3}$$

expressed in terms of predefined tangentially continuous vector interpolation functions τ and unknown coefficients h. The functions τ can
be of either the tetrahedral or the covariant projection hexahedral variety
(Chapter 7, Sections 5 and 6 respectively). Here, unlike in the waveguide
problem where the working vector variable can usefully be split into its
longitudinal and transverse components, no simplification is possible and
Eq. (11.3) must be used as it stands. However the construction of the
finite element matrix equation

$$\mathbf{A}\mathbf{h} + k_0^2 \mathbf{B}\mathbf{h} = 0, \tag{11.4}$$

is straightforward, where $\mathbf{h}$ is the vector of unknown coefficients in Eq.
(11.3). The equations

$$A_{ij}^e = \int_\Omega \nabla \times \tau_i^e \cdot \epsilon_r^{-1} \nabla \times \tau_j^e \, d\Omega \tag{11.5}$$

$$B_{ij}^e = \int_\Omega \tau_i^e \cdot \mu_r \tau_j^e \, d\Omega \tag{11.6}$$

define the matrices relating to an individual finite element e, whilst the
global matrices $\mathbf{A}$ and $\mathbf{B}$ are assembled from the element arrays in the
usual fashion. It would be usual to assume ϵ_r and μ_r to be constant within
any given element, whereas the formulation is valid and presents no
particular difficulty for anisotropic materials, where the relative consti-
tuent parameters assume a tensor form.

The matrices in Eq. (11.4) are sparse and the extraction of eigenvalues
k_{0m} and eigenvectors $\mathbf{h}_m$, corresponding to the cavity resonant frequencies
and the associated field patterns respectively, is a standard problem in
numerical analysis (Davies, Fernandez and Philippou, 1982). Whilst no
spurious modes are encountered using the above formulation, it may be
observed that $\nabla \times \mathbf{H} = 0$, $k_0 = 0$ is a valid, though not very useful, solution of Eq. (11.1). Consequently, depending upon the particular eigen-
solver used, such zero eigenvalues may turn up; however in a
computation purely to determine the properties of an isolated cavity,
there is no difficulty in recognizing and discarding such modes.

The method described here can be extended to deterministic frequency-domain problems, for instance cases where a cavity containing
lossy materials is externally driven. The formulation given here is certainly valid for situations where either or both of ϵ and μ are complex,
representing a dissipative medium. Furthermore the presence of a patch
with known, nonzero transverse fields, arising say from a waveguide

excitation, is easily taken into account in applying the essential boundary constraint. For **H** as the working variable (say) this is

$$\mathbf{H} \times \mathbf{1_n} = \mathbf{P}, \tag{11.7}$$

where **P** is the prescribed transverse field. Then as a result of the pre-scribed values of h_j which occur, a right-hand side to Eq. (11.4) appears,

$$\mathbf{Ah} + k_0^2 \mathbf{Bh} = \mathbf{p}, \tag{11.8}$$

so that at a given frequency, a solution

$$\mathbf{h} = [\mathbf{A} + k_0^2 \mathbf{B}]^{-1} \mathbf{p} \tag{11.9}$$

can be extracted by standard sparse matrix manipulations. In principle the nonlinear problem arising in the analysis of microwave cavity indus-trial and domestic ovens can be solved in this way. Here the dielectric properties of the material being heated in the oven depend critically upon its temperature, whilst the latter may be determined from a separate concurrent numerical computation depending upon the time average field set by Eq. (11.9). In practice, microwave oven cavities operate over-moded such that a sizable number of the discrete modes expected appear together. In these circumstances, workers in this field (Dibben and Metaxas, 1994) have found it preferable to work in the time domain using methods not dissimilar to those detailed for the nonlinear eddy-currents problem described in Section 6.

12.   Further reading

Bíró and Preis (1989) give an excellent account of the options open for the solution of eddy-current problems by the finite element method using vector potential **A**. The reader may wish to consult this paper for information relating to configurations more general than the simple arrangements discussed here. The paper by Bossavit and Vérité (1983) describing an eddy-current solver employing vector edge elements is the first account ever of such finite elements being successfully used in the solution of an electromagnetic problem. Kriezis *et al.* (1992) give a comprehensive bibliography relating to all aspects of eddy currents, including finite element formulations and solutions.

The book by Ramo, Whinnery and Van Duzer (1994), now in its third edition, can be regarded as the standard reference, at a level appropriate to this text, relating to device-oriented high-frequency electromagnetics.

The reivew by Rahman, Fernandez and Davies (1991) and the update by Davies (1993) together represent a comprehensive general coverage of the treatment of waveguides and cavities by the finite element method.

13. Bibliography

Bíró, O. and Preis, K. (1989), 'On the use of magnetic vector potential in the finite element analysis of three-dimensional eddy currents', *IEEE Transactions on Magnetics*, **25**, pp. 3145–59.

Bossavit, A. and Vérité, J.C. (1983), 'The "Trifou" code: Solving the 3-D eddy-current problem by using H as state variable', *IEEE Transactions on Magnetics*, **19**, pp. 2465–70.

Csendes, Z.J. and Silvester, P. (1970), 'Numerical solution of dielectric waveguides', *IEEE Transactions on Microwave Theory and Techniques*, **MTT-18**, pp. 1124–31.

Davies, J.B. (1993), 'Finite element analysis of waveguides and cavities – a review', *IEEE Transactions on Magnetics*, **29**, pp. 1578–83.

Davies, J.B., Fernandez, F.A. and Philippou, G.Y. (1982), 'Finite element analysis of all modes in cavities with circular symmetry', *IEEE Transactions on Microwave Theory and Techniques*, **MTT-30**, pp. 1975–80.

Dibben, D.C. and Metaxas, A.C. (1994), 'Finite element time domain analysis of multimode applicators using edge elements', *Journal of Microwave Power and Electromagnetic Energy*, **29**(4), pp. 242–51.

Dillon, B.M. and Webb, J.P. (1994), 'A comparison of formulations for the vector finite element analysis of waveguides', *IEEE Transactions on Microwave Theory and Techniques*, **42**(2), pp. 308–16

Ferrari, R.L. (1991), 'Finite element solution of time-harmonic modal fields in periodic structures', *Electronics Letters*, **27**(1), pp. 33–4.

Kettunen, L. and Turner, L.R. (1992), 'How to define the minimum set of equations for edge elements', *International Journal of Applied Electromagnetics in Materials*, **3**, pp. 47–53.

Kriezis E.E., Tsiboukis, T.D., Panas, S.M. and Tegopoulos, J.A. (1992), 'Eddy currents: Theory and applications', *Proceedings of the IEEE*, **80**, pp. 1559–89.

Lu, Y. and Fernandez, F.A. (1993), 'An efficient finite element solution of inhomogeneous anisotropic and lossy dielectric waveguides', *IEEE Transactions on Microwave Theory and Techniques*, **41**, pp. 1215–23.

Lu, Y. and Fernandez, F.A. (1994), 'Vector finite element analysis of integrated optical waveguides', *IEEE Transactions on Magnetics*, **30**, pp. 3116–19.

Mias, C. (1995), '*Finite element modelling of the electromagnetic behaviour of spatially periodic structures*', PhD dissertation, University of Cambridge.

Rahman, B.M.A., Fernandez, F.A. and Davies, J.B. (1991), 'Review of finite element methods for microwave and optical waveguides, *Proceedings of the IEEE*, **79**, pp. 1442–8.

Ramo, S., Whinnery, J.R. and Van Duzer, T. (1994), *Fields and waves in communication electronics*, 3rd edn. New York. Wiley: xix + 844.

Webb, J.P. (1993), 'Edge elements and what they can do for you', *IEEE Transactions on Magnetics*, **29**, pp. 1460–5.

Webb, J.P. and Forghani, B. (1990), 'A scalar-vector method for 3D eddy current problems using edge elements', *IEEE Transactions on Magnetics*, **26**, pp. 2367–9.

Wood, W.L. (1990), *Practical time-stepping schemes*, Oxford: Clarendon Press, ix + 373 pp.

9

Unbounded radiation and scattering

1. Introduction

Radiation and scattering of high-frequency electromagnetic energy arises in many important physical circumstances, particularly those relating to communications and radar systems. The need to design and improve such systems has given rise to much complex and sophisticated numerical analysis, mostly tailored to suit particular geometries. Integral methods have been used extensively and may be applied quite effectively to obtain numerical results for geometries involving just highly conducting elements, since for such cases the appropriate Green's functions are readily available and only the conducting surfaces need be discretized. Dense but relatively small matrices turn up in the resulting linear algebra; the techniques used are often described as 'moment methods', see Chapter 6. Purely integral techniques can be extended to situations with the presence of uniform dielectric or magnetic materials, whereas the dense matrices generated tend to become large and the procedures unwieldy. The differential-operator finite element approaches to the solution of high-frequency electromagnetic problems are well suited to cases where there is inhomogeneous material present, resulting in relatively large but sparse and banded matrices, easily handled by standard linear algebra routines. However invariably problems involve an inhomogeneous region which is limited in extent, bordered by free space constituting an external region which reaches to infinity. It is thus a logical further step to develop hybrid finite element methods, incorporating discretized boundary operations dealing with the uniform external region, for solving general radiation and scattering problems.

1.1 *The general problem*

A general configuration which typifies both antenna and scattering problems is shown in Fig. 9.1. A region Ω, having an arbitrarily-

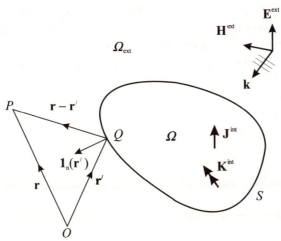

Fig 9.1 General radiation/scattering configuration. The region Ω, bounded by surface S, may contain both internal sources and material with constitutive constants $\mu_0\mu_r(\mathbf{r})$, $\epsilon_0\epsilon_r(\mathbf{r})$. The external region Ω_{ext} is free space, characterized by μ_0, ϵ_0, and may contain radiation source fields.

placed closed boundary S with outward normal vector unit $\mathbf{1}_n$, contains all of the problem-specified nonuniform linear material characterized by relative permittivity ϵ_r and permeability μ_r. The region may also include equivalent internal electric and magnetic currents $\mathbf{J}^{int}$ and $\mathbf{K}^{int}$ respectively representing sources of magnetic and electric field, see Kong (1986); such an inclusion is appropriate if it is required to model a transmitting antenna array. The rest of space outside Ω, denoted by Ω_{ext}, contains no material medium and extends indefinitely. There may however be impressed external incident fields $\mathbf{E}^{ext}$, $\mathbf{H}^{ext}$; these are the sources appropriate when scattering problems are being considered. The total resultant field, either $\mathbf{E}$ or $\mathbf{H}$, is sought everywhere; when one is determined, the other follows from Maxwell's equations. This general arrangement has been considered by Jin, Volakis and Collins (1991).

2.   Finite element treatment of the inhomogeneous region

Maxwell's curl equations in complex phasor form for frequency ω (Chapter 3) are restated here to include the equivalent internal source terms:

$$\nabla \times \mathbf{E} = -j\omega\mu_0\mu_r\mathbf{H} - \mathbf{K}^{int}, \tag{2.1}$$

$$\nabla \times \mathbf{H} = j\omega\epsilon_0\epsilon_r\mathbf{E} + \mathbf{J}^{int}. \tag{2.2}$$

Note that in Eq. (2.2), no specific term **J** representing induced current has been included. Since linear materials are to be assumed everywhere, any secondary current arising from the primary source excitation can be taken into account by means of a complex relative permittivity ϵ_r with imaginary part $j\epsilon_{ri} = -j\sigma/\omega\epsilon_0$, where σ represents an equivalent high-frequency conductivity.

Assume initially that the hybrid finite element analysis is to be carried out employing the E-variable. Eliminating **H** between Eqs. (2.1) and (2.2) gives

$$\nabla \times (\mu_r^{-1}\nabla \times \mathbf{E}) - k^2\epsilon_r\mathbf{E} + jk\eta\mathbf{J}^{int} + \nabla \times (\mu_r^{-1}\mathbf{K}^{int}) = 0, \qquad (2.3)$$

where $k^2 = \omega^2\mu_0\epsilon_0$ and $\eta^2 = \mu_0/\epsilon_0$.

2.1   *Weighted residual procedure*

Equation (2.3) may be put into the weak form either by variational means or using the strictly equivalent Galerkin weighted residuals; the latter route (Chapter 3, Sections 5.5–5.7) is taken here. Thus take the scalar product of Eq. (2.3) with some vector weighting function **W** and integrate over the volume Ω, applying the vector integration-by-parts identity

$$\int_\Omega \mathbf{W} \cdot \{\nabla \times (\mu_r^{-1}\nabla \times \mathbf{E})\} \, d\Omega = \int_\Omega \mu_r^{-1}(\nabla \times \mathbf{W}) \cdot (\nabla \times \mathbf{E}) \, d\Omega$$
$$- \oint_S \mu_r^{-1}\mathbf{W} \times (\nabla \times \mathbf{E}) \cdot \mathbf{1}_n \, dS,$$
$$(2.4)$$

see Appendix 2, Eq. (2.5), to establish the weighted residual

$$R = \int_\Omega \{\mu_r^{-1}(\nabla \times \mathbf{W}) \cdot (\nabla \times \mathbf{E}) - k^2\epsilon_r\mathbf{W} \cdot \mathbf{E}\} \, d\Omega$$
$$+ \int_\Omega \{jk\eta\mathbf{W} \cdot \mathbf{J}^{int} + \mathbf{W} \cdot \nabla \times (\mu_r^{-1}\mathbf{K}^{int})\} \, d\Omega$$
$$+ jk\eta \oint_S \mathbf{W} \cdot (\mathbf{H} \times \mathbf{1}_n) \, dS. \qquad (2.5)$$

The surface integral in Eq. (2.5) arises from the corresponding term in Eq. (2.4) after some vector manipulation and a substitution using the Maxwell equation (2.1). Note that on S, by definition, the source terms are zero. This surface integral component of the weighted residual will turn out to have critical significance in coupling the finite element region Ω to the boundary integral region Ω_{ext}.

2.2   *Galerkin finite elements using vector interpolation functions*

The residual expression Eq. (2.5) may be used to set up finite element matrix equations. For the vector variable here, the preferred procedure is to expand **E** in terms of tangentially continuous vector interpolation functions τ_j (see Chapter 7, Sections 5–7). This rules out the possibility of spurious corruptions in the numerical solution and easily allows the handling of material property discontinuities within Ω. Let the finite element expansion be

$$\mathbf{E} = \sum_{j=1}^{N} e_j \tau_j, \tag{2.6}$$

where the e_j represent (mostly) unknown expansion coefficients. For simplicity here it is assumed that the lowest possible order vector expansion is used; if tetrahedral elements are chosen, these will be the order 0.5. Nedelec–Bossavit ones (Chapter 7, Section 5.3), if bricks, the covariant projection elements of mixed order $m = 0$, $n = 1$ with straight sides (Chapter 7, Section 6). The significant property of such vector elements is that whilst τ_j may vary over the edge j to which it belongs, its component along that edge is constant and of magnitude unity; moreover the tangential part of τ_j strictly vanishes on any edge where $i \neq j$. In the Galerkin form of the weighted residual process, the weights **W** are chosen from the same collection of functions τ.

First consider some of the possibilities for S, other than it being the boundary between Ω and Ω_{ext}, the residual Eq. (2.5) being the basis for setting up the finite element equations. Suppose part of S happens to be an electric wall (a perfect conductor or an appropriate plane of symmetry), that is to say the field **E** there has to be normal to S whilst **H** lies in that surface. Provided **E** is constrained such that its components tangential to S vanish, the weighted residual process requires that in choosing the **W**'s from the τ's, those τ_j which belong to degrees of freedom specified as zero by an essential boundary constraint on S should be ignored, **W** being set to zero instead. The surface integral in Eq. (2.5) over the relevant portion of S (call this integral I_S say) vanishes and does not enter into the finite element matrix assembly. Alternatively, if a portion of S represents an electromagnetically driven patch, such as a waveguide port, then it may be assumed that the tangential **E** is known there, to be prescribed as such in the trial function. Again the corresponding τ_j is ignored and the weight function instead set to zero; I_S vanishes once more whereas the prescribed **E**-components enter into the finite element matrix assembly to make up a driving function. In either case the possibility of any finite element linear algebraic equation being set up trying to solve for an already-prescribed field component is thereby ruled out. Finally if S is a magnetic wall, $H \times 1_n = 0$ on S, such as at the surface of an infinitely permeable magnetic material or an appropriate plane of symmetry, Eq. (2.5) again shows that I_S vanishes; this time no particular constraint need be placed upon **E** or the corresponding τ's. The possibilities discussed here show that when the region Ω represents some real antenna or scatterer, there is considerable flexibility for modelling its detailed internal structure.

It now remains to set the residual defined by Eq. (2.5) to zero and recover from it the finite element matrices corresponding to the inhomogeneous region Ω. In Eq. (2.6) N is the total number of degrees of freedom, including those on S, corresponding to the vector **E** in either a tetrahedral or 'brick' element discretization of Ω. The unknown vector **H** represented in the surface integral portion of the residual expression, Eq. (2.5), is expanded similarly using the same edge-based representation for (say) N_S element edges laying in the surface S. Such an edge expansion is particularly appropriate here, since although there may be components of **H** normal to S, there is no reference to such in the residual expression Eq. (2.5). As already remarked, to set up the Galerkin weighted residual procedure equivalent to the variational process, the weight functions are chosen to be the same as the expansion functions. Thus substituting $\mathbf{W} = \tau_i$ (or $\mathbf{W} = 0$, if i corresponds to a constrained edge), into the weighted residual expression, Eq. (2.5), for $i = 1, \ldots, N$ a matrix relation

$$\mathbf{A}^E \mathbf{e} + \mathbf{B}^E \mathbf{h}_S = \mathbf{c}^E \qquad (2.7)$$

ensues, the superscript E indicating that the **E**-field has been chosen as the main variable of Eq. (2.5), whilst **e** and $\mathbf{h}_S$ correspond to vector arrays of edge coefficients respectively of length N and N_S. The matrices in Eq. (2.7) are, by inspection from Eq. (2.5)

$$A_{ij}^E = \int_\Omega \{\mu_r^{-1}(\nabla \times \tau_i) \cdot \nabla \times \tau_j) - k^2 \epsilon_r \tau_i \cdot \tau_j\} \, d\Omega,$$

$$i \text{ and } j = 1, \ldots, N, \quad (2.8)$$

$$B_{ij}^E = jk\eta \oint_S \tau_i \cdot (\tau_j \times 1_n) \, dS, \quad i = 1, \ldots, N, \; j = 1, \ldots, N_S, \quad (2.9)$$

$$c_i^E = -\int_\Omega \tau_i \cdot \{jk\eta \mathbf{J}^{int} + \nabla \times (\mu_r^{-1} \mathbf{K}^{int})\} \, d\Omega,$$

$$i = 1, \ldots, N. \quad (2.10)$$

Now Eqs. (2.8)–(2.10) are written as if each τ_i were a significant function of position throughout the whole space Ω. However being a vector generalization of the 'pyramid' functions of Section 5.7, Chapter 3, any τ_i is in fact nonzero only in the geometrical finite elements common to the *i*th edge. Furthermore the source terms $\mathbf{J}^{int}$ and $\mathbf{K}^{int}$ will ordinarily be non-zero only in a few finite elements. Thus it turns out that the matrices $\mathbf{A}$, $\mathbf{B}$ and the vector $\mathbf{c}$ contain many zeros and may be efficiently assembled in the usual way, element by element. In particular, if i in the surface integral Eq. (2.9) represents an edge not in S, then τ_i is zero in the integrand; counting the N_S edges first, the indexes $i > N_S$ thus generate rows exclusively of zeros in $\mathbf{B}$. They are formally retained merely in order to ensure a consistent number of rows in the matrix equation Eq. (2.7).

2.3   *Dual magnetic formulation*

In cases where it is advantageous to work with magnetic fields, the dual of Eq. (2.5) is given by

$$R = \int_{\Omega} \{\epsilon_r^{-1}(\nabla \times \mathbf{W}) \cdot (\nabla \times \mathbf{H}) - k^2 \mu_r \mathbf{W} \cdot \mathbf{H}\} \, d\Omega$$

$$+ \int_{\Omega} \left\{ \frac{jk}{\eta} \mathbf{W} \cdot \mathbf{K}^{int} - \mathbf{W} \cdot \nabla \times (\epsilon_r^{-1} \mathbf{J}^{int}) \right\} d\Omega$$

$$- \frac{jk}{\eta} \oint_S \mathbf{W} \cdot (\mathbf{E} \times \mathbf{1}_n) \, dS. \tag{2.11}$$

This similarly leads to volume and surface finite element discretized duals $\mathbf{h}$ and $\mathbf{e}_S$ connected by the matrix equation

$$\mathbf{A}^M \mathbf{h} + \mathbf{B}^M \mathbf{e}_S = \mathbf{c}^M, \tag{2.12}$$

where the dual matrix elements are

$$A_{ij}^M = \int_{\Omega} \{\epsilon_r^{-1}(\nabla \times \tau_i) \cdot (\nabla \times \tau_j) - k^2 \mu_r \tau_i \cdot \tau_j\} \, d\Omega,$$

$$i \text{ and } j = 1, \ldots, N, \tag{2.13}$$

$$B_{ij}^M = -\frac{jk}{\eta} \oint_S \tau_i \cdot (\tau_j \times \mathbf{1}_n) \, dS, \quad i = 1, \ldots, N, \, j = 1, \ldots, N_S \tag{2.14}$$

$$c_i^M = \int_{\Omega} \tau_i \cdot \left\{ -\frac{jk}{\eta} \mathbf{K}^{int} + \nabla \times (\epsilon_r^{-1} \mathbf{J}^{int}) \right\} d\Omega, \, i = 1, \ldots, N. \tag{2.15}$$

Notice that the matrices $\mathbf{B}^E$ and $\mathbf{B}^M$ are only trivially different.

3.   **Boundary integral treatment of the exterior region**

Taking stock of the results reached in the previous section, it may be noted that the edge coefficients h_{Si}, $i = 1, , N_S$, associated with the finite element boundary surface S, are linearly independent of the corresponding boundary subset of the edge coefficients e_{Si}. Thus there are not enough equations to solve for the unknown coefficients either in Eq. (2.7) or its dual, Eq. (2.12); the N_S more equations required may be set up by a boundary element treatment of the exterior region. Such a treatment invokes Huygens' principle, that the field solution in a region such as Ω_{ext} is entirely determined by the tangential fields specified over the bounding surface of that region, here effectively the surface S. The bounding surface for Ω_{ext} properly includes a portion at infinity but it contributes nothing to the Huygens field, because the Sommerfield radiation condition holds there.

3.1   *Mathematical statement of Huygens' principle*

Consider the radiation/scattering problem here for which the region Ω, bounded by a surface S, contains all of the relevant materials and internal field sources. The rest of the infinite problem may be represented by Ω_{ext} bounded by $S_{\text{ext}} \cup S_\infty$, with S_{ext} abutting closely up to S, whereas S_∞ is a boundary at infinity. It may be assumed that Ω_{ext} comprises of free space; assume that it contains neither sources nor, for the moment, any fields of external origin whatsoever. Then its governing equation, for say the **E**-variable, expressed in terms of $\mathbf{r}'$ is

$$\nabla' \times \nabla' \times \mathbf{E}(\mathbf{r}') - k_0^2 \mathbf{E}(\mathbf{r}') = 0 \qquad \text{for } \mathbf{r}' \in \Omega_{\text{ext}}. \tag{3.1}$$

Multiplying Eq. (3.1) by the dyadic Green's function $\mathfrak{G}$ (see Appendix 2, Section 4) and integrating with respect to $\mathbf{r}'$ over Ω_{ext} gives

$$\int_{\Omega_{\text{ext}}} \mathfrak{G}(\mathbf{r}, \mathbf{r}') \cdot \left\{ \nabla' \times \nabla' \times \mathbf{E}(\mathbf{r}') - k_0^2 \mathbf{E}(\mathbf{r}') \right\} d\Omega' = 0. \tag{3.2}$$

Similarly multiplying the $\mathbf{r}'$-version of Eq. (4.11) in Appendix 2 by $\mathbf{E}(\mathbf{r}')$ and integrating similarly results in

$$\int_{\Omega_{\text{ext}}} \mathbf{E}(\mathbf{r}') \cdot \left\{ \nabla' \times \nabla' \times \mathfrak{G}(\mathbf{r}, \mathbf{r}') - k_0^2 \mathfrak{G}(\mathbf{r}, \mathbf{r}') \right\} d\Omega'$$
$$= \int_{\Omega_{\text{ext}}} \mathbf{E}(\mathbf{r}') \cdot \mathbb{I}\delta(\mathbf{r} - \mathbf{r}') d\Omega', \tag{3.3}$$

where $\mathbb{I}$ is the unit dyadic. Subtract Eq. (3.3) from Eq. (3.2), after noting that the right-hand side of Eq. (3.3) is simply $\mathbf{E}(\mathbf{r})$ whilst $\mathfrak{G}$ is a symmetric dyadic so that $\mathfrak{G} \cdot \mathbf{E} = \mathbf{E} \cdot \mathfrak{G}$, to give

$$-\mathbf{E}(\mathbf{r}) = \int_{\Omega_{\text{ext}}} \left\{ \mathfrak{G}(\mathbf{r}, \mathbf{r}') \cdot \nabla' \times \nabla' \times \mathbf{E}(\mathbf{r}') \right.$$
$$\left. - \mathbf{E}(\mathbf{r}') \cdot \nabla' \times \nabla' \times \mathfrak{G}(\mathbf{r}, \mathbf{r}') \right\} d\Omega'. \tag{3.4}$$

Now apply the vector form of Green's theorem, Eq. (3.5) of Appendix 2, to Eq. (3.4). The result is

$$\mathbf{E}(\mathbf{r}) = \oint_{S_{ext}} \{ \mathfrak{G}(\mathbf{r},\mathbf{r}') \times \nabla' \times \mathbf{E}(\mathbf{r}')$$

$$- \mathbf{E}(\mathbf{r}') \times \nabla' \times \mathfrak{G}(\mathbf{r},\mathbf{r}') \} \cdot \mathbf{1}'_{n,ext} \, dS. \tag{3.5}$$

The omission of S_∞ from the range of the surface integral in Eq. (3.5) is justified on the grounds that $\mathbf{E}$ and $\mathfrak{G}$ vanish in the limit at infinity as $1/r$ according to the Sommerfeld radiation law. Examination of the integral omitted then reveals by inspection that it also vanishes in that limit. The external field $\mathbf{E}^{ext}(\mathbf{r})$, that which would have existed everywhere had the region Ω been empty, temporarily ignored in the analysis, may now be inserted in the obvious place, added to the right-hand side of Eq. (3.5). Further, using the Maxwell electric curl equation to substitute for $\nabla' \times \mathbf{E}(\mathbf{r}')$, Eq. (3.5) may be manipulated to be expressed as

$$\mathbf{E}(\mathbf{r}) = \mathbf{E}^{ext}(\mathbf{r}) + \oint_S (\nabla \times \mathfrak{G}(\mathbf{r},\mathbf{r}') \cdot \{ \mathbf{1}'_n \times \mathbf{E}_S(\mathbf{r}')$$

$$- jk\eta\mathfrak{G}(\mathbf{r},\mathbf{r}') \cdot \{ \mathbf{1}'_n \times \mathbf{H}_S(\mathbf{r}') \}) \, dS'. \tag{3.6}$$

Here $\mathbf{1}'_{n,ext}$, the normal vector assigned to S_{ext} at $\mathbf{r}'$, has been replaced by $-\mathbf{1}'_n$, referring to the coincident surface S. Notice that the unprimed operator $\nabla\times$ replaces $\nabla'\times$, also with reversal of the order of $\mathbf{E}$ and $\mathfrak{G}$; this is allowable since both $\mathfrak{G}$ and $\nabla \times \mathfrak{G}$ are symmetrical in $\mathbf{r}$ and $\mathbf{r}'$. Since $\mathbf{E}_S$ and $\mathbf{H}_S$ would in any case remain continuous between S and S_{ext}, the materials enclosed within Ω may extend right up to the boundary S. The dual of Eq. 3.6 is

$$\mathbf{H}(\mathbf{r}) = \mathbf{H}^{ext}(\mathbf{r}) + \oint_S \left(\nabla \times \mathfrak{G}(\mathbf{r},\mathbf{r}') \cdot \{ \mathbf{1}'_n \times \mathbf{H}_S(\mathbf{r}') \} \right.$$

$$\left. + \frac{jk}{\eta} \mathfrak{G}(\mathbf{r},\mathbf{r}') \cdot \{ \mathbf{1}'_n \times \mathbf{E}_S(\mathbf{r}') \} \right) dS'. \tag{3.7}$$

In all, the significance of the results here is seen to be that the radiation due to a region Ω containing materials and sources may be calculated from a knowledge of just the tangential fields $\mathbf{E}_S$ and $\mathbf{H}_S$ over the surface S enclosing Ω. The test point $\mathbf{r}$ may be placed upon S and a further cross-product throughout with $\mathbf{1}_n(\mathbf{r})$ taken, to yield an integral equation relating $\mathbf{E}_S$ and $\mathbf{H}_S$; then, because $\mathfrak{G}$ ultimately depends upon $1/|\mathbf{r} - \mathbf{r}'|$, there is a singularity at $\mathbf{r} = \mathbf{r}'$ which fortunately is integrable. In some instances it may be convenient to replace $\mathbf{1}'_n \times \mathbf{H}_S(\mathbf{r}')$ and $\mathbf{1}'_n \times \mathbf{E}_S(\mathbf{r}')$ by surface electric and magnetic currents $\mathbf{J}_S(\mathbf{r}')$ and $-\mathbf{K}_S(\mathbf{r})$ respectively. Equations (3.6) and (3.7) are in fact alternative mathematical statements of

Huygens' principle. The fields $\mathbf{H}_S(\mathbf{r}')$ and $\mathbf{E}_S(\mathbf{r}')$, or the corresponding currents $\mathbf{J}_S(\mathbf{r}')$ and $-\mathbf{K}_S(\mathbf{r}')$ may be regarded as equivalent sources replacing the real sources enclosed within the body Ω. The field notation is retained here since $\mathbf{E}_S$ and $\mathbf{H}_S$ need to appear as variables, being a subset of the general vector fields belonging to Ω.

Thus Eqs. (3.6) and (3.7) with $\mathbf{r}$ placed on S represent the basic boundary integral formulation which enables the infinitely extending free-space exterior region to be connected to the inhomogeneous region Ω; the latter may be modelled by the finite element method without incurring the discretization of an infinite volume of space. Because the scalar Green's function $G(\mathbf{r}, \mathbf{r}')$ embedded in the dyadic is acted upon by an operator which includes second degree differential components, the singularity involves terms $0(1/|\mathbf{r} - \mathbf{r}'|^3)$. Further analytical manipulation is desirable to reduce this singularity to $0(1/|\mathbf{r} - \mathbf{r}'|)$, so as to render it integrable without too much difficulty.

4.   The hybrid FEM/BEM procedure

The general configuration is pursued a little further before considering some special cases. There are several combinations of the dual routes which in principle may be followed in order to establish a set of matrix equations yielding the hybrid FEM/BEM (finite element method/ boundary element method) field solution; assume say that it is required to establish the E-vector overall. Either of the dual Huygens integral equations, Eqs. (3.6) or (3.7), with $\mathbf{r}$ on S, may be used to eliminate the unknown edge expansion vector array $[\mathbf{h}_S]$ arising in the finite element equation for the interior region Ω, Eq. (2.7). It remains to discretize the Huygens equations so as to be able to do this. Write Eq. (3.6) in a more compact form as

$$-\mathbf{E} + \mathbf{L}_{e1}^S (\mathbf{E}_S \times \mathbf{1}_n') + \mathbf{L}_{e2}^S (\mathbf{H}_S \times \mathbf{1}_n') + \mathbf{E}^{\text{ext}} = 0, \qquad (4.1)$$

where $\mathbf{L}_{e1,2}^S$ are operators; with $\mathbf{r}$ lying upon S, taking the cross-product with the unit normal of S yields

$$-\mathbf{E}_S \times \mathbf{1}_n + \mathbf{L}_{e1}^S (\mathbf{E}_S \times \mathbf{1}_n') \times \mathbf{1}_n + \mathbf{L}_{e2}^S (\mathbf{H}_S \times \mathbf{1}_n') \times \mathbf{1}_n + \mathbf{E}_S^{\text{ext}}$$
$$\times \mathbf{1}_n = 0. \qquad (4.2)$$

Here is a continuous representation relating $\mathbf{E}_S$ and $\mathbf{H}_S$ in the subspace of the surface S, amenable to the weighted residual operation to give a residual

$$R_S = -\oint_S \mathbf{W}_S \cdot (\mathbf{E}_S \times \mathbf{1}_n)\, dS + \oint_S \mathbf{W}_S \cdot \mathbf{L}^S_{e1}(\mathbf{E}_S \times \mathbf{1}'_n) \times \mathbf{1}_n\, dS$$

$$+ \oint_S \mathbf{W}_S \cdot \mathbf{L}^S_{e2}(\mathbf{H}_S \times \mathbf{1}'_n) \times \mathbf{1}_n\, dS + \oint_S \mathbf{W}_S \cdot (\mathbf{E}^{\text{ext}}_S \times \mathbf{1}_n)\, dS$$

$$(4.3)$$

for any weight function $\mathbf{W}_S$ in the S-subspace. Clearly R_S vanishes if you have the correct $\mathbf{E}_S$ and $\mathbf{H}_S$. The Galerkin procedure is invoked by expanding $\mathbf{E}_S$ and $\mathbf{H}_S$ in terms of the edge expansion functions lying in S, say τ_j, $j = 1, \ldots, N_S$ and choosing $\mathbf{W}_S$ successively to be the same vector functions τ_i, $i = 1, \ldots, N_S$. Also multiplying throughout by the factor jk/η to get a tidy result, the resulting matrix equation is found to be

$$\mathbf{B}^M \mathbf{e}_S + \mathbf{P}^E \mathbf{e}_S + \mathbf{Q}^E \mathbf{h}_S + \mathbf{y}^E = 0, \qquad (4.4)$$

where the $\mathbf{B}^M$ has already been encountered in the dual finite element discretization of Ω, see Eqs. (2.12) and (2.14), whilst

$$P^E_{ij} = j\frac{k}{\eta} \oint_S \tau_i \cdot \left\{ \mathbf{L}^S_{e1}(\tau_j \times \mathbf{1}_n) \times \mathbf{1}_n \right\}\, dS, \qquad (4.5)$$

$$Q^E_{ij} = \frac{jk}{\eta} \oint_S \tau_i \cdot \left\{ \mathbf{L}^S_{e2}(\tau_j \times \mathbf{1}_n) \times \mathbf{1}_n \right\}\, dS, \qquad (4.6)$$

$$y^E_i = \frac{jk}{\eta} \oint_S \tau_i \cdot (\mathbf{E}^{\text{ext}} \times \mathbf{1}_n)\, dS. \qquad (4.7)$$

In Eq. (2.14) j is taken over the range $1, \ldots, N_S$ whereas i is formally taken beyond N_S to N, filling out $\mathbf{B}^M$ with zeros so as to maintain consistency of matrix dimensions in Eq. (2.7). Clearly the same can be done here. It has to be assumed that some means of dealing with the singularity arising from the dyadic Green's function can be found. The dual Huygens equation Eq. (3.2) gives in the same way

$$\mathbf{B}^E \mathbf{h}_S + \mathbf{P}^M \mathbf{h}_S + \mathbf{Q}^M \mathbf{e}_S + \mathbf{y}^M = 0, \qquad (4.8)$$

where $\mathbf{B}^E$ has also appeared before, see Eqs. (2.7) and (2.9), whilst $\mathbf{P}^M$, $\mathbf{Q}^M$ and $\mathbf{y}^M$ may be obtained from Eqs. (4.5)–(4.7) respectively by replacing η by $-1/\eta$, $\mathbf{L}^S_{e1,2}$ by the corresponding operator $\mathbf{L}^S_{h1,2}$ defined by Eq. (3.2) and $\mathbf{E}^{\text{ext}}$ by $\mathbf{H}^{\text{ext}}$. It is clear that in principle, now the vector $\mathbf{e}_S$, representing the arbitrary-boundary transverse magnetic field, can be eliminated between Eq. (2.7) and either Eq. (4.4) or Eq. (4.8) so as to allow the full vector array of electric field edge coefficients $\mathbf{e}$ to be determined; alternatively the dual formulation can similarly be used to get the full $\mathbf{h}$. Thus the field solution within Ω may be evaluated in a finite element approximation. Other radiation/scattering parameters, as appropriate, may then be worked out from the solution without any particular difficulty.

4.1   *Radiation from a cavity-backed aperture in a conducting half-space*

The configuration of Fig. 9.2, considered by Jin, Volakis and Collins (1991), represents a somewhat simplified version of the general problem considered above. Choose **E** as the working variable. Evidently there is no particular difficulty in setting up the principle matrix relationship, Eq. (2.7); it remains to construct the auxiliary matrix relationship necessary so as to eliminate the aperture magnetic field $\mathbf{H}_S$ from the principle equation. Either of the Huygens boundary integrals Eqs. (3.6) or (3.7) could be used for this elimination; in the following subsection it will be assumed that Eq. (3.7) has been selected.

· 4.2   *Simplification using the equivalence principle*

In the case here of a cavity-backed aperture in a perfectly conducting half-plane, a much simpler form of the boundary integral equation becomes valid. The simplification arises from a combination of the equivalence principle and image theory, bearing in mind the general feature enlarged upon by Harrington (1961), that there are many possible equivalences for any given real situation. The simplification may be understood by considering the sequence, illustrated in Fig. 9.3, of three configurations which are equivalent for the free-space right-hand half-plane. Figure 9.3(*a*) illustrates the actual setup. In Fig. 9.3(*b*), the cavity region is shown replaced by a perfect conductor bounded by the aperture plane together with a surface magnetic source current $\mathbf{K}_S = \mathbf{E}_S \times \mathbf{1}_n$. There is no electric source current, so the $\mathbf{H}_S$ term disappears from Eq. (3.7). There is now no field inside the cavity but this is of no consequence. The aperture area, now considered as a perfect conductor, is compatible with the rest of the boundary plane, so that any exter-

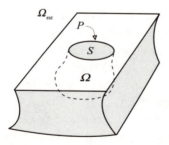

Fig 9.2 Cavity-backed aperture S in a conducting half-space bounded by a plane P.

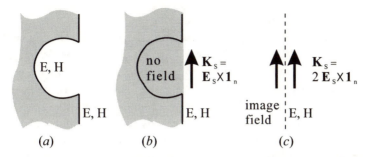

Fig 9.3 Equivalence principle applied to a cavity-backed aperture in a conducting half-space. (*a*) Physical configuration; (*b*) cavity replaced by a perfect conductor carrying surface magnetic current $\mathbf{K}_S$ in the aperture plane; (*c*) the perfectly conducting half-space replaced by an image magnetic current.

nal field excitation is simply dealt with. In Fig. 9.3(*c*) the perfectly conducting plane in the aperture is replaced by an image magnetic current, which by the usual rules is identical with the source and in this case coincident with it, giving a revised magnetic current of $\mathbf{K}_S = 2\mathbf{E}_S \times \mathbf{1}_n$ together with an inconsequential image field inside the cavity. Now it is a simple matter also to add in the effects of an external source field. Thus Eq. (3.7) simplifies to

$$\mathbf{H}(\mathbf{r}) = \mathbf{H}^{ext} + \mathbf{H}^{ref} - \frac{2jk}{\eta} \oint_S \mathbb{G}(\mathbf{r},\mathbf{r}') \cdot \{\mathbf{E}_S(\mathbf{r}') \times \mathbf{1}_n'\} \, dS', \qquad (4.9)$$

where $\mathbf{H}^{ref}$ represents the magnetic field which would result from the incident field $\mathbf{H}^{ext}$ as a result of reflection in the conducting ground plane *without* the aperture. Now placing $\mathbf{r}$ upon S, taking the cross-product throughout with $-jk\eta\mathbf{1}_n$, then the dot-product with an appropriate $\mathbf{W}_S(\mathbf{r})$ and finally integrating over S gives, with some rearrangement,

$$
\begin{aligned}
R_S = jk\eta &\int_S \mathbf{W}_S(\mathbf{r}) \cdot \{\mathbf{H}_S(\mathbf{r}) \times \mathbf{1}_n\} \, dS \\
&- 2k^2 \int_S dS \int_S dS' \{\mathbf{W}_S(\mathbf{r}) \times \mathbf{1}_n\} \cdot (\mathbb{G}(\mathbf{r},\mathbf{r}') \cdot \{\mathbf{E}_S(\mathbf{r}') \times \mathbf{1}_n'\}) \\
&- 2jk\eta \int_S \mathbf{W}_S(\mathbf{r}) \cdot (\mathbf{H}^{ext} \times \mathbf{1}_n) \, dS = 0, \qquad (4.10)
\end{aligned}
$$

where from simple plane wave theory, the relationship $\mathbf{H}^{ref} \times \mathbf{1}_n = \mathbf{H}^{ert} \times \mathbf{1}_n$ at the ground plane has been utilized. Note that since $\mathbf{E}_S$ must vanish on the perfect conductors, the S-integrations here are just over the aperture plane. The closed surface integral sign $\oint$ has been replaced by just the open surface integral $\int$ to indicate this.

4.3   *Manipulating the Green's function*

Before Eq. (4.10) can be utilized, the awkward, highly singular terms in $|\mathbf{r} - \mathbf{r}'|$ raised to the powers -2 and -3 need to be circumvented. The full expression for the dyadic $\mathfrak{G}$, Appendix 2, Eq. (4.12), reveals that the difficulty is caused by the $\nabla\nabla$-operator. As employed here, the ∇-operator relates to the variable $\mathbf{r}$, but when acting upon the scalar Green's kernel G, Appendix 2, Eq. (4.5), it is readily seen that if ∇' is considered to be the corresponding operator with respect to $\mathbf{r}'$ then $\nabla' = -\nabla$. Thus the critical part of Eq. (4.10) is the term

$$A = 2 \int_S \mathrm{d}S \int_S \mathrm{d}S' [\mathbf{W}_S(\mathbf{r}) \times \mathbf{1}_n] \cdot (\nabla\nabla' G(\mathbf{r},\mathbf{r}') \cdot \{\mathbf{E}_S(\mathbf{r}') \times \mathbf{1}'_n\}),$$

$$(4.11)$$

which arises from putting in $\mathfrak{G}(\mathbf{r},\mathbf{r}')$ explicitly, changing from $\nabla\nabla$ to $-\nabla\nabla'$ and cancelling the k^2 coefficient. The vector theorem Eq. (2.3) of Appendix 2 can be rewritten in two-dimensional form applied to the aperture S and its perimeter P as

$$\int_S \nabla u \cdot \mathbf{F} \, \mathrm{d}S = -\int_S u\nabla \cdot \mathbf{F} \, \mathrm{d}S + \oint_P u\mathbf{F} \cdot \mathbf{1}_{np} \, \mathrm{d}l,$$

$$(4.12)$$

where $\mathbf{1}_{np}$ is the unit normal, lying in the aperture plane, to an element of length $\mathrm{d}l$ of the perimeter P. The theorem may be applied twice to Eq. (4.11), first associating $\mathbf{F}$ with $\mathbf{E}_S(\mathbf{r}') \times \mathbf{1}_n$ and then with $\mathbf{W}_S(\mathbf{r}) \times \mathbf{1}_n$. In the first instance $\mathbf{F} \cdot \mathbf{1}_{np} = (\mathbf{E}_S \times \mathbf{1}'_n) \cdot \mathbf{1}_{np} = \mathbf{E}_S \cdot (\mathbf{1}'_n \times \mathbf{1}_{np}) = -\mathbf{E}_S \cdot \mathbf{1}_t$ where $\mathbf{1}_t$ is the unit tangential vector to P. Since $\mathbf{E}_S$ must be normal to the aperture edge, $\mathbf{F} \cdot \mathbf{1}_{np}$ vanishes. In a Galerkin analysis, $\mathbf{W}_S$ is chosen from the same vector family as $\mathbf{E}_S$, so $\mathbf{F} \cdot \mathbf{1}_{np}$ vanishes also in the second instance. Thus

$$A = 2 \int_S \mathrm{d}S \int_S \mathrm{d}S' G(\mathbf{r},\mathbf{r}')\nabla \cdot \{\mathbf{W}_S(\mathbf{r}) \times \mathbf{1}_n\}\nabla' \cdot \{\mathbf{E}_S(\mathbf{r}') \times \mathbf{1}'_n\},$$

$$(4.13)$$

so that Eq. (4.10) may be rewritten as

$$R_S = jk\eta \int_S \mathbf{W}_S(\mathbf{r}) \cdot \{\mathbf{H}_S(\mathbf{r}) \times \mathbf{1}_n\} \, \mathrm{d}S$$

$$+ 2 \int_S \mathrm{d}S \int_S \mathrm{d}S' G(\mathbf{r},\mathbf{r}')\nabla \cdot \{\mathbf{W}_S(\mathbf{r}) \times \mathbf{1}_n\}\nabla' \cdot \{\mathbf{E}_S(\mathbf{r}') \times \mathbf{1}'_n\}$$

$$- 2jk\eta \int_S \mathbf{W}_S(\mathbf{r}) \cdot (\mathbf{H}^{\text{ext}} \times \mathbf{1}_n) \, \mathrm{d}S.$$

$$(4.14)$$

4.4   *The boundary integral matrices*

Equation (4.14) leads to the matrix relation

$$\mathbf{B}^{\mathrm{E}}\mathbf{h}_{\mathrm{S}} + 2\mathbf{Q}^{\mathrm{M}}\mathbf{e}_{\mathrm{S}} + 2\mathbf{y}^{\mathrm{M}} = 0, \tag{4.15}$$

where $\mathbf{B}^{\mathrm{E}}$ and $\mathbf{y}^{\mathrm{M}}$ have appeared before, specifically (writing Eq. (2.9) again)

$$B_{ij}^{\mathrm{E}} = jk\eta \int_{S} \boldsymbol{\tau}_i \cdot (\boldsymbol{\tau}_j \times \mathbf{1}_{\mathrm{n}})\, \mathrm{d}S, \tag{4.16}$$

and, setting down explicitly the dual of Eq. (4.7),

$$y_i^{\mathrm{M}} = -jk\eta \int_{S} \boldsymbol{\tau}_i \cdot (\mathbf{H}^{\mathrm{ext}} \times \mathbf{1}_{\mathrm{n}})\, \mathrm{d}S, \tag{4.17}$$

whilst now

$$Q_{ij} = -k^2 \int_{S} \mathrm{d}S \int_{S} \mathrm{d}S' G(\mathbf{r}, \mathbf{r}')\{\boldsymbol{\tau}_i(\mathbf{r}) \times \mathbf{1}_{\mathrm{n}}\} \cdot \{\boldsymbol{\tau}_j(\mathbf{r}' \times \mathbf{1}_{\mathrm{n}}')$$
$$+ \int_{S} \mathrm{d}S \int_{S} \mathrm{d}S' G(\mathbf{r}, \mathbf{r}')\nabla \cdot \{\boldsymbol{\tau}_i(\mathbf{r}) \times \mathbf{1}_{\mathrm{n}}\}\nabla' \cdot \{\boldsymbol{\tau}_j(\mathbf{r}') \times \mathbf{1}_{\mathrm{n}}'\}. \tag{4.18}$$

Having chosen an advantageous route through the boundary integral maze, there is no need to solve Eq. (4.15) for the unknown edge coefficient vector $\mathbf{h}_{\mathrm{S}}$ or to work out $\mathbf{B}^{\mathrm{E}}$. The array term $\mathbf{B}^{\mathrm{E}}\mathbf{h}_{\mathrm{S}}$ can be eliminated entirely between Eq. (4.15) and Eq. (2.7), to give

$$\mathbf{A}^{\mathrm{E}}\mathbf{e} - 2\mathbf{Q}^{\mathrm{M}}\mathbf{e}_{\mathrm{S}} = \mathbf{c}^{\mathrm{E}} + 2\mathbf{y}^{\mathrm{M}}, \tag{4.19}$$

which may be inverted immediately for the electric field edge-coefficients $\mathbf{e}$, including of course the subset $\mathbf{e}_{\mathrm{S}}$.

The evaluation of individual array elements above presents no particular difficulty and proceeds, according to the type of element being employed, as for the scalar problems dealt with in previous chapters. In general Gaussian quadrature will be used; however for the singular $i = j$ case of Eq. (4.18) the substitution

$$G(\mathbf{r}, \mathbf{r}') = \left\{\frac{\exp(-jk|\mathbf{r} - \mathbf{r}'|)}{|\mathbf{r} - \mathbf{r}'|} - \frac{1}{|\mathbf{r} - \mathbf{r}'|}\right\} + \frac{1}{|\mathbf{r} - \mathbf{r}'|} \tag{4.20}$$

is made. The term within curly brackets is nonsingular and may be handled in the computation of Q_{ii} by Gaussian quadrature, whereas an analytical procedure, following the lines developed in Chapter 5 for the two-dimensional situation there, can be followed to deal with the remaining part of the Green's function. However, whilst the matrix $\mathbf{A}^{\mathrm{E}}$ has the usual sparse and banded properties, $\mathbf{Q}^{\mathrm{M}}$ is dense and the latter's $O(N_S^2)$

storage requirement may cause difficulty if a direct solution of Eq. (4.19) is attempted.

4.5   *Discrete convolution and FFT for aperture surface-integration*

If a conjugate gradient (CG) iterative method of solution (Chapter 10, Section 4) is applied to Eq. (4.19), it may be observed that the matrices on the left-hand side of that equation never themselves need be worked out. All that ever is required is the matrix product $\mathbf{A}^E \mathbf{e}_k - 2\mathbf{Q}^M \mathbf{e}_{Sk}$, where $\mathbf{e}_k$, including its subset on S, represents one of the successively computed CG iterative estimates of the solution-vector $\mathbf{e}_k$. Evaluating $\mathbf{A}^E \mathbf{e}_k$ by means of storing the sparse, banded array $\mathbf{A}^E$ and a direct matrix multiplication represents no particular problem, but in view of its dense character, other means of working out $\mathbf{Q}^M \mathbf{e}_{Sk}$ should be considered. An alternative fast Fourier transform (FFT) method, used by Jin, Volakis and Collins (1991), exploits the fact that when first-order edge elements (preferably the brick variety) are employed so as to give a uniform rectangular mesh of points in the aperture, the latter matrix product can be put into a form representing a discrete convolution. This can be evaluated by performing a discrete (fast) Fourier transform and then making use of the fact that the Fourier transform of a convolution

$$h(x) = \int_0^\infty f(x')g(x - x')\,\mathrm{d}x' \tag{4.21}$$

is the simple product of the transforms of f and g made separately. The procedure has been described by Brigham (1988) as a 'shortcut by the long way round'. Only $O(N_S)$ storage is required, whilst the number of multiplications performed is of order $N_S \log_2 N_S$ compared with the N_S^2 of the straight matrix operation. The result may be expressed as

$$\sum_{p=0}^{P-1}\sum_{q=0}^{Q-1} Q_{m-p,n-q}^M e_{Sm,n}$$

$$= \mathrm{FFT}^{-1}\{\mathrm{FFT}(Q_{m,n}^M)\mathrm{FFT}(e_{Sm,n})\}. \tag{4.22}$$

Here the convolution and transforms are two-dimensional, so a two-dimensional FFT operator is implied. The integers P and Q represent the size of the rectangular mesh employed to represent the aperture, whilst p, q and m, n are summation and location indexes respectively corresponding to discrete coordinates within the mesh. Such coordinates are defined by the rectangular finite element subdivision of the aperture. In practice the mesh defined by P and Q must be chosen to extend over linear dimensions at least twice as great as the actual area of the aperture, the extra space being filled out with zeros; such a choice avoids the well-known aliasing problems associated with finite discrete convolutions. The CG-FFT technique may be applied to other situations provided the boundary integral can be represented as a convolution.

5.   Absorbing boundary conditions for vector wavefields

As has already been seen, the essence of the open-boundary problem lies in coping with, in one or other of the dual residual expressions, the surface integral term over the fictitious surface S bounding the finite element region. It is easy to see that the requirement is to ensure that the nonreflection of waves meeting S is simulated. To avoid confusion, for the present just consider the **E**-variable residual, Eq. (2.5); the dual alternatives to the analysis below merely involve the usual simple exchanges of variables and parameters.

Assume that S is spherical and lies in free space as close as possible to the inhomogeneous region of the problem. Then $\mathbf{1}_n$ may be replaced by the unit radial vector $\mathbf{1}_r$ whilst the surface integral term in Eq. (2.5) can be written

$$I = \oint_S \mathbf{W} \cdot (\mathbf{1}_r \times \nabla \times \mathbf{E}) \, dS. \tag{5.1}$$

As already has been discussed in the previous sections, the components of $\nabla \times \mathbf{E}$ transverse to S, equivalent to $\mathbf{H}_S$, are unavailable in the main finite element matrix equations for the volume Ω and so some means of relating them to the components $\mathbf{E}_S$ must be found. The Green's function method for doing this reproduces the boundary transparency property exactly, but is global in its application to S and so leads to dense matrices with $O(N^2)$ storage requirements. In some circumstances it is preferable to use one or other of the asymptotically correct 'absorbing' boundary conditions (ABC's) in order to truncate the open space of a radiation or scattering problem. The ABC's are applied locally in differential form to the boundary and so give rise to sparse matrices. Mathematically ABC's arise from the observation that a radiation field, $\mathbf{E}$ say, consistently travelling outwards and coming from sources placed at finite distances from some arbitrarily placed origin, can always be represented in spherical coordinates by

$$\mathbf{E}(r, \theta, \phi) = e^{-jkr} \sum_{n=0}^{\infty} \frac{\mathbf{E}_n(\theta, \phi)}{r^{n+1}} \tag{5.2}$$

in the free space surrounding the sources. This is plausible enough but not entirely obvious, since these are waves appearing to arise from sources at the origin whereas the proposition includes sources anywhere not at infinity; however the matter has been rigorously proved. Accepting the validity of vector series expansions such as Eq. (5.2), a general Nth order operator $B_N(\mathbf{u})$ has been constructed by Webb and Kanellopoulos (1989) such that acting upon the series, the result is exactly zero if it terminates at $1/r^N$. On the other hand if the series goes on beyond that term, the result approximates to zero neglecting terms of order $1/r^{2N+1}$. Importantly, B_N includes the curl operator in its construction in such a way that, by invoking the Maxwell electric curl equation, it can relate the unavailable transverse magnetic field $\mathbf{H}_S$ implied in the open-boundary integral expression, Eq. (4.21), to the accessible $\mathbf{E}_S$. Then, for instance, the discretized form of the radiation problem, Eq. (2.7), becomes soluble.

5.1   *Derivation of the vector absorbing boundary condition*

In order to follow through the derivation, it is appropriate to note the rule for taking the curl of a product,

$$\nabla \times (\psi \mathbf{u}) = \nabla \psi \times \mathbf{u} + \psi \nabla \times \mathbf{u}; \tag{5.3}$$

and to remind the reader of the expressions for grad and curl in spherical coordinates (r, θ, ϕ):

$$\nabla \psi = \mathbf{1}_r \frac{\partial \psi}{\partial r} + \mathbf{1}_\theta \frac{1}{r} \frac{\partial \psi}{\partial \theta} + \mathbf{1}_\phi \frac{1}{r \sin \theta} \frac{\partial \psi}{\partial \phi} \tag{5.4}$$

$$\begin{aligned}
\nabla \times \mathbf{u} = &\mathbf{1}_r \frac{1}{r \sin \theta} \left\{ \frac{\partial}{\partial \theta} (u_\phi \sin \theta) - \frac{\partial u_\theta}{\partial \phi} \right\} \\
&+ \mathbf{1}_\theta \frac{1}{r} \left\{ \frac{1}{\sin \theta} \frac{\partial u_r}{\partial \phi} - \frac{\partial}{\partial r} (r u_\phi) \right\} \\
&+ \mathbf{1}_\phi \frac{1}{r} \left\{ \frac{\partial}{\partial r} (r u_\theta) - \frac{\partial u_r}{\partial \theta} \right\}.
\end{aligned} \tag{5.5}$$

Define a differential operator

$$L_N(\mathbf{u}) = \mathbf{1}_r \times \nabla \times \mathbf{u} - \left(jk + \frac{N}{r} \right) \mathbf{u}, \qquad N = 0, 1, 2, \ldots \tag{5.6}$$

Now S is a spherical surface of radius r, so that there are E-vector components $\mathbf{1}_r E_{nr}$ and $\mathbf{E}_{nt} = \mathbf{1}_\theta E_{n\theta} + \mathbf{1}_\phi E_{n\phi}$ normal and tangential to S respectively. Then applying the operator (5.6) to the series expansion of $\mathbf{E}$, using the vector formulae (5.4) and (5.5) it is found that

$$L_N \left\{ e^{-jkr} \frac{\mathbf{E}_{nt}(\theta, \phi)}{r^{n+1}} \right\} = (n - N) e^{-jkr} \frac{\mathbf{E}_{nt}(\theta, \phi)}{r^{n+2}} \tag{5.7}$$

and, for $N \geq 0, n \geq 0$,

$$L_N\left(\nabla_t\left\{e^{-jkr}\frac{E_{nr}(\theta,\phi)}{r^{n+1}}\right\}\right) = (n+1-N)\nabla_t\left[e^{-jkr}\frac{E_{nr}(\theta,\phi)}{r^{n+2}}\right],$$

$$(5.8)$$

where ∇_t denotes the tangential (θ,ϕ) part of the gradient operator. Notice that after the L_N-operation, because of the particular choice of the second, multiplicative term in the operator here, all θ- and ϕ-derivatives of the E_n components have cancelled out. In each case above the operation is seen to have the effect of multiplying its argument by i/r, where i is an integer which can be zero. It is this property which enables the ABC vector operator.

$$B_N(\mathbf{u}) = L_{N-1}^N(\mathbf{u}_t) + sL_N^{N-1}(\nabla_t u_r) \qquad (5.9)$$

to be constructed, where an integer superscript i attached to L means 'apply L to its argument i times' whilst s is an arbitrary number. Using Eqs. (5.7) and (5.8) repeatedly, it is seen by inspection that

$$B_N\left\{e^{-jkr}\frac{E_n(\theta,\phi)}{r^{n+1}}\right\} = (n+1-N)(n+2-N)\dots ne^{-jkr}\frac{E_{nt}(\theta,\phi)}{r^{n+1+N}}$$

$$+ s(n+1-N)(n+2-N)\dots(n-1)\nabla_t$$

$$\left\{e^{-jkr}\frac{E_{nr}(\theta,\phi)}{r^{n+N}}\right\},$$

$$(5.10)$$

which evidently vanishes for $n = 1, 2, \dots, N-1$. Because all radiation can be considered as if arising from oscillating poles, the radial field component is expected to have an inverse-square-law component but no inverse-linear one, so that E_{0r} vanishes in all cases (this can be justified rigorously). Thus the operator B_N also vanishes for $n = 0$ and does indeed annihilate the first N terms of the radiation function series expansion. Noting, from Eq. (5.4), that ∇_t in the spherical coordinates used here includes a factor $1/r$, it is seen that for $n > N$, the terms left over after application of the ABC operator are proportional to $1/r^{n+N+1}$. Thus when acting upon a radiation function $\mathbf{H}$, it is confirmed that the operator does not indeed have the property that

$$B_N(\mathbf{H}) = O\left(\frac{1}{r^{2N+1}}\right). \qquad (5.11)$$

5.2   *Practical forms for the boundary operators*

The general form for B_N written explicitly is rather complicated; however the practical operators are the first- and second-order ones. From Eq. (5.9) it is seen that for the first-order case

$$B_1(\mathbf{E}) = L_0^1(\mathbf{E}_t) + sL_1^0(\nabla_t E_r)$$
$$= \mathbf{1}_r \times \nabla \times \mathbf{E}_t - jk\mathbf{E}_t + s\nabla_t E_r. \tag{5.12}$$

Using the spherical coordinates vector calculus expressions Eqs. (5.4) and (5.5), it is simple to show that the above first-order ABC expression may be recast as

$$B_1(\mathbf{E}) = \mathbf{1}_r \times \nabla \times \mathbf{E} - jk\mathbf{E}_t + (s - 1)\nabla_t E_r. \tag{5.13}$$

Physically this is saying that, correct to $O(1/r^3)$, locally at any point on S there are plane waves proceeding outwards. Moreover if there is any radial component of $\mathbf{E}$, locally it does not change with either of the transverse spherical coordinates θ or ϕ. If the arbitrary number s is set equal to unity, Eq. (5.12) corresponds to the Sommerfeld radiation condition, see Kong (1986). The second-order ABC as expressed by Eq. (5.9) is

$$B_2(\mathbf{E}) = L_1^2(\mathbf{E}_t) + sL_2^1(\nabla_t E_r), \tag{5.14}$$

where L_1 and L_2 are given by Eq. (5.6). After some algebra and utilizing the fact that $\mathbf{E}$ satisfies the vector wave equation, the second-order ABC can be put into the useful form

$$B_2(\mathbf{E}) = -2(\mathbf{1}_r \times \nabla \times \mathbf{E} - jk\mathbf{E}_t)\left(jk + \frac{1}{r}\right) + \nabla \times \{\mathbf{1}_r(\nabla \times \mathbf{E})_r\}$$
$$(s - 1)\nabla_t(\nabla_t \cdot \mathbf{E}_t) + (2 - s)jk\nabla_t E_r. \tag{5.15}$$

5.3   *ABC's applied to dipole radiation*

The fields for an oscillating (Hertzian) dipole of strength Idz at the origin and pointing in the z-direction ($\theta = 0$) are, see for instance Kong (1986), exactly

$$H_\phi = Ae^{-jkr}\left(\frac{jk}{r} + \frac{1}{r^2}\right)\sin\theta, \tag{5.16}$$

$$E_r = 2\eta Ae^{-jkr}\left(\frac{1}{r^2} + \frac{1}{jkr^3}\right)\cos\theta, \tag{5.17}$$

$$E_\theta = \eta Ae^{-jkr}\left(\frac{jk}{r} + \frac{1}{r^2} + \frac{1}{jkr^3}\right)\sin\theta, \tag{5.18}$$

where $A = Idz/(4\pi)$. It is of interest to substitue these practical fields into the expressions for the first- and second-order ABC's to confirm properties which have been ascribed to the radiation operators. Noting that the spherical coordinate representations here take the form $(E_r, E_\theta, 0)$ and $(0, 0, H_\phi)$, first of all the boundary operators are simplified accordingly. It is found that

$$\mathbf{B}_1(\mathbf{H}) = jk\left(\frac{E_\theta}{\eta} - H_\phi\right)\mathbf{1}_\phi, \tag{5.19}$$

$$\mathbf{B}_1(\mathbf{E}) = jk(\eta H_\phi - E_\theta)\mathbf{1}_\theta + (s-1)\frac{1}{r}\frac{\partial E_r}{\partial\theta}\mathbf{1}_\theta, \tag{5.20}$$

$$\mathbf{B}_2(\mathbf{H}) = -2\left(jk + \frac{1}{r}\right)jk\left(\frac{E_\theta}{\eta} - H_\phi\right)\mathbf{1}_\phi - \frac{jk}{\eta^r}\frac{\partial E_r}{\partial\theta}\mathbf{1}_\phi \tag{5.21}$$

$$\mathbf{B}_2(\mathbf{E}) = -2\left(jk\frac{1}{r}\right)jk(\eta H_\phi - E_\theta)\mathbf{1}_\theta$$
$$+ (s-1)\frac{1}{r^2}\frac{\partial}{\partial\theta}\frac{1}{\sin\theta}\frac{\partial}{\partial\theta}(E_\theta\sin\theta)\mathbf{1}_\theta + (2-s)jk\frac{1}{r}\frac{\partial E_r}{\partial\theta}\mathbf{1}_\theta, \tag{5.22}$$

whence substituting from Eqs. (5.16)–(5.18) gives

$$\mathbf{B}_1(\mathbf{H}) = Ae^{-jkr}\frac{\sin\theta}{jkr^3}\mathbf{1}_\phi, \tag{5.23}$$

$$\mathbf{B}_1(\mathbf{E}) = -\eta Ae^{-jkr}\left\{\frac{2s-1}{r^3} + \frac{2(s-1)}{jkr^4}\right\}\sin\theta\mathbf{1}_\theta, \tag{5.24}$$

$$\mathbf{B}_2(\mathbf{H}) = 0, \tag{5.25}$$

$$\mathbf{B}_2(\mathbf{E}) = -\eta Ae^{-jkr}\frac{2(s-1)\sin\theta}{jkr^5}\mathbf{1}_\theta. \tag{5.26}$$

The results are exactly as predicted:

■ The $1/r$ term is absent in E_r (H_r is identically zero) showing that E_{0r} and H_{0r} do in fact vanish.
■ The results of both $\mathbf{B}_1$ operations are $O(1/r^3)$.
■ The $\mathbf{B}_2$ operation on the dipole electric field gives a remainder $O(1/r^5)$, however the magnetic field expression terminates in $1/r^2$ requiring that $\mathbf{B}_2(\mathbf{H})$ shall vanish, which indeed it does.

5.4   *Application of the second-order ABC to scattering problems*
    The use of absorbing boundary conditions is particularly appropriate for scattering problems involving large and complex bodies, where it becomes difficult to arrange suitable artificial boundaries over which to apply Green's function integrals. It is usually advantageous to separate the total field into its incident and scattered parts, solving for the latter as a dependent variable, thus avoiding the loss of accuracy sometimes encountered through near-cancellation of the two. Let

$$\mathbf{E} = \mathbf{E}^{\text{ext}} + \mathbf{E}^{\text{sca}}, \tag{5.27}$$

$$\mathbf{H} = \mathbf{H}^{\text{ext}} + \mathbf{H}^{\text{sca}}. \tag{5.28}$$

The superscript 'ext', as used previously, denotes a field arising from some external agency, it being understood that such a field extends into the space occupied by the scatterer as if the scattering material were not there; the superscript 'sca' denotes the extra field which arises due to the presence of that material. The total field satisfies Maxwell's equations pertaining to the real distribution of materials, so substituting the two parts together into the electric curl equation gives

$$\nabla \times \mathbf{E}^{\text{ext}} + \nabla \times \mathbf{E}^{\text{sca}} = -\mathrm{j}\omega\mu_0\mu_r(\mathbf{H}^{\text{ext}} + \mathbf{H}^{\text{sca}}). \qquad (5.29)$$

The incident field satisfies the free-space Maxwell equations, whence substituting $\nabla \times \mathbf{E}^{\text{ext}} = -\mathrm{j}\omega\mu_0\mathbf{H}^{\text{ext}}$ into Eq. (5.28) gives

$$\nabla \times \mathbf{E}^{\text{sca}} + \mathrm{j}\omega\mu_0\mu_r\mathbf{H}^{\text{sca}} = -\mathrm{j}\omega\mu_0(\mu_r - 1)\mathbf{H}^{\text{ext}}. \qquad (5.30)$$

Recalling Eq. (2.1), Maxwell's electric curl equation with the inclusion of a magnetic current source term, a fictitious magnetic source current $\mathbf{K}^{\text{ext}} = \mathrm{j}\omega\mu_0(\mu_r - 1)\mathbf{H}^{\text{ext}}$ arising inside the scatterer can be used to replace the externally incident magnetic field giving the very convenient Maxwell form

$$\nabla \times \mathbf{E}^{\text{sca}} + \mathrm{j}\omega\mu_0\mu_r\mathbf{H}^{\text{sca}} = -\mathbf{K}^{\text{ext}}. \qquad (5.31)$$

Similarly

$$\nabla \times \mathbf{H}^{\text{sca}} - \mathrm{j}\omega\epsilon_0\epsilon_r\mathbf{E}^{\text{sca}} = \mathrm{j}\omega\epsilon_0(\epsilon_r - 1)\mathbf{E}^{\text{ext}}, \qquad (5.32)$$

$$= \mathbf{J}^{\text{ext}}, \qquad (5.33)$$

a known electric current source replacing the incident electric field. It is now allowable to proceed as if there were just these fictitious internal sources and no external fields, to which can be added any real internal sources that may be present.

Evidently the residual equation now valid instead of Eq. (2.5) is

$$R = \int_\Omega \{\mu_r^{-1}(\nabla \times \mathbf{W}^{\text{sca}}).(\nabla \times \mathbf{E}^{\text{sca}}) - k^2\epsilon_r\mathbf{W}^{\text{sca}} \cdot \mathbf{E}^{\text{sca}}\} \, \mathrm{d}\Omega$$

$$+ \int_\Omega \{\mathrm{j}k\eta\mathbf{W}^{\text{sca}} \cdot \mathbf{J}^{\text{ext}} + \mathbf{W}^{\text{sca}} \cdot \nabla \times (\mu_r^{-1}\mathbf{K}^{\text{ext}})\} \, \mathrm{d}\Omega$$

$$- \oint_S \mathbf{W}^{\text{sca}} \cdot (\nabla \times \mathbf{E}^{\text{sca}} \times \mathbf{1}_n) \, \mathrm{d}S. \qquad (5.34)$$

Here the surface integral term $\mathrm{j}k\eta\mathbf{H}$ in Eq. (2.5) has been put back to its Maxwell equation equivalent of $-\nabla \times \mathbf{E}$. Now it is simple to apply an ABC condition, since the fields in the latter refer to outward-travelling waves, here that is $\mathbf{E}^{\text{sca}}$ alone. The second-order ABC is the lowest worth considering and, setting $\mathbf{B}_2(\mathbf{E}^{\text{sca}}) = 0$, may be written

$$1_r \times \nabla \times \mathbf{E}^{\text{sca}} = jk\mathbf{E}^{\text{sca}} + \frac{1}{2(jk + 1/r)} \{\nabla \times 1_r(\nabla \times \mathbf{E}^{\text{sca}})r$$

$$+ (s-1)\nabla_t(\nabla_t \cdot \mathbf{E}_t^{\text{sca}}) + (2-s)jk\nabla_t E_r^{\text{sca}}\}. \quad (5.35)$$

Noting that the surface S is spherical, $1_r = 1_n$, the expression Eq. (5.35) is suitable for substitution into the residual Eq. (5.34). The substitution yields a residual expression, alternative to Eq. (2.5), written entirely in terms of the *accessible* E-components already included in the tangentially continuous vector finite element representation of the volume Ω, therefore avoiding use of the boundary integral process.

Now consider the choice of the arbitrary parameter s. If $s = 2$ is chosen so as to eliminate the last term in Eq. (5.35), the resulting finite element matrices becomes symmetric.[1] This is so because in the residual Eq. (5.34) with its integration over the spherical surface S, an integration by parts can be effected transferring one of the ∇ operators in each of the second and third terms of the right-hand side of Eq. (5.34) to the weight function itself. A similar operation applied to the last term in Eq. (5.34) is not possible, so it is fortunate that opportunity presents itself of selecting $s = 2$ to get rid of it. The final form for the residual thus becomes

$$R = \int_\Omega \{\mu_r^{-1}(\nabla \times \mathbf{W}^{\text{sca}}) \cdot (\nabla \times \mathbf{E}^{\text{sca}}) - k^2\epsilon_r \mathbf{W}^{\text{sca}} \cdot \mathbf{E}^{\text{sca}}\} \, d\Omega$$

$$+ \int_\Omega \{jk\eta\mathbf{W}^{\text{sca}} \cdot \mathbf{J}^{\text{ext}} + \mathbf{W}^{\text{sca}} \cdot \nabla \times (\mu_r^{-1}\mathbf{K}^{\text{ext}})\} \, d\Omega$$

$$+ jk \oint_S \mathbf{W}_t^{\text{sca}} \cdot \mathbf{E}_t^{\text{sca}} \, dS$$

$$+ \frac{1}{2\left(jk + \dfrac{1}{r}\right)} \oint_S \{(\nabla \times \mathbf{W}^{\text{sca}})_r \cdot (\nabla \times \mathbf{E}^{\text{sca}})_r$$

$$+ (\nabla_t \cdot \mathbf{W}^{\text{sca}})(\nabla_t \cdot \mathbf{E}_t^{\text{sca}})\} \, dS. \quad (5.36)$$

Note that in performing the integrations by parts over the closed sphere S, the line integrals normally appearing after such integrations are absent. Completion of the finite element treatment now follows without difficulty. The treatment here requires the inhomogeneous object being analysed to be surrounded by a fictitious spherical surface which lies entirely external to the object. Some objects may nearly conform to a spherical surface and so are very suitable for the vector ABC treatment described here. Other more extreme shapes will require the modelling of a

---

[1] Less systematic derivations of $\mathbf{B}_2$, earlier than the one by Webb and Kanellopoulos (1989) given here, found the case $s = 1$ producing troublesome nonsymmetric matrices.

considerable volume of empty space in the finite element region. It is allowable to let $r \to \infty$ in a vector ABC, say Eq. (5.15), thus corresponding to a locally plane surface. It may be possible to construct a more nearly conformal fictitious hexahedral surface around the object, in which case the piecewise application of the planar ABC to the hexahedron facets is permissible.

6.   Fictitious absorbers

The last alternative discussed for dealing with the radiation open boundary is perhaps the simplest and most obvious method. The finite element procedures described here all work quite satisfactorily with complex μ and ϵ representing lossy materials. Why not clad the open boundary with judiciously chosen layers of such materials, conveniently terminated by a perfect conductor, to simulate the anechoic chamber used in practical antenna and scattering evaluations? If a plane wave is incident obliquely upon a planar junction between materials i and j with constitutive constants μ_i, ϵ_i and μ_j, ϵ_j respectively, then the reflection coefficient is readily calculable from simple electromagnetic theory. For polarized waves with incident and refracted angles θ_i, θ_j respectively (Fig. 9.4) the reflection and transmission coefficients are, see Kong (1986),

$$\rho_{ij} = \frac{1 - p_{ij}}{1 - p_{ij}}, \tag{6.1}$$

$$\tau_{ij} = \frac{2}{1 + p_{ij}}, \tag{6.2}$$

where with

$$k_{ij} = \omega(\mu_{ij}\epsilon_{ij})^{1/2} \tag{6.3}$$

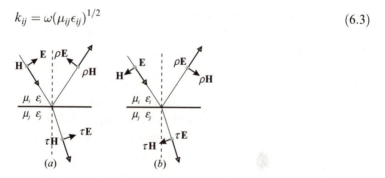

Fig 9.4   Reflection of plane waves from a material interface. (*a*) TE polarization. (*b*) TM polarization.

and (Snell's law)

$$\sin \theta_j = \frac{k_i}{k_j} \sin \theta_i, \tag{6.4}$$

then

$$p_{ij} = \frac{\mu_i k_i \cos \theta_i}{\mu_j k_j \cos \theta_j}, \tag{6.5}$$

for TE incidence, as in Fig. 9.4(a), and

$$p_{ij} = \frac{\epsilon_i k_i \cos \theta_j}{\epsilon_j k_j \cos \theta_i}, \tag{6.6}$$

for TM incidence, Fig. 9.4(b). The formulae Eqs. (6.1)–(6.6) are ordinarily applied to lossless media for real angles θ, but they remain valid if any of the μ_j, ϵ_j and θ are complex. They can be applied repeatedly to a multilayered medium $i = 1, 2, \ldots, n$, abutting free space, material 0, and terminating with a perfect conductor, material $n+1$, as shown in Fig. 9.5. The resulting wave amplitudes from the multiple reflections are summed, taking into account the changes in phase arising from the spatial displacement of each reflecting interface. Thus the overall free-space reflection coefficient, ρ_0 say, can be worked out for a real θ_0 in rather a complicated but straightforward calculation. Choosing the thicknesses $t_1, t_2, \ldots, t_n$ arbitrarily for a smallish number of layers, the closed form ρ_0 can be optimized for minimum magnitude over a wide range of real incidence angles θ_0 in terms of the complex values μ_j and ϵ_j.

Jin, Volakis and Liepa (1992) did this for a TE-incidence two-dimensional scattering problem, taking three layers each of thickness 0.05λ with complex constitutive constants

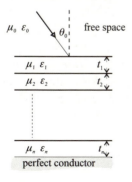

Fig 9.5   General layered fictitious absorber.

$$\left.\begin{array}{l} \epsilon_{r1} = (-0.1140202, -1.413109), \quad \mu_{r1} = (-0.9256691, 0.07614109), \\ \epsilon_{r2} = (0.05407664, 0.1946043), \quad \mu_{r2} = (0.3195902, 0.2922617), \\ \epsilon_{r3} = (-1.053359, 0.3214443), \quad \mu_{r3} = (-0.02842726, -5.400275). \end{array}\right\}$$

$$(6.7)$$

The reflection coefficient was negligible over a range of θ_0 from 0 to 70 degrees and compared favourably with results from a second-order two-dimensional treatment using ABC's. It was pointed out that although the optimized data is frequency sensitive and only valid over a small bandwidth, expressing the thicknesses in terms of the free-space wavelength λ allows the actual dimensions to be scaled accordingly in a frequency domain analysis. The separation of cases into TE and TM incidence is meaningful in a two-dimensional analysis; the TM case here can be set up invoking duality. In a three-dimensional analysis the optimization would have to extend over both polarizations.

7.   Further reading

The book by Kong (1986) serves as an excellent reference for more advanced topics in electromagnetics relevant to scattering and radiation. The reader should note that Kong chooses the physicists' optional minus in the phasor wave factor $\exp\{\pm j(\omega t - kz)\}$, whereas this book adopts the engineers' plus sign. Thus many of the results there need a change of sign to conform with our convention. Another widely used standard textbook in this area, though of earlier vintage, is by Harrington (1961). Much pioneering work on the application of finite elements methods to the problem of radiation and scattering has been done by J.L. Volakis and his collaborators. A good deal of this is summarized in the *IEEE Magazine* paper by Jin, Volakis and Collins (1991). The arguments supporting the statements of the weighted residual rules used freely here in setting up the vector finite elements are given more fully by Ferrari and Naidu (1990). Tai (1994) gives a classic account of the dyadic Green's function fundamental to the hybrid FEM/BEM method. The book by Brigham (1988) is a well-known standard reference for the fast fourier transform. The paper by Webb and Kanellopoulos (1989) represents the first satisfactory account of a vector ABC suitable for three-dimensional vector finite element application.

8.   Bibliography
Brigham, E.O. (1988), *The Fast Fourier Transform and its Applications.* Englewood Cliffs: Prentice-Hall International. xi+ 448 pp.

Ferrari, R.L. and Naidu, R.L. (1990), 'Finite-element modelling of high-frequency electromagnetic problems with material discontinuities', *IEE Proceedings*, **137**, Pt. A, pp. 314–20.

Harrington, R.F. (1961), *Time-harmonic Electromagnetic Fields*. New York: McGraw-Hill. xi + 480 pp.

Jin, J.-M., Volakis, J.L. and Collins, J.D. (1991), 'A finite-element – boundary-integral method for scattering and radiation by two- and three-dimensional structures'. *IEEE Antennas and Propagation Magazine*, **33**, pp. 22–32.

Jin, J.-M., Volakis, J.L. and Liepa, V. (1992), 'Fictitious absorber for truncating finite element meshes in scattering', *IEE Proceedings*, **139**, Pt. H, pp. 472–75.

Kong, J.A. (1986), *Electromagnetic Wave Theory*. New York: Wiley. xi + 696 pp.

Tai, C.-T. (1994), *Dyadic Green's Functions in Electromagnetic Theory*. 2nd edn. New York: IEEE Press. xiii + 343 pp.

Webb, J.P. and Kanellopoulos, V.N. (1989), 'Absorbing boundary conditions for the finite element solution of the vector wave equation', *Microwave and Optical Technology Letters* **2**, pp. 370–2.

10

Numerical solution of finite element equations

1.   Introduction

In the finite element method, integral- or differential-equation problems are solved by substituting an equivalent problem of matrix algebra, in the expectation that the techniques of numerical mathematics will allow fairly direct solution of the matrix problem through the use of digitial computers. Corresponding to the boundary-value problems or the Fredholm integral equations arising in electromagnetics, there then results a matrix equation of the form

$$(\mathbf{A} + p\mathbf{B})\mathbf{x} = \mathbf{y}, \tag{1.1}$$

where $\mathbf{A}$ and $\mathbf{B}$ are matrices, $\mathbf{x}$ is the unknown vector of coefficients, and $\mathbf{y}$ is a known vector, while p is a scalar parameter. This equation has a nontrivial solution if $\mathbf{A}$ is a positive definite matrix and either (1) p is known, and $\mathbf{y}$ does not vanish, or (2) p is unknown, and $\mathbf{y}$ is identically zero. Problems of the former class are usually called *deterministic*; they produce a single solution vector $\mathbf{x}$ (provided the value of p does not match one of the possible solutions of the second class). The second class are known as *eigenvalue* problems; they possess as many solutions or *eigenvectors* $\mathbf{x}$ as the matrix has rows, each corresponding to one *eigenvalue* p, so that each solution in fact consists of a scalar-vector pair $(p_k, \mathbf{x}_k)$. It may appear at first glance that in this case the system of N simultaneous equations yields $N + 1$ answers, the N components of $\mathbf{x}$ and the value of p; however, such is not the case. Since equations of this class are always homogeneous, solutions to eigenvalue problems are unique only up to a multiplicative constant: if $\mathbf{x}$ is a solution, so is $k\mathbf{x}$, where k is any number. Hence the solution vector $\mathbf{x}$ really only contains $N - 1$ independent degrees of freedom.

This chapter briefly surveys the most common techniques in present-day use for solving deterministic problems arising in finite element analysis. The solution of finite element equations, like much else in the finite element art, is a highly specialized area; the novice is well advised to use clever programs already written by someone else, not to attempt to write his own. This is true even for the relatively simple task of solving deterministic problems; it is doubly true for eigenvalue problems! The object here is merely to give the reader some idea of what might be found in currently available programs.

2.   Gaussian elimination methods

Many present-day finite element analysis programs rely on one version or another of the Gaussian triangular decomposition technique for solving the large systems of simultaneous equations commonly encountered, and even where they do not, alternative techniques borrow large segments of Gaussian methodology. The small programs given in the utility package of Appendix 4 rely on triangular decomposition exclusively.

2.1   *Triangular decomposition*

In its basic form, the Gaussian elimination method differs very little from that generally learned in school under the name of *successive elimination*: one eliminates one variable at a time until only one equation in one unknown is left. When that one has been determined, the remainder are found by taking the equations in the reverse order, each time substituting the already known values. These two phases of the process are generallay called *forward elimination* and *back substitution*. In large-scale finite element programs, the triangular decomposition method is made to operate in the first instance on the coefficient matrix alone. To illustrate, suppose **S** is a symmetric, positive definite matrix, as many finite element matrices are. It can be shown that such a matrix may always be written as the product of two triangular matrices,

$$\mathbf{S} = \mathbf{L}\mathbf{L}^{\mathrm{T}}. \tag{2.1}$$

Here, **L** is a lower triangular matrix; it has only zero elements above and to the right of its principal diagonal. (The superscript T denotes transposition, so that $\mathbf{L}^{\mathrm{T}}$ is upper triangular.) The process of computing the lower triangular matrix **L** is called *triangular factorization*, or *triangular decomposition*; **L** is known as the lower triangular factor of **S**. The requirement that **S** must be a positive definite matrix corresponds physically to the requirement that stored energy must be positive for any and all excitations. This requirement is certainly met by physical systems encountered in practice.

Suppose now it is required to solve the matrix equation, which might typically arise from finite element analysis,

$$\mathbf{Sx} = \mathbf{y}. \tag{2.2}$$

To do so, $\mathbf{S}$ is factored as in Eq. (2.1), and (2.2) is rewritten as the pair of equations

$$\mathbf{Lz} = \mathbf{y} \tag{2.3}$$

$$\mathbf{L}^{\mathrm{T}}\mathbf{x} = \mathbf{z}, \tag{2.4}$$

where $\mathbf{z}$ is an auxiliary vector. Because $\mathbf{L}$ is triangular, row k of $\mathbf{L}$ can contain nonzero entries only in its first k columns. In particular, the first row of $\mathbf{L}$ contains only one nonzero element, so that the first component of $\mathbf{z}$ can be computed immediately. The second row of $\mathbf{L}$ contains only two nonzero elements, hence it relates only the first two components of $\mathbf{z}$; but since the first component of $\mathbf{z}$ is known, the second may now be computed. Continuing, the system of equations (2.3) is quicky solved for $\mathbf{z}$. Since $\mathbf{L}^{\mathrm{T}}$ is also triangular, a similar process is used for solving Eq. (2.4) to find $\mathbf{x}$; only the sequence of operations is reversed because $\mathbf{L}^{\mathrm{T}}$ is upper rather than lower triangular.

Clearly, the key to equation solving lies in triangular decomposition of the coefficient matrix $\mathbf{S}$. The necessary steps are readily deduced by examining the desired result. Written out in detail, Eq. (2.1) requires that

$$S_{ik} = \sum_{j=1}^{\min(i,k)} L_{ij}L_{kj}. \tag{2.5}$$

Note that the summation only extends to the lower of i or k, for all elements beyond the diagonal of $\mathbf{L}$ must vanish. For the diagonal elements themselves, Eq. (2.5) may be solved to yield

$$L_{ii} = \sqrt{S_{ii} - \sum_{j=1}^{i-1} L_{ij}^2}. \tag{2.6}$$

The first row of $\mathbf{L}$ only contains a diagonal element, which is given by

$$L_{11} = \sqrt{S_{11}}. \tag{2.7}$$

In the second (and every subsequent) row, the summation of Eq. (2.5) contains no more terms than the column number k. Working across the rows in their natural sequence, only one new unknown appears in each row. The method is thus easily applied, and may be summarized in the following prescription:

(a) Set

$$L_{11} = \sqrt{S_{11}}. \tag{2.7}$$

(b) In each row i, compute the off-diagonal element in each column k by

$$L_{ik} = \frac{1}{L_{kk}} \left(S_{ik} - \sum_{j=1}^{i-1} L_{ij} L_{kj} \right) \tag{2.8}$$

and the diagonal element by

$$L_{ii} = \sqrt{S_{ii} - \sum_{j=1}^{i-1} L_{ij}^2}. \tag{2.6}$$

until all of **L** has been calculated.

To illustrate the procedure, consider the symmetric positive definite matrix

$$\mathbf{S} = \begin{bmatrix} 4 & 4 & 4 & & & & & \\ 4 & 5 & 2 & & & & & \\ 4 & 2 & 9 & 1 & 2 & 1 & & \\ & & 1 & 10 & -7 & 1 & & \\ & & 2 & -7 & 14 & & 2 & 1 \\ & & 1 & 1 & & 9 & -8 & -2 \\ & & & & 2 & -8 & 9 & \\ & & & & 1 & -2 & & 6 \end{bmatrix}. \tag{2.9}$$

To improve clarity of presentation, all zero entries have been left blank in **S**. By using the technique described, the matrix (2.9) is easily shown to have as its triangular factor

$$\mathbf{L} = \begin{bmatrix} 2 & & & & & & & \\ 2 & 1 & & & & & & \\ 2 & -2 & 1 & & & & & \\ & & 1 & 3 & & & & \\ & & 2 & -3 & 1 & & & \\ & & 1 & & -2 & 2 & & \\ & & & & 2 & -2 & 1 & \\ & & & & 1 & & -2 & 1 \end{bmatrix}. \tag{2.10}$$

Correctness of the decomposition may of course be verified easily by multiplying as indicated in Eq. (2.1).

To complete the example, suppose the right-hand vector **y** to be given by

$$\mathbf{y}^T = \begin{bmatrix} 4 & 1 & 10 & -18 & 16 & 8 & -9 & 1 \end{bmatrix}. \tag{2.11}$$

Solving Eq. (2.3) by the process of forward elimination described above yields the auxiliary vector **z** as

$$\mathbf{z}^T = [2 \quad -3 \quad 0 \quad -6 \quad -2 \quad 2 \quad -1 \quad 1]. \tag{2.12}$$

The solution vector **x** may then be recovered by the process of back substitution, ie., by solving Eq. (2.4):

$$\mathbf{x}^T = [-5 \quad 3 \quad 3 \quad -3 \quad -1 \quad 2 \quad 1 \quad 1 \quad]. \tag{2.13}$$

It should be evident that the process of triangular decomposition need only be performed once for any given matrix, even if several different right-hand sides are of interest. Of course, the forward elimination and back substitution will have to be performed separately for each one. Happily, as will be shown next, the relative amount of computational work involved in the elimination and back substitution process is not large.

2.2   *Time and storage for decomposition*

To assess the efficacy of the triangular decomposition method, some assessment of the total time and memory requirements is necessary. An estimate of both will be given here for dense (fully populated) matrices; refinements useful for sparse matrices will be developed subsequently.

The process of triangular decomposition proceeds step-by-step, computing one new entry of the triangular factor **L** at each step. To find element k in row i of **L**, k arithmetic operations are required, where an *operation* is defined as the combination of one multiplication or division and one addition or subtraction, along with such integer operations as may be necessary on array subscripts. The number $M(i)$ of operations needed to compute row i of the triangular decompose is the sum of operations required for the elements of that row,

$$M(i) = \sum_{k=1}^{i} k = i(i+1)/2, \tag{2.14}$$

plus one square root operation. The work required for the entire matrix is the sum of the amounts needed for the individual rows, or

$$M = \sum_{k=1}^{N} k(k+1)/2 = \frac{1}{6}N(N+1)(N+2), \tag{2.15}$$

plus N square roots. This amount of work of course covers the decomposition only; nothing is included for actually solving equations. The forward elimination needed for solving with one right-hand side uses up an additional E operations,

$$E = \sum_{k=1}^{N} k = N(N + 1)/2. \tag{2.16}$$

Exactly the same number of operations is needed for back substitution. Thus the amount of work involved in forward elimination and back substitution is, for all practical purposes, equal to the work required for one matrix–vector multiplication, which requires exactly N^2 operations.

Where several right-hand sides are to be solved for with the same coefficient matrix — for example, in calculating the magnetic field in a particular set of coils for several different sets of current values — it may at first glance seem tempting to compute the inverse of the coefficient matrix, and then to multiply each new right-hand side by the inverse. This approach cannot save computing time, and it frequently wastes computer memory, for the following reasons. The most economic procedure for computing the inverse of a matrix is to perform triangular decomposition, and then to solve N times, each right-hand side being one column taken from the unit matrix. While these solutions can be carried out a little more economically than the general case (operations with known zeros can be eliminated by clever programming), the cost of computing an explicit inverse must always be higher than that of triangular decomposition because it actually begins by carrying out a triangular decomposition and then performs a few other operations besides. Subsequent matrix–vector multiplications entail the same cost as elimination and back substitution. Hence the invert-and-multiply strategy can never be advantageous as compared to decompose-eliminate-backsubstitute.

The computer memory required to calculate and store the explicit inverse of a matrix can never be less than that needed for the triangular factors, and may often amount to considerably more. Consider again Eq. (2.8), which prescribes the operations required to compute element L_{ik} of the triangular factor $\mathbf{L}$. Suppose element S_{i1} of the original coefficient matrix $\mathbf{S}$ is zero. Examination of (2.8) shows that the corresponding element of $\mathbf{L}$ is then zero also. Further, if $S_{i2}, S_{i3}, \ldots, S_{ik}$ are all zero, then no nonzero terms can appear in the summation in (2.8); the corresponding elements of $\mathbf{L}$ must vanish. Of course, this argument does not hold true for any and all zero elements of $\mathbf{S}$. If there are any nonzero entries to the left of S_{ik}, the summation in (2.8) does not in general vanish, and the corresponding entry in $\mathbf{L}$ is very likely not to vanish either. Thus one may conclude that the leftmost nonzero element in any given row of **L** will in general correspond to the leftmost nonzero element in the same row of **S**; but from that element up to the diagonal, **L** may be more fully populated than **S**.

Since the leftmost zero elements of the rows of **L** correspond exactly to the leftmost zero elements in **S**, finite element programs are often arranged to avoid storing the leading zeros. Considerable economies of storage may be effected in this way. An explicit inverse of **S**, on the other hand, is full, with no zero entries in the general case. Computing and storing the inverse requires space for a fully populated matrix, while substantially less storage often suffices for the triangular factors. Thus the storage requirement for **L** is bounded above by the storage requirement of the inverse; there cannot occur any case in which computing the inverse can be better from the point of view of storage. Consequently, practical finite element programs never compute and store inverses. They generally either perform triangular decomposition, then eliminate and back substitute; or else they combine the elimination and back substitution with the decomposition so that the triangular factors are not explicitly stored.

2.3   *A Choleski decomposition program*

To illustrate the techniques employed in equation solving, a simple Choleski decomposition program may prove instructive. Program `eqsolv` in the utility program package (see Appendix 4) is in fact the program called by all sample finite element programs in this book. It is not optimal in any sense, it does not even exploit matrix symmetry to economize on memory. However, it is probably easy to read and understand.

After the COMMON block which allows for access to all relevant variables in the calling program, subroutine `eqsolv` proceeds with triangular decomposition. The Choleski algorithm is used, in exactly the form stated above, with special treatment of the initial matrix row to avoid zero or negative subscripts. Subroutine `chlsky` fills both upper and lower halves of the matrix storage area with the upper and lower triangular factors. This approach is admittedly wasteful of storage, so it is never used in practical finite element programs. However, the more sophisticated storage arrangements, such as the band or profile methods, render the program indexing more complicated. They are therefore often much more difficult to read and understand.

After the Choleski decomposition has been performed, `eqsolv` carries out forward elimination and back substitution. In this process, the right-hand side originally furnished is overwritten with the solution. `eqsolv` performs triangular decomposition, forward elimination, and back substitution in a single program, but it is quite usual to subdivide equation-solving programs instead into two or three subroutines. Removing the triangular decomposition to a separate program allows solution with several different right-hand sides, without repeating the decomposition. The penalty paid for doing so is negligible in computer time and space.

3.   Compact storage of matrices

Most useful two-dimensional field problems can be modelled by matrix equations in 100–10 000 variables, while three-dimensional vector fields involving complicated geometric structures may require as many as 50 000–500 000 variables. With present-day computers the solution of a hundred simultaneous algebraic equations is a simple matter, regardless of the structure of the coefficient matrix. Thousands of equations can be handled easily, provided the coefficient matrix contains many zeros and the program is so organized as to take advantage of them. With full coefficient matrices, containing few or no zero elements, even 1000 equations is a large number, for the storage of the coefficient matrix alone will require about four megabytes of computer memory!

As a general rule, coefficient matrices derived from integral equations are full or nearly full. On the other hand, discretization of differential equations by means of finite elements tends to produce sparse matrices because any one nodal variable will be directly connected only to nodal variables which appear in the same finite element. Hence, the number of nonzero entries per matrix row generally depends on the type of element employed, and has little to do with the type of problem. Thus, very large systems of simultaneous equations arising from discretization of differential equations are usually also very sparse. This section briefly examines the conventional methods for exploiting matrix sparsity to economize on both storage and computing time.

3.1   *Band and profile storage*

When triangular decomposition of a matrix is performed, it is usually found that the lower triangular factor **L** is denser (contains fewer zeros) than the original matrix. As noted above, the leftmost nonzero entry in any given row of **L** must in general coincide with the leftmost nonzero entry in the corresponding row of the original matrix **S**. To put the matter another way, any left-edge zeros in **S** are preserved in decomposition. Zeros to the right of the leftmost nonzero element, on the other hand, are not in general preserved; they are said to fill in. Of course, any particular matrix element may fortuitously turn out to be zero because of numerical cancellation. Such zeros are sometimes called *computed* zeros. The zeros of interest here result from the topological structure of the matrix, and they remain zero in any matrix with the same structure, independently of numerical values; they are referred to as *structural* zeros. By way of example, consider the matrix **S** of Eq. (2.9). The S_{65} and S_{87} entries in this matrix are zero. But comparison with (2.10) shows that they have filled in during the process of decomposition. On the other hand, the L_{64} and L_{86} entries in **L** are zero, as the result of accidental numerical cancellation. To verify this point, it suffices to decompose another matrix of exactly the same structure as **S**, but with different numerical values. The L_{64} and L_{86} entries of **L** do not in general remain zero, while the left-edge zeros are preserved.

Equation-solving programs generally need to reserve computer memory space for both **S** and **L**. There is clearly no need to reserve space for the left-edge zeros in **S** and **L**, for they are known to remain zero throughout the entire computation. Therefore, it is usual to arrange for storage of matrices in one of several compacted forms. Two of the most common arrangements are band matrix storage and profile storage.

Let **S** be a symmetric matrix. Suppose that the leftmost nonzero entry in row i occurs in column $N(i)$. The half-bandwidth of M of **S** is defined as

$$M = 1 + \max_{i}\{i - n(i)\}, \tag{3.1}$$

the maximum being taken over all i, $1 \leq i \leq N$. Storage in banded form is arranged by storing exactly M numbers for each matrix row. For example, in row k, the first entry stored would be that belonging to column $k - M + 1$, the last one that belonging to column k (the diagonal element). The missing elements in the first few rows are usually simply filled in with zeros. For example, the matrix **S** of Eq. (2.9) can be stored in a single array as

$$\text{band}(\mathbf{S}) = \begin{bmatrix} 0 & 0 & 0 & 4 \\ 0 & 0 & 4 & 5 \\ 0 & 4 & 2 & 9 \\ & & 1 & 10 \\ & 2 & -7 & 14 \\ 1 & 1 & & 9 \\ & 2 & -8 & 9 \\ 1 & -2 & & 6 \end{bmatrix}. \tag{3.2}$$

In this representation, all zeros contained in the original matrix have been left blank for the sake of clarity. However, also for the sake of clarity, the zeros artifically introduced to fill the leading array elements have been entered explicitly. Even for the quite small matrix of this example, the storage required in banded form is smaller than that needed for full storage, requiring 32 locations (including the artificial zeros) as against 36 (taking advantage of matrix symmetry). For large matrices, the difference can be very large indeed.

The advantage of banded storage is simplicity: only one new item of information is required, the half-bandwidth M. But this form of storage may still be relatively inefficient because many rows are likely to contain at least some structural zeros. Indeed, in the form (3.2) four structural zeros are stored. These are avoided by the so-called profile storage arrangement. In this arrangement, only those members of each row are stored which fall between the leftmost nonzero entry and the diagonal. The matrix of (2.9) is then stored as a single numeric string:

$$\text{prof}(\mathbf{S}) = [4 \quad 4 \quad 5 \quad 4 \quad 2 \quad 9 \quad 1 \quad 10 \quad 2 \quad -7 \quad 14$$
$$\qquad\qquad 1 \quad 1 \quad 0 \quad 9 \quad 2 \quad -8 \quad 9 \quad 1 \quad -2 \quad 0 \quad 6]. \tag{3.3}$$

The proper indices of these values can be recovered only if the position of the leftmost nonzero entry in each row is known. One simple fashion to keep track of it is to store the leftmost nonzero entry locations of $N(i)$, as an integer string:

$$\text{left}(\mathbf{S}) = [1 \quad 1 \quad 1 \quad 3 \quad 3 \quad 3 \quad 5 \quad 5]. \tag{3.4}$$

In the case used in this example, the storage requirement for $\mathbf{S}$ is now reduced to 22 locations, plus the array of leftmost locations $N(i)$. As compared to full matrix storage, the reduction is likely to be quite remarkable, in the case of large matrices with a few fully populated or nearly fully populated rows.

One further method of memory conservation is available to the analyst. Careful reexamination of Eqs. (2.5)–(2.8) will show that once the member L_{ik} of $\mathbf{L}$ has been computed, the corresponding entry S_{ik} of $\mathbf{S}$ is never used again. Consequently, there is no need to store both $\mathbf{S}$ and $\mathbf{L}$: it is sufficient to allocate space for one array only, to deposit $\mathbf{S}$ in this space initially, and then to overwrite $\mathbf{S}$ with $\mathbf{L}$ as decomposition proceeds. Should $\mathbf{S}$ be required for some purpose at a later time, it can be read into memory again from some external storage medium.

The triangular decomposition, forward elimination, and back substitution processes are unaltered in principle when matrices are stored in compact forms. However, the change in data storage implies that the details of subscript computation (i.e., the array indexing in the computer programs) must be modified also. Since there is no need to perform vacuous operations on the left-edge zeros, the summations in Eqs. (2.6) and (2.8) may be altered to run, not from $j = 1$ (the left edge of the full matrix) but from the first nonzero matrix entry. Thus, (2.8) should be replaced by

$$L_{ik} = \frac{1}{L_{kk}} \left(S_{ik} - \sum_{j=J}^{i-1} L_{ij} L_{kj} \right)$$ (3.5)

where the lower limit J of the summation is given by

$$J = i - M$$ (3.6)

for band-stored matrices, because no computations should be executed outside the band edge. For profile-stored matrices, no calculation should take place if either factor lies beyond the left edge of its row, so that

$$J = \max\{n(i), n(k)\}.$$ (3.7)

The lower summation limit for Eq. (2.6) of course must be amended correspondingly.

In performing triangular decomposition of a band-stored matrix, each off-diagonal element of the lower triangular factor is calculated by a single application of Eq. (3.5), and each diagonal element by

$$L_{ii} = \sqrt{S_{ii} - \sum_{j=J}^{i-1} L_{ij}^2}.$$ (3.8)

It is important to observe that the entries of row i are computed using entries from the immediately preceding rows, but never from rows numbered lower than $i - M + 1$. Consequently, there is no need to house rows $1, 2, \dots, i - M$ in the immediate-access (core) memory of the computer while work on row i proceeds; only the rows from $i - M + 1$ to i are required. That is to say, the immediate-access memory actually necessary is just that sufficient to house the in-band elements of $M + 1$ rows, or a total of $(M + 1)(M + 2)/2$ numbers. The bandwidth of a sparse matrix is often very much smaller than its total number N of rows. Many large-scale finite element analysis programs therefore use so-called out-of-core banded solvers, which house the matrix $\mathbf{S}$ (in banded storage) on disk, tape or other external storage media. Triangular decomposition is then performed by computing one or more rows of $\mathbf{L}$, writing them to an output tape or disk, then moving on the remaining rows in immediate-access memory, and reading in one or more additional rows of $\mathbf{S}$. The matrices $\mathbf{S}$ and $\mathbf{L}$ are thus passed through memory, with only $M + 1$ (or fewer) rows ever concurrently resident. Matrices of orders 5000–250 000, with bandwidths of 500–2500 or so, are often treated in this way. A matrix of order 50 000 with $M = 1000$ requires about 1 000 000 words of immediate-access memory for such out-of-core triangular decomposition and subsequent equation solving, a downright modest demand for present-day computers. Storage of the matrix itself takes about 50 000 000 words of storage in banded form, but would require 2 500 000 000 words if written out without exploitation of sparsity — a gross waste of storage, since 98% of that space would be filled with zeros!

3.2   *Structure of finite element matrices*

Most practical finite element programs exploit matrix sparsity in order to keep both computing time and memory requirements within reasonable bounds. Band matrix and profile methods store all zeros within the band or profile, but reject those outside the band. Keeping the matrix bandwidth small is therefore a matter of prime concern.

Rarely, if ever, do random node numberings on finite elements lead to good matrix sparsity patterns. In most practical programs, nodes are assumed to be randomly numbered at the outset, and some systematic renumbering scheme is then applied to produce an improved sparsity pattern. The only known technique guaranteed to produce the true minimum bandwidth is that of enumerating all possible numberings. Unfortunately, in an N-variable problem there are $N!$ possible numberings, so that the enumerative technique is quite impractical. On the other hand, there do exist several methods which produce nearly minimal bandwidths or profiles in most problems.

Renumbering methods take no account of the numeric values of matrix entries; they are concerned only with matrix structure. To exhibit structure in a form independent of the numeric values, and without regard to the numbering of variables, it is convenient to give the coefficient matrix a graphical representation. Since every nonzero matrix entry S_{ij} represents a direct connection between variables i and j, one may depict the matrix as a set of linking connections between variables. As an example, Fig. 10.1 shows a structural representation of Eq. (2.9). Here the variables are shown as graph nodes (circles), while the nonzero matrix entries are indicated by lines linking the variables. For first-order triangular elements, such a graphical representation has exactly the appearance of the finite element mesh itself.

To make the representation of Fig. 10.1 complete, the element values could be entered along each link. They have been omitted, since only the structure of the matrix is of interest, not the numerical values. In fact, the variable numbers could well be omitted too, since their numbering is essentially arbitrary anyway.

In computer storage, the structure graph of a matrix may be represented by a set of integer pairs $E = (i,j)$, each denoting a matrix element and identified by the pair of variable numbers (i.e., row and column

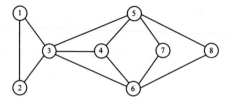

Fig 10.1   Representation, as an undirected graph, of the symmetric matrix of Eq. (2.9).

numbers). Each matrix element corresponds to an edge (a linking line) in the graph. The half-bandwidth M of a matrix is then given by the maximal node number difference $i - j$, plus one, taken over all the matrix elements E. That is,

$$M = 1 + \max_{E} |i - j|. \tag{3.9}$$

Similarly, the matrix *profile*, the total storage P required for the matrix in profile storage, is given by

$$P = \sum_{E} (|i - j| + 1). \tag{3.10}$$

Good node numberings are obviously those which minimize either the bandwidth M or the profile P, depending on which form of storage is contemplated. A good numbering for one purpose is not necessarily good for the other, since a few very large differences $|i - j|$ do not affect the profile P very much, but are ruinous for banded storage. Nevertheless, orderings that produce good bandwidths also give good profiles in a surprising number of practical cases.

Many node numbering methods begin by classifying all the variables into sets called *levels*. A starting variable, say K, is first chosen and assigned to level $L(0)$. Next, level $L(i)$ is created by taking all the variables which are directly connected to variables of level $L(i - 1)$ but which have not been assigned to any of levels $L(0), L(1), \ldots, L(i - 1)$. Each and every variable can belong to only one level. For example, starting with variable 1, the levels for the left graph of Fig. 10.2 are

$$\left.\begin{aligned} L(0) &= [\,1\,] \\ L(1) &= [\,2 \quad 3\,] \\ L(2) &= [\,4 \quad 5 \quad 6\,] \\ L(3) &= [\,7 \quad 8\,]. \end{aligned}\right\} \tag{3.11}$$

There are altogether four levels; in other words, the shortest path between node 1 and the most distant node of the graph requires traversing three

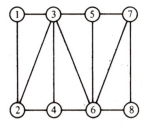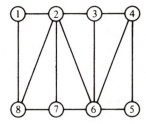

Fig 10.2   Two possible node (variable) numberings, which do not yield the same matrix bandwidth.

links. Indeed, it is always possible to reach a node in $L(i)$ by traversing exactly i links from the starting node because $L(i)$ is defined as the set of all nodes directly connected to (and hence one step away from) all nodes in $L(i-1)$. The number of levels is dependent on the choice of starting node. For the left half of Fig. 10.2, but this time starting with variable 3, the level structure is quickly obtained as

$$\left.\begin{aligned}
L(0) &= [\,3\,] \\
L(1) &= [\,1 \quad 2 \quad 4 \quad 5 \quad 6\,] \\
L(2) &= [\,7 \quad 8\,].
\end{aligned}\right\} \tag{3.12}$$

The total number of levels, as well as the grouping of variables within the levels, depends on the starting node. If all possible starting nodes are tried, and the highest-numbered level reached for any of them is $L(K)$, the matrix is said to have a *topological diameter* of K.

3.3   *Variable-numbering techniques*

There exists a substantial number of specific renumbering techniques for exploiting sparsity. No practical method known to date is guaranteed to produce the truly optimal numbering, but several come fairly close in most cases. This area is still developing, so it is not appropriate to give a detailed treatment of the available methods here; but since all the available methods are based essentially on examination of a graphical representation of matrix structure, certain common principles can be stated.

In Fig. 10.2, the numbering on the left is superior to that on the right because the variable number differences $|i-j|$ are smaller. Correspondingly, the nonzero matrix elements cluster more tightly around the principal diagonal. Good numberings are usually such that nodes with widely different numbers are widely separated. Such separation is obtained if all nodes in level $L(i)$ are numbered before those, in level $L(i+1)$. In other words, once level $L(0)$ has been created by assigning the number 1 to the starting node, the members of level $L(1)$ should be numbered next, then those of $L(2), \ldots,$ until no more levels are left. While the general principle is simple enough, it leaves open two important questions: how to choose the starting node, and how to number the members of each level. One quite effective technique chooses the node with the fewest connections to all others as the starting node. A level structure is then constructed, starting from that node. Next, within each level those members are numbered first which are directly connected to the smallest number of members of the next level. The rationale for doing so is that the longer one delays numbering a particular node or variable, the smaller the index differences $|i - j|$ between it and the members of the next level; numbering the most strongly connected variables last thus implies that the largest number of connecting links will have small differences $|i - j|$. While this argument is essentially probabilistic and gives no guarantees as to quality of numbering, it does work quite well in a large number of cases.

To take the case of Fig. 10.2 again, with the level structure as in Eqs. (3.11), the nodes have the following number of connections to nodes of the next higher level:

node:	1	2	3	4	5	6	7	8
level:	0	1	1	2	2	2	3	3
connections:	2	1	3	0	1	2	0	0

It is readily verified that the left-hand numbering shown in Fig. 10.2 is precisely the numbering given by the above method.

From Eqs. (3.11) and (3.12) it should be clear that the best numberings are likely to result from those choices of starting point which yield many levels, each containing only a few nodes. This observation leads to another group of methods, in which the matrix structure is examined to find its topological diameter, and the starting node is chosen at one end of a true diameter. While the amount of work required to find the new variable numbering is generally increased by the extra operations needed to find the diameter, the additional effort is usually amply repaid by decreased costs of actually solving equations afterwards.

4.   Iterative and semi-iterative solvers

The family of *iterative* solution methods, widely used in the early days of computing as an alternative to elimination methods, has regained popularity in recent years. In iterative methods, stepwise corrections are added to a known (but wrong) solution, thereby creating a sequence of approximate solutions which converge to the correct solution. In other words,

$$\mathbf{Ax} = \mathbf{b} \tag{4.1}$$

is solved by creating a sequence of corrections $\delta\mathbf{x}_0, \delta\mathbf{x}_1, \ldots, \delta\mathbf{x}_k$ which yield a convergent succession of approximate solutions by

$$\mathbf{x}_{k+1} = \mathbf{x}_k + \delta\mathbf{x}_k \tag{4.2}$$

beginning from some arbitrarily chosen initial value $\mathbf{x}_0$. If the initial estimate is not far wrong, the desired accuracy of approximation may be reached quickly, so iterative methods are able to make good use of any prior knowledge about the solution. Alternatively, they are able to trade accuracy for computing time. In effect, they create better approximate solutions out of bad ones, so iterative solvers should really be called *solution improvers*.

A closely related group of methods, usually called *semi-iterative*, has the interesting property of guaranteed convergence after a finite and known number of steps. Although in a strict sense they are deterministic rather than iterative, they usually converge long before the theoretical number of steps has been executed. In practice, they are therefore not viewed as a distinct class from iterative methods.

4.1   *Steepest descent iteration*

One simple but attractive iteration is the method of steepest descent. In this scheme, solving equations is viewed as equivalent to minimizing the quadratic form $F(x)$, defined by

$$F(\mathbf{z}) = \mathbf{z}^{\mathrm{T}}\mathbf{Az} - 2\mathbf{z}^{\mathrm{T}}\mathbf{b} \tag{4.3}$$

for any arbitrary vector $\mathbf{z}$. If the matrix $\mathbf{A}$ is positive definite, $F(z)$ has a minimum at the true solution $\mathbf{z} = \mathbf{x}$. To seek a minimum of F, some arbitrary starting point $\mathbf{x}_0$ is selected. There, and at every subsequent trial solution $\mathbf{x}_k$, the greatest immediate reduction in F is sought by choosing $\mathbf{x}_{k+1}$ so that the difference between solutions $\delta\mathbf{x}_k$ is normal to the locus of constant F, i.e., so $\delta\mathbf{x}_k$ points in the direction of steepest descent in F. The next trial solution $\mathbf{x}_{k+1}$ is then established by proceeding to the lowest possible value of F in that direction.

Implementing the steepest descent method in practice is easy enough. Suppose $\mathbf{u}$ is a short vector in such a direction that $F(\mathbf{x}_k) = F(\mathbf{x}_k + \mathbf{u})$; in other words, $\mathbf{u}$ is a vector tangent to the constant-F surface that passes through the point $\mathbf{x}_k$. By Eq. (4.3),

$$F(\mathbf{x}_k + \mathbf{u}) = (\mathbf{x}_k + \mathbf{u})^{\mathrm{T}}\mathbf{A}(\mathbf{x}_k + \mathbf{u}) - 2(\mathbf{x}_k + \mathbf{u})^{\mathrm{T}}\mathbf{b}. \tag{4.4}$$

Provided **u** is small, the second-order quantity $\mathbf{u}^T\mathbf{A}\mathbf{u}$ may be neglected, so

$$F(\mathbf{x}_k + \mathbf{u}) = F(\mathbf{x}_k) + \mathbf{u}^T\mathbf{A}\mathbf{x}_k + \mathbf{x}_k^T\mathbf{A}\mathbf{u} - 2\mathbf{u}^T\mathbf{b}. \tag{4.5}$$

But $F(\mathbf{x}_k) = F(\mathbf{x}_k + \mathbf{u})$. Equation (4.5) may therefore be written

$$\mathbf{u}^T(\mathbf{A}\mathbf{x}_k - \mathbf{b}) + (\mathbf{A}\mathbf{x}_k - \mathbf{b})^T\mathbf{u} = 0. \tag{4.6}$$

The residual vector

$$\mathbf{r}_k = \mathbf{A}\mathbf{x}_k - \mathbf{b} \tag{4.7}$$

is thus orthogonal to any small displacement **u** whose addition does not alter F. Conversely, the steepest descent direction is the direction of $\mathbf{r}_k$. Once the residual vector $\mathbf{r}_k$ is known, minimizing F in the direction of steepest descent is easy. This minimum must lie somewhere along a line that starts at the point $\mathbf{x}_k$ and proceeds in the residual direction. It therefore lies at a point $\mathbf{x}_k + \alpha_k\mathbf{r}_k$, where α_k is the scalar multiplier that minimizes the quadratic form $F(\mathbf{x}_k + \alpha_k\mathbf{r}_k)$. Now

$$\begin{aligned} F(\mathbf{x}_{k+1}) &= F(\mathbf{x}_k + \alpha_k\mathbf{r}_k) \\ &= F(\mathbf{x}_k) + \alpha_k\mathbf{r}_k^T\mathbf{A}\mathbf{x}_k + \alpha_k\mathbf{x}_k^T\mathbf{A}\mathbf{r}_k + \alpha_k^2\mathbf{r}_k^T\mathbf{A}\mathbf{r}_k + 2\alpha_k\mathbf{r}_k^T\mathbf{b} \end{aligned} \tag{4.8}$$

so that the usual requirement for minimizing F,

$$\frac{d}{d\alpha_k}F(\mathbf{x}_k + \alpha_k\mathbf{r}_k) = 0, \tag{4.9}$$

becomes in this particular case

$$\mathbf{r}_k^T\mathbf{A}\mathbf{x}_k + \alpha_k\mathbf{r}_k^T\mathbf{A}\mathbf{r}_k - \mathbf{r}_k^T\mathbf{b} = 0 \tag{4.10}$$

and yields the result

$$\alpha_k = \frac{\mathbf{r}_k^T\mathbf{r}_k}{\mathbf{r}_k^T\mathbf{A}\mathbf{r}_k}. \tag{4.11}$$

Finding α_k does not require much computation; it takes one matrix-vector multiplication to compute the product $\mathbf{A}\mathbf{r}_k$, then two vector inner products.

The great economy of memory characteristic of iterative techniques is well illustrated by the steepest descent method, which only requires storage for **A** and two vectors $\mathbf{x}_k$ and $\mathbf{r}_k$. However, the matrix **A** is never modified in the course of iteration — it is only ever used in computing matrix vector products — so it can be kept in very compact storage; further, it can be stored on a secondary medium such as tape or disk, with only one row retrieved into working memory at a time. Many other iterative methods can also be made to work well with the coefficient matrix on disk and only a few vectors in random-access storage. The primary storage required is then proportional to the number of variables, and very large problems can be handled in modest amounts of fast-access memory. Consequently, such methods are well suited to dealing with enormous matrices. Iterative methods were popular in the early days of computing, when memory was at a premium, for just this reason. They have become popular again in recent years as three-dimensional solutions have grown practical despite their gluttonous appetite for memory.

4.2   *The conjugate gradient method*

The conjugate gradient or *CG* method is an improved version of the steepest descent method. The main weakness of the steepest descent method is that the successive corrections $\delta\mathbf{x}_k$, each taken in the direction of the current residual $\mathbf{r}_k$, are not mutually orthogonal. Error eliminated at one step is therefore likely to be reintroduced at another. The conjugate gradient method avoids this difficulty; it computes successive solution estimates $\mathbf{x}_k$ by

$$\mathbf{x}_{k+1} = \mathbf{x}_k + \alpha_k \mathbf{p}_k. \tag{4.12}$$

where the direction vectors $\mathbf{p}_k$ are selected so as to make successive residuals orthogonal to each other,

$$\mathbf{r}_k^T \mathbf{r}_{k+1} = 0. \tag{4.13}$$

As in the steepest descent method, the parameter α_k is chosen to make the greatest possible improvement at each step. Every trial solution $\mathbf{x}_{k+1}$ is constructed from its predecessor $\mathbf{x}_k$ by adding a correction step, so that

$$\begin{aligned} F(\mathbf{x}_{k+1}) &= F(\mathbf{x}_k + \alpha_k \mathbf{p}_k) \\ &= F(\mathbf{x}_k) + \alpha_k \mathbf{p}_k^T \mathbf{A}\mathbf{x}_k + \alpha_k \mathbf{x}_k^T \mathbf{A}\mathbf{p}_k + \alpha_k^2 \mathbf{p}_k^T \mathbf{A}\mathbf{p}_k + 2\alpha_k \mathbf{p}_k^T \mathbf{b}. \end{aligned} \tag{4.14}$$

Minimization of F then requires that

$$\mathbf{p}_k^T \mathbf{A}\mathbf{x}_k + \alpha_k \mathbf{p}_k^T \mathbf{A}\mathbf{p}_k - \mathbf{p}_k^T \mathbf{b} = 0. \tag{4.15}$$

Solving for α_k,

$$\alpha_k = \frac{\mathbf{p}_k^T \mathbf{r}_k}{\mathbf{p}_k^T \mathbf{A}\mathbf{p}_k}. \tag{4.16}$$

This prescription is valid for minimization along any direction $\mathbf{p}_k$, no particular choice of $\mathbf{p}_k$ having yet been made and no orthogonality requirement having been imposed. It is interesting to note that successive vectors $\mathbf{r}_k$ do not need to be computed explicitly at each step but can be found by a reduction relation. Multiplying Eq. (4.12) by $\mathbf{A}$ and subtracting $\mathbf{b}$ on both sides,

$$\mathbf{r}_{k+1} = \mathbf{r}_k + \alpha_k \mathbf{A} \mathbf{p}_k. \tag{4.17}$$

Finding residuals by this recursive approach is computationally cheap because the vector $\mathbf{A} \mathbf{p}_k$ needs to be computed anyway.

If the orthogonality requirement of Eq. (4.13) holds, then premultiplying Eq. (4.17) by $\mathbf{r}_k$ on both sides leads directly to an alternative expression for α_k,

$$\alpha_k = \frac{\mathbf{r}_k^{\mathrm{T}} \mathbf{r}_k}{\mathbf{r}_k^{\mathrm{T}} \mathbf{A} \mathbf{p}_k}. \tag{4.18}$$

This value of the multiplier α_k ensures that successive residuals $\mathbf{r}_k$ are mutually orthogonal, while α_k as given by Eq. (4.16) ensures that F is minimized at each step. Clearly, the direction vectors $\mathbf{p}_k$ must be chosen in such a way that the two prescriptions yield the same value of α_k. Relation (4.17) implies a useful orthogonality property that may not be immediately apparent. Rewriting and collecting terms, Eq. (4.15) becomes

$$\mathbf{p}_k^{\mathrm{T}} (\mathbf{A} \mathbf{x}_k - \mathbf{b}) + \alpha_k \mathbf{p}_k^{\mathrm{T}} \mathbf{A} \mathbf{p}_k = 0, \tag{4.19}$$

or, by Eq. (4.17),

$$\mathbf{p}_k^{\mathrm{T}} \mathbf{r}_{k+1} = 0. \tag{4.20}$$

This results suggests a simple way of creating a search direction $\mathbf{p}_{k+1}$. Begin by taking the residual $\mathbf{r}_{k+1}$, as would be done in the steepest descent method, and add a component orthogonal to $\mathbf{r}_{k+1}$ just large enough to satisfy Eq. (4.12). Then

$$\mathbf{p}_{k+1} = \mathbf{r}_{k+1} + \beta_k \mathbf{p}_k, \tag{4.21}$$

where the scaling constant β_k remains to be determined. Premultiplying both sides by $\mathbf{p}_k^{\mathrm{T}} \mathbf{A}$, Eq. (4.21) becomes

$$\mathbf{p}_k^{\mathrm{T}} \mathbf{A} \mathbf{p}_{k+1} = \mathbf{p}_k^{\mathrm{T}} \mathbf{A} \mathbf{r}_{k+1} + \beta_k \mathbf{p}_k^{\mathrm{T}} \mathbf{A} \mathbf{p}_k, \tag{4.22}$$

and the multiplier β_k is given by

$$\beta_k = \frac{\mathbf{p}_k^{\mathrm{T}} \mathbf{A} \mathbf{r}_{k+1} - \mathbf{p}_k^{\mathrm{T}} \mathbf{A} \mathbf{p}_{k+1}}{\mathbf{p}_k^{\mathrm{T}} \mathbf{A} \mathbf{p}_k}. \tag{4.23}$$

Enforcing orthogonality of residuals as in Eq. (4.12) is equivalent to requiring the values of α_k as given by Eqs. (4.16) and (4.18) to be equal. To determine the conditions of equality, premultiply Eq. (4.21) by $\mathbf{r}_{k+1}^{\mathrm{T}}$:

$$\mathbf{r}_{k+1}^{T}\mathbf{p}_{k+1} = \mathbf{r}_{k+1}^{T}\mathbf{r}_{k+1} + \beta_{k}\mathbf{r}_{k+1}^{T}\mathbf{p}_{k}. \tag{4.24}$$

By Eq. (4.20), the last term on the right must vanish, leaving

$$\mathbf{r}_{k+1}^{T}\mathbf{r}_{k+1} = \mathbf{r}_{k+1}^{T}\mathbf{p}_{k+1} \tag{4.25}$$

which implies

$$\mathbf{r}_{k}^{T}\mathbf{r}_{k} = \mathbf{r}_{k}^{T}\mathbf{p}_{k}. \tag{4.26}$$

Using this substitution, Eq. (4.18) becomes

$$\alpha_{k} = \frac{\mathbf{p}_{k}^{T}\mathbf{r}_{k}}{\mathbf{r}_{k}^{T}\mathbf{A}\mathbf{p}_{k}}. \tag{4.27}$$

Substitute the expression (4.21) for the residual $\mathbf{r}_{k}$ in the denominator:

$$\alpha_{k} = \frac{\mathbf{p}_{k}^{T}\mathbf{r}_{k}}{\mathbf{p}_{k}^{T}\mathbf{A}\mathbf{p}_{k} - \beta_{k-1}\mathbf{p}_{k-1}^{T}\mathbf{A}\mathbf{p}_{k}}. \tag{4.28}$$

If $\mathbf{p}_{k}$ and $\mathbf{p}_{k+1}$ are made to be conjugate to each other with respect to the matrix $\mathbf{A}$ — in other words, if

$$\mathbf{p}_{k+1}^{T}\mathbf{A}\mathbf{p}_{k} = 0, \tag{4.29}$$

then the second term in the denominator of Eq. (4.17) vanishes and the two values of α_{k} are identical. This requirement is met by setting

$$\beta_{k} = -\frac{\mathbf{p}_{k}^{T}\mathbf{A}\mathbf{r}_{k+1}}{\mathbf{p}_{k}^{T}\mathbf{A}\mathbf{p}_{k}}. \tag{4.30}$$

Equation (4.29) is what gives the conjugate gradient method its name: the method seeks minima along *gradient* lines chosen to have *conjugate* directions.

To summarize, the conjugate gradient method chooses an initial estimate $\mathbf{x}_{0}$ more or less arbitrarily, computes the associated residual $\mathbf{r}_{0}$ and chooses the initial search direction $\mathbf{p}_{0}$ to coincide with $\mathbf{r}_{0}$:

$$\mathbf{r}_{0} = \mathbf{A}\mathbf{x}_{0} - \mathbf{b}, \tag{4.31}$$
$$\mathbf{p}_{0} = \mathbf{r}_{0}. \tag{4.32}$$

It then computes a succession of residuals and search directions by the recursive steps

$$\alpha_k = \frac{\mathbf{p}_k^{\mathrm{T}} \mathbf{r}_k}{\mathbf{r}_k^{\mathrm{T}} \mathbf{A} \mathbf{p}_k},$$

(4.27)

$$\mathbf{x}_{k+1} = \mathbf{x}_k + \alpha_k \mathbf{p}_k,$$

(4.12)

$$\mathbf{r}_{k+1} = \mathbf{r}_k + \alpha_k \mathbf{A} \mathbf{p}_k,$$

(4.17)

$$\beta_k = -\frac{\mathbf{p}_k^{\mathrm{T}} \mathbf{A} \mathbf{r}_{k+1}}{\mathbf{p}_k^{\mathrm{T}} \mathbf{A} \mathbf{p}_k},$$

(4.30)

$$\mathbf{p}_{k+1} = \mathbf{r}_{k+1} + \beta_k \mathbf{p}_k.$$

(4.21)

In this process, error is removed in one independent search direction at a time and not reintroduced subsequently, except perhaps through round-off error accumulation. After N steps there is no direction left in which correction is still required,

$$\mathbf{x}_N = \mathbf{x}.$$

(4.33)

The resulting solution is therefore the exact and only solution $\mathbf{x}$ of the original matrix equation.

In practical cases, adequate solution accuracy is often reached before all N steps have been taken. The conjugate gradient method is therefore termed a *semi-iterative* method; it requires monitoring for convergence just like a true iterative method, but it is certain to complete its task fully in a predeterminable, finite number of steps just like an explicit (Gaussian) method.

4.3   *Preconditioned conjugate gradients*

The conjugate gradient method converges with a speed strongly dependent on the eigenvalue spectrum of the matrix $\mathbf{A}$. If $\mathbf{A}$ has only a few distinct eigenvalues, or has many closely clustered eigenvalues, the conjugate gradient algorithm converges quickly, reaching a very good estimate of the solution $\mathbf{x}$ in far fewer than N steps. Unfortunately, matrices arising from field problems usually have very wide eigenvalue spectra; they have to, if the matrix equations are to be good models of electromagnetic continuum problems with widely distributed resonant frequencies or characteristic time constants! This observation suggests modifying, or *preconditioning*, the matrix $\mathbf{A}$ to hasten convergence. The central idea is simple enough. A positive definite and symmetric precon-ditioning matrix $\mathbf{B}$ is chosen and the conjugate gradient process is carried out using a matrix $\mathbf{B}\mathbf{A}\mathbf{B}^{\mathrm{T}}$ instead of $\mathbf{A}$. After convergence has been achieved, the solution of the original problem is recovered from the modified one. If $\mathbf{B}$ is chosen so that $\mathbf{B}\mathbf{A}\mathbf{B}^{\mathrm{T}}$ has closely clustered eigen-values, the modified process will converge much more quickly than the original one. The modified procedure takes very little extra work; Eq. (4.21) is replaced by

$$\mathbf{p}_{k+1} = \mathbf{Br}_{k+1} + \beta_k \mathbf{p}_k, \tag{4.34}$$

and the scaling factor β_k is now calculated by

$$\beta_k = -\frac{\mathbf{p}_k^{\mathrm{T}} \mathbf{ABr}_{k+1}}{\mathbf{p}_k^{\mathrm{T}} \mathbf{Ap}_k}. \tag{4.35}$$

Note that preconditioning leaves all the orthogonality properties untouched, so convergence in N iterations is still guaranteed. From the point of view of the matrix equation the generalized method is equivalent to rewriting $\mathbf{Ax} = \mathbf{b}$ as

$$(\mathbf{BAB}^{\mathrm{T}})(\mathbf{B}^{-\mathrm{T}}\mathbf{x}) = \mathbf{Bb}. \tag{4.36}$$

The idea now is to solve for the vector $\mathbf{y} = \mathbf{B}^{\mathrm{T}}\mathbf{x}$ using the modified right-hand side and coefficient matrix of Eq. (4.36), then to recover $\mathbf{x}$ from $\mathbf{y}$.

To illustrate the potential effectiveness of preconditioning, suppose $\mathbf{B}$ is chosen to be the inverse of the Choleski factor of $\mathbf{A}$,

$$\mathbf{A} = \mathbf{LL}^{\mathrm{T}}, \tag{4.37}$$
$$\mathbf{B} = \mathbf{L}^{-\mathrm{T}}. \tag{4.38}$$

The coefficient matrix $\mathbf{BAB}^{\mathrm{T}}$ now becomes the unit matrix and the conjugate gradient iteration converges in just one step. Of course, this ideal selection for $\mathbf{B}$ is impractical; if $\mathbf{L}$ were known, there would be no point in iterating!

Where the original matrix $\mathbf{A}$ is sparse, the *incomplete* Choleski decomposition is an inexpensive way of constructing a sparse, lower triangular, preconditioning matrix $\mathbf{C}$ with low memory requirements. It is based on a bold stroke of imagination: to keep computing costs down, many elements of the Choleski factors are forced to zero so they need neither storage nor subsequent computing time. The resulting decomposition is at best approximate; but that does not matter much, since poor accuracy in decomposition only means that a larger number of conjugate gradient steps will be required. The incomplete Choleski decomposition begins by preassigning a sparsity pattern to the lower triangular matrix $\mathbf{C}$. Commonly, $\mathbf{C}$ is made to have the same topology as $\mathbf{A}$; whatever specialized techniques may be adopted for storing large and sparse $\mathbf{A}$ are then equally applicable to $\mathbf{C}$. (In particular, any special indexing arrays valid for $\mathbf{A}$ are also valid for $\mathbf{C}$.) In other words, a lower triangular matrix $\mathbf{C}$ is sought such that

$$C_{ij} = 0 \quad \text{if } A_{ij} = 0, \text{ or } j > i. \tag{4.39}$$

The decomposition itself is carried out in exactly the standard Choleski manner, with the very important exception that the sparsity pattern is maintained by insisting on the preassigned zero values wherever they are used in the computation. Since the zeros are artificially introduced, the decomposition is no longer exact,

$$\mathbf{A} = \mathbf{C}\mathbf{C}^{\mathsf{T}} + \mathbf{E}, \tag{4.40}$$

where $\mathbf{E}$ is an error matrix. The preconditioning matrix is then the inverse of $\mathbf{C}$,

$$\mathbf{B} = \mathbf{C}^{-1}. \tag{4.41}$$

The effect of preconditioning may be appreciated by examining the matrix product $\mathbf{B}\mathbf{A}\mathbf{B}^{\mathsf{T}}$:

$$\mathbf{B}\mathbf{A}\mathbf{B}^{\mathsf{T}} = \mathbf{I} + \mathbf{C}^{-1}\mathbf{E}\mathbf{C}^{-\mathsf{T}}. \tag{4.42}$$

where $\mathbf{I}$ is the unit diagonal matrix. If $\mathbf{E}$ is small the eigenvalue structure of the preconditioned matrix will be close to that of the identity matrix, or at least closer to it than $\mathbf{A}$ itself.

The incomplete Choleski decomposition may fail because square roots of negative numbers appear as diagonal elements. This problem can be handled in a systematic fashion. Let $\mathbf{A}(c)$ be defined as

$$\mathbf{A}(c) = \mathbf{A} + c\mathbf{D}, \tag{4.43}$$

where $\mathbf{D}$ contains all the diagonal elements of $\mathbf{A}$,

$$\mathbf{D} = \text{diag}\,[A_{ii}] \tag{4.44}$$

and c is some nonnegative real number. Clearly, $\mathbf{A}(0) = \mathbf{A}$. For larger values of c, the eigenvalues of matrices in this family occur in clusters much like those of $\mathbf{A}$, but their values are shifted farther to the right. Since the amount of eigenvalue shift increases with c, the eigenvalues of $\mathbf{A}(c)$ can always be shifted far enough to the right to ensure that no negative diagonal elements are encountered in the incomplete decomposition. This observation suggests that decomposition of $\mathbf{A}(0)$ should be attempted first; should it fail, the attempt is repeated with a sequence of growing values of c until success is achieved.

When the eigenvalue spectrum of $\mathbf{A}$ is modified by shifting, large shift parameters c prevent decomposition failure while small values minimize the amount of deviation from the original $\mathbf{A}$ and therefore produce the best preconditioning matrices. The shift parameter c should therefore be chosen as small as possible while still guaranteeing successful decomposition. This value cannot be determined in advance, but it is known to be bounded by zero from below and by a row norm of $\mathbf{A}$ from above:

$$0 \leq c < \max_i \sum_j |A_{ij}|. \tag{4.45}$$

Although trivial, the lower bound is relatively tight; the upper bound is quite loose. A useful strategy is to try decomposition at the lower bound first. If it is unsuccessful, a geometric mean (golden section) search is made between the upper and lower bounds to determine the smallest acceptable value of c. Geometric mean searching is impossible, however, if one of the bounds is zero. A workable approach is therefore to set the lower bound at the *machine epsilon* ϵ_M of the computer being used, i.e., at the number equal to the smallest increment the machine is able to distinguish from unity:

$$\epsilon_M = \inf_\epsilon \epsilon, \forall \epsilon \in |1 + \epsilon| - 1 \neq 0. \tag{4.46}$$

A good approximation to the machine epsilon can be found by examining a sequence of trial values, beginning with $\epsilon = 1$ and successively dividing by two. Values of about 10^{-7} should be expected with conventional 32-bit arithmetic. While it may seem wasteful to restart decomposition afresh after each failure, the incomplete Choleski decomposition is only a small part of the total computing effort required so some small inefficiency in it can be tolerated readily.

When the matrix **A** arises from finite element discretization of physical boundary-value problems, incomplete Choleski decomposition usually produces modified matrices $\mathbf{BAB}^\mathrm{T}$ having all but $\sqrt{N}$ of their eigenvalues in a small neighbourhood of unity, typically between 0.9 and 1.1. This curious phenomenon has been observed in a large number of cases and has been documented in the literature by numerous researchers. It occurs consistently enough to hint at some underlying fundamental reason, though none is as yet known. Whatever its cause, it leads to convergence of the conjugate gradient method in about $\sqrt{N}$ steps, for all the remaining eigenvalues lie very close to each other so all the residual components corresponding to them are eliminated in only a few conjugate gradient steps.

5.   Further reading

There are many good textbooks on the solution of simultaneous equations, to which the reader may refer with profit. The mathematical substance of Gaussian elimination, as well as many other useful methods, is well started by Watkins (1991). Both their mathematical properties and the algorithmic details are given in by Golub and Van Loan (1989), whose treatment may at times be more demanding than Watkins', but whose book is also filled with good practical advice. Carey and Oden (1984) present both sound mathematics and advice born of experience, and blend the matrix mathematics with a sound treatment of finite element mesh generation.

Finite element matrices have many peculiar properties so that these quite general works may not suffice. Thus, finite element textbooks in the structural and mechanical engineering areas may be consulted; although the physical problems there quite often differ profoundly from those of electromagnetics, the structure of the finite element equations is frequently the same. The book by Bathe and Wilson (1980) is specifically concerned with numerical methods applicable to finite elements, and is therefore to be recommended. Irons and Ahmad (1980) is less mathematically oriented but evaluates techniques and suggests ways of avoiding pitfalls based on practical experience. Rao's (1989) treatment is expository and may be easier for a first reading than the more mathematical works. Jin (1993) describes applications to problems of electrical engineering.

The various renumbering methods now available are less well documented in textbook literature, so that recourse must be had to the periodical literature. The first, and still popular, method is that of Cuthill and McKee (1969), which was superseded in due course by the similar, but more effective, algorithm of Gibbs, Poole and Stockmeyer (1976). A very fast method, which has not yet been extensively implemented, is that described by George and Liu (1980).

Conjugate gradient methods have been known since the middle 1950s but have gained prominence only in the past decade. Beckman (1960) gives an early, but short and very readable, account of the conjugate gradient method itself. Manteuffel (1980) treats the incomplete Choleski preconditioning method in considerable mathematical detail, with sufficient attention to its programming aspects.

As matrix sizes grow, parallel processing is assuming increasing importance in matrix handling. Although this is a fast developing area, the book by Modi (1988) describes the fundamentals well and will probably remain worth reading for many years.

6.   Bibliography

Bathe, K.-J. and Wilson, E. (1980), *Numerical Methods in Finite Element Analysis*. New York: McGraw-Hill. xv + 529 pp.

Beckman, F.S. (1960), 'The solution of linear equations by the conjugate gradient method', in *Mathematical Methods for Digital Computers*. Vol. **1**, pp. 62–72. New York: John Wiley.

Carey, G.F. and Oden, J.T. (1984), *Finite Elements: Computational Aspects.*
(Vol. 3 of the Texas Finite Element Series.) Englewood Cliffs: Prentice-Hall.
x + 350 pp.

Cuthill, E. and McKee, J. (1969), 'Reducing the bandwidth of sparse
symmetric matrices', *Proceedings of the 24th National Conference of the
Association for Computing Machinery* (ACM Publication P-69), pp. 157–72.
New York: Association for Computing Machinery.

George, A. and Liu, J.W.H. (1980), 'A minimal storage implementation of the
minimum degree algorithm', *Society for Industrial and Applied Mathematics
Journal on Numerical Analysis*, **17**, pp. 282–99.

Gibbs, N.E., Poole, W.G., Jr. and Stockmeyer, P.K. (1976), 'An algorithm for
reducing bandwidth and profile of a sparse matrix', *Society for Industrial
and Applied Mathematics Journal on Numerical Analysis*, **13**, pp. 236–50.

Golub, G.H. and Van Loan, C.F. (1989), *Matrix computations.* 2nd ed.
Baltimore: Johns Hopkins University Press. xix + 642 pp.

Irons, B. and Ahmad, S. (1980), *Techniques of Finite Elements.* Chichester: Ellis
Horwood. 529 pp.

Jin, J.-M. (1993), *The Finite Element Method in Electromagnetics.* New York:
John Wiley. xix + 442 pp.

Manteuffel, T.A. (1980), 'An incomplete factorization technique for positive
definite linear systems'. *Mathematics of Computation*, **34**, pp. 473–97.

Modi, J.J. (1988), *Parallel algorithms and matrix computation.* New York:
Oxford University Press. xi + 260 pp.

Rao, S.S. (1989), *The Finite Element Method in Engineering.* 2nd edn. Oxford:
Pergamon. xxvi + 643 pp.

Watkins, D.S. (1991), *Fundamentals of matrix computations.* New York: John
Wiley. xiii + 449 pp.

Appendix 1

Calculations on simplex elements

The matrix element calculations given in Chapter 4, and the corresponding three-dimensional work of Chapter 9, rely on two mathematical facts: an integration formula for polynomials such as occur in Eq. (3.11) of Chapter 4, and the trigonometric identity of Eq. (3.21) of the same chapter. Since the reader may be interested in how these key facts were arrived at, a short derivation is given here.

1. Integration in homogeneous coordinates

In computing element matrices for triangular elements, it is necessary to evaluate the definite integral $I(i,j,k)$,

$$I(i,j,k) = \int \zeta_1^i \zeta_2^j \zeta_3^k \, \frac{d\Omega}{A}. \tag{1.1}$$

This task is best accomplished by noting that in any area integration, the area element may be written in whatever coordinates may be convenient, provided the Jacobian of the coordinate transformation is included,

$$d\zeta_1 d\zeta_2 = \frac{\partial(\zeta_1, \zeta_2)}{\partial(x, y)} \, dx \, dy. \tag{1.2}$$

Since the transformation between Cartesian and triangle coordinates is given by

$$\zeta_1 = (a_1 + b_1 x + c_1 y)/(2A), \tag{1.3}$$
$$\zeta_2 = (a_2 + b_2 x + c_2 y)/(2A), \tag{1.4}$$

where A is the triangle area, the Jacobian is readily evaluated:

$$\frac{\partial(\zeta_1, \zeta_2)}{\partial(x, y)} = \left(\frac{1}{2A}\right)^2 \begin{vmatrix} b_1 & c_1 \\ b_2 & c_2 \end{vmatrix} = \frac{1}{2A}. \tag{1.5}$$

The integral $I(i,j,k)$ may therefore be written in the form of an iterated integral

$$I(i,j,k) = 2\int_0^1 \int_0^{1-\zeta_1} \zeta_1^i \zeta_2^j (1-\zeta_1-\zeta_2)^k \, d\zeta_2 \, d\zeta_1. \tag{1.6}$$

Integrating by parts, one obtains

$$I(i,j,k) = \frac{2k}{j+1}\int_0^1 \int_0^{1-\zeta_1} \zeta_1^i \zeta_2^{j+1}(1-\zeta_1-\zeta_2)^{k-1} \, d\zeta_2 \, d\zeta_1 \tag{1.7}$$

and hence

$$I(i,j,k) = \frac{j!k!}{(j+k)!} I(i,j+1,k-1). \tag{1.8}$$

Applying Eq. (1.8) repeatedly, there then results

$$I(i,j,k) = \frac{j!k!}{(j+k)!} I(i,j+k,0) \tag{1.9}$$

and

$$I(i,j,k) = \frac{i!j!k!}{(i+j+k)!} I(0,i+j+k,0). \tag{1.10}$$

But only one of the triangle coordinates actually appears in the integral $I(0,i+j+k,0)$ on the right of Eq. (1.10). Evaluation is therefore straightforward, and yields

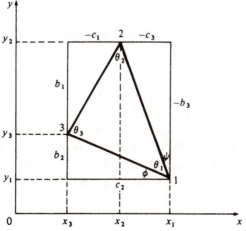

A1.1 Geometry of an arbitrary triangle, for proof of the cotangent identity of Chapter 4.

$$I(i,j,k) = \frac{i!j!k!2!}{(i+j+k+2)!} \tag{1.11}$$

as given in Chapter 4.

A similar method is applicable to tetrahedra, the main difference being simply that integration is carried out over four homogeneous coordinates rather than three.

2.   The cotangent identity

Equation (3.21) of Chapter 4 is readily derived from basic trigonometric considerations. Let a triangle be given, and let it be circumscribed by a rectangle whose sides are aligned with the Cartesian axes, as in Fig. A1.1. The three vertices of the triangle subdivide the rectangle sides into segments given by the numbers

$$b_i = y_i - y_{i-1}, \tag{2.1}$$
$$c_i = x_{i-1} - x. \tag{2.2}$$

These are the same as Eqs. (3.17) and (3.18) of Chapter 4. From Fig. A1.1, it is evident that

$$\cot\theta_1 = \cot\left(\frac{\pi}{2} - \phi - \psi\right), \tag{2.3}$$

which may be written

$$\cot\theta_1 = \frac{\tan\phi + \tan\psi}{1 - \tan\phi\tan\psi} \tag{2.4}$$

or, substituting the ratios of lengths for the tangents of included angles,

$$\cot\theta_1 = -\frac{b_2b_3 + c_2c_3}{b_2c_3 - c_2b_3}. \tag{2.5}$$

The denominator may now be recognized as being twice the triangle area,

$$\cot\theta_1 = -\frac{b_2b_3 + c_2c_3}{2A}, \tag{2.6}$$

so that the first part of Eq. (3.21) of Chapter 4, with $k = 1$, has been obtained. Similar results for $k = 2$ and $k = 3$ result on interchange of vertex numbering, or directly from the same geometric figure.

The second part of Eq. (3.21) of Chapter 4 is obtainable by direct algebraic means by adding two distinct included angle cotangents, expressed as in Eq. (2.4) above, and collecting terms.

Appendix 2

Integration by parts, Green's theorems and Green's functions

1.   Scalar one-dimensional formula

Ordinary integration by parts concerns a scalar variable in one dimension, expressed as follows:

$$\int_{x_1}^{x_2} u\frac{dv}{dx}\, dx = uv\Big|_{x_1}^{x_2} - \int_{x_1}^{x_2} \frac{du}{dx} u\, dx. \tag{1.1}$$

Some generalizations of this operation have been used in arguments given in the main text. They derive from the divergence theorem of vector calculus, here assumed as a starting point.

2.   Vector integration by parts

The divergence theorem applied to the vector $u\mathbf{F}$, u and $\mathbf{F}$ being sufficiently well-behaved functions of position, may be written as

$$\int_{\Omega} \nabla \cdot (u\mathbf{F})\, d\Omega = \oint_{S} u\mathbf{F} \cdot \mathbf{1}_n\, dS. \tag{2.1}$$

Here S is a closed surface with outward normal $\mathbf{1}_n$ and Ω is the volume contained within S. Using the divergence of products rule

$$\nabla \cdot (u\mathbf{F}) = (\nabla u) \cdot \mathbf{F} + u\nabla \cdot \mathbf{F} \tag{2.2}$$

in Eq. (2.1) and rearranging gives

$$\int_{\Omega} u\nabla \cdot \mathbf{F}\, d\Omega = \oint_{S} u\mathbf{F} \cdot \mathbf{1}_n\, dS - \int_{\Omega} (\nabla u) \cdot \mathbf{F}\, d\Omega. \tag{2.3}$$

Equation (2.3) bears more than a superficial resemblance to Eq. (1.1) and the operation which it represents may be described as *vector integration by parts*. Note that at the expense of introducing a surface integral, the differential operator ∇ is transferred from the vector $\mathbf{F}$ to the scalar u.

Another useful vector integration by parts formula, couched entirely in terms of vector variables, arises from application of the divergence of cross-products rule, here stated as

$$\nabla \cdot (\mathbf{P} \times \mathbf{F}) = (\nabla \times \mathbf{P}) \cdot \mathbf{F} - \mathbf{P} \cdot (\nabla \times \mathbf{F}), \tag{2.4}$$

where $\mathbf{P}$ and $\mathbf{F}$ are vector functions of position. Integrating Eq. (2.4) over the volume Ω and applying the divergence theorem to the left-hand side gives, after rearranging terms, the formula

$$\int_\Omega \mathbf{P} \cdot (\nabla \times \mathbf{F}) \, d\Omega = \int_\Omega (\nabla \times \mathbf{P}) \cdot \mathbf{F} \, d\Omega - \oint_S (\mathbf{P} \times \mathbf{F}) \cdot \mathbf{1}_n \, dS.$$

$$\tag{2.5}$$

Again, the operator ∇ is transferred from one variable to the other at the price of a surface integral.

3.   Green's theorems

If the substitution $\mathbf{F} = \nabla u$ is made in Eq. (2.3) there results one of the well-known Green's formulae,

$$\int_\Omega u \nabla^2 v \, d\Omega = \oint_S u(\nabla v) \cdot \mathbf{1}_n \, dS - \int_\Omega (\nabla u) \cdot (\nabla v) \, d\Omega. \tag{3.1}$$

Equation (3.1) is often used in its two-dimensional form

$$\int_S u \nabla^2 v \, dS = \oint_C u(\nabla v) \cdot \mathbf{1}_n \, dC - \int_S (\nabla u) \cdot (\nabla v) \, dS, \tag{3.2}$$

where C is a plane curve and S the area within it. It should be noted that $(\nabla v) \cdot \mathbf{1}_n$ is synonymous with the normal gradient $\partial v / \partial n$ on the bounding surface or curve. Interchanging u and v in Eq. (3.1) and then subtracting the result from the original equation gives

$$\int_S (u \nabla^2 v - v \nabla^2 u) \, dS = \oint_C \{u(\nabla v) - v(\nabla u)\} \cdot \mathbf{1}_n \, dC. \tag{3.3}$$

This is the formula which is generally described as 'Green's theorem'.

Making the substitution $\mathbf{F} = \nabla \times \mathbf{Q}$ into Eq. (2.5) gives the vector equivalent to Eq. (3.1)

$$\int_\Omega \mathbf{P} \cdot (\nabla \times \nabla \times \mathbf{Q}) \, d\Omega$$

$$= \int_\Omega (\nabla \times \mathbf{P}) \cdot (\nabla \times \mathbf{Q}) \, d\Omega - \oint_S (\mathbf{P} \times \nabla \times \mathbf{Q}) \cdot \mathbf{1}_n \, dS. \tag{3.4}$$

Interchanging $\mathbf{P}$ and $\mathbf{Q}$ in Eq. (3.4) then subtracting the result from the original now gives a vector form of Green's theorem,

$$\int_\Omega \{\mathbf{P}\cdot(\nabla \times \nabla \times \mathbf{Q}) - \mathbf{Q}\cdot(\nabla \times \nabla \times \mathbf{P})\}\,d\Omega$$

$$= -\oint_S \{(\mathbf{P}\times\nabla\times\mathbf{Q}) - (\mathbf{Q}\times\nabla\times\mathbf{P})\}\cdot\mathbf{1}_n\,dS. \tag{3.5}$$

The perhaps unexpected minus sign in front of the right-hand term of Eq. (3.5) falls into place on recalling the vector identity defining the Laplacian operator when it acts upon a vector,

$$\nabla^2\mathbf{P} = \nabla(\nabla\cdot\mathbf{P}) - \nabla\times\nabla\times\mathbf{P}. \tag{3.6}$$

4. Green's functions for the free-space wave equations

Consider the complex phasor inhomogeneous scalar wave equation

$$\nabla^2 V + k_0^2 V = -\frac{\rho(\mathbf{r})}{\epsilon_0}, \tag{4.1}$$

where $k_0^2 = \omega^2\mu_0\epsilon_0$, which describes the scalar electric potential in free space due to an oscillating distribution of charge $\rho(\mathbf{r})\exp(j\omega t)$ contained within a volume Ω. Using a three-dimensional delta function $\delta(\mathbf{r} - \mathbf{r}')$ gives, by definition,

$$\rho(\mathbf{r}) = \int_\Omega \rho(\mathbf{r}')\delta(\mathbf{r} - \mathbf{r}')\,d\Omega. \tag{4.2}$$

Suppose a solution of Eq. (4.1) is sought in the form

$$V(\mathbf{r}) = \frac{1}{\epsilon_0}\int_\Omega G(\mathbf{r},\mathbf{r}')\rho(\mathbf{r}')\,d\Omega'. \tag{4.3}$$

Then it is clear that the *scalar Green's function* $G(\mathbf{r},\mathbf{r}')$ has to satisfy the scalar Helmholtz equation

$$\nabla^2 G + k_0^2 G = -\delta(\mathbf{r} - \mathbf{r}'). \tag{4.4}$$

Assuming $\mathbf{r}$ to be the independent space variable and $\mathbf{r}'$ to be some fixed point in space it is a straightforward matter to check that

$$G(\mathbf{r},\mathbf{r}') = \frac{\exp(-jk_0|\mathbf{r} - \mathbf{r}'|)}{4\pi|\mathbf{r} - \mathbf{r}'|} \tag{4.5}$$

is a solution of the homogeneous version of Eq. (4.4) obtained if $\mathbf{r} - \mathbf{r}' \neq 0$. A loose argument then goes on to say that as $\mathbf{r} \to \mathbf{r}'$, the wave term $k_0^2 V$ in Eq. (4.1) may be neglected and one is looking for the spherically symmetric static potential solution corresponding to unit charge at the point $\mathbf{r}'$. This evidently is what you have if $k_0 = 0$,

$\rho = 1$ in Eqs. (4.1) and (4.3). Using similar arguments in a two-dimensional analysis it turns out that

$$G(\rho, \rho') = -\frac{j}{4} H_0^{(2)}(k_0|\rho - \rho'|), \tag{4.6}$$

where ρ is the cylindrical radius vector and $H_0^{(2)}$ is the zero-order Hankel function of the second kind.

With these preliminaries a more general *dyadic Green's function* $\mathbb{G}(\mathbf{r}, \mathbf{r}')$ may be constructed, to be used in solving vector field problems in a similar fashion to the simple kernel $G(\mathbf{r}, \mathbf{r}')$ used in scalar situations. Briefly, a 'dyadic' may be described as a 3×3 matrix operator $\mathbb{D}$ such that acting upon a vector $\mathbf{A}$, the operation $\mathbb{D} \cdot \mathbf{A}$ produces another vector, oriented in a different direction from the first and of different magnitude but in a fashion independent of the coordinate system chosen. Consider the curl curl Maxwell equation describing the vector electric field due to a radiating source current distribution in otherwise empty space,

$$\nabla \times \nabla \times \mathbf{E}(\mathbf{r}) - k_0^2 \mathbf{E}(\mathbf{r}) = -j\omega\mu_0 \mathbf{J}^S(\mathbf{r}). \tag{4.7}$$

In general $\mathbf{E}$ and $\mathbf{J}^S$ will not lie in the same direction, thus the relation between the two is expected to be a dyadic integral expression

$$\mathbf{E}(\mathbf{r}) = -j\omega\mu_0 \int_\Omega \mathbb{G}(\mathbf{r}, \mathbf{r}') \cdot \mathbf{J}^S(\mathbf{r}') \, d\Omega', \tag{4.8}$$

where Ω encompasses all of the current sources. Now using the three-dimensional delta function $\delta(\mathbf{r} - \mathbf{r}')$, the right-hand side of Eq. (4.7) can be written

$$-j\omega\mu_0 \mathbf{J}^S(\mathbf{r}) = -j\omega\mu_0 \int_\Omega \delta(\mathbf{r} - \mathbf{r}')\mathbb{I} \cdot \mathbf{J}^S(\mathbf{r}') \, d\Omega' \tag{4.9}$$

where $\mathbb{I}$ is the unit dyadic

$$\mathbb{I} = \begin{bmatrix} 1 & 0 & 0 \\ 0 & 1 & 0 \\ 0 & 0 & 1 \end{bmatrix}. \tag{4.10}$$

Substituting Eqs. (4.8) and (4.9) into Eq. (4.7), it becomes evident that the equation for the dyadic Green's function is

$$\nabla \times \nabla \times \mathbb{G}(\mathbf{r}, \mathbf{r}') - k_0^2 \mathbb{G}(\mathbf{r}, \mathbf{r}') = \mathbb{I}\delta(\mathbf{r} - \mathbf{r}'), \tag{4.11}$$

where the ∇ operator refers to differentiation with respect to the unprimed space-variable $\mathbf{r}$. It may be confirmed by direct substitution that the solution to Eq. (4.11) is

$$\mathfrak{G}(\mathbf{r}, \mathbf{r}') = \left[\mathbb{1} + \frac{1}{k_0^2} \nabla \nabla \right] G(\mathbf{r}, \mathbf{r}'), \tag{4.12}$$

where $G(\mathbf{r}, \mathbf{r}')$ is the scalar free-space Green's function satisfying the Helmholtz equation, Eq. (4.4). Because of the symmetry of the equations in $\mathbf{r}$ and $\mathbf{r}'$, if required the operator $\nabla \times \nabla$ may be replaced by $\nabla' \times \nabla'$, the prime here signifying differentiation with respect to $\mathbf{r}'$.

Appendix 3

Simplex element tables

1.   Introduction

This appendix gives numerical values of the fundamental element matrices for line segments, triangles, and tetrahedra of moderate orders. The $\mathbf{C}^{(1)}$ (embedding), $\mathbf{D}^{(1)}$ (differentiation), $\mathbf{T}$ (metric) and $\mathbf{P}$ (projection) matrices are given in all cases. For line segments, they are augmented by the $\mathbf{S}$ matrices. For triangles and tetrahedra $\mathbf{S}$ matrices cannot be given, so the $\mathbf{Q}^{(1)}$ matrices are presented instead for triangles, the $\mathbf{Q}^{(12)}$ matrices for tetrahedra. The lowest orders of matrix are given here in printed form. They are also available in machine-readable files, far preferable when any but the smallest matrices must be dealt with, because verifying and proofreading numeric matrices is a difficult task even for a second-order tetrahedron. All matrices are given in the form of integer quotients, so they are known exactly and all word-length considerations disappear from the presentation; they may, of course, reappear in particular computing systems if any of the integers exceeds the available machine word-length. All matrices have been normalized to a common denominator, so it suffices to tabulate the matrix of numerators and to specify the single common denominator separately.

The printed tables only cover the first few (very few) matrix orders; experience shows that manual entry and proofreading of large tables borders on the impracticable, so the reader is strongly advised to use the machine-readable files for higher orders. These tables are organized by dimensionality first, matrix order thereafter; they thus begin by listing all the matrices for a line segment of first order, then of second, and so on; triangles follow the line segments, and tetrahedra follow the triangles. Matrices are given in their natural form, save for one exception: where the natural form of the matrix has more columns than rows (e.g., $\mathbf{P}$ and

$\mathbf{C}^{(1)}$), they are printed in transposed form so as to avoid breaking matrix rows at line ends in printing.

The machine-readable files are organized with every matrix in a separate file, so it may be found and used without reference to the others. Every matrix listing begins with a line that identifies the matrix, in a fixed format:

1	6	12	TTRIA1.DAT
order	*lines to follow*	*common divisor*	*file name*

The *file names* adhere to a common structure. The first character identifies the matrix (e.g., T); the next four identify the simplex (LINE, TRIA, TETR for line segments, triangles and tetrahedra, respectively). The numeral that follows gives the order of polynomial approximation, identical to the *order* given on the left.

Following this header line, each matrix element is listed on a separate line in the file, in the format

1	1	2	1.66666666666667e − 001
1	2	1	8.33333333333333e − 002
1	...		

Here the first two integers identify the row and column numbers of the matrix element. Matrix elements are given both as integer fractions and in floating-point (double precision) form; the latter is often convenient in program construction, while the former guarantees exact values independently of the precision or machine word-length. The floating-point number at the right of each line is the element value. When expressed as a fraction, the third integer in the row gives its numerator, while the common denominator for all elements in the matrix is given in the header line. In the example above, T_{12} is in the second line of the table, and has the value $1/12$, or $8.333\ldots \times 10^{-2}$.

Some known matrix properties are exploited to save space. Matrices **T** **S**, $\mathbf{Q}^{(1)}$ and $\mathbf{Q}^{(12)}$ are known to be fairly dense but symmetric; only their upper triangle elements are given. Matrices **C** and $\mathbf{D}^{(1)}$ are not symmetric, but contain many zeros, so only their nonzero elements appear in the files. Matrix **P** unfortunately is neither symmetric nor particularly sparse, so there is little choice but to tabulate it in its entirety. In all cases, the second number in the header line gives the total number of tabulated matrix elements that follow the header.

2.   Line-segment matrices

The line-segment matrices, to order 4, follow: $\mathbf{T}, \mathbf{S}, \mathbf{C}^{(1)T}, \mathbf{P}^T$ $\mathbf{D}^{(1)}$. Each matrix is given by listing on the first line its name, and the common denominator for all its elements; the rest of the matrix then appears in its natural printed form (except for possible transposition, as indicated above). All these as well as others are available in machine-readable form on the laboratory diskette.

Order 1

```
T    6              S    1
     2    1              1   -1
     1    2             -1    1

C    4              P    3
     4    1              2   -1
     0    1              2    2
     0    0             -1    2

D    1
     1    1
     0    0
```

Order 2

```
T   30              S    3
     4    2   -1         7   -8    1
     2   16    2        -8   16   -8
    -1    2    4         1   -8    7

C   27              P   80
    27    4   -1        62   -5   18
     0   16    8        54   45  -54
     0   -2    2       -54   45   54
     0    0    0        18   -5   62

D    1
     3    1   -1
     0    2    4
     0    0    0
```

Order 3

```
T 1680                      S    40
   128    99   -36    19         148  -189    54   -13
    99   648   -81   -36        -189   432  -297    54
   -36   -81   648    99          54  -297   432  -189
    19   -36    99   128         -13    54  -189   148

C  512                      P  2835
   512    45   -16     5        2403  -122    53  -432
     0   405   144   -27        1728  1888  -352  1728
     0   -81   144   135       -2592  1368  1368 -2592
     0    15   -16    15        1728  -352  1888  1728
     0     0     0     0        -432    53  -122  2403

D    2
    11     2    -1     2
     0     9     6    -9
     0     0     6    18
     0     0     0     0
```

Order 4

```
T   5670
      292      296    -174      56     -29
      296     1792    -384     256      56
     -174     -384    1872    -384    -174
       56      256    -384    1792     296
      -29       56    -174     296     292

S    945
     4925    -6848    3048   -1472     347
    -6848    16640  -14208    5888   -1472
     3048   -14208   22320  -14208    3048
    -1472     5888  -14208   16640   -6848
      347    -1472    3048   -6848    4925

C   3125
     3125      176     -63      28     -11
        0     2816     672    -192      64
        0     -704    1512    1008    -176
        0      256    -288     448     704
        0      -44      42     -42      44
        0        0       0       0       0

P 774144
   694144   -29017    9072     -89   80000
   400000   581675  -75600   13675 -400000
  -800000   291950  453600  -84050  800000
   800000   -84050  453600  291950 -800000
  -400000    13675  -75600  581675  400000
    80000      -89    9072  -29017  694144

D      3
       25        3      -1       1      -3
        0       22       8      -6      16
        0        0      18      18     -36
        0        0       0      12      48
        0        0       0       0       0
```

3. Triangle matrices

Triangle matrices are tabulated similarly to the line-segment matrices, except that the $\mathbf{Q}^{(1)}$ matrices appear where the $\mathbf{S}$ matrices were given for line segments. Thus the order of presentation is

Order 1

```
T   12              Q   2
     2    1    1         0    0    0
     1    2    1         0    1   -1
     1    1    2         0   -1    1

C    4              P   5
     4    1    1         2   -1   -1
     0    1    0         3    3   -1
     0    0    1         3   -1    3
     0    0    0        -1    2   -1
     0    0    0        -1    3    3
     0    0    0        -1   -1    2

D    1
     1    1    1
     0    0    0
     0    0    0
```

Order 2

```
T    180
      6      0      0     -1     -4     -1
      0     32     16      0     16     -4
      0     16     32     -4     16      0
     -1      0     -4      6      0     -1
     -4     16     16      0     32      0
     -1     -4      0     -1      0      6

Q      6
      0      0      0      0      0      0
      0      8     -8      0      0      0
      0     -8      8      0      0      0
      0      0      0      3     -4      1
      0      0      0     -4      8     -4
      0      0      0      1     -4      3

C     27
     27      4      4     -1     -1     -1
      0     16      0      8      4      0
      0      0     16      0      4      8
      0     -2      0      2     -1      0
      0      0      0      0      4      0
      0      0     -2      0     -1      2
      0      0      0      0      0      0
      0      0      0      0      0      0
      0      0      0      0      0      0
      0      0      0      0      0      0

P    280
    136     -4     -4     36     36     36
    216    126     -9   -144    -54     36
    216     -9    126     36    -54   -144
   -144    126    -54    216     -9     36
   -144    126    126   -144    126   -144
   -144    -54    126     36     -9    216
     36     -4     36    136     -4     36
     36     -9    -54    216    126   -144
     36    -54     -9   -144    126    216
     36     36     -4     36     -4    136

D      1
      3      1      1     -1     -1     -1
      0      2      0      4      2      0
      0      0      2      0      2      4
      0      0      0      0      0      0
      0      0      0      0      0      0
      0      0      0      0      0      0
```

Order 3

```
T 6720
     76     18     18      0     36      0     11     27     27     11
     18    540    270   -189    162   -135      0   -135    -54     27
     18    270    540   -135    162   -189     27    -54   -135      0
      0   -189   -135    540    162    -54     18    270   -135     27
     36    162    162    162   1944    162     36    162    162     36
      0   -135   -189    -54    162    540     27   -135    270     18
     11      0     27     18     36     27     76     18      0     11
     27   -135    -54    270    162   -135     18    540   -189      0
     27    -54   -135   -135    162    270      0   -189    540     18
     11     27      0     27     36     18     11      0     18     76
```

Q 80

0	0	0	0	0	0	0	0	0	0
0	135	-135	-27	0	27	3	0	0	-3
0	-135	135	27	0	-27	-3	0	0	3
0	-27	27	135	-162	27	3	0	0	-3
0	0	0	-162	324	-162	0	0	0	0
0	27	-27	27	-162	135	-3	0	0	3
0	3	-3	3	0	-3	34	-54	27	-7
0	0	0	0	0	0	-54	135	-108	27
0	0	0	0	0	0	27	-108	135	-54
0	-3	3	-3	0	3	-7	27	-54	34

C 512

512	45	45	-16	-16	-16	5	5	5	5
0	405	0	144	72	0	-27	-18	-9	0
0	0	405	0	72	144	0	-9	-18	-27
0	-81	0	144	-36	0	135	36	-9	0
0	0	0	0	216	0	0	108	108	0
0	0	-81	0	-36	144	0	-9	36	135
0	15	0	-16	10	0	15	-8	5	0
0	0	0	0	-18	0	0	36	-18	0
0	0	0	0	-18	0	0	-18	36	0
0	0	15	0	10	-16	0	5	-8	15
0	0	0	0	0	0	0	0	0	0
0	0	0	0	0	0	0	0	0	0
0	0	0	0	0	0	0	0	0	0
0	0	0	0	0	0	0	0	0	0
0	0	0	0	0	0	0	0	0	0

P 5103

2943	-2	-2	-7	-7	-7	-432	-432	-432	-432
4320	2704	-176	-256	-112	32	2160	1296	432	-432
4320	-176	2704	32	-112	-256	-432	432	1296	2160
-4320	2232	-1008	2232	-84	288	-4320	-1008	288	-432
-4320	2544	2544	-672	2016	-672	2160	-1872	-1872	2160
-4320	-1008	2232	288	-84	2232	-432	288	-1008	-4320
2160	-256	1296	2704	-112	432	4320	-176	32	-432
2160	-672	-1872	2544	2016	-1872	-4320	2544	-672	2160
2160	-1872	-672	-1872	2016	2544	2160	-672	2544	-4320
2160	1296	-256	432	-112	2704	-432	32	-176	4320
-432	-7	-432	-2	-7	-432	2943	-2	-7	-432
-432	32	432	-176	-112	1296	4320	2704	-256	2160
-432	288	288	-1008	-84	-1008	-4320	2232	2232	-4320
-432	432	32	1296	-112	-176	2160	-256	2704	4320
-432	-432	-7	-432	-7	-2	-432	-7	-2	2943

D 2

11	2	2	-1	-1	-1	2	2	2	2
0	9	0	6	3	0	-9	-6	-3	0
0	0	9	0	3	6	0	-3	-6	-9
0	0	0	6	0	0	18	6	0	0
0	0	0	0	6	0	0	12	12	0
0	0	0	0	0	6	0	0	6	18
0	0	0	0	0	0	0	0	0	0
0	0	0	0	0	0	0	0	0	0
0	0	0	0	0	0	0	0	0	0
0	0	0	0	0	0	0	0	0	0
0	0	0	0	0	0	0	0	0	0

$\mathbf{T}, \mathbf{Q}^{(1)}, \mathbf{C}^{(1)T}, \mathbf{P}^T, \mathbf{D}^{(1)}.$

4.   Tetrahedron matrices

Order 1

T 20

					Q	2			
2	1	1	1			1	-1	0	0
1	2	1	1			-1	1	0	0
1	1	2	1			0	0	0	0
1	1	1	2			0	0	0	0

C 4

					P	15			
4	1	1	1			3	-2	-2	-2
0	1	0	0			8	8	-2	-2
0	0	1	0			8	-2	8	-2
0	0	0	1			8	-2	-2	8
0	0	0	0			-2	3	-2	-2
0	0	0	0			-2	8	8	-2
0	0	0	0			-2	8	-2	8
0	0	0	0			-2	-2	3	-2
0	0	0	0			-2	-2	8	8
0	0	0	0			-2	-2	-2	3

D 1

1	1	1	1
0	0	0	0
0	0	0	0
0	0	0	0

Order 2

T 420

6	-4	-4	-4	1	-6	-6	1	-6	1
-4	32	16	16	-4	16	16	-6	8	-6
-4	16	32	16	-6	16	8	-4	16	-6
-4	16	16	32	-6	8	16	-6	16	-4
1	-4	-6	-6	6	-4	-4	1	-6	1
-6	16	16	8	-4	32	16	-4	16	-6
-6	16	8	16	-4	16	32	-6	16	-4
1	-6	-4	-6	1	-4	-6	6	-4	1
-6	8	16	16	-6	16	16	-4	32	-4
1	-6	-6	-4	1	-6	-4	1	-4	6

Q 10

3	-4	-1	-1	1	1	1	0	0	0
-4	8	0	0	-4	0	0	0	0	0
-1	0	8	4	1	-8	-4	0	0	0
-1	0	4	8	1	-4	-8	0	0	0
1	-4	1	1	3	-1	-1	0	0	0
1	0	-8	-4	-1	8	4	0	0	0
1	0	-4	-8	-1	4	8	0	0	0
0	0	0	0	0	0	0	0	0	0
0	0	0	0	0	0	0	0	0	0
0	0	0	0	0	0	0	0	0	0

C   27

27	4	4	4	-1	-1	-1	-1	-1	-1
0	16	0	0	8	4	4	0	0	0
0	0	16	0	0	4	0	8	4	0
0	0	0	16	0	0	4	0	4	8
0	-2	0	0	2	-1	-1	0	0	0
0	0	0	0	0	4	0	0	0	0
0	0	0	0	0	0	4	0	0	0
0	0	-2	0	0	-1	0	2	-1	0
0	0	0	0	0	0	0	0	4	0
0	0	0	-2	0	0	-1	0	-1	2
0	0	0	0	0	0	0	0	0	0
0	0	0	0	0	0	0	0	0	0
0	0	0	0	0	0	0	0	0	0
0	0	0	0	0	0	0	0	0	0
0	0	0	0	0	0	0	0	0	0
0	0	0	0	0	0	0	0	0	0
0	0	0	0	0	0	0	0	0	0
0	0	0	0	0	0	0	0	0	0
0	0	0	0	0	0	0	0	0	0
0	0	0	0	0	0	0	0	0	0

P   448

88	-1	-1	-1	36	36	36	36	36	36
360	153	9	9	-180	-72	-72	36	36	36
360	9	153	9	36	-72	36	-180	-72	36
360	9	9	153	36	36	-72	36	-72	-180
-180	153	-72	-72	360	9	9	36	36	36
-180	198	198	-72	-180	198	-72	-180	-72	36
-180	198	-72	198	-180	-72	198	36	-72	-180
-180	-72	153	-72	36	9	36	360	9	36
-180	-72	198	198	36	-72	-72	-180	198	-180
-180	-72	-72	153	36	36	9	36	9	360
36	-1	36	36	88	-1	-1	36	36	36
36	9	-72	36	360	153	9	-180	-72	36
36	9	36	-72	360	9	153	36	-72	-180
36	-72	9	36	-180	153	-72	360	9	36
36	-72	-72	-72	-180	198	198	-180	198	-180
36	-72	36	9	-180	-72	153	36	9	360
36	36	-1	36	36	-1	36	88	-1	36
36	36	9	-72	36	9	-72	360	153	-180
36	36	-72	9	36	-72	9	-180	153	360
36	36	36	-1	36	36	-1	36	36	88

D   1

3	1	1	1	-1	-1	-1	-1	-1	-1
0	2	0	0	4	2	2	0	0	0
0	0	2	0	0	2	0	4	2	0
0	0	0	2	0	0	2	0	2	4
0	0	0	0	0	0	0	0	0	0
0	0	0	0	0	0	0	0	0	0
0	0	0	0	0	0	0	0	0	0

Tables of the tetrahedron matrices are organized similarly to the line segments and triangles, with the sole exception that the matrix $\mathbf{Q}^{(12)}$ appears in place of $\mathbf{S}$ or $\mathbf{Q}^{(1)}$.

Appendix 4

Utility programs and style notes

1. Introduction

This Appendix lists several generally useful subprograms for matrix manipulation. It also clarifies some points of programming style as followed throughout this book. Most of the subprograms have no particularly close association with the material of any one chapter, but they are called upon to perform necessary services by programs listed in the several chapters. It therefore seems appropriate to group them in a single listing.

2. Input-data arrangements

Nearly all the programs in this book accept input data in roughly similar formats. Typically, a list of nodes is followed by a list of elements, which is in turn followed by a list of constraints. The lists are separated by lines blank except for a slant character / in the leading character position. This technique is based on the special meaning Fortran 77 assigns to the / character in list-directed input–output. It leads to program constructs a little different from C, Pascal, or many other languages, and therefore merits a brief discussion.

List-directed input–output is free-format. Numbers may be placed anywhere on the line, only their sequence need be correct. When input is requested, as many full lines are read in as is necessary to ensure that *the input list has been fully satisfied*, even if that means reading several lines. Since Fortran r e a d statements always read full lines, a blank line will not be visible at all. To force termination of reading before the list is satisfied, Fortran 77 recognizes the / character to signify *stop input*. Any variables on the input list but not yet read are then not read — they keep the values they had prior to the input request. This mechanism provides a clean way of terminating an input stream of unknown length. The following demonstration program reads integers i new and places them in an array list. After each item is read, a copy of it is maintained in variable iold. After the each input request, the newly read inew is compared with iold; if they do not differ, the program concludes that no reading has taken place and an end-of-list / character must have been encountered. Of course, this scheme only works if the item being compared is necessarily unique (e.g., a node number).

```
C*********************************************************
C      Demonstration of the Fortran 77 null-read character /.
C*********************************************************
      parameter (NULL=0, LONG=4)
      Dimension list(LONG)
   10 continue
C
      i = 0
      iold = NULL
   20 read (*,*, end=90) inew
      if (inew .ne. iold) then
        write (*,*) inew
        i = i + 1
        list(i) = inew
        iold = inew
        go to 20
      else
        write (*,100) i
      endif
C
      go to 10
   90 stop
  100   format (' Length =', i2, '. Next list please!')
      end
```

It may be instructive to examine this short program closely, particularly the dozen lines in the middle that constitute the actual list-reading loop. A small input file and the corresponding output are shown in Fig. A4.1. Note that the blank line (after 7901) is ignored, and that data placement within each line is unimportant.

In actual finite element programs the entities to be read and compared are generally more complex, e.g., nodes or finite elements. However, the program structure remains in all cases as it is sketched here, only the individual operations become more complicated.

693
  501
438
    384
/
   7901
 	
8053
/
3114
	693
	501
	438
	384
	Length = 4. Next list please!
	7901
	8053
	Length = 2. Next list please!
	3114

Fig A4.1 Input and output files for demonstration program. Line ends have no special significance, / does.

3.   Utility programs

When used in conjunction with finite element programs, the subroutines of this Appendix are best treated as a utility library. Under most Fortran systems, this means compiling the object modules and archiving them in a library, which can then be called upon as the various programs may require.

All the programs listed here are written in Fortran 77 and have been verified for strict compliance with the Fortran 77 standard. They should therefore compile under any reasonable Fortran 77 compiler. In addition, the authors hope that the listings are sufficiently commented to permit modification in an informed and systematic fashion, should modification indeed be required.

```fortran
C****************************************************************
C
      Subroutine eqsolv(s, x, y, n, maxn)
C
C****************************************************************
C      Copyright (c) 1995  P.P. Silvester and R.L. Ferrari
C****************************************************************
C     Solves the system of equations with square symmetric co-
C     efficient matrix s,  in n variables but dimensions maxn,
C                          s * x  =  y
C     Procedure:   Cholesky decomposition, forward elimination
C     and back substitution.  s and y are not preserved.
C================================================================
      dimension s(maxn,maxn), x(maxn), y(maxn)
C
C     Cholesky decomposition replaces s by triangular factors
      call chlsky(s, s, n, maxn)
C        then forward eliminate and back substitute.
      call fwdelm(s, y, y, n, maxn)
      call backsb(s, y, x, n, maxn)
      return
      end
C
C****************************************************************
C
      Subroutine chlsky(a, fl, n, maxn)
C
C****************************************************************
C      Copyright (c) 1995  P.P. Silvester and R.L. Ferrari
C****************************************************************
      dimension a(maxn,maxn), fl(maxn,maxn)
C
      do 70 j = 1,n
        fl(j,j) = a(j,j)
        do 20 k = 1,j-1
          fl(j,j) = fl(j,j) - fl(j,k)*fl(j,k)
   20     continue
        if (fl(j,j) .gt. 0.) then
          fl(j,j) = sqrt(fl(j,j))
          do 50 I = j+1,n
            fl(i,j) = a(i,j)
            do 40 k = 1,j-1
              fl(i,j) = fl(i,j) - fl(i,k)*fl(j,k)
   40         continue
            fl(i,j) = fl(i,j) / fl(j,j)
            fl(j,i) = fl(i,j)
   50       continue
        else
          call errexc('CHLSKY', j)
        endif
   70   continue
      return
      end
C
```

```
C***************************************************************
C
      Subroutine fwdelm(fl, b, y, n, maxn)
C
C***************************************************************
C     Copyright (c) 1995  P.P. Silvester and R.L. Ferrari
C***************************************************************
C     Forward elimination of triangular system  (fl)*y = b
C===============================================================
      dimension fl(maxn,maxn), b(maxn), y(maxn)
C
      y(1) = b(1) / fl(1,1)
      do 60 I = 2,n
        y(i) = b(i)
        do 40 j = 1,i-1
          y(i) = y(i) - fl(i,j)*y(j)
   40     continue
        y(i) = y(i) / fl(i,i)
   60   continue
      return
      end
C
C***************************************************************
C
      Subroutine backsb(u, b, x, n, maxn)
C
C***************************************************************
C     Copyright (c) 1995  P.P. Silvester and R.L. Ferrari
C***************************************************************
C     Back substitution of triangular system  u*x = b
C===============================================================
      dimension u(maxn,maxn), b(maxn), x(maxn)
C
      x(n) = b(n) / u(n,n)
      do 60 I = n-1,1,-1
        x(i) = b(i)
        do 40 j = i+1,n
          x(i) = x(i) - u(i,j)*x(j)
   40     continue
        x(i) = x(i) / u(i,i)
   60   continue
      return
      end
C
C***************************************************************
C
      Subroutine vecini(a, nr)
C
C***************************************************************
C
C     Initializes vector a, of nr entries, to all zeros.
C===============================================================
      dimension a(nr)
C===============================================================
C
      do 10 i = 1,nr
        a(i) = 0.
   10   continue
      return
      end
C
```

```
C************************************************************
C
      Subroutine matini(a, nr, nc)
C
C************************************************************
C
C     Initializes  matrix a, of nr rows and nc columns, to all
C     zero entries.
C============================================================
      dimension a(nr, nc)
C============================================================
C
      do 20 i = 1,nr
        do 10 j = 1,nc
          a(i,j) = 0.
   10     continue
   20   continue
      return
      end
C
C************************************************************
C
      Function locate(i,j)
C
C************************************************************
C
C     Returns linear storage-mode  location index of the (i,j)
C     element of  a symmetric matrix  whose lower  triangle is
C     stored by rows in a linear array.
C
      if (j .lt. i) then
        locate = (i*(i-1))/2+j
      else
        locate = (j*(j-1))/2+i
      endif
      return
      end
C
C************************************************************
C
      Subroutine errexc(prog, ierr)
C
C************************************************************
C
C     Exception handler for serious errors: prints the program
C     name and error numbers, and halts execution.
C
      Character*(*) prog
      write (*,100) prog, ierr
C     stop 911
      return
  100 format(1x/' Stopped in routine ', a6, ', error ',
     *        'number ', i5)
      end
```

Appendix 5

Laboratory problems and exercises

1. Introduction

This appendix suggests some laboratory work to accompany *Finite Elements for Electrical Engineers*. As a convenience to readers, all the computer programs listed in the book are available through the Internet (by anonymous FTP) and the authors are also prepared to furnish them on an MS−DOS diskette. This appendix could therefore be viewed as a user's manual to accompany the programs. While it mainly deals with how they might best be used in an MS−DOS environment, the program source code is written in standard Fortran 77, is not restricted to MS−DOS and will work equally well under various Unix, VMS or other computing systems.

Note: The programs and related data in this book are copyright material, like all other parts of the book. The files provided at the anonymous FTP site, and on the MS−DOS diskette obtainable from the authors, are provided there solely as a convenience for the reader; they remain copyright materials and must be treated on the same basis as this book, except that a security backup copy may be made. In other words, they may be used for personal study but they may not be sold, rented, or reproduced, without the express written permission of the copyright holders. Written permission must be obtained for the programs, tables, data or auxiliary files to be incorporated in any collections, anthologies or software systems. However, they may be duplicated and given to any rightful owner of a copy of *Finite Elements for Electrical Engineers* (Silvester and Ferrari; 3rd edition, Cambridge University Press 1996), provided that: (1) the recipient understands and agrees to these distribution restrictions, (2) the copy is complete, including all programs, data, auxiliary files and copyright notices, and (3) no charge of any kind is made.

2. **Laboratory working procedures**

The authors have found the following laboratory procedures convenient and believe that others may find them useful too.

2.1 *Background requirements*

To make best use of the material provided, some acquaintance with the Fortran programming language will obviously be necessary. The programs furnished here are written in standard Fortran 77 and make no use of any operating system dependent features. Thus any standard textbook on the Fortran language should serve as a reference and any good Fortran 77 compiler should do equally well.

Because the programs are all in a standard dialect of the Fortran language, there should be no difficulty installing and running them under any reasonable operating system. The remarks to follow here are mainly directed to MS–DOS and assume some familiarity with the principal system commands in it: how to navigate around directories, how to use a text editor, how to run a Fortran compiler, and how to use common system utilities. The Microsoft Fortran 77 compiler is suitable, as is the Prospero Software *Profort* compiler. Excellent compilers are also available from Watcom Software and Lahey. Two public-domain Fortran 77 compilers (g77 and BCF77) are available from Internet download sites. All programs in this book have been tested with the Microsoft and Profort compilers. BCF77 (version 1.3) appears to have difficulties with the / end-of-input marker but otherwise runs well. The g77 compiler (from the Free Software Foundation) appears to be good, but not all the programs given here have been tested with it. Under the various Unix operating systems, the time-tested f77 works well; an alternative method is to use f2c (available by anonymous FTP – it too comes from the Free Software Foundation) to translate the Fortran code into C, then to use the local C compiler. The translated C code is rarely readable, but it compiles properly.

The executable modules provided on the diskette and by anonymous FTP are MS–DOS programs; they will of course not execute under any other operating system.

2.2 *Installation*

The files as furnished on diskette are organized in four main directories: FORTRAN, EXECUTBL, SIMPLXLO, SIMPLXHI. The latter two contain tables of simplex elements, of low and high orders respectively. Directory EXECUTBL contains compiled, executable MS–DOS programs. Directory FORTRAN in its turn contains even subdirectories. Each of these houses the Fortran 77 source code, sample data sets and corresponding output for one of the code collections at the end of a chapter in this book. The directory structure is as shown in Fig. A5.1.

2.3   *Working procedure*

Directory *WORK* shown in Fig. A5.1 does not exist on the available diskette; it is suggested that the user create it. The files on diskette are then installed by simply copying them into the tree structure shown.

When first starting out to compile and link programs, rather than just running the ready-made ones, an object version of the finite element utility package that furnishes general-purpose functions (file reading and writing, matrix manipulation, etc.) will be required. Copy the utility package FEUTILIT.FOR into the *WORK* directory and compile it. The source copy in *WORK* may be deleted immediately, it will no longer be required; and anyway there is still a copy of it in directory FORTRAN.

Most of the work is probably best done by copying the contents of a whole subdirectory (e.g., LOSSYLIN) into the *WORK* directory and performing all the required manipulations there. In this way there is no danger of altering the original code inadvertently.

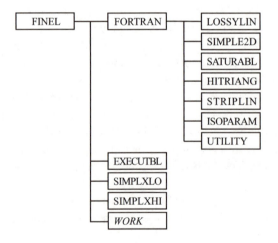

Fig A5.1   Directory structure of the programs on disk.

2.4   *Missing files*

The most frequent difficulty encountered by beginners is a complaint from the linker about missing files. Each one of the laboratory programs consists of a main program and several subroutines. Most of these call other subroutines from the general utility package, and in a few cases from other finite element programs. When programs are linked to make up the executable code, each and every one of the parts must be present in directory *WORK*. If it is not, the linker will issue a diagnostic message to indicate that one or more of the required modules could not be found in the list of files given to the linker.

On most occasions when missing object modules are encountered, the matter is simple: one of the required pieces really is missing (the utility package is an easy one to forget). Some linkers are willing to produce an executable module even in the absence of one or two program parts. This is sometimes useful if the missing pieces supply what are essentially cosmetic improvements. However, there really is not much point even attempting such partial linking of finite element programs.

3. Laboratory exercises

The following laboratory problem sets generally begin with easy tasks, sometimes verging on the trivial, but difficulty escalates toward the end of each problem set.

3.1 *One-dimensional finite elements*

Program LOSSYLIN solves the problem of a lossy direct-current transmission line. Set up this program so it will run properly, and try at least some of the following tasks. Note that, as furnished, the program reads in data in the following format:

2.	1.	0.5	*Applied volts* 2.0, $r = 1, g = 0.5$;
4	1.5		*4 elements in first* 1.5 *of length,*
3	0.6		*3 elements in next* 0.6 *units,*
			end of input.

The above, and several other data sets, appear in files provided on the laboratory diskette (with filename extension .DAT).

(1) Run the LOSSYLIN program with the data file(s) provided on the laboratory diskette, and ensure it works, as indicated by the correct voltage distribution along the line. Results corresponding to at least one input data file are provided on the diskette also (corresponding file names, with extensions .DAT and .OUT).

(2) How can total power loss be determined? How can the input impedance of the line, as seen at the source terminals, be found? (There are at least two distinct ways.) Can these calcula-

tions be made on the external data only, without altering the program itself?

(3) Does any significant accuracy difference result from using smaller elements near the sending end? Near the receiving end?

(4) The transmission line modelled by the LOSSYLIN program is terminated in a resistor of R_t ohms, not left open circuited. How can the program be modified to take a termination into account? Carry out the modification and show how the voltage varies along the length of this line. How does it behave if $R_t = \sqrt{r/g}$?

(5) (*Advanced problem.*) Modify the transmission-line program so that the resistivity and conductivity can be specified for sections of line, not merely for the whole line. Apply the modified program to a transmission line of length L, made of conductors having a resistance of $2r$ (ohms per unit length) from the sending end to its middle, then of conductors with resistance $r/2$ from the middle to the receiving end; $g = 1/r$ throughout. Use the modified program to determine how the voltage varies along the line, as compared to a line with uniform resistance r along its whole length.

3.2   *First-order triangular elements*

The program SIMPLE2D as furnished on the laboratory diskette solves Poisson's equation on a triangular finite element mesh, using first-order (piecewise-flat) elements. Like the LOSSYLIN program, it requires some parts of the FEUTILIT utility function library in order to work. Set up this program and try at least some of the following tasks.

(1) Run SIMPLE2D using the data shown in this book (and reproduced on the diskette). Ensure the program works properly, comparing results with those shown in the book.

(2) Using SIMPLE2D, find the electric potential distribution in a coaxial line made up of a circular inner conductor of radius a inside a square outer conductor of side $3a$. (Modelling only one-eighth of the line will save computational effort.) The circular boundary may be approximated by a regular polygon; 16, 32 or 64 sides might prove adequate to model a circle. What are the correct boundary conditions to use?

(3) How can one compute the capacitance per unit length of coaxial line, starting from a potential solution such as provided by SIMPLE2D? (There are at least two perfectly good ways.) Can this computation be made separately from SIMPLE2D,

using its output, or is it necessary to modify SIMPLE2D? Find the capacitance of the circle-in-square line.

(4) (*Advanced problem.*) Repeat the coaxial-line problem several times, with increasing numbers of elements. Do the capacitance results converge to a definite solution? Is it analytically known? Determine the law of error behaviour by curve-fitting, e.g., by plotting the error on log–log paper.

3.3   *First-order elements for saturable materials*

The program SATURMAT as furnished on the laboratory diskette solves a nonlinear Poisson's equation on a mesh of first-order (piecewise-planar) triangular elements. Like the other programs, it requires some parts of the FEUTILIT utility function library in order to work. Set up this program and try at least some of the following tasks.

(1) Run SATURMAT using the data shown by Silvester and Ferrari (and reproduced on the diskette). Ensure the program works properly, comparing results with those shown in the book.

(2) Using SATURMAT, find the magnetic vector potential (i.e., calculate the flux distribution) in a large iron block containing two square holes filled with current-carrying conductors of opposite polarity; the holes are 1 cm on a side, with their sides parallel, and are separated by a gap of 2 cm. Compute the flux distribution for at least three current values, choosing one so that the material is unsaturated, one so that the material is heavily saturated, and one or more part-way between. What is the difference in the potential distributions?

(3) How can the stored energy (per unit length) be determined in the structure described above? Can this computation be done without modifying SATURMAT? Set up and carry out an appropriate computation.

(4) As the SATURMAT program is listed here, its Newton iteration is initially slowed by an empirically set variable damper. Investigate the effect of this variable by trying several different settings. How many iterations does the Newton process require if started undamped, i.e., setting damper = 1.0?

(5) (*Advanced problem.*) How can the leakage flux of a magnetic device be estimated without counting lines in a flux plot? Devise an appropriate method of circulation and implement it in SATURMAT.

3.4   *High-order elements*
The matrices tabulated in this book include many, but not all, the matrices useful in finite element analysis. The following problems explore some of the matrix properties and involve computation of some matrices not given here.

(1) Show, both theoretically and by computation, that all the elements of any **T** matrix sum to exactly 1.

(2) Show, both theoretically and by computation, that all the elements in any one row or column of any **Q** matrix sum to exactly zero.

(3) Find the **Q** matrices for one-dimensional elements from the corresponding **D** matrices.

(4) Find the **Q** matrices for tetrahedral elements from the corresponding **D** matrices.

(5) (*Advanced problem.*) Test computationally whether the projection and embedding matrices commute. Show theoretically that the computed results are correct.

(6) (*Advanced problem.*) Find the element matrices necessary for solving axisymmetric scalar potential problems.

(7) (*Advanced problem.*) Find the element matrices necessary for solving axisymmetric vector potential problems in which the potential vector only possesses an azimuthal component (as would be the case for a solenoidal coil).

(8) (*Advanced problem.*) Problem (2) above implies that the null-space of **Q** contains functions of a certain kind. What kind? What other functions may be expected in that null-space?

3.5   *High-order triangular element program*
The high-order triangular element program solves Poisson's equation on a triangular finite element mesh, using high-order (piecewise-polynomial) elements. Its general organization is very similar to SIMPLE2D in most other respects. Set up this program and try at least some of the following tasks. Note that in a high-order triangle the nodes must be given to the program by rows, beginning at a vertex. For example, the element shown in Fig. A5.2 will have the node list ordered in a data line as 15 21 12 10 13 22; an equally valid arrangement would be 10 13 21 22 12 15.

This program resides in file HITRIA2D.FOR, except for the element matrices themselves which appear in HI2DTRIM.FOR. Like the other finite element programs, it also requires some utility routines from the FEUTILIT library.

(1) Verify that the program works properly, by repeating one of the problems solved earlier using SIMPLE2D.

(2) A square conductor of uniform finite conductivity and side length $2a$ carries a current known to vary slowly enough to ensure its uniform distribution. This conductor is placed in a square tube, of side length $3a$, made of superconductive (or very highly conductive) material which will not be significantly penetrated by magnetic flux; the two tubes effectively form a single coaxial go-and-return circuit. Model this problem with triangular elements and solve it.

(3) How can the inductance of a go-and-return circuit, such as that described, be found from the field solution? (There are at least two perfectly good ways to do so.) Can a method be found that allows entirely external computation, without significant internal modification of HI2DMAIN.FOR? Implement this or any reasonable method and compute the inductance of the square cable.

(4) (*Advanced problem.*) Keeping the number of finite element nodes approximately constant, solve the square cable problem using first-order elements, then second-order, then higher-orders. Is it possible to deduce from the solutions how the error varies with element order? With the number of nodes? An analytic solution is not easily found; but the inductance determined with the finest mesh of the highest-order elements available is probably very accurate indeed.

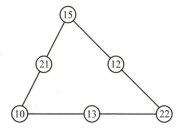

Fig A5.2 A second-order triangular element. To suit the computer programs, the nodes must be listed in parallel rows.

3.6   *Strip line integral equation*

Program STRIPLIN solves the problem of a strip transmission line parallel to a ground plane. Set up this program and try at least some of the following tasks. Note that, as furnished, the program reads in data in the format illustrated by the sample data file provided:

	0.5	*Spacing from ground plane* $= 0.5$
4	1.5	*4 elements in first 1.5 of width*
3	0.6	*3 elements in next 0.6 of width*
		end of input data.

(1) Run the STRIPLIN program with the data provided, and ensure it works correctly, as indicated by the correct charge distribution across the line.

(2) Run the STRIPLIN program to determine capacitance per unit length with several different element subdivisions. Determine a law of error variation, if possible — e.g., by plotting on log–log paper.

(3) How can the electric potential V be found at an arbitrary point (x, y) which is not on the conductor? Extend the program to permit such computation.

(4) Devise a way of computing potential at an arbitrary point (x, y) by means of an external program, i.e., without modifying STRIPLIN at all. Construct such a program.

(5) Construct a solution to the problem of two parallel coplanar flat strip lines with no ground plane, by modifying STRIPLIN appropriately.

(6) (*Advanced problem.*) Find the charge distribution on, and the capacitance between, two parallel coplanar flat strips on the face of an infinite dielectric half-space of permittivity ϵ_1.

(7) (*Advanced problem.*) The classical microstrip line is a flat strip conductor of width w on the face of an infinitely wide dielectric sheet of thickness h and permittivity ϵ_1, backed by a conductive ground plane. Find its capacitance per unit length, the TEM-mode characteristic impedance, and the propagation velocity of a TEM wave.

3.7   *Isoparametric curvilinear elements*

Program ISO8EL.FOR is not really a program, but a subroutine to compute the element matrices for an isoparametric 8-noded rectangle. This subroutine can be fitted into the HI2DMAIN program, thereby converting it into `ISO8MAIN.FOR`, a program able to use both isoparametric and triangular straight-sided elements. Inclusion of this subroutine is easy because the data structure within `HI2DMAIN` is intentionally set up to accommodate new types of element. Note: with provision for *both* high-order triangles and isoparametrics, the total program code may exceed 64K (depending on the compiler used). If working under `MS−DOS`, either use a compiler memory model that permits larger code segments, or remove the triangular elements from the program.

Data for the isoparametric elements follow the format established for high-order triangles. Node numbering on an isoparametric element must always be counterclockwise, starting at a corner. For example, in Fig. A5.3, the nodes may be listed in the order 2 3 14 6 11 7 9 4; but 11 7 9 4 2 3 14 6 is an equally acceptable numbering sequence.

Begin by modifying the program and verifying that it works, then attempt at least some of the further work.

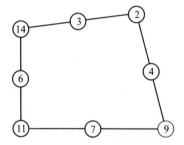

Fig A5.3   An eight-noded isoparametric element. The nodes must be listed counterclockwise, starting at a corner node.

(1) Make the necessary modifications to `HI2DMAIN` to incorporate `ISO8ELEM`. Test to verify that the program works properly, by comparing with the field results for Laplace's equation $\nabla^2 \phi = 0$ in a square region of side length 1 whose top and bottom edges are at potentials $+1$ and -1 respectively, the other two sides being left at their natural boundary conditions (i.e., $\partial\phi/\partial n = 0$).

(2) A rectangular coaxial line consists of a square conductor with side length 2 inside a square conductor of side 3. Solve Laplace's equation $\nabla^2 \phi = 0$ in the dielectric of this cable, using a single quadrilateral element.

(3) A circular coaxial line has inner and outer radii of $2a$ and $3a$ respectively. Find the potential distribution in such a line by modelling a radial portion of it (say one-eighth) by curvilinear

isoparametric elements. Compare with the known analytic solution and with any triangular-element solutions computed earlier.

(4) How can the capacitance (per unit length) of a coaxial line be determined from the potential solution computed on curvilinear isoparametric elements? Can such a calculation be carried out externally, i.e., without modifying the field solution program extensively, or at all? Implement a method for finding capacitance and check it by calculating the approximate capacitance of the circular coaxial line.

Index